THERMAL SYSTEM DESIGN AND SIMULATION

THERMAL SYSTEM DESIGN AND SIMULATION

P.L. DHAR

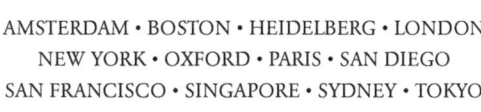

AMSTERDAM • BOSTON • HEIDELBERG • LONDON
NEW YORK • OXFORD • PARIS • SAN DIEGO
SAN FRANCISCO • SINGAPORE • SYDNEY • TOKYO

Academic Press is an imprint of Elsevier

Academic Press is an imprint of Elsevier
50 Hampshire Street, 5th Floor, Cambridge, MA 02139, United States
The Boulevard, Langford Lane, Kidlington, Oxford OX5 1GB, United Kingdom

Notices
Knowledge and best practice in this field are constantly changing. As new research and experience broaden
our understanding, changes in research methods, professional practices, or medical treatment may become
necessary.

Practitioners and researchers must always rely on their own experience and knowledge in evaluating and
using any information, methods, compounds, or experiments described herein. In using such information
or methods they should be mindful of their own safety and the safety of others, including parties for
whom they have a professional responsibility.

To the fullest extent of the law, neither the Publisher nor the authors, contributors, or editors, assume any
liability for any injury and/or damage to persons or property as a matter of products liability, negligence or
otherwise, or from any use or operation of any methods, products, instructions, or ideas contained in the
material herein.

Library of Congress Cataloging-in-Publication Data
A catalog record for this book is available from the Library of Congress

British Library Cataloguing-in-Publication Data
A catalogue record for this book is available from the British Library

ISBN 978-0-12-809449-5

For information on all Academic Press publications
visit our website at https://www.elsevier.com/

Working together
to grow libraries in
developing countries

www.elsevier.com • www.bookaid.org

Acquisition Editor: Lisa Reading
Editorial Project Manager: Maria Convey
Production Project Manager: Susan Li
Designer: Mark Rogers

Typeset by SPi Global, India

CONTENTS

PREFACE

The methodology of design of engineering systems has, over the last three decades, undergone a major shift—from an art, where the experience of the designer was of paramount importance, to a science with rigorous optimization procedures choosing the best possible design from among a wide variety of designs generated by a software. Thermal systems, because of their complexity, badly needed such a tool to help the designers. This book documents the procedures that have been developed over these years to bring about this transition.

I had the privilege of introducing a new course "Thermal System Simulation and Design" for postgraduate students of Thermal Engineering at IIT Delhi about 25 years ago. It was primarily a sharing of my own research work, and of my other research students at IIT Delhi. Gradually as the course took shape, its scope was widened based on the feedback from the students. At that time there was just one book available on this topic, namely "Design of Thermal Systems" by Prof. W.F. Stoecker. This excellent book was the recommended text book for the course. As the course developed it was noticed that many of the concepts that we had developed, like use of information flow diagrams to simplify the simulation procedures, novel optimization methods suitable for thermal system simulation, etc., were not adequately covered in this or even the other text "Design and Optimization of Thermal Systems" by Prof. Y. Jaluria, which was published in 1998. During the course of our research work we had developed the detailed simulation procedures for various equipment used in refrigeration systems, in thermal power plants, and novel desiccant-based cooling systems. The only references available with the students for studying these were the original research papers or the PhD theses and the research monograph on "Computer Simulation and Optimization of Refrigeration Systems" written by me and my former research student Dr G.R. Saraf. As similar courses were introduced at many other institutes, the need arose for an up-to-date text which incorporated all the topics discussed in these courses. This book is written in response to that need.

I had got the feedback that my earlier monograph on refrigeration systems design was being referred by the designers in the industry, and so the present book has been written keeping their needs also in view.

The book has been enriched by the feedback of hundreds of students—many of them professional engineers from the industry—who took this course. My former colleagues Prof. M.R. Ravi and Prof. Sangeeta Kohli, who have also taught the course many times over these years, gave many valuable suggestions. Many thanks to all of them! Thanks are also due to Mr. Harsh Sharma who painstakingly typed the first draft of the book in LaTeX software.

I hope the book will be useful to both the students and the practicing engineers interested in this exciting field of great practical importance. Any suggestions for its improvement would be gratefully acknowledged.

P.L. Dhar

CHAPTER 1

Introduction

The design of thermal systems—be these power producing systems, like a thermal power plant, or power absorbing systems, like a central air-conditioning plant—has traditionally been carried out using thumb rules based largely on experience. Over the years, consortiums of experienced engineers have been formed by reputed publishers, as well as well-established professional societies like ASME and ASHRAE, to produce authoritative handbooks to help design safe and functional systems. These handbooks give valuable information on the likely range of values of important design parameters, but the final choice of the values of various design parameters for a typical application still rests with the design engineer. Any engineer would naturally wish that his choice of design parameters should result in an "optimum design"—that which best satisfies the requirement of his clients—be it minimization of the cost, weight, or floor area, or the maximization of the efficiency, coefficient of performance, etc. This demands that a large number of alternative designs be obtained and evaluated with respect to the optimization criterion. Traditionally even to arrive at a workable design of a thermal system, with its numerous interconnected components, has been such a laborious task that rarely any attempt was made to arrive at the "optimum design." Many companies did modify their equipment designs progressively, on the basis of the performance of earlier models, and thus slowly the designs were improved over many generations of "models." However, over the last three decades, the possibility of doing high speed computation through desktop computers has made computer-aided design commercially viable. The designers can thus aim at obtaining optimum designs of thermal systems. Increasing competition and rapidly increasing costs of energy have, in fact, made this task imperative.

To appreciate the difference between a "feasible design" and the "optimal design" let us consider a simple, commonly encountered problem: design of a heat exchanger, say, for recovering waste heat from the flue gases leaving a diesel generator (DG) set to produce steam for use in a laboratory. The first step in the design process would be to specify the amount, the temperature, and pressure of the steam required in the laboratory. Let us assume, for the sake of illustration, that the laboratory requires 5 kg/min of dry saturated steam at a pressure of 2 bar. Knowing the minimum temperature of water in winter, say 10°C, we can calculate the amount of heat transfer that should occur in the heat exchanger. The mass flow rate and the temperature of the hot gases exiting from

Thermal System Design and Simulation
http://dx.doi.org/10.1016/B978-0-12-809449-5.00001-2

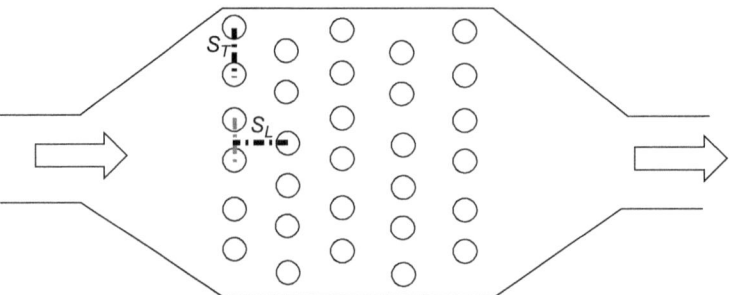

Fig. 1.1 Fin tube heat exchanger for recovering heat from DG set exhaust.

the DG set at the design conditions would be known from its performance data. Now, using this data we need to design a suitable heat exchanger for this application.

The first step in design would be to choose the type of heat exchanger, from among various possibilities [1]. This is usually done by the design engineer on the basis of his experience, keeping in view various systemic constraints, like the type of fluids, their fouling potential and cleanability, the permissible pressure drops, consequences of leakage, etc. In this case, say, we choose tube fin heat exchanger to minimize the pressure drop on the gas side. Next we have to choose the tube diameter and the fin height. This is usually done on the basis of prevalent "good industrial practices." Suppose we choose tubes of base diameter 16 mm, ID of 14 mm, with integral fins on the outside of 2 mm height spaced on the tube with a pitch of 4 mm. We then have to decide on the configuration of the tube bank, the longitudinal and lateral pitches, S_T and S_L, and the size of the duct through which the hot gases would flow, as shown in Fig. 1.1.

These are again chosen based on the past experience with this type of heat recovery units. Thus the two pitches S_T and S_L, could be chosen, typically as 32 and 30 mm, respectively. The sizing of the duct will have to be done to ensure that the gas side velocity is within acceptable limits[1] and all the tubes can be accommodated within a reasonable length. We could choose a gas velocity over the tube bundle, say of 10 m/s, and design the exhaust gas duct accordingly. We then choose a "reasonable" value of velocity of water in the tubes, say 1.5 m/s (usually water velocity is limited to 3 m/s to minimize erosion), and decide on the number of the tubes needed in a single pass of the heat exchanger to achieve the required mass flow rate of water.

Having fixed the basic configurations and the crucial design variables, the heat transfer calculations can be done. We estimate the heat transfer coefficient of the gas

---

[1] If it is too low, the gas side heat transfer coefficient would be very small, and if it is too high, the gas side pressure drop would be high and that would influence the performance of the IC engine of the DG set through increase in its exhaust pressure.

side and the water side by using appropriate correlations from the literature; find the overall heat transfer coefficient and the effective mean temperature difference between the two streams, and then estimate the heat transfer area needed. The length of the heat exchanger tubes can then be calculated. We thus have a feasible design of the heat exchanger which will deliver the required amount of steam.

It is obvious from the above description of the design process that we could have obtained many other "feasible" designs by changing the values of the variables like tube diameter, fin pitch, fin height, velocity of water, velocity of hot gases, longitudinal and lateral pitches of the tube bundle, etc., assumed above on the basis of "good industrial practices." For each change in the value of a design variable, a different design of the heat exchanger would be obtained. We could then choose between them based on a suitable optimization criterion. The optimization criterion could, for example, be minimization of the initial cost, or of the pressure drop on the gas side (since that would influence the performance of the IC engine of the DG set) or a suitable combination of the two. Thus we could, for example, apportion a cost to the steam produced and to the reduction in DG set output caused by gas side pressure drop in the heat exchanger, and combine these suitably with the initial cost of the heat exchanger to evolve a composite function which gives the net increase in profit during the entire projected life of the DG set, due to incorporation of the heat recovery system. The heat exchanger design which maximizes this profit could be chosen as the "optimum design." Another approach could be to constrain the pressure drop on the gas side to be less than a specified value and then minimize the initial cost of the heat exchanger; yet another approach being to maximize the second law efficiency of the heat exchanger.

Now, since the number of possible feasible designs is extremely large, we need to devise suitable strategy to narrow down the search domain in consonance with the chosen objective to arrive at the "optimum design" with minimum computational effort. This necessitates use of mathematical/numerical optimization techniques to generate alternative designs which are likely to be "better" than the initial design. Further, to assess whether the designs so "generated" are feasible, we need a procedure to determine the performance of the system based on such a design. This requires computer-based system simulation procedures. System simulation, in turn, needs comprehensive procedures to predict the performance of each component of the system and a methodology to integrate these procedures in tune with their actual interconnection in the system. Thus a comprehensive computer program to obtain optimal design of a thermal system would essentially be an optimization algorithm to maximize/minimize the objective function subject to the constraints that it provides the required thermal performance[2] without compromising on other nonthermal performance measures like

---

[2] Which is predicted with the help of the system simulation program.

long life, safety, permissible wear and tear, noise, etc.[3] More often than not, the optimization algorithm is a search procedure to generate progressively "better" designs, which need to be checked for feasibility by using the system simulation procedure.

The system simulation procedures are also quite useful in their own right. They enable us to assess the influence of various operating parameters on the performance of the system or its components without actually conducting laboratory tests which are often very expensive and time consuming. We can thus do a sensitivity analysis to identify the relative influence of various components and/or the operating conditions on the overall system performance. Most thermal systems rarely operate on the design conditions, and the system simulation procedures can help us predict their "off-design" performance. This is often of great help in designing suitable control strategies to ensure safe and "optimal" operation even under off-design operating conditions.

1.1 OUTLINE OF THE BOOK

As indicated by the title of the book, its focus is on simulation and design of thermal systems. By the term "system" we imply a collection of components with interrelated performance, and by "simulation" we mean predicting the performance of a system for a given set of input conditions. Thermal systems generally involve transfer of heat and work, often through fluids moving through various components. Thus analysis of the performance of thermal systems demands a through knowledge of the fundamentals of thermodynamics, heat and mass transfer, and fluid mechanics. We shall review these very briefly in Chapter 3.

Most of the equipments used in a thermal system involve heat and mass transfer. For such heat/mass exchangers it is often possible to develop detailed models of process simulation. However, for some important equipment like multistage turbines and compressors, detailed thermofluid modeling is extremely complicated. Often, while simulating thermal systems using such turbomachines, and other equipments where process models would be too cumbersome to incorporate in the system simulation program, the performance curves of such equipments are taken from the manufacturer's catalogs, and converted into equations for ease of use in the computer programs. We also need to fit suitable equations into the thermodynamic property data. Equation fitting is usually done using the "least squares technique." The art and science (or rather mathematics!) of fitting equations into data are discussed in Chapter 2, along with other mathematical techniques needed in thermal system simulation. These include commonly used methods for solution of simultaneous algebraic equations (both linear and nonlinear) and differential equations, basic concepts of Laplace transformation, etc. A brief discussion on engineering economics is also included in this chapter, focusing

---

[3] These nonthermal considerations are often incorporated through constraints on variables, see also Chapter 11.

mainly on the basic principles of converting complex optimization objectives[4] into financial terms.

The detailed process models of typical equipment used in thermal systems like heat and mass exchangers, various types of compressors, gasifier, etc., are presented in Chapter 4. Most thermal systems use atmosphere as a heat source or a sink. Since the environmental conditions (like temperature, humidity content, solar radiation intensity, etc.) are inherently uncertain, and can only be predicted with certain probabilities, a rigorous study of the performance of thermal systems should incorporate this probabilistic description. Such an approach is termed as stochastic simulation, and shall be only briefly discussed in the last chapter of the book. In Chapter 4 the focus is on deterministic simulation, where it is assumed that the input variables are precisely specified. Further, though the focus of the book is primarily on thermal systems like power producing systems (IC engines, steam power plants, etc.) and cold producing systems (like refrigeration and air-conditioning equipment), in this chapter we have also illustrated the application of basic laws of thermodynamics and heat-mass transfer to develop process models for other applications involving transfer of heat and mass like cooling of electronic equipment, manufacturing processes, heat treatment, dehydration of foods, etc.

System simulation involves integration of the models/equations for predicting the performance of various components into a comprehensive procedure which ensures that various conservation equations (like those for mass, momentum, and energy conservation) are satisfied. From a mathematical perspective system simulation involves solution of simultaneous "equations," mostly nonlinear, representing the performance of its components. Many a times these "equations" are actually detailed process models of the equipment wherein various variables are intricately related, usually through differential equations. To evolve a suitable system simulation strategy, the component simulation models are represented in the form of information flow diagrams which indicate the minimum input variables necessary to obtain the desired output information. By suitably choosing the input and output variables, and combining these information flow diagrams judiciously, it is possible to evolve strategies which significantly reduce the computational effort for system simulation. The concept of information flow diagram and its utility in system simulation are discussed in Chapter 5. A few comprehensive case studies of some common thermal systems, illustrating the use of all these concepts are presented in Chapter 6.

The main focus of the book is on optimum design of thermal systems. As discussed briefly above, the objectives of optimum design, the constraints and the design variables, can all vary depending upon the specific requirements of the situation and the designer's preferences. Thus, for example, the objective function could be economic, like

---

[4] Like minimization of the total cost of a plant taking into account both the initial capital cost and the running cost spread over its entire life time.

minimizing the initial investment, or life cycle cost; or thermodynamic, like maximizing some index of system performance (like the thermal efficiency or the coefficient of performance, COP), or minimizing the exergy loss. While ideally all the components of a system should be designed together, a situation on the ground may not permit change in the design of some of the components, and so the design variables would have to be appropriately constrained. The different approaches of formulating the optimal design problems and identifying the design variables and the constraints are discussed in Chapter 7.

The techniques of optimization can be broadly divided into two categories, namely analytical techniques and numerical (or search) techniques. There exists a whole range of analytical and numerical techniques for optimization of various types of functions. A brief discussion of a few of these, which are suitable for nonlinear objective functions commonly encountered while optimizing thermal systems, is presented in Chapter 8. In Chapter 9 firstly a few simple examples of thermodynamic optimization and optimum design of some typical equipment used in thermal systems are presented. These are followed by three comprehensive case studies of the optimum design of the waste heat recovery boiler of a combined cycle power plant, the optimum design of a refrigeration system, and a liquid desiccant-based air-conditioning system.

Design of thermal systems is conventionally done on the basis of their desired steady-state performance. However, in actual operation these systems rarely operate on steady state since the "load" on the system and the environmental conditions are changing continuously. The behavior of a system under transient conditions is therefore of great importance, especially to ensure that performance is not impaired greatly and the system safety is not threatened. A simplified approach at analyzing the dynamic performance of typical thermal systems, and the control systems for these is presented in Chapter 10.

Besides the thermal design, there are many other considerations like material selection, mechanical strength, erosion, noise, etc., which play a crucial role in deciding the final design of most equipment. These are often incorporated as constraints in the optimization of the thermal design and therefore a thermal engineer should have a working knowledge about these factors. A brief discussion on these issues is presented in Chapter 11. With increasing awareness about need to conserve energy, stochastic considerations are also becoming important. Thus while usually a thermal power plant design is optimized for a fixed power output, stochastic approach would require taking into account a probabilistic description of the power demand. A brief presentation on these, as also on some other emerging trends in thermal system design, like use of proprietary softwares, is also included in this last chapter of the book.

REFERENCE

[1] R.K. Shah, D.P. Sekulic, Fundamentals of Heat Exchanger Design, John Wiley & Sons, Inc., Hoboken, NJ, 2003.

CHAPTER 2

Mathematical Background

The core objective of simulating a thermal system is to develop its mathematical model using which it should be possible to predict its performance at different operating conditions. As we shall see in the next chapter, this mathematical model could be in the form of linear/nonlinear algebraic or differential equations. Often, we also need to develop algebraic equations to represent the experimental data on component performance and also various thermodynamic properties of the working substance. Thus knowledge of techniques to solve these equations is a prerequisite for system simulation. In this chapter we shall present a brief overview of the mathematical techniques for fitting equations into data, as also the methods commonly used to solve simultaneous linear and nonlinear algebraic and differential equations. Since in most practical situations we need to take recourse to computers, focus will be only on the methods which are suitable for computerization.

Dynamic simulation of each component invariably involves differential equations and when we have a number of these components interacting with each other, the analysis is greatly facilitated by the use of Laplace transformation. Accordingly a brief presentation on its basic concepts and its utility in solving differential equations is included in this chapter.

For simulating the performance of various thermal equipment, we have to use empirical correlations for predicting heat-transfer coefficients and friction factors. These correlations based on experimental data have inherent uncertainty which, in complex flow situations like boiling, can be as high as 10% to 20%. This can significantly influence the results of simulation. Therefore a brief discussion on analysis of uncertainty and its propagation is also included in this chapter.

Economics is often the key factor in optimal design. Both the initial capital cost, and the cost of "running" and "maintaining" the system over its life time need to be considered. Further the "worth" of money keeps on falling with passage of time due to inflation, interest rates, etc. Thus evolving a suitable economic objective function for optimal design is a complex task. This is briefly outlined in the last section of this chapter.

2.1 LINEAR ALGEBRAIC EQUATIONS

In its most general form, such a set of n-equations can be written as

$$a_{11}\, x_1 + a_{12}\, x_2 + \cdots + a_{1n}\, x_n = b_1$$
$$a_{21}\, x_1 + a_{22}\, x_2 + \cdots + a_{2n}\, x_n = b_2 \tag{2.1}$$
$$\vdots$$
$$a_{n1}\, x_1 + a_{n2}\, x_2 + \cdots + a_{nn}\, x_n = b_n$$

This can also be written in matrix form as

$$[A][X] = [B] \tag{2.2}$$

$$\text{where} \quad [A] = \begin{bmatrix} a_{11} & a_{12} & \cdots & a_{1n} \\ a_{21} & a_{22} & \cdots & a_{2n} \\ \vdots & \vdots & \vdots & \vdots \\ a_{n1} & a_{n2} & \cdots & a_{nn} \end{bmatrix} \tag{2.3}$$

$$\text{where} \quad [X] = \begin{bmatrix} x_1 \\ x_2 \\ \vdots \\ x_n \end{bmatrix} \quad \text{and} \quad [B] = \begin{bmatrix} b_1 \\ b_2 \\ \vdots \\ b_n \end{bmatrix} \tag{2.4}$$

Multiplying Eq. (2.2) on both sides by the inverse of matrix $[A]$, that is $[A^{-1}]$, we get

$$[X] = [A^{-1}][B] \tag{2.5}$$

Thus the solution of a set of linear equations can be obtained by finding the inverse of the coefficient matrix $[A]$. However, direct determination of the inverse of a matrix, for $n > 3$, is an extremely cumbersome task. A computationally more efficient, and popular, method is the *Gaussian elimination* method. It is a direct extension of the familiar method of solving two simultaneous equations by eliminating one unknown from an equation with the help of the other equation, and is illustrated below by solving three simultaneous equations, eliminating one variable at a time.

Equations

$$2x_1 + 3x_2 - x_3 = 39 \tag{2.6}$$

$$10x_1 - 2x_2 + 5x_3 = -25 \tag{2.7}$$

$$-7x_1 + 5x_2 + 7x_3 = 1 \tag{2.8}$$

Step 1: Eliminate coefficient of x_1, from Eqs. (2.7), (2.8) by using Eq. (2.6). For this we first multiply Eq. (2.6) by 5 (on both sides) and subtract Eq. (2.7) from it. Similarly we multiply Eq. (2.6) by 3.5 and add Eq. (2.8) to it. This gives

$$2x_1 + 3x_2 - x_3 = 39 \tag{2.9}$$

$$17x_2 - 10x_3 = 220 \tag{2.10}$$

$$15.5x_2 + 3.5x_3 = 137.5 \tag{2.11}$$

Step 2: Eliminate coefficient of x_2 from Eq. (2.11) by using Eq. (2.10). For this, we multiply this equation by $\left(\frac{15.5}{17}\right)$ and subtract Eq. (2.11) from it. This gives

$$2x_1 + 3x_2 - x_3 = 39 \tag{2.12}$$

$$17x_2 - 10x_3 = 220 \tag{2.13}$$

$$-12.6176x_3 = 63.088 \tag{2.14}$$

Eq. (2.14) gives $x_3 = -5$.
Substituting it in Eq. (2.13) we get

$$x_2 = 10$$

Substituting values of x_2 and x_3 in Eq. (2.12), we get

$$x_1 = 2.0$$

Thus the solution to the three equations is $x_1 = 2$, $x_2 = 10$, and $x_3 = -5$.

This method can be easily computerized by combining the $[A]$ and $[B]$ matrices into an augmented matrix $[A|B]$ and performing elementary row transformations to convert it into upper triangular form. Thus augmented matrix for these equations would be

$$\begin{bmatrix} 2 & 3 & -1 & | & 39 \\ 10 & -2 & 5 & | & -25 \\ -7 & 5 & 7 & | & 1 \end{bmatrix}$$

Then row transformations could be done sequentially as done above. Thus in the first attempt we use the transformation rule [Row A × coeff. of x_1 in Row 1 − Row 1 × coeff. of x_1 in Row A] to bring zero's in the first column in all the rows except the first row. This transforms the matrix to

$$\begin{bmatrix} 2 & 3 & -1 & 39 \\ 0 & -34 & 20 & -440 \\ 0 & 31 & 7 & 275 \end{bmatrix}$$

Similarly to transform the third row, we use the rule: [Row 3 × coeff. of x_2 in Row 2 − Row 2 × coeff. of x_2 in Row 1] to get

$$\begin{bmatrix} 2 & 3 & -1 & 39 \\ 0 & -34 & 20 & -440 \\ 0 & 0 & -858 & 4290 \end{bmatrix}$$

Thus from the last row we get

$$x_3 = \frac{4290}{-858} = -5$$

Substituting in the second row, we get

$$x_2 = \frac{-440 + 100}{-34} = 10$$

Substituting these values in the first row we get as before

$$x_1 = 2.0$$

2.1.1 Difficulties Encountered in Gaussian Elimination

During implementation of the simple Gaussian elimination, sometimes difficulties are encountered which may prematurely terminate the solution procedure (due to encountering zero coefficient on a diagonal element) or introduce large round-off errors. For exhaustive treatment the student is advised to refer to specialist books [1, 2]. Here a few practical tips are mentioned.

Thus if the term on the diagonal of the coefficient matrix is zero, then it is not possible to eliminate that variable from the remaining equations. This difficulty can be easily overcome by just interchanging the rows of the augmented matrix, that is by using another equation which has a nonzero coefficient for that variable as the pivot equation to eliminate that variable from the remaining equations. Of course, if after triangularization we find zero on the diagonal of the last equation, this implies that the equation is not independent of the set of remaining equations. Thus no solution is possible.

Difficulties may also be encountered if the pivot element though nonzero, is very small in comparison to the new coefficients in column vector. This can give rise to large round-off errors, especially during hand calculations. Such problems can be handled by pivoting, that is, rearranging the equations to put the coefficient of largest magnitude on the diagonal. Large roundoff errors also occur when the coefficients of different equations greatly differ from each other. This can happen due to differing units, for example, in equations relating the thermocouple output to temperature, the output would be in millivolts while the temperature may be in hundreds on Kelvin scale. In such situations it helps to do scaling, for example, by dividing each equations by the largest coefficient in it, so that the maximum coefficient of all the equations is of the order of 1. Often a combination of scaling and pivoting is necessary to get accurate solutions, as illustrated in following example.

Example 2.1

Solve the following set of equations both without and with pivoting, taking four significant digits during the calculations.

$$0.0001x_1 + 3x_2 - x_3 = -2.995$$
$$x_1 + 9x_2 - 7x_3 = 25$$
$$0.2x_1 - 15x_2 + x_3 = 9$$

Solution
Without Pivoting

Carrying out the triangularization, as before, we get the equations:

$$0.0001x_1 + 3x_2 - x_3 = -2.995$$
$$-2.9991x_2 - 0.9993x_3 = -2.9975$$
$$-0.0010x_3 = -0.0038$$

which give the solution as:

$$x_3 = 3.8, \quad x_2 = 0.2667, \quad \text{and} \quad x_1 = 49$$

With Pivoting

The equations are rearranged as

$$x_1 + 9x_2 - 7x_3 = 25$$
$$0.0001x_1 + 3x_2 - x_3 = -2.995$$
$$0.2x_1 - 15x_2 + x_3 = 9$$

On triangularization, we get

$$x_1 + 9x_2 - 7x_3 = 25$$
$$-2.9991x_2 + 0.9993x_3 = 2.9975$$
$$9.5904x_3 = 38.3616$$

which give the solution as

$$x_3 = 4, \quad x_2 = 0.3333, \quad \text{and} \quad x_1 = 50.0003$$

which is quite near the exact solution of

$$x_3 = 4, \quad x_2 = 1/3, \quad \text{and} \quad x_1 = 50$$

The students are encouraged to solve these equations retaining only three significant digits during the calculations and verify that without pivoting the solution obtained is $x_3 = 6, x_2 = 1$, and $x_1 = 50$; while with pivoting the solution is $x_3 = 4.002, x_2 = 0.334$, and $x_1 = 50.008$.

Example 2.2

Solve the following set of simultaneous equations using simple Gaussian elimination.

$$3.1x_1 + 225x_2 + 100x_3 = 475.5$$
$$210x_1 - 85x_2 + x_3 = 880.1$$
$$2x_1 - 5x_2 + 4x_3 = 0.4$$

Solve these again after scaling and compare the results with those obtained earlier.

Solution

The augmented matrix for the problem is

$$\begin{bmatrix} 3.1 & 225 & 100 & 475.5 \\ 210 & -85 & 1 & 880.1 \\ 2 & -5 & 4 & 0.4 \end{bmatrix}$$

To bring out the advantage of scaling, we first solve these equations carrying only four significant digits.

Triangularization of the above matrix by Gaussian elimination gives

$$\begin{bmatrix} 3.1 & 225 & 100 & 475.5 \\ 0 & 47{,}510 & 20{,}100 & 97{,}120 \\ 0 & 0 & 444{,}000 & 80{,}000 \end{bmatrix}$$

The solution as obtained by back substitution is

$$x_3 = \frac{80{,}000}{444{,}000} = 0.1802$$

$$x_2 = \frac{97{,}120 - 20{,}100 \times 0.1802}{47{,}510} = 1.968$$

$$x_1 = \frac{475.5 - 100 \times 0.1802 - 225 \times 1.968}{3.1}$$
$$= 4.742$$

which is quite different from the exact solution of $x_3 = 0.1$, $x_2 = 2.0$, and $x_1 = 5.0$.

If we scale the first two equations by dividing each by 100 on both sides, we get

$$0.031x_1 + 2.25x_2 + x_3 = 4.755$$
$$2.1x_1 - 0.85x_2 + 0.01x_3 = 8.801$$
$$2x_1 - 5x_2 + 4x_3 = 0.4$$

Now all the coefficients are of similar order of magnitude, and we can solve the equations by Gaussian elimination. However, before doing so we interchange the first two equations so that the diagonal element has the largest value in that column. The new augmented matrix is

$$\begin{bmatrix} 2.1 & -0.85 & 0.01 & 8.801 \\ 0.031 & 2.25 & 1 & 4.755 \\ 2 & -5 & 4 & 0.4 \end{bmatrix}$$

After triangularization we get (retaining only four significant digits during calculations)

$$\begin{bmatrix} 2.1 & -0.85 & 0.01 & 8.801 \\ 0 & -4.751 & -2.1 & -9.713 \\ 0 & 0 & -58.29 & -5.83 \end{bmatrix}$$

which gives

$$x_3 = \frac{-5.83}{-58.29} = 0.1000$$

$$x_2 = \frac{-9.713 + 2.1 \times 0.1}{-4.751} = 2.000$$

$$x_1 = \frac{8.801 - 0.01 \times 0.1 + 0.85 \times 2}{2.1} = 5.00$$

Thus we get the exact solution after scaling and pivoting even when the calculations are done retaining only four significant digits.

2.2 NONLINEAR ALGEBRAIC EQUATIONS

Before discussing procedures for solving a set of nonlinear equations, we shall review, in brief, the methods for solving a single nonlinear equation.

It is obvious that iterations would be necessary, and one of the most popular methods of iteration is the Newton-Raphson method (NRM). Here, starting from an estimate x_i of the root, the next estimate x_{i+1} is found as:

$$x_{i+1} = x_i - \frac{f(x_i)}{f'(x_i)} \tag{2.15}$$

where $f(x) = 0$ is equation to be solved, $f(x_i)$ the value of function $f(x)$ at the initial estimate, and $f'(x_i)$ is the value of the gradient of function $f(x)$ evaluated at $x = x_i$. Fig. 2.1 gives a graphical representations of Eq. (2.15) and shows how through successive iterations we approach the correct solution.

This method works quite well if the initial guess is near the correct solution and the values of the gradient do not change very rapidly. Fig. 2.2 indicates a few cases where the method would not be able to drive the iterations toward the correct solution. In the view of this difficulty, methods have been devised to ensure that the iterations converge

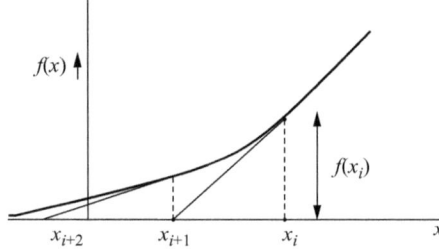

Fig. 2.1 Newton-Raphson method.

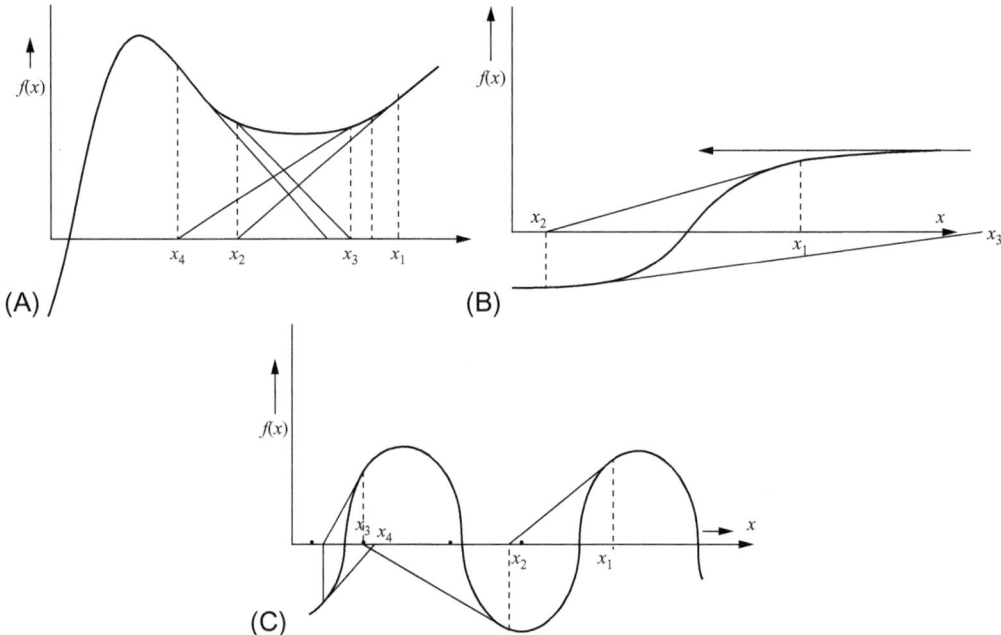

Fig. 2.2 A few cases illustrating poor convergence of Newton-Raphson iterations.

to the solution. It has been found that if the solution is first bracketed to lie within two limiting values corresponding to which the function values are of opposite signs, most iterative methods would then converge to the correct solution. In case of equations encountered in thermal system simulation and design, usually it is possible to identify such limiting values of the variable on the basis of physical considerations. Otherwise, we could arbitrarily take a "very small" and a "very large" value of the variable and increase the gap between them till the values of $f(x)$ at the two ends are of opposite sign. Once the solution is bracketed, its correct value can be found by using any iterative procedure.

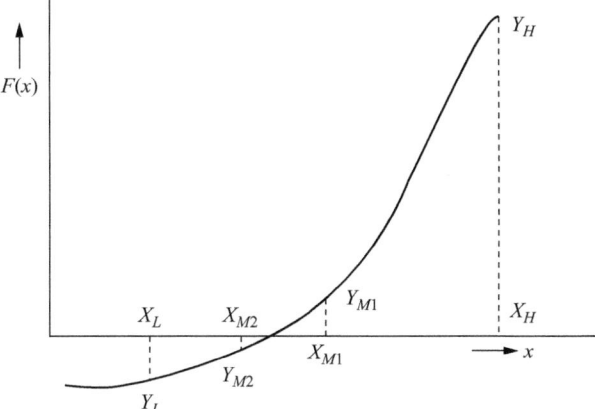

Fig. 2.3 Bisection method.

A simple, and yet effective procedure is the bisection technique. If x_L and x_H denote the two limiting values of x for which the function values $f(x_L)$ and $f(x_H)$ are of opposite sign, then the value of the function is evaluated at $x_{m1} = (x_L + x_H)/2$ and $f(x_{M1})$ compared with $f(x_L)$ and $f(x_H)$. Keeping x_{M1} as one of the new limiting values, the other limiting value is chosen as x_L or x_H whichever has the function value of sign opposite to that of $f(x_m)$ (see Fig. 2.3).

Thus in the case illustrated in Fig. 2.3, the initial set of limiting values is x_L, x_H, and after first bisection, the next set is (x_L, x_{M1}). In the next iteration the bisection point is x_{M2}, and on the basis of the values of $f(x_L)$ and $f(x_{M1})$, the next interval bracketing the solution is (x_{M2}, x_{M1}). It should be clear from Fig. 2.3 that algorithm would converge to the correct solution within a few iterations.

An algorithm, which converges faster than the successive bisection procedure outlined above, is the method of false position, or the *regula-falsi* method. It makes use of the values of the function at the two limiting values to estimate the value of the root of the equation. This is done by replacing the actual curve by a straight line as shown graphically in Fig. 2.4. Since this replacement gives a false position of the root the method is known as *regula-falsi* or the method of false position.

Mathematically it involves locating an estimate of the root by linear interpolation between the two points already known, namely x_L and x_H (Fig. 2.4). Thus the new value ("false position") of the root is

$$x_{FP1} = \frac{x_H f(x_L) - x_L f(x_H)}{f(x_L) - f(x_H)} \tag{2.16}$$

As before, for the next iteration we choose the limiting values as x_{FP1} and x_H since the values of the function for these two points are of opposite sign, and find the new estimate as:

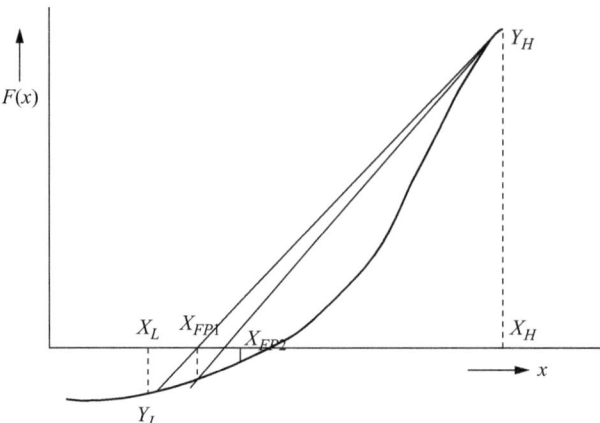

Fig. 2.4 Regula-falsi method.

$$x_{FP2} = \frac{x_H\, f(x_{FP1}) - x_{FP1}\, f(x_H)}{f(x_{FP1}) - f(x_H)}$$

The iteration is carried forward till the correct solution is obtained (within desired accuracy). In most practical situations the *regula-falsi* method gives faster convergence than the bisection method.

Example 2.3

The effectiveness ϵ of a cross-flow heat exchanger is given by the equation

$$\epsilon = 1 - \exp\left[\frac{1}{C_r}(N)^{0.22}\{\exp[-C_r(N)^{0.78}] - 1\}\right]$$

where N, called the number of transfer units, is a factor dependent on heat exchanger size and heat-transfer coefficients, and C_r is the ratio of heat capacities of the two fluids.

Determine the value of N of the heat exchanger which has an effectiveness of 0.7 with $C_r = 0.8$.

Solution

Substituting the given values of C_r and ϵ in the above equation we get the nonlinear equation for N whose solution could be found by any of the methods mentioned above.

We firstly write the equation:

$$f(N) = 0.3 - \exp[1.25N^{0.22}\{\exp[-0.8N^{0.78}] - 1\}] = 0$$

Let us first try using NRM. Differentiating the equation we get

$$f'(N) = -(1.25 \times 0.22N^{-0.78}\{\exp[-0.8N^{0.78}] - 1\} + 1.25N^{0.22}$$
$$\cdot \{-0.8 \times 0.78N^{-0.22}\exp[-0.8N^{0.78}]\})$$
$$\cdot \exp[1.25N^{0.22}\{\exp[-0.8N^{0.78}] - 1\}]$$

Taking the value of $N = 0.5$, as the initial assumption we calculate newer estimates using Eq. (2.15). The results are given in following table:

Iteration no.	N_i	$f(N_i)$	$f'(N_i)$
1	0.5	−0.37053	0.44615
2	1.330501	−0.1312	0.183758
3	2.044488	−0.03244	0.1035
4	2.35789	−0.00331	0.083381
5	2.397603	−4.3E−05	0.081233
6	2.398131	−7.4E−09	0.081205

Thus the method converges to the correct solution in six iterations.

Bracketing the Solution

We take two limiting trial values of the variable as 0.5 and 5.0. The values of the function for these values are

$$f(0.5) = -0.37053$$
$$f(5.0) - +0.112422$$

We could now try bisection method. The calculation steps are given in following table:

x_i	$f(x_i)$	x_i	$f(x_i)$
0.5	−0.37053	5.0	+0.112422
		2.75	+0.025607
1.625	−0.08347		
2.1875	−0.01836		
		2.46875	0.005615
2.328125	−0.00582		
		2.398437	2.48E−05
2.363281	−0.00286		
2.381359	−0.00137		
2.389898	−0.00067		
2.3941675	−0.00032		
2.396302	−0.00015		
2.397365	−6.2E−05		
2.397901	−1.9E−05	2.398169	3.08E−06
2.398035	−7.8E−06		
2.398102	−2.4E−06	2.398135	3.15 E−07
2.398119	−9.8E−07		
2.398127	−3.3E−07		

Once again we obtain the correct solution but in a larger number of steps than in the NRM. We will also find the solution by the *regula-falsi* approach, using Eq. (2.16) to advance the iterations. The details of the iterations are given in following table:

x_i	$f(x_i)$	x_i	$f(x_i)$
0.5	−0.37053	5.0	+0.112422
		3.952	+0.082989
		3.320	+0.057306
		2.9422	+0.037451
		2.718	+0.023502
		2.5857	+0.014348
		2.5079	+0.008604
		2.4623	+0.005104
		2.4356	+0.003006
		2.4200	+0.001763
		2.4109	+0.001033
		2.40558	+0.000603
		2.40248	+0.000353
		2.40067	+0.000206
		2.39961	+0.00012
		2.39899	6.7E−05
		2.39864	4.13E−05
		2.39843	2.43E−05
		2.39830	1.37E−05
		2.39823	8.03E−06
		2.398188	4.62E−06
		2.398164	2.67E−06
		2.398150	1.53E−06
		2.398142	8.83E−07
		2.398137	4.77E−07
		2.398135	3.15E−07
		2.398133	1.52E−07
		2.398132	7.13E−08

Here too, we do reach the exact solution, but in much larger number of iterations.[a]

A comparison of the iterations involved in each method indicates that NRM converges most rapidly, arriving at the correct solution in just six iterations while other methods need 18–26 iterations. This is generally true for most functions, if the initial value is near the correct solution. However, if the initial guess is far from the correct solution, difficulties can arise. In the present problem too, the students are advised to verify that for a stating value of $N = 6.0$, the NRM would predict the next estimate of N as −2.137, which is an infeasible value and further calculations can not be done in view of exponents of N involved in the function. The method thus stops after one iteration without reaching the solution. We also note that in both the bisection and the *regula-falsi* methods, as the estimate approaches the correct solution, the "rate of convergence" shows down considerably. In order to illustrate this, we kept our "accuracy" requirements high, demanding that the function value should be of very low, of the order of 10^{-8}. In case function values of the order of 10^{-5} are acceptable, the number of iterations is reduced considerably. It is therefore of great importance to select the requirement of "accuracy" of final result properly especially in practical situations involving solution of many simultaneous nonlinear equations.

[a] This difference in the number of iterations does not translate into similar difference in the computational time since in every iterations of the NRM an additional computation of the gradient of the function is needed. This may even offset the advantage accruing from smaller number of iterations, especially in cases, like the present example, where the gradient function is more complex involving many mathematical operations.

NRM can be easily extended to solve simultaneous nonlinear equations. Thus considering a set of n-equations, in n-variables $x_1, x_2, \ldots, x_n$, written in a general form as:

$$f_1(x_1, x_2, \ldots, x_n) = 0$$
$$f_2(x_1, x_2, \ldots, x_n) = 0$$
$$\vdots$$
$$f_n(x_1, x_2, \ldots, x_n) = 0 \tag{2.17}$$

if $[X]^i = (x_1^i, x_2^i, \ldots, x_n^i)$ is the set of values of the variables assumed/obtained at iteration i, the next estimate of the solution as provided by NRM is

$$[X]^{i+1} = (x_1^{i+1}, x_2^{i+1}, \ldots, x_n^{i+1})$$

where

$$x_1^{i+1} = (x_1^i + \Delta x_1^i), \quad x_2^{i+1} = x_2^i + \Delta x_2^i,$$
$$x_3^{i+1} = (x_3^i + \Delta x_3^i), \ldots, x_n^{i+1} = x_n^i + \Delta x_n^i \tag{2.18}$$

and the increments are determined from solution of n-linear equations:

$$\begin{bmatrix} \dfrac{\partial f_1}{\partial x_1} & \dfrac{\partial f_1}{\partial x_2} & \cdots & \dfrac{\partial f_1}{\partial x_n} \\ \dfrac{\partial f_2}{\partial x_1} & \dfrac{\partial f_2}{\partial x_2} & \cdots & \dfrac{\partial f_2}{\partial x_n} \\ \vdots & \vdots & \vdots & \vdots \\ \dfrac{\partial f_n}{\partial x_1} & \dfrac{\partial f_n}{\partial x_2} & \cdots & \dfrac{\partial f_n}{\partial x_n} \end{bmatrix} \begin{bmatrix} \Delta x_1^i \\ \Delta x_2^i \\ \vdots \\ \Delta x_n^i \end{bmatrix} = - \begin{bmatrix} f_1 \\ f_2 \\ \vdots \\ f_n \end{bmatrix}_i \tag{2.19}$$

The subscript i appended to the matrices in Eq. (2.19) indicates all the values in the matrices are calculated for the set of values of variables $(x_1^i, x_2^i, \ldots, x_n^i)$.

This method converges rapidly if one starts from an estimate close to the correct solution, otherwise the method may diverge and not reach the solution. Moreover, the number of function calculations needed at each step is quite high, namely $n^2 + n$. In many cases encountered in thermal system simulation, direct estimation of the partial derivative in Eq. (2.19) may not be possible. This increases the computations since these derivatives have then to be calculated numerically. Due to errors inherent in the numerical calculation of the partial derivatives, the probability of divergence of the solution is increased. A more robust method for solution of a system of complex nonlinear equations has been proposed by Warner [3]. This is especially suited for equations where it is difficult to find the partial derivatives of various functions. As this method has been widely used in thermal system simulation and is not discussed much in popular text books, we shall present the method in full detail in the following section.

2.2.1 Warner's Method

To develop the theoretical foundation of this method, let us presume that we have an estimate $[X]^i$ for the solution to Eq. (2.17). On substituting the values of the variables, namely $x_1^i, x_2^i, \ldots, x_n^i$, we can calculate the values of the functions on the RHS of all the equations. Let these be indicated as the residual vector

$$[R]^i \equiv (R_1^i, R_2^i, R_3^i, \ldots, R_n^i)$$

with

$$R_1^i = f_1(x_1^i, x_2^i, \ldots, x_n^i)$$

$$R_2^i = f_2(x_1^i, x_2^i, \ldots, x_n^i)$$

$$\vdots$$

$$R_n^i = f_n(x_1^i, x_2^i, \ldots, x_n^i) \tag{2.20}$$

Now, if we assume that the above equations have a unique solution[1] then it should be possible to invert the functional relationship and express the variables $[x]^i$ in terms of the $[R]^i$ as

$$x_1^i = \phi_1(R_1^i, R_2^i, \ldots, R_n^i)$$

$$x_2^i = \phi_2(R_1^i, R_2^i, \ldots, R_n^i)$$

$$\vdots$$

$$x_n^i = \phi_n(R_1^i, R_2^i, \ldots, R_n^i) \tag{2.21}$$

If the assumed values $[x]^i$ are not very far from the correct solution $[C]$ we can expand above functional relationships using Taylor's series expansion:

---

[1] In most practical situations involving thermal systems it is usually true.

$$x_1^i = c_1 + \left(\frac{\partial \phi_1}{\partial R_1}\right)_i (R_1^i - R_{1,c}) + \left(\frac{\partial \phi_1}{\partial R_2}\right)_i (R_2^i - R_{2c}) + \cdots$$

$$x_2^i = c_2 + \left(\frac{\partial \phi_2}{\partial R_1}\right)_i (R_1^i - R_{1,c}) + \left(\frac{\partial \phi_2}{\partial R_2}\right)_i (R_2^i - R_{2c}) + \cdots$$

$$\vdots$$

$$x_n^i = c_n + \left(\frac{\partial \phi_n}{\partial R_1}\right)_i (R_1^i - R_{1,c}) + \left(\frac{\partial \phi_n}{\partial R_2}\right)_i (R_2^i - R_{2c}) + \cdots \quad (2.22)$$

Since $[C]$ is the vector of correct solution to Eq. (2.17) it follows that

$$R_{1c} = 0, \ R_{2c} = 0, \ \ldots, \ R_{nc} = 0$$

Hence the above set of equations can be expressed in matrix form as

$$[x_1^i, x_2^i, \ldots, x_n^i] = [R_1^i, R_2^i, R_3^i, \ldots, R_n^i \ 1]
\begin{bmatrix}
\frac{\partial \phi_1}{\partial R_1} & \frac{\partial \phi_2}{\partial R_1} & \cdots & \frac{\partial \phi_n}{\partial R_1} \\
\frac{\partial \phi_1}{\partial R_2} & \frac{\partial \phi_2}{\partial R_2} & \cdots & \frac{\partial \phi_n}{\partial R_2} \\
\vdots & \vdots & \vdots & \vdots \\
\frac{\partial \phi_1}{\partial R_n} & \frac{\partial \phi_2}{\partial R_n} & \cdots & \frac{\partial \phi_n}{\partial R_n} \\
c_1 & c_2 & \cdots & c_n
\end{bmatrix} \quad (2.23)$$

Now if the values of the residuals $[R]^i$ are calculated for $n+1$ sets of values of variables $[X]^i$, we can get $n+1$ equations of the above type. These can be combined and written in matrix notation as

$$\begin{bmatrix}
x_1^1 & x_2^1 & \cdots & x_n^1 \\
x_1^2 & x_2^2 & \cdots & x_n^2 \\
\vdots & \vdots & \vdots & \vdots \\
x_1^{n+1} & x_2^{n+1} & \cdots & x_n^{n+1}
\end{bmatrix} =
\begin{bmatrix}
R_1^1 & R_2^1 & \cdots & R_n^1 & 1 \\
R_1^2 & R_2^2 & \cdots & R_n^2 & 1 \\
\vdots & \vdots & \vdots & \vdots & \vdots \\
R_1^{n+1} & R_2^{n+1} & \cdots & R_n^{n+1} & 1
\end{bmatrix}
\begin{bmatrix}
\frac{\partial \phi_1}{\partial R_1} & \frac{\partial \phi_2}{\partial R_1} & \cdots & \frac{\partial \phi_n}{\partial R_1} \\
\frac{\partial \phi_1}{\partial R_2} & \frac{\partial \phi_2}{\partial R_2} & \cdots & \frac{\partial \phi_n}{\partial R_2} \\
\vdots & \vdots & \vdots & \vdots \\
\frac{\partial \phi_1}{\partial R_n} & \frac{\partial \phi_2}{\partial R_n} & \cdots & \left(\frac{\partial \phi_n}{\partial R_n}\right) \\
c_1 & c_2 & \cdots & c_n
\end{bmatrix}$$

$$(2.24)$$

or in more compact from as

$$[\chi] = [\rho][\psi] \quad (2.25)$$

where

$$[\chi] =
\begin{bmatrix}
x_1^1 & x_2^1 & \cdots & x_n^1 \\
x_1^2 & x_2^2 & \cdots & x_n^2 \\
\vdots & \vdots & \vdots & \vdots \\
x_1^{n+1} & x_2^{n+1} & \cdots & x_n^{n+1}
\end{bmatrix} \quad (2.26)$$

$$[\rho] = \begin{bmatrix} R_1^1 & R_2^1 & \cdots & R_n^1 & 1 \\ R_1^2 & R_2^2 & \cdots & R_n^2 & 1 \\ \vdots & \vdots & \vdots & \vdots & 1 \\ R_1^{n+1} & R_2^{n+1} & \cdots & R_n^{n+1} & 1 \end{bmatrix} \tag{2.27}$$

$$[\psi] = \begin{bmatrix} \dfrac{\partial \phi_1}{\partial R_1} & \dfrac{\partial \phi_2}{\partial R_1} & \cdots & \dfrac{\partial \phi_n}{\partial R_1} \\ \dfrac{\partial \phi_1}{\partial R_2} & \dfrac{\partial \phi_2}{\partial R_2} & \cdots & \dfrac{\partial \phi_n}{\partial R_2} \\ \vdots & \vdots & \vdots & \vdots \\ \dfrac{\partial \phi_1}{\partial R_n} & \dfrac{\partial \phi_2}{\partial R_n} & \cdots & \left(\dfrac{\partial \phi_n}{\partial R_n}\right) \\ c_1 & c_2 & \cdots & c_n \end{bmatrix} \tag{2.28}$$

Eq. (2.24) can be solved for matrix $[\psi]$ as

$$[\psi] = [\rho]^{-1}[\chi] \tag{2.29}$$

The last row of the matrix ψ gives a better estimate of the correct solution, that is $(c_1, c_2, \ldots, c_n)$. Thus starting from $(n+1)$ arbitrary sets of values of the variables, a better estimate of the correct solution can be obtained.

In the next step the "poorest" set of assumed values, out of the initial $(n+1)$ sets, is replaced by the newly obtained estimate and the procedure continued. Identification of the "poorest" set is done on the basis of a comparison of the "total error" for each set defined as

$$T_i = \sum_{j=1}^{n} |R_j^i| \quad \text{for } i = 1, 2, \ldots, n+1 \tag{2.30}$$

If the "poorest" set is the kth set, with values of variables $x_1^k, x_2^k, \ldots, x_n^k$ then various terms in Eq. (2.23) corresponding to this set are replaced by the newly calculated values $(c_1, c_2, \ldots, c_n)$ and the corresponding values of the residuals. A new improved estimate for the correct solution is then obtained by solving Eq. (2.24) again. The procedure is continued till one of the values of T_i becomes less than a prespecified value. The corresponding set of variables is then the required solution.

An important advantage of this method is that it does not entail calculation of any partial derivatives and relies only on the function calculations. Thus computational effort in each iteration is much smaller than in the NRM. Of course, its rate of convergence to the correct solution is slower than in NRM but this also helps in avoiding heavy "oscillations" that are often encountered with NRM.

We shall illustrate the method by a simple example.

Example 2.4

In a vapor compression refrigeration plant the heat transfer in the condenser (Q_c), the cooling in the evaporator (Q_e) as also the power consumption in the compressor (P) are mainly dependent on the evaporator and the condenser temperatures, namely T_e and T_c. The experimental data have been fitted into following equations:

$$Q_e = 12.5 + 0.2T_e + 0.001T_e^2 - 0.05T_c - 0.0005T_c^2$$
$$Q_c = 5.5 + 0.2T_c + 0.001T_c^2$$
$$P = 3 + 0.01T_c + 0.005T_cT_e + 0.06T_e$$

where T_c and T_e are in °C and Q_c, Q_e, and P are in kW.

Determine the values of T_c and T_e for the case when the power consumption is 5.427 kW.

Solution

We know in a vapor compression refrigeration plant, the overall energy balance demands that

$$Q_e + P = Q_c$$

This gives an equation in T_e and T_c as

$$12.5 + 0.2T_e + 0.001T_e^2 - 0.05T_c - 0.0005T_c^2 + 3 + 0.01T_c$$
$$+ 0.005T_cT_e + 0.06T_e = 5.5 + 0.2T_c + 0.001T_c^2$$

which on rearranging becomes

$$10 + 0.26T_e + 0.001T_e^2 - 0.24T_c - 0.0015T_c^2 + 0.005T_cT_e = 0$$

Let us call this as $f_1(T_e, T_c) = 0$

The second equation arises from the prescribed value of compressor power:

$$3 + 0.01T_c + 0.005T_cT_e + 0.06T_e = 5.427$$

which can be rearranged as

$$f_2(T_e, T_c) = 0.01T_c + 0.005T_cT_e + 0.06T_e - 2.427 = 0$$

We shall solve these two nonlinear equations for two variables T_c and T_e by both the methods discussed above.

Newton-Raphson Method

We first write expressions for the four partial derivatives of Eq. (2.19)

$$\frac{\partial f_1}{\partial T_e} = 0.26 + 0.002T_e + 0.005T_c$$

$$\frac{\partial f_1}{\partial T_c} = -0.24 - 0.003T_c + 0.005T_e$$

$$\frac{\partial f_2}{\partial T_e} = 0.005\,T_c + 0.06$$

$$\frac{\partial f_2}{\partial T_c} = 0.01 + 0.005\,T_e$$

As a first guess, we assume $T_e = 0°C$, $T_c = 40°C$.

Substituting these values in above equations, the next set of values can be found using Eqs. (2.18), (2.19). Eq. (2.19) gives

$$\begin{bmatrix} 0.46 & -0.36 \\ 0.26 & 0.01 \end{bmatrix} \begin{bmatrix} \Delta T_e \\ \Delta T_c \end{bmatrix} = - \begin{bmatrix} -2.0 \\ -2.027 \end{bmatrix}$$

These two linear equations can be solved using any of the methods discussed in Section 3.1 to get

$$\Delta T_e = 7.635 \quad \Delta T_c = 4.198$$

This gives the next set of estimates of the variables T_e and T_c as

$$T_e = 7.635°C \quad T_c = 40 + 4.198 = 44.198°C$$

Using these values, we recalculate the partial derivatives and the function values. Eq. (2.19) now becomes

$$\begin{bmatrix} 0.49626 & -0.33442 \\ 0.28099 & 0.48175 \end{bmatrix} \begin{bmatrix} \Delta T_e \\ \Delta T_c \end{bmatrix} = - \begin{bmatrix} -0.192937 \\ -0.160339 \end{bmatrix}$$

whose solution is

$$\Delta T_e = -0.53374 \quad \Delta T_c = -0.21511$$

The new set of estimates is

$$T_e = 7.102°C \quad T_c = 43.983°C$$

When we calculate the function values for these temperatures we get

$$f_1 = 0.001118 \quad \text{and} \quad f_2 = 0.000786$$

since these values are quite small, we can terminate the iterations and take solution as

$$T_e = 7.1°C \quad \text{and} \quad T_c = 43.98°C.$$

Warner's Method

Here, we have to find the function values for three sets of assumptions for T_c and T_e. We take these as $(0, 40), (5, 40)$, and $(5, 45)$, and calculate the values of f_1 and f_2 for each of these. This gives us the residual matrix $[\rho]$ as

$$[\rho] = \begin{bmatrix} -2 & -2.027 & 1 \\ 0.325 & -0.727 & 1 \\ -1.3875 & -0.552 & 1 \end{bmatrix}$$

We find the inverse of this matrix using online matrix calculator [4] and substitute it in Eq. (2.29) to get

$$[\psi] = \begin{bmatrix} -0.0595 & 0.56266 & -0.50317 \\ -0.058221 & -0.20824 & 0.79045 \\ -0.40393 & 0.66575 & -0.73818 \end{bmatrix} \begin{bmatrix} 0 & 40 \\ 5 & 40 \\ 5 & 45 \end{bmatrix} = \begin{bmatrix} 0.29745 & -2.51625 \\ 2.91105 & 3.95225 \\ 7.01965 & 43.69090 \end{bmatrix}$$

This gives the next set of assumptions as

$$T_e = 7.02 \quad \text{and} \quad T_c = 43.69$$

Further we find the magnitude of "total" error for each of the three sets of values used in above calculations, in accordance with Eq. (2.30). The errors are

$$T_1 = 2 + 2.027 = 4.027$$

$$T_2 = 0.325 + 0.727 = 1.052$$

$$T_3 = 1.3875 + 0.552 = 1.9395$$

T_1 being the largest, we replace the first set of values of T_e and T_c by the new estimates, and find the values of f_1 and f_2 for these values. The reconstituted $[\rho]$ matrix is

$$[\rho] = \begin{bmatrix} 0.059175 & -0.03538 & 1 \\ 0.325 & -0.727 & 1 \\ -1.3875 & 0.552 & 1 \end{bmatrix}$$

And the new $[\psi]$ matrix is

$$[\psi] = \begin{bmatrix} 0.15379 & 0.45402 & -.60781 \\ 1.50499 & -1.27138 & -0.23361 \\ 1.04415 & -0.07185 & 0.02770 \end{bmatrix} \begin{bmatrix} 7.02 & 43.69 \\ 5 & 40 \\ 5 & 45 \end{bmatrix} = \begin{bmatrix} 0.31066 & -2.47156 \\ 3.04008 & 4.38536 \\ 7.10918 & 43.99141 \end{bmatrix}$$

Thus the next estimate of the correct solution is

$$T_e = 7.109 \quad T_c = 43.99$$

for which the function values are

$$f_1 = 0.001886 \quad \text{and} \quad f_2 = 0.00311$$

Since f_2 is still significant we do another iteration. The "total error" values for this set are

$$T_1 = 0.059175 + 0.03538 = 0.094555$$

$$T_2 = 0.325 + 0.727 = 1.052$$

$$T_3 = 1.3875 + 0.552 = 1.9375$$

Clearly T_3 is the largest and so we replace the third set of values of the variables with the new estimate found above. The new $[\rho]$ matrix is

$$[\rho] = \begin{bmatrix} 0.059175 & -0.03538 & 1 \\ 0.325 & -0.727 & 1 \\ 0.001886 & 0.00311 & 1 \end{bmatrix}$$

and the new ψ matrix is

$$[\psi] = \begin{bmatrix} 24.8416 & -1.3096 & -23.532 \\ 10.99378 & -1.94923 & -9.04455 \\ -0.08104 & 0.00853 & 1.07251 \end{bmatrix} \begin{bmatrix} 7.02 & 43.69 \\ 5 & 40 \\ 7.10918 & 43.9914 \end{bmatrix} = \begin{bmatrix} 0.54681 & -2.26012 \\ 3.13085 & 4.46663 \\ 7.09842 & 43.98178 \end{bmatrix}$$

This gives the new estimate of correct solution as $T_e = 7.0984$ and $T_c = 43.9818$. The values of the functions for these values of the variables are

$$f_1 = -0.00024 \quad \text{and} \quad f_2 = -0.00027$$

which are quite small.

Thus the solution is $T_e = 7.0984°C$ and $T_c = 43.9818°C$.

Comment
It can be seen that both the methods converge to the correct solution quite quickly. The Warner's method needs one more iteration than NRM, but there is only one set of function calculation at each iteration while NRM needs calculation of all the partial derivatives as also the function values.

In the above problem the functions were polynomials and for such "well behaved" functions the convergence of iterations is very quick. However, if the functions have exponential and/or logarithmic terms—and these are often encountered in thermal systems—the gradients can change very sharply and convergence becomes quite difficult. We shall illustrate this in Example 2.5.

Example 2.5
A simulation problem is finally reduced to solution of following two nonlinear equations:

$$f_1(x_1, x_2) = 1.0 + \ln(x_1/x_2) + x_1^{0.5} = 0$$
$$f_2(x_1, x_2) = e^{x_1} - e^{x_2} + 5 = 0$$

Determine the values of x_1 and x_2.

Solution
Let us start with initial guess values as $x_1 = 3$ and $x_2 = 3$.

We first use NRM. The partial derivatives are

$$\frac{\partial f_1}{\partial x_1} = \frac{1}{x_1} + 0.5x_1^{-0.5}$$

$$\frac{\partial f_1}{\partial x_2} = -\frac{1}{x_2}$$

$$\frac{\partial f_2}{\partial x_1} = e^{x_1} \quad \frac{\partial f_2}{\partial x_2} = -e^{x_2}$$

At the initial guess we have

$$f_1 = 2.732051 \quad \text{and} \quad f_2 = 5.0$$

and the gradient matrix is

$$\begin{bmatrix} 0.622008 & -0.3333 \\ 20.0855 & -20.0855 \end{bmatrix}$$

Thus Eq. (2.19) becomes

$$\begin{bmatrix} 0.622008 & -0.3333 \\ 20.0855 & -20.0855 \end{bmatrix} \begin{bmatrix} \Delta x_1 \\ \Delta x_2 \end{bmatrix} = -\begin{bmatrix} -2.732051 \\ 5.0 \end{bmatrix}$$

whose solution is

$$\Delta x_1 = -9.17659$$
$$\Delta x_2 = -8.92764$$

Thus new estimates of the solution are

$$\Delta x_1 = 3 - 9.17659 = -6.17659$$
$$\Delta x_2 = 3 - 8.92764 = -5.92764$$

On substituting these values in the above expressions we find that the value of the function f_1 can not be calculated since it involves taking the square root of a negative number. Hence NRM comes to a halt.

We can either start from another set of guesses for x_1 and x_2 or use the approach of *partial substitution*. This involves reducing the extent of change in the initial values, that is, taking

$$x_j^{1+1} = x_j^i + \lambda \Delta x_j \quad \text{(for } j = 1, 2\text{)}$$

where λ is a factor ($0 < \lambda < 1$), known as relaxation factor. The value of λ should be suitably chosen to avoid the kind of difficulty encountered here. Here negative values of x_1 or x_2 are not admissible in view of the presence of the logarithmic and square root terms in the first function. Clearly this demands that $(3 - 8.92764 \times \lambda)$ and $(3 - 9.17659 \times \lambda)$ should be positive numbers. Taking $\lambda = 0.3$, we get the new estimates as

$$x_1 = 3 - 9.17659 \times 0.3 \simeq 0.2470$$
$$x_2 = 3 - 8.92764 \times 0.3 \simeq 0.3217$$

Starting from these values all the calculations can be repeated, to get the various terms in Eq. (2.19). This gives

$$\begin{bmatrix} 5.054638 & -3.10849 \\ 1.280179 & -1.37947 \end{bmatrix} \begin{bmatrix} \Delta x_1 \\ \Delta x_2 \end{bmatrix} = -\begin{bmatrix} 1.23276 \\ 4.900708 \end{bmatrix}$$

Solving these equations we get

$$\Delta x_1 = +4.52116$$
$$\Delta x_2 = +7.74834$$

The new values of variables are

$$x_1 = 0.2470 + 4.52116 = 4.76816$$
$$x_2 = 0.3217 + 7.74834 = 8.07004$$

Substituting these values and repeating above steps we can perform further iterations. The results are summarized in Table 2.1.

For the sake of comparing the two methods, we shall solve this problem using Warner's method also taking the initial three sets of values as:

$$(3, 3)(\text{the same as in NRM}), (3, 2) \text{ and } (2, 4).$$

The iterations are shown in Table 2.2.

It is clear that further iterations would change the "new estimate" only marginally. Since the "total error" with the new estimate is much larger than the rest, we should discard this and replace it by another set of values. Moreover, it is clear from the nature of functions (see, e.g., f_2) that $x_2 > x_1$. Keeping this in view we modify the other sets too, and carry out further iterations as indicated below.

$$[\chi] \qquad [\rho] \qquad [\rho]^{-1} \qquad \psi = [\rho]^{-1}[\chi]$$

$$\begin{bmatrix} 2 & 3 \\ 3 & 4 \\ 2.5 & 3.5 \end{bmatrix} \begin{bmatrix} 2.008 & -7.696 & 1 \\ 2.444 & -29.51 & 1 \\ 2.245 & -15.93 & 1 \end{bmatrix} \begin{bmatrix} -8.595 & -5.212 & 13.807 \\ -0.1259 & -0.15 & 0.276 \\ 17.29 & 9.31 & -25.6 \end{bmatrix} \begin{bmatrix} 1.6919 & 1.6919 \\ -0.0120 & -0.01203 \\ -1.489 & -0.489 \end{bmatrix}$$

Remarks

We note that Warner's method is predicting the next estimate as $x_1 = -1.489$ and $x_2 = -0.489$. These values being negative are inadmissible. As in NRM, we resort to "partial substitution" of the "poorest set" by this new set; the poorest set being the one with largest magnitude of errors defined by Eq. (2.30). Here it is the set number 2 with $x_1 = 3$ and $x_2 = 4$. For "partial substitution" we take $\lambda = 0.5$, which gives the new estimate as

$$x_1 = \frac{3 - 1.489}{2} = 0.755 \quad x_2 = \frac{4 - 0.489}{2} = 1.755$$

Subsequent iterations are given in following table.

$$\text{Function values at the final estimate:} \tag{2.31}$$
$$f_1 = 0.000288$$
$$f_2 = -0.00016$$
$$\text{Thus the correct solution is reached} \tag{2.32}$$
$$x_1 = 0.3726$$
$$x_2 = 1.8643$$

Table 2.1 Iterations for solution by NRM

x_1	x_2	Partial derivative	Function values	Δx_1	Δx_2	New x_1	New x_2
3.0	3.0	$\begin{bmatrix} 0.622008 & -0.3333 \\ 20.0855 & -20.0855 \end{bmatrix}$	$\begin{bmatrix} 2.732051 \\ 5.0 \end{bmatrix}$	-9.1765	-8.9276	$-6.1765^{\star}$	$-5.9276^{\star}$
			$\star$ Partial substitution $\lambda = 0.3$				
0.2470	0.3217	$\begin{bmatrix} 5.0546 & -3.1085 \\ 1.2802 & -1.3795 \end{bmatrix}$	$\begin{bmatrix} 1.23276 \\ 4.9007 \end{bmatrix}$	4.521	7.748	4.768	8.07
4.768	8.07	$\begin{bmatrix} 0.438714 & -0.12392 \\ 117.6836 & -3197.1 \end{bmatrix}$	$\begin{bmatrix} 2.65735 \\ -3074.4 \end{bmatrix}$	-6.3975	-1.2091	$-1.63^{\star}$	6.86
			$\star$ Partial substitution $\lambda = 0.7$				
0.2897	7.2236	$\begin{bmatrix} 4.3808 & -0.1384 \\ 1.336 & -1371.4 \end{bmatrix}$	$\begin{bmatrix} -1.678 \\ -1365.08 \end{bmatrix}$	0.3558	-0.9961	0.6455	6.2275
0.6455	6.2275	$\begin{bmatrix} 2.1715 & -0.1606 \\ 1.907 & -506.49 \end{bmatrix}$	$\begin{bmatrix} -0.4633 \\ -499.58 \end{bmatrix}$	0.138	-0.983	0.7835	5.2445
0.7835	5.2445	$\begin{bmatrix} 1.8412 & -0.1907 \\ 2.1891 & -189.52 \end{bmatrix}$	$\begin{bmatrix} -0.01601 \\ -182.332 \end{bmatrix}$	-0.0916	-0.963	0.6919	4.2815
0.6919	4.2815	$\begin{bmatrix} 2.0464 & -0.2336 \\ 1.9975 & -72.35 \end{bmatrix}$	$\begin{bmatrix} -0.009188 \\ -65.35 \end{bmatrix}$	-0.1078	-0.907	0.5841	3.3745
0.5841	3.3745	$\begin{bmatrix} 2.3663 & -0.2963 \\ 1.7934 & -29.21 \end{bmatrix}$	$\begin{bmatrix} 0.01033 \\ -22.416 \end{bmatrix}$	-0.1012	-0.7736	0.4829	2.6009
0.4829	2.6009	$\begin{bmatrix} 2.7903 & -0.3845 \\ 1.6208 & -13.48 \end{bmatrix}$	$\begin{bmatrix} -0.0111 \\ -6.855 \end{bmatrix}$	-0.0753	-0.518	0.4076	2.083
0.4076	2.083	$\begin{bmatrix} 3.2366 & -0.4801 \\ 1.5032 & -8.0285 \end{bmatrix}$	$\begin{bmatrix} -0.00716 \\ -1.5253 \end{bmatrix}$	-0.0313	-0.1958	0.3763	1.8871
0.3763	1.8871	$\begin{bmatrix} 3.4725 & -0.5299 \\ 1.4569 & -6.6002 \end{bmatrix}$	$\begin{bmatrix} -0.001023 \\ -0.14332 \end{bmatrix}$	-0.004	-0.023	0.3723	1.8641
0.3723	1.8641	Function values: $f_1 = -0.00067; f_2 = 0.00094$					

Table 2.2 Iterations of Warner's method

χ	ρ	$[\rho]^{-1}$	$\psi = [\rho]^{-1}[\chi]$	Remarks
$\begin{bmatrix} 3 & 3 \\ 3 & 2 \\ 2 & 4 \end{bmatrix}$	$\begin{bmatrix} 2.732 & 5 & 1 \\ 3.1375 & 17.696 & 1 \\ 1.721 & -42.2 & 1 \end{bmatrix}$	$\begin{bmatrix} -9.5014 & 7.4874 & 2.01398 \\ 0.2247 & -0.1604 & -0.0643 \\ 25.8342 & -19.6536 & -5.1806 \end{bmatrix}$	$\begin{bmatrix} -2.014 & -5.473 \\ 0.0643 & 0.096 \\ 8.18 & 17.47 \end{bmatrix}$	Replace third set
$\begin{bmatrix} 3 & 3 \\ 3 & 2 \\ 8.18 & 17.47 \end{bmatrix}$	$\begin{bmatrix} 2.732 & 5 & 1 \\ 3.1375 & 17.696 & 1 \\ 3.101 & -3.9E07 & 1 \end{bmatrix}$	$\begin{bmatrix} -2.4661 & 2.4661 & 0.0 \\ 0.0 & 0.0 & 0.0 \\ 7.3736 & -6.7374 & 0.0 \end{bmatrix}$	$\begin{bmatrix} 0.0 & -2.466 \\ 0.0 & 0.0 \\ 3.0 & 9.737 \end{bmatrix}$	Replace third set
$\begin{bmatrix} 3 & 3 \\ 3 & 2 \\ 3.0 & 9.737 \end{bmatrix}$	$\begin{bmatrix} 2.732 & 5 & 1 \\ 3.1375 & 17.696 & 1 \\ 1.5547 & -16{,}907.6 & 1 \end{bmatrix}$	$\begin{bmatrix} -2.4733 & 2.4715 & 0.00186 \\ 0.00023 & -0.00017 & -0.00006 \\ 7.756 & -6.7512 & -0.0047 \end{bmatrix}$	$\begin{bmatrix} 0.0003 & -2.4589 \\ 0.0 & -0.00023 \\ 3.0 & 9.719 \end{bmatrix}$	Replace third set
$\begin{bmatrix} 3 & 3 \\ 3 & 2 \\ 3.0 & 9.719 \end{bmatrix}$	$\begin{bmatrix} 2.732 & 5 & 1 \\ 3.1375 & 17.696 & 1 \\ 1.5566 & -16{,}605.5 & 1 \end{bmatrix}$	$\begin{bmatrix} -2.47346 & 2.47157 & 0.00189 \\ 0.00024 & -0.00017 & -0.00006 \\ 7.756 & -6.7515 & -0.00486 \end{bmatrix}$	$\begin{bmatrix} 0.000 & -2.4581 \\ 0.0003 & -0.0002 \\ 3.0 & 9.7188 \end{bmatrix}$	Values repeated, see text.
$\begin{bmatrix} 2 & 3 \\ 0.755 & 1.755 \\ 2.5 & 3.5 \end{bmatrix}$	$\begin{bmatrix} 2.0087 & -7.6965 & 1 \\ 1.0254 & 1.3432 & 1 \\ 2.2446 & -15.933 & 1 \end{bmatrix}$	$\begin{bmatrix} 2.8958 & -1.3805 & -1.515 \\ 0.20435 & -0.03954 & -0.1648 \\ -3.244 & 2.4687 & 1.7753 \end{bmatrix}$	$\begin{bmatrix} 0.961 & 0.961 \\ -0.03318 & -0.03318 \\ -0.1859 & 0.8141 \end{bmatrix}$	x_1-ve, do partial substitution; replace set 3
$\begin{bmatrix} 2 & 3 \\ 0.755 & 1.755 \\ 1.157 & 2.152 \end{bmatrix}$	$\begin{bmatrix} 2.0087 & -7.6965 & 1 \\ 1.0254 & 1.3432 & 1 \\ 1.4551 & -0.4217 & 1 \end{bmatrix}$	$\begin{bmatrix} -0.822 & -3.3866 & 4.2086 \\ -0.200 & -0.2577 & 0.4577 \\ 1.1118 & 4.819 & -4.9308 \end{bmatrix}$	$\begin{bmatrix} 0.6684 & 0.6473 \\ -0.065 & -0.067 \\ 0.157 & 1.182 \end{bmatrix}$	Replace set 1
$\begin{bmatrix} 0.157 & 1.182 \\ 0.755 & 1.755 \\ 1.157 & 2.152 \end{bmatrix}$	$\begin{bmatrix} -0.6225 & 2.909 & 1 \\ 1.0254 & 1.3432 & 1 \\ 1.4551 & -0.4217 & 1 \end{bmatrix}$	$\begin{bmatrix} -0.78917 & 1.48848 & -0.69931 \\ -0.19202 & 0.92847 & -0.73645 \\ 1.0673 & -1.7743 & 1.707 \end{bmatrix}$	$\begin{bmatrix} 0.1908 & 0.17457 \\ -0.1812 & -0.1823 \\ 0.803 & 1.8211 \end{bmatrix}$	Replace set 1

$$\begin{bmatrix} 0.803 & 1.8211 \\ 0.755 & 1.755 \\ 1.157 & 2.152 \end{bmatrix} \quad \begin{bmatrix} 1.07726 & 1.05358 & 1 \\ 1.0254 & 1.3442 & 1 \\ 1.4551 & -0.4217 & 1 \end{bmatrix} \quad \begin{bmatrix} -53.04 & 44.312 & 8.7293 \\ -12.906 & 11.348 & 1.5577 \\ 71.736 & -59.612 & -11.045 \end{bmatrix} \quad \begin{bmatrix} 0.96321 & -0.04047 \\ 0.00672 & -0.23467 \\ -0.24169 & 2.11196 \end{bmatrix}$$

x_1-ve, do partial substitution; replace set 2

$$\begin{bmatrix} 0.803 & 1.8211 \\ 0.2566 & 1.9335 \\ 1.157 & 2.152 \end{bmatrix} \quad \begin{bmatrix} 1.07726 & 1.05358 & 1 \\ -0.513 & -0.6211 & 1 \\ 1.4551 & -0.4217 & 1 \end{bmatrix} \quad \begin{bmatrix} -0.06695 & -0.49526 & 0.56221 \\ 0.66071 & -0.12684 & -0.53387 \\ 0.37602 & 0.66715 & -0.04317 \end{bmatrix} \quad \begin{bmatrix} 0.46963 & 0.13037 \\ -0.11968 & -0.19091 \\ 0.42319 & 1.8818 \end{bmatrix}$$

Replace set 1

$$\begin{bmatrix} 0.42319 & 1.8818 \\ 0.2566 & 1.9335 \\ 1.157 & 2.152 \end{bmatrix} \quad \begin{bmatrix} 0.15837 & -0.03849 & 1 \\ -0.513 & -0.6211 & 1 \\ 1.4551 & -0.4217 & 1 \end{bmatrix} \quad \begin{bmatrix} -0.19692 & -0.37836 & 0.57529 \\ 1.94334 & -1.28041 & -0.66293 \\ 1.10599 & 0.01064 & -0.11662 \end{bmatrix} \quad \begin{bmatrix} 0.48519 & 0.1359 \\ -0.27316 & -0.24532 \\ 0.33584 & 1.85086 \end{bmatrix}$$

Replace set 3

$$\begin{bmatrix} 0.42319 & 1.8818 \\ 0.2566 & 1.9335 \\ 0.33584 & 1.85086 \end{bmatrix} \quad \begin{bmatrix} 0.15838 & -0.03849 & 1 \\ -0.513 & -0.6211 & 1 \\ -0.12725 & 0.03382 & 1 \end{bmatrix} \quad \begin{bmatrix} 3.04671 & -0.33641 & -2.7103 \\ -1.79451 & -1.32875 & 3.12326 \\ 0.44839 & 0.00214 & 0.54947 \end{bmatrix} \quad \begin{bmatrix} 0.29279 & 0.06646 \\ -0.05146 & -0.16533 \\ 0.37484 & 1.86491 \end{bmatrix}$$

Replace set 2

$$\begin{bmatrix} 0.42319 & 1.8818 \\ 0.37484 & 1.86491 \\ 0.33584 & 1.85086 \end{bmatrix} \quad \begin{bmatrix} 0.15838 & -0.03849 & 1 \\ 0.00778 & -0.00047 & 1 \\ -0.12725 & 0.03382 & 1 \end{bmatrix} \quad \begin{bmatrix} 1122.68 & -2367.33 & 1244.65 \\ 4420.59 & -9350.64 & 4930.05 \\ -6.6613 & 15.0326 & -7.3713 \end{bmatrix} \quad \begin{bmatrix} 5.73993 & 1.4746 \\ 21.46396 & 5.3967 \\ 0.34025 & 1.856 \end{bmatrix}$$

Replace set 1

$$\begin{bmatrix} 0.34025 & 1.85599 \\ 0.37484 & 1.86491 \\ 0.33584 & 1.85086 \end{bmatrix} \quad \begin{bmatrix} -0.11318 & 0.00727 & 1 \\ 0.00778 & -0.00047 & 1 \\ -0.12725 & 0.03382 & 1 \end{bmatrix} \quad \begin{bmatrix} -11.05126 & 8.5570 & 2.494 \\ -43.5148 & 4.5341 & 38.981 \\ 0.06557 & 0.93552 & -0.0011 \end{bmatrix} \quad \begin{bmatrix} 0.285 & 0.0635 \\ -0.0151 & -0.1595 \\ 0.3726 & 1.8643 \end{bmatrix}$$

Solution

The above example demonstrates the difficulties encountered in solving highly nonlinear equations and the methods employed to achieve convergence even when the intermediate steps give inadmissible estimates. This usually happens when our initially assumed values are far away from the correct solution. When the number of equations (and consequently the number of variables) is large, even one guesstimate being far off can upset the iterations. This can happen both with NRM and Warner's method, though the former is more susceptible to predicting wild values at intermediate steps.

In many practical problems it is possible to identify realistic limits to the values of the variables and a common strategy to contain divergence is to bring the trial values within the limits (usually by partial substitution) whenever these are crossed.

2.3 EQUATION FITTING

More often than not, engineering data are presented in the form of tables. The manufacturers usually indicate the performance data of their equipment—fans, turbines, evaporators, or even complete refrigeration plants—in the form of series of tables. Thermodynamic properties of various fluids have also been conventionally presented in tabular forms, mainly because of the ease of their use in manual calculations. Simple interpolation techniques—usually linear, sometime logarithmic—are employed to estimate the properties at intermediate values. However, in the computer calculations—which are invariably necessary in thermal system simulation—storing, retrieving, and interpolation from tabular data are not very efficient. If equations are developed which represent the tabulated data accurately the computational process becomes extremely efficient.

The first step in fitting an equation to data is to identify the nature of equation—should it be a straight line, a polynomial (of which degree), or a more complex nonlinear equation? An insight into the interrelationships between the data on physical grounds does help. Thus if we have the saturation pressure-temperature data for any fluid, it is very likely that a relationship of the kind $\ln P = A + B/T$ would be a good fit, as is suggested by Clasius-Clayperon equation of thermodynamics. In case the number of independent variables is 1 or 2, graphical plotting of the data can help identify the equation likely to fit the data well. Fig. 2.5 indicates a few commonly encountered curves and the corresponding equations.

Having decided upon the "type" of equation to be fitted into the data, the next step is to find the "unknown" coefficients of the equation. If the number of unknown coefficients equals the number of data points then our task is quite straight forward. Thus, for example, suppose we want to fit a polynomial $y = a + bx + cx^2 + dx^3$ into four data points (x_1, y_1), (x_2, y_2), (x_3, y_3), (x_4, y_4). Then clearly, values of the four coefficients are the solutions to the following linear equations:

$$y_1 = a + bx_1 + cx_1^2 + dx_1^3$$
$$y_2 = a + bx_2 + cx_2^2 + dx_2^3$$

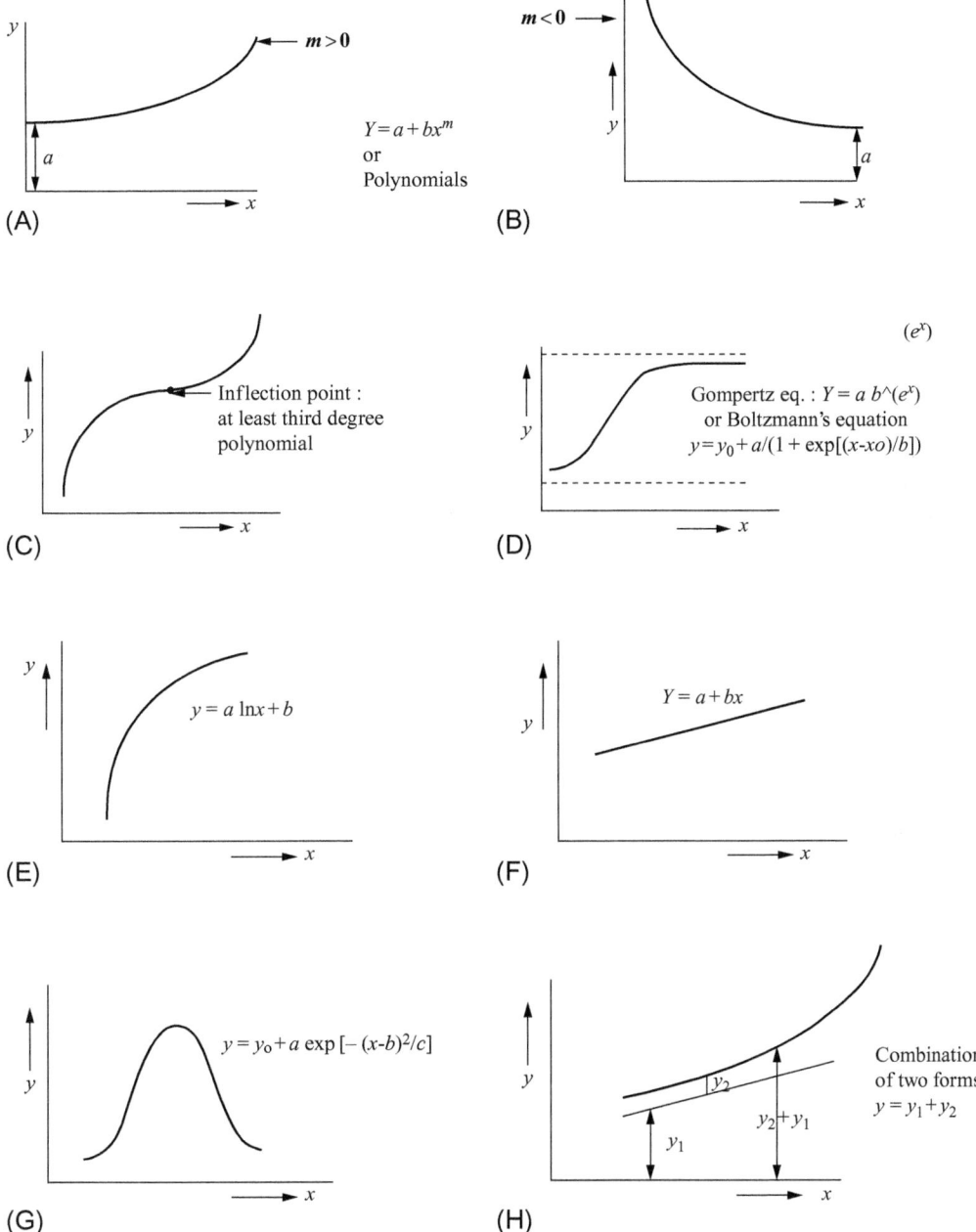

Fig. 2.5 Choosing equation for fitting into data.

$$y_3 = a + bx_3 + cx_3^2 + dx_3^3$$
$$y_4 = a + bx_4 + cx_4^2 + dx_4^3$$

In case the equations are nonlinear in coefficients, we end up with the four nonlinear equations to be solved, for example, with the equation. $y_1 = a+bx_1+cx_1^d$ the coefficients have to be found by solving the following nonlinear equations:

$$y_1 = a + bx_1 + cx_1^d$$
$$y_2 = a + bx_2 + cx_2^d$$
$$y_3 = a + bx_3 + cx_3^d$$
$$y_4 = a + bx_4 + cx_4^d$$

In either case we can get a solution to the four equations in four variables using any of the methods explained in Sections 2.1 and 2.2. However, in case the number of data points is larger than the unknown coefficients—and this is usually the case—then the values of the coefficients are so chosen that the "overall error" in predicting the values of the dependent variable at all the data points is minimal. The most commonly adopted measure of the "overall error" is the sum of squares of the difference between the calculated and the given values of the dependent variable, for all the N data point, that is

$$E = \sum_{i=1}^{N} e_i^2 = \sum_{i=1}^{N} (y_{i,data} - y_{i,calculated})^2 \tag{2.33}$$

The set of coefficients which minimizes the value of E are determined, and accepted as the "best fit" into the data. This method is called least squares regression.

This criterion has a number of advantages including the fact that it gives a unique result for a given set of data. Further, in most practical cases, the data being fitted are experimental data. Certain measurement errors are bound to creep into the these. It can be demonstrated that if these measurement errors have a normal distribution then least squares regression would give the best (i.e., statistically the most likely) estimates of the coefficients.

The values of the coefficients which would minimize the least squares error can be determined by using the basic methodology of differential calculus. We shall develop below the procedure for generalized least squares method for linear regression.

2.3.1 Generalized Linear Regression

The most general form of relationship between the dependent variable y and the independent variables x_i can be written as

$$\phi(y) = a_0 + a_1 \, \phi_1(\vec{X}) + a_2 \, \phi_2(\vec{X}) + \cdots + a_n \, \phi_n(\vec{X}) \tag{2.34}$$

where $\vec{X}$ represents the variable vector $(x_1, x_2, \ldots, x_m)$.

The error function E of Eq. (2.33) can therefore be written as

$$E = \sum_{i=1}^{N} e_i^2 = \sum_{i=1}^{N} \left[\phi(y)_i - a_0 - a_1\phi_1(\vec{X})_i - a_2\phi_2(\vec{X})_i \cdots a_n\phi_n(\vec{X})_i \right]^2 \qquad (2.35)$$

To find the values of the coefficients $a_0, a_1, a_2, \ldots, a_n$ which minimize E, we take partial derivatives of E with respect to each of these and set them equal to zero. This gives, for partial differentiation with respect to a_0:

$$\sum_{i=1}^{N} 2 \left[\phi(y)_i - a_0 - a_1\phi_1(\vec{X})_i - a_2\phi_2(\vec{X})_i - \cdots - a_n\phi_n(\vec{X})_i \right] (-1) = 0$$

$$\text{or} \quad a_0 \, N + a_1 \sum_{i=1}^{N} \phi_1(\vec{X})_i + a_2 \sum_{i=1}^{N} \phi_2(\vec{X})_i + \cdots \qquad (2.36)$$

$$\cdots + a_n \sum_{i=1}^{N} \phi_1(\vec{X})_i = \sum_{i=1}^{N} \phi(y)_i \qquad (2.37)$$

Similarly differentiating with respect to other coefficients, namely $a_1, a_2, \ldots, a_n$ we get

$$a_0 \sum \phi_1(\vec{X})_i + a_1 \sum \phi_1(\vec{X})_i\phi_1(\vec{X})_i + a_2 \sum \phi_1(\vec{X})_i\phi_2(\vec{X})_i$$
$$\cdots + a_n \sum \phi_1(\vec{X})_i\phi_n(\vec{X})_i = \sum \phi_1(\vec{X})_i\phi_1(y)_i$$

$$a_0 \sum \phi_2(\vec{X})_i + a_1 \sum \phi_2(\vec{X})_i\phi_1(\vec{X})_i + a_2 \sum \phi_2(\vec{X})_i\phi_2(\vec{X})_i$$
$$\cdots + a_n \sum \phi_2(\vec{X})_i\phi_n(\vec{X})_i = \sum \phi_2(\vec{X})_i\phi_1(y)_i$$

$$\vdots$$

$$a_0 \sum \phi_n(\vec{X})_i + a_1 \sum \phi_n(\vec{X})_i\phi_1(\vec{X})_i + a_2 \sum \phi_n(\vec{X})_i\phi_2(\vec{X})_i$$
$$\cdots + a_n \sum \phi_n(\vec{X})_i\phi_n(\vec{X})_i = \sum \phi_n(\vec{X})_i\phi_1(y)_i$$

These equations can be written in matrix form as:

$$\begin{bmatrix} N & \Sigma\phi_1 & \Sigma\phi_2 & \cdots & \Sigma\phi_n \\ \Sigma\phi_1 & \Sigma\phi_1\phi_1 & \Sigma\phi_1\phi_2 & \cdots & \Sigma\phi_1\phi_n \\ \Sigma\phi_2 & \Sigma\phi_2\phi_1 & \Sigma\phi_2\phi_2 & \cdots & \Sigma\phi_2\phi_n \\ \vdots & \vdots & \vdots & \vdots & \vdots \\ \Sigma\phi_n & \Sigma\phi_n\phi_1 & \Sigma\phi_n\phi_2 & \cdots & \Sigma\phi_n\phi_n \end{bmatrix} \begin{bmatrix} a_0 \\ a_1 \\ a_2 \\ \vdots \\ a_n \end{bmatrix} = \begin{bmatrix} \Sigma\phi(y) \\ \Sigma\phi_1\phi(y) \\ \Sigma\phi_2\phi(y) \\ \vdots \\ \Sigma\phi_n\phi(y) \end{bmatrix} \qquad (2.38)$$

This is a set of $n + 1$ linear equations in $n + 1$ unknowns, namely the coefficients a_0, a_1, a_2, a_n, which can be solved by using the methods discussed in Section 2.1.

The accuracy with which the equation would be able to predict the data depends on the extent to which the equation captures the actual relationship between the dependent variable (y) and the independent variables ($\vec{X}$), as also the extent of "measurement errors" in the data itself. Quite often even a few "rouge" data points can upset the regression process considerably and one needs to be cognizant of that.

There are many statistical measures which are used to "quantify" the goodness of fit. Some of the most common ones are enumerated below:

(a) Arithmetic mean fractional deviation

$$f_1 = \frac{\Sigma f_i}{N} \tag{2.39}$$

where $f_i = \frac{(y_{i,data} - y_{i,calculated})}{y_{i,data}}$ for nonzero values of the data. Otherwise for

$$\text{for} \quad y_{i,data} = 0 \quad f_i = (y_{i,data} - y_{i,calculated})$$

It is a measure of centering or average accuracy of the prediction.

(b) Mean absolute fractional deviation

$$f_2 = \frac{\Sigma |f_i|}{N} \tag{2.40}$$

(c) Standard deviation

$$f_3 = \sqrt{\frac{1}{N-1}[\Sigma(f_i - f_1)^2]} \tag{2.41}$$

(d) Root mean square fractional deviation

$$f_4 = \sqrt{\frac{(\Sigma f_i^2)}{N}} \tag{2.42}$$

f_2, f_3, f_4 provides measures of scatter or lack of precision in prediction.

Standard softwares are available which enable us to do least squares regression for commonly used equations involving linear, polynomial, power law, and logarithmic relationship between two variables. For nonstandard relationships and situations involving multiple variables, one has to formulate Eq. (2.38) and solve this to get the values of the coefficients.

Example 2.6

Table 2.3 gives the refrigeration capacity (Q_e) of a reciprocating compressor as a function of the evaporator and condensing temperatures (T_e and T_c, respectively). Using the method of least squares regression, fit the following equation into the data:

$$Q = a_0 + a_1 T_c + a_2 T_e + a_3 T_c T_e$$

Determine the values of various measures of the goodness of fit.

Table 2.3 Refrigeration capacity in kW

Evap. temperature (T_e) (°C)	Cond. temperature (T_c) (°C)		
	25	35	45
0	152.7	117.1	81.0
5	182.9	141.9	101.3
10	215.4	170.7	126.5

Solution

Expressing the given equation in terms of generalized functional relationship of Eq. (2.34) we can write the following equivalence:

$$\phi(y) = Q \quad \phi_1(\vec{X}) = T_c \quad \phi_2(\vec{X}) = T_e \quad \phi_3(\vec{X}) = T_c T_e$$

Various terms needed in Eq. (2.38) can now be calculated using the given data. The detailed calculation can be conveniently done using software Excel. The results are given in Tables 2.4A and 2.4B.

Using the coefficients as calculated in Tables 2.4A and 2.4B, Eq. (2.38) becomes

$$\begin{bmatrix} 9 & 315 & 45 & 1575 \\ 315 & 11{,}625 & 1575 & 58{,}125 \\ 45 & 1575 & 375 & 13{,}125 \\ 1575 & 58{,}125 & 13{,}125 & 48{,}375 \end{bmatrix} \begin{bmatrix} a_0 \\ a_1 \\ a_2 \\ a_3 \end{bmatrix} = \begin{bmatrix} 1289.5 \\ 42{,}710.5 \\ 7256.5 \\ 241{,}007.5 \end{bmatrix}$$

Solving these simultaneous equations we get

$$a_0 = 242.5444 \quad a_1 = -3.606667$$
$$a_2 = 8.40333 \quad a_3 = -0.08600$$

The refrigeration capacity calculated by substituting these values of coefficients in regression equation, and the calculation of various "goodness-of-fit" measures are shown in Table 2.4A.

It can be seen that the chosen equation represents the data reasonably well. Extremely low f_1 values indicate good average accuracy of prediction, but the f_2 value is not so low, still the predicted values are, on an average, only about ±0.08% away from the actual values. The other indicators of the scatter are also of similar magnitude. If higher accuracy of prediction is desired, additional terms need to be added in the regression equation.

Table 2.4A Detailed calculations for Example 2.6

φ1	φ2	φ(y)	φ3	φ1φ1	φ1φ2	φ1φ3	φ2φ1	φ2φ2	φ2φ3	φ3φ1	φ3φ2	φ3φ3
25	0	152.7	0	625	0	0	0	0	0	0	0	0
35	0	117.1	0	1225	0	0	0	0	0	0	0	0
45	0	81	0	2025	0	0	0	0	0	0	0	0
25	5	182.9	125	625	125	3125	125	25	625	3125	625	15,625
35	5	141.9	175	1225	175	6125	175	25	875	6125	875	30,625
45	5	101.3	225	2025	225	10,125	225	25	1125	10,125	1125	50,625
25	10	215.4	250	625	250	6250	250	100	2500	6250	2500	62,500
35	10	170.7	350	1225	350	12,250	350	100	3500	12,250	3500	122,500
45	10	126.5	450	2025	450	20,250	450	100	4500	20,250	4500	202,500
						Summation of various column values						
315	45	1289.5	1575	11,625	1575	58,125	1575	375	13,125	58,125	13,125	484,375

Table 2.4B Detailed calculations for Example 2.6…(contd)

φ1φY	φ2φY	φ3φY	y_pred	f1	f2	f3	f4
3817.5	0	0	152.377765	0.002110249	0.002110249	4.12116E−06	4.45315E−06
4098.5	0	0	116.311095	0.00673702	0.00673702	4.43134E−05	4.53874E−05
3645	0	0	80.244425	0.009328086	0.009328086	8.55237E−05	8.70132E−05
4572.5	914.5	22,862.5	183.644415	−0.004070066	0.004070066	1.72246E−05	1.65654E−05
4966.5	709.5	24,832.5	143.277745	−0.009709267	0.009709267	9.58334E−05	9.42699E−05
4558.5	506.5	22,792.5	102.911075	−0.015903998	0.015903998	0.000255494	0.000252937
5385	2154	53,850	214.911065	0.002269893	0.002269893	4.79482E−06	5.15242E−06
5974.5	1707	59,745	170.244395	0.002669039	0.002669039	6.70216E−06	7.12377E−06
5692.5	1265	56,925	125.577725	0.007290711	0.007290711	5.19917E−05	5.31545E−05
		Summation of various column values [col 1–3], average of col 5–8 values					
42,710.5	7256.5	241,007.5	—	8.01853E−05	0.006676481	0.008411295	0.00793065

2.3.2 Nonlinear Regression

Sometimes there is a need to fit nonlinear equations like

$$y = a_0 + a_1 x^{a_2} + a_3 e^{-a_4 x}$$

into the given (x, y) data. The basic philosophy of choosing the coefficients of a regression equation is to minimize the sum of squares of the "errors," Eq. (2.33) is applicable here too. However, if we follow the procedure outlined in Section 2.3.1, the resulting equations would be highly nonlinear and difficult to solve. Instead of this "direct" method two indirect approaches are often recommended. The first involves linearizing the original equation using Taylor's series expansion (curtailed after first derivative) around the estimates and using the least squares method of Section 2.3.1 to get new estimates of the parameters that would minimize the overall "error" term of Eq. (2.33). However, this method is often not stable and can oscillate widely, especially when the initial guess of parameter values is not near the correct solution.

The second approach consists of minimizing Eq. (2.33) using nonlinear optimization techniques discussed in Chapter 8. This is usually the most successful approach toward proper estimation of the unknown parameters of the nonlinear equation, and we shall discuss it there.

2.4 DIFFERENTIAL EQUATIONS

Differential equations—both ordinary and partial—are often encountered in simulating the performance of various thermal equipment like heat exchangers, finned surfaces, spray cooling devices, etc. In most practical cases the equations involved are quite complex and analytical solutions are not available. However, in some simple cases, the equations can be reduced to standard forms, for example, the second-order ordinary differential equation with constant coefficients, encountered frequently is

$$\frac{d^2 y}{dx^2} + a \frac{dy}{dx} + by = 0 \tag{2.43}$$

(also written for brevity as $y'' + ay' + by = 0$) for which the analytical solution can be written as

$$y = \exp\left(\frac{-ax}{2}\right) \cdot \phi(x)$$

The value of function $\phi(x)$ depends on the magnitude of the term $(a^2 - 4b)$, as given below:

$$\lambda^2 = a^2 - 4b > 0 \qquad \phi(x) = C_1 \exp\left(\frac{\lambda x}{2}\right) + C_2 \exp\left(-\frac{\lambda x}{2}\right)$$

$$\lambda^2 = 4b - a^2 > 0 \qquad \phi(x) = C_1 \sin\left(\frac{\lambda x}{2}\right) + C_2 \cos\left(\frac{\lambda x}{2}\right)$$

$$a^2 - 4b = 0 \qquad \phi(x) = C_1 x + C_2 \tag{2.44}$$

where C_1 and C_2 are arbitrary constants whose values have to be chosen so as to satisfy the prescribed boundary conditions. The specific equations for which analytical solutions are available can be seen from standard book or from internet.[2]

For numerical solution of differential equations, the basic approach is to represent the derivatives by using finite difference approximations. These approximations are derived from the Taylor's series expansion of a function $f(x)$ about a point x_i:

$$f(x_{i+1}) = f(x_i + h) = f(x_i) + h f'(x) + \frac{h^2}{2} f''(x) + \cdots \tag{2.45}$$

If we omit the terms involving gradients higher than $f'(x)$, we get from Eq. (2.45):

$$f'(x_i) = \frac{f(x_i + h) - f(x_i)}{h} + O(h) \tag{2.46}$$

This is termed as the forward difference approximation of the derivative, and has truncation errors of the order of the step size h.

In a similar manner we can get a backward difference approximation as

$$f'(x_1) = \frac{f(x_i) - f(x_{i-1})}{h} + O(h) \tag{2.47}$$

Further, if we make judicious use of both backward and forward differences, it can be easily proved by using Taylor's series expansion that a more accurate estimate of the gradient is the centered difference approximation:

$$f'(x_i) = \frac{f(x_{i+1}) - f(x_{i-1})}{2h} + O(h^2) \tag{2.48}$$

Proceeding in a similar manner even more accurate representations of the first derivative as also finite difference formulation for higher derivatives can be obtained. Some commonly used formulae are given below:

$$f'(x_i) = \frac{f(x_{i+1}) - f(x_{i-1})}{2h} + O(h^2) \tag{2.49}$$

$$f'(x_i) = \frac{-f(x_{i+2}) + 8f(x_{i+1}) - 8f(x_{i-1}) + f(x_{i-2})}{12h} + O(h^4) \tag{2.50}$$

$$f''(x_i) = \frac{f(x_{i+1}) - 2f(x_i) + f(x_{i-1})}{h^2} + O(h^2) \tag{2.51}$$

---

[2] See, for example, http://eqworld.ipmnet.ru/en/solutions/ode.htm.

$$f''(x_i) = \frac{-f(x_{i+2}) + 16f(x_{i+1}) - 30f(x_i) + 16f(x_{i-1}) - f(x_{i-2})}{12h^2} + O(h^4) \quad (2.52)$$

$$f'''(x_i) = \frac{f(x_{i+2}) - 2f(x_{i+1}) + 2f(x_{i-1}) - f(x_{i-2})}{2h^3} + O(h^2) \quad (2.53)$$

The techniques of numerical solution of differential equations are based on such finite difference approximations for the derivatives.

2.4.1 Single-Step Methods

This is a class of methods where to find the value of the independent variable y_{i+1} at x_{i+1} only the information at single previous point x_i is used. The simplest of these methods is the Euler's method which is based on the forward difference approximation of the derivative. Thus for the differential equation

$$\frac{dy}{dx} = f(x, y) \quad (2.54)$$

using the forward difference formula, Eq. (2.46), we get

$$y_{i+1} = y_i + h f(x_i, y_i) \quad (2.55)$$

Thus, starting from a known initial value y_0 at $x = x_0$, the values of y at other values of x, at intervals of h the step size, can be found.

The modified Euler's method involves three-step calculations to improve the accuracy:

(i) Predict y_{i+1} using Eq. (2.55). Call it predictor: y_{i+1}^p

(ii) Calculate the gradient $f(x_{i+1}, y_{i+1}^p)$ using Eq. (2.54)

(iii) Correct the first estimate obtained in step (i) using the average of the slopes at (x_i, y_i) and that calculated in step (ii) above, that is

$$y_{i+1} = y_i + h \cdot \frac{f(x_i, y_i) + f(x_{i+1}, y_{i+1}^p)}{2} \quad (2.56)$$

This is termed as the corrector, and the y_{i+1} value thus obtained is a better estimate than that obtained using the basic Euler's method. It can be shown that while the overall truncation error in Euler's method is of the order of $O(h)$, in modified Euler's method this error is of the order $O(h^2)$.

This three-step approach is termed as a predictor-corrector approach and is commonly employed in multistep methods discussed later.

Even more accurate methods to solve ordinary differential equations have been developed. One of the most popular methods is the classical Runge-Kutta method which has a truncation error if the order $O(h^4)$. The procedure for solving Eq. (2.54) using this method is given below.

$$y_{i+1} = y_i + \frac{h}{6}(k_1 + 2k_2 + 2k_3 + k_4) \tag{2.57}$$

where

$$k_1 = f(x_i, y_i) \tag{2.58}$$

$$k_2 = f\left(x_i + \frac{h}{2}, y_i + \frac{k_1}{2}h\right) \tag{2.59}$$

$$k_3 = f\left(x_i + \frac{h}{2}, y_i + \frac{k_2}{2}h\right) \tag{2.60}$$

$$k_4 = f(x_i + h, y_i + k_3 h) \tag{2.61}$$

By suitably incorporating additional terms in Eq. (2.57), the accuracy can be further increased. Details of such formulae can be seen in standard texts on numerical methods [1, 2].

2.4.2 Multistep Methods

As we solve a differential equation using any of the single-step methods mentioned above, after a few steps we have information on (y_i, x_i, f_i) values for say, $i = 1, 2, 3, 4$. By using this information judiciously it should be possible to fit a polynomial $p(x)$ between f and x. We could then solve for y by analytically integrating this polynomial $p(x)$, since from Eq. (2.54) it follows that:

$$y_{i+1} - y_i = \int_{x_i}^{x_{i+1}} f(x, y)dx = \int_{x_i}^{x_{i+1}} p(x)dx$$

Many formulae have been devised, based on this approach, differing in the order of polynomial fitted. Two of the most popular methods are briefly explained below.

Adams-Bashforth Method

$$y_{i+1} = y_i + \frac{h}{24}(55f_i - 59f_{i-1} + 37f_{i-2} - 9f_{i-3}) \tag{2.62}$$

Adams-Moulton Method
  (i) Use Eq. (2.62) as a predictor of y_{i+1}. Call it y_{i+1}^p
  (ii) Calculate the value of function f at

$$x = x_{i+1}, \ y = y_{i+1}^p \quad \text{call it } f_{i+1}^p \tag{2.63}$$

(iii) Calculate the corrected value of y_{i+1} as

$$y_{i+1} = y_i + \frac{h}{24}(9f_{i+1}^p + 19f_i - 5f_{i-1} + f_{i-2}) \tag{2.64}$$

Both these methods have overall error of the order $0(h^4)$.

To use this method it is necessary to solve the ordinary differential equation (ODE) for first few steps using any of the single-step methods, since both Eqs. (2.62), (2.65) need previous function values to march the solution ahead. Thus Runge-Kutta method could be employed to get four or more points and then one could switch over to any of these multistep methods. These methods are faster than the Runge-Kutta fourth-order method since only two new values need to be calculated at each step. The difference between the "predicted" and the "corrected" values gives an idea of the likely magnitude of the truncation error. If it is unacceptable, the step size h can even be reduced (usually halved), and if it is extremely small, h can even be increased (usually doubled) to speed up the calculations. When the step size is halved, at the switch-over point we need to interpolate for the values in between the known points. This can be done using the following formulae:

$$y_{i-1/2} = \frac{1}{128}[35y_i + 140y_{i-1} - 70y_{i-2} + 28y_{i-3} - 5y_{i-4}] \tag{2.65}$$

$$y_{i-3/2} = \frac{1}{64}[-y_i + 24y_{i-1} + 54y_{i-2} - 16y_{i-3} + 3y_{i-4}] \tag{2.66}$$

Example 2.7
Solve the differential equation

$$\frac{dy}{dx} + y = (x+1)/2$$

given at $x = 0, y = -1$.

Solution
We shall solve the equation by various methods discussed above to bring out their relative merits, using a step size of $h = 0.1$. We first rewrite the equation in the form of Eq. (2.54) as:

$$\frac{dy}{dx} = \frac{(x+1)}{2} - y = f(x, y)$$

$$\text{Thus} \quad f(x, y) = \frac{x+1}{2} - y$$

Euler's Method

Calculations in accordance with Eq. (2.55) are given in following table.

i	x_i	y_i	$f(x,y)$	y_{i+1}
1	0	−1	1.5	−0.85
2	0.1	−0.85	1.4	−0.71
3	0.2	−0.71	1.31	−0.579
4	0.3	−0.579	1.229	−0.4561
5	0.4	−0.4561	1.1561	−0.34049
6	0.5	−0.34049	1.09049	−0.23144
7	0.6	−0.23144	1.031441	−0.1283
8	0.7	−0.1283	0.978297	−0.03047
9	0.8	−0.03047	0.930467	+0.06258
10	0.9	+0.06258	0.88742	0.151322
11	1.0	0.151322	0.848678	0.236189
12	1.1	0.236189	0.813811	0.31757
13	1.2	0.31757	0.78243	0.395813

Modified Euler's Method

Using Eq. (2.56), taking values of $f(x_i, y_i)$ and y_{i+1}^p from Eq. (2.55).

i	x_i	y_i	$f(x_i, y_i)$	y_{i+1}^p	$f(x_{i+1}, y_{i+1}^p)$	y_{i+1}^C
1	0	−1	1.5	−0.85	1.4	−0.855
2	0.1	−0.855	1.405	−0.7145	1.3145	−0.71903
3	0.2	−0.71903	1.319025	−0.58712	1.237123	−0.59122
4	0.3	−0.59122	1.241218	−0.4671	1.167096	−0.4708
5	0.4	−0.4708	1.170802	−0.35372	1.103722	−0.35708
6	0.5	−0.35708	1.107076	−0.24637	1.046368	−0.2494
7	0.6	−0.2494	1.049404	−0.14446	0.994463	−0.14721
8	0.7	−0.14721	0.99721	−0.04749	0.947489	−0.04998
9	0.8	−0.04998	0.949975	+0.045022	0.904978	+0.042772
10	0.9	+0.042772	0.907228	0.133495	0.866505	0.131459
11	1.0	0.131459	0.868541	0.218313	0.831687	0.21647
12	1.1	0.21647	0.83353	0.299823	0.800127	0.298156
13	1.2	0.298156	0.801844	0.37834	0.12166	0.344331

Runge-Kutta Method

The solution is done using Eq. (2.57) where various gradients in this equation are calculated using Eqs. (2.58)–(2.61).

i	x_i	y_i	k_1	k_2	k_3	k_4	y_{i+1}
1	0	−1	1.5	1.45	1.4525	1.40475	−0.85484
2	0.1	−0.85484	1.404838	1.359596	1.361858	1.318652	−0.71873
3	0.2	−0.71873	1.318731	1.277794	1.279841	1.240747	−0.59082
4	0.3	−0.59082	1.240818	1.203778	1.20563	1.170255	−0.47032
5	0.4	−0.47032	1.17032	1.136804	1.13848	1.106472	−0.35653
6	0.5	−0.35653	1.106531	1.076204	1.0777121	1.048759	−0.24881
7	0.6	−0.24881	1.048812	1.021371	1.022743	0.996538	−0.14659
8	0.7	−0.14659	0.996586	0.971756	0.972998	0.949286	−0.04933
9	0.8	−0.04933	0.949329	0.926863	0.927986	0.906531	+0.04343
10	0.9	+0.04343	0.90657	0.886241	0.887258	0.867844	+0.13212
11	1.0	0.13212	0.86788	0.849486	0.850405	0.832839	+0.217129
12	1.1	0.217129	0.832871	0.816228	0.81706	0.801165	0.298805
13	1.2	0.298805	0.801195	0.786135	0.786888	0.772506	0.377468

Adams-Bashforth Method

We take the first four values of $y's$ from the Runge-Kutta method and then calculate subsequent values of y using Eq. (2.62).

i	x_i	y_i	$f(x, y)$	y_{i+1}, e.g., 3.59
1	0	−1	1.5	
2	0.1	−0.85484	1.404838	
3	0.2	−0.71873	1.318731	
4	0.3	−0.59082	1.240818	−0.47032
5	0.4	−0.47032	1.170323	−0.35654
6	0.5	−0.35654	1.106535	−0.24882
7	0.6	−0.24882	1.048818	−0.14659
8	0.7	−0.14659	0.996593	−0.04934
9	0.8	−0.04934	0.949338	0.04342
10	0.9	0.04342	0.90658	0.13211
11	1.0	0.13211	0.86789	0.217118
12	1.1	0.217118	0.832882	0.298795
13	1.2	0.298795	0.801205	0.377457

Adams-Moulton Predictor-Corrector Method

i	x_i	y_i	$f(y_i, x_i)$	y_{i+1}^{P}, e.g., 3.60	f_{i+1}^{P}	y_{i+1}^{C}, e.g., 3.61
1	0	−1	1.5	−		
2	0.1	−0.85484	1.404837	−		
3	0.2	−0.71873	1.318731	−		
4	0.3	−0.59082	1.240818	−0.47032	1.170323	−0.47032
5	0.4	−0.47032	1.17032	−0.35653	1.106533	−0.35653
6	0.5	−0.35653	1.10653	−0.24881	1.048813	−0.24881
7	0.6	−0.24881	1.048811	−0.14659	0.996587	−0.14658
8	0.7	−0.14658	0.996584	−0.04933	0.94933	−0.04933
9	0.8	−0.04933	0.949328	−0.04343	0.90657	+0.043431
10	0.9	−0.043431	0.906569	+0.13212	0.86788	0.132122
11	1.0	0.132122	0.867878	0.217129	0.832871	0.21713
12	1.1	0.21713	0.83287	0.298806	0.801194	0.298807
13	1.2	0.298807	0.801193	0.377468	0.772532	0.377469

The following table shows a comparison of the results obtained by these numerical methods with the exact analytical solution. It can be seen that all the fourth-order methods, the Runge-Kutta, Adams–Bashforth, and Adams–Moulton methods give extremely good agreement with the analytical solution. Euler's method predicts values of y having the largest difference from the correct solution; the modified Euler's method being much better.

i	x_i	y_i (Anal)	y_1 (Euler)	y_i (Mod-Eul)	y_i (Runge-Kutta)	y_i (Adams-Bashforth)	y_i (Adam Moulton)
1	0	−1	−1	−1	−1		
2	0.1	−0.85484	−0.85	−0.855	−0.85484		
3	0.2	−0.71873	−0.71	−0.71903	−0.71873		
4	0.3	−0.59082	−0.579	0.59122	−0.59082		
5	0.4	−0.47032	−0.4561	−0.4708	−0.47032	−0.47032	−0.47032
6	0.5	−0.35653	−0.34049	−0.35708	−0.35653	−0.35654	−0.35653
7	0.6	−0.24881	−0.23144	−0.2494	−0.24881	−0.24882	−0.24881
8	0.7	−0.14659	−0.1283	−0.14721	−0.14659	−0.14659	−0.14658
9	0.8	−0.04933	−0.3047	−0.04998	−0.04933	−0.04934	−0.04933
10	0.9	0.04343	0.06258	+0.042772	0.04343	+0.04342	+0.043431
11	1.0	0.132121	0.151322	+0.131459	0.13212	0.13211	0.132122
12	1.1	0.217129	0.236189	0.21647	0.217129	0.217118	0.21713
13	1.2	0.298806	0.31757	+0.298156	0.298805	0.298795	0.298807

2.4.3 Systems of Equations

While simulating thermal systems we often come across higher order differential equations, as also systems of such equations to be solved simultaneously. Any higher order differential equation can always be converted into a system of first-order equations. Thus the differential equation

$$a\frac{d^3 x}{dt^3} + b\frac{d^2 x}{dt^2} + c\frac{dx}{dt} = f(x, t) \tag{2.67}$$

can be transformed into three first-order equations by defining

$$y = \frac{dx}{dt} \quad \text{and} \quad z = \frac{dy}{dt} = \frac{d^2 x}{dt^2} \cdots \tag{2.68}$$

as

$$a\frac{dz}{dt} + bz + cy = f(x, t) \tag{2.69}$$

Thus the original third-order differential equation (2.67) is transformed into three first-order equations in x, y, and z, namely Eqs. (2.68), (2.69).

All the methods discussed in Section 2.4.1 can be easily extended to a system of equations taking care to follow appropriate sequence in calculations.

In simple one-step methods like the Euler's method, the only care needed is to advance every equation by one step before beginning the next step. In Runge-Kutta methods the values of each gradient terms (i.e., k_i) should be found for all the variables before advancing to the next k_i. We use previous k-value in incrementing the function values and the value of h to increment the independent variable. Similarly in predictor-corrector methods, we determine the predictor values for all the variables first, and then the corrected values; only thereafter we advance to the next step and repeat the calculations. We shall illustrate this with a simple example.

Example 2.8

Solve the following second-order differential equation numerically.

$$\frac{d^2x}{dt^2} + 4x = 4t^2 - 36\cos 4t + 2$$

given at $t = 1$ $x = 0.857664$ and $\frac{dx}{dt} = 9.417043$.

Solution

We first convert it into a system of two first-order equations by defining

$$\frac{dx}{dt} = y$$

The given equation becomes

$$\frac{dy}{dt} = (4t^2 - 36\cos 4t + 2 - 4x)$$

Let us solve it first using the simple Euler's method, which gives (Eq. 2.55)

$$x_{i+1} = x_i + h \cdot (y)_i$$
$$y_{i+1} = y_i + h(4t^2 - 36\cos 4t + 2 - 4x)_i$$

The results are given in following table.

t	x	y	t	x	y
1	0.857664	9.417043	1.6	7.109438	0.826851
1.1	1.799368	12.02709	1.7	7.192123	−4.36839
1.2	3.002078	13.09775	1.8	6.755284	−9.01907
1.3	4.311852	12.35792	1.9	5.853377	−12.4152
1.4	5.547644	9.822517	2.0	4.611852	−14.0171
1.5	6.529896	5.795422			

We shall now solve these equations using the fourth-order Runge-Kutta method (Eqs. 2.57–2.61):

$$x_{i+1} = x_i + \frac{h}{6}(k_{1,x} + 2k_{2,x} + 2k_{3,x} + k_{4,x})$$

$$y_{i+1} = y_i + \frac{h}{6}(k_{1,y} + 2k_{2,y} + 2k_{3,y} + k_{4,y})$$

The first function values ($k_{1,x}$ and $k_{1,y}$) are calculated at the starting point, that is, $k_{1,x} = y_i$ and $k_{1,y} = 4t^2 - 36\cos 4t + 2 - 4x_i$. Here t is given the value at t_i.

Next the gradients $k_{2,x}$ and $k_{2,y}$ have to be found for value of $t = t_i + h/2, x_0 = x_i + \frac{h}{2}k_{1,x}$, and $y = y_i + \frac{h}{2}k_{1,y}$ (Eq. 2.59). This gives

$$k_{2,x} = y_i + \frac{h}{2}k_{1,y}$$

$$k_{2,y} = 4\left(t_i + \frac{h}{2}\right)^2 - 36\cos\left(4\left(t_i + \frac{h}{2}\right)\right) + 2 - 4\left(x_i + \frac{h}{2}k_{1x}\right)$$

Now we find the gradients k_{3x} and $k_{3,y}$ for values of

$$t = t_i + \frac{h}{2}; \quad x = x_i + \frac{h}{2}k_{2x}; \quad y = y_i + \frac{h}{2}k_{2,y}$$

This gives for our problem, using Eq. (2.60)

$$k_{3,x} = y_i + \frac{h}{2}k_{2y}$$

$$k_{3,y} = 4\left(t_i + \frac{h}{2}\right)^2 - 36\cos\left(4\left(t_i + \frac{h}{2}\right)\right) + 2 - 4\left(x_i + \frac{h}{2}k_{2,x}\right)$$

Similarly using Eq. (2.61), we can calculate $k_{4,x}$ and $k_{4,y}$ as

$$k_{4,x} = y_i + h \cdot k_{3y}$$

$$k_{4,y} = 4(t_i + h)^2 - 36\cos(4(t_i + h)) + 2 - 4(x_i + hk_{3,x})$$

Substituting these values in the equations for x_{i+1} and y_{i+1}, the solution can be advanced by one step. The calculations for the present problem done using Excel software are summarized in following table.

Detailed calculations for Example 2.8

t	x	y	k1x	k1y	k2x	k2y	k3x	k3y	k4x	k4y
1	0.857664	9.417043	9.4170	26.1005	10.7221	18.7453	10.3543	18.4843	11.2655	10.3316
1.1	1.904919	11.26523	11.2652	10.2843	11.7794	1.4548	11.3380	1.3519	11.4004	−7.5448
1.2	3.05326	11.40445	11.4044	−7.6030	11.0243	−16.4558	10.5817	−16.3797	9.7665	−24.5523
1.3	4.12631	9.77401	9.7740	−24.6118	8.5434	−32.019	8.1731	−31.7729	6.5967	−37.8548
1.4	4.95637	6.60650	6.6065	−37.9058	4.7112	−42.6155	4.4757	−42.2364	2.3829	−45.1819
1.5	5.41242	2.39331	2.3933	−45.2158	0.1325	−46.3939	0.0736	−45.9417	−2.2009	−45.1938
1.6	5.42250	−2.19137	−2.1914	−45.2047	−4.4516	−42.5701	−4.3199	−42.1181	−6.4032	−37.7004
1.7	4.98688	−6.39606	−6.3961	−37.6858	−8.2803	−31.5588	−7.974	−31.1819	−9.5142	−23.6986
1.8	4.17989	−9.51049	−9.5105	−23.6602	−10.6935	−14.9152	−10.2562	−14.6786	−10.9783	−5.2224
1.9	3.14009	−10.97832	−10.9783	−5.16571	−11.2366	4.9029	−10.7332	4.9546	−10.4829	14.9709
2	2.050076	−10.48632								

Comparing the predictions of the two methods with the exact analytical solution (see following table) we find that the Runge-Kutta method gives remarkably accurate agreement with the correct solution.

Comparison of Solutions

t	x			$y = dx/dt$		
	Analytical	Euler	Runge-Kutta	Analytical	Euler	Runge-Kutta
1	0.857664	0.857664	0.857664	9.417043	9.417043	9.417043
1.1	1.904944	1.799368	1.904919	11.26522	12.02709	11.2652
1.2	3.053423	3.002078	3.05326	11.4044	13.09775	11.40445
1.3	4.126553	4.311852	4.126307	9.773901	12.35792	9.77400
1.4	4.956674	5.547644	4.956369	6.60631	9.822517	6.606504
1.5	5.412751	6.529896	5.412423	2.393016	5.795422	2.39331
1.6	5.422806	7.109438	5.422502	−2.19177	0.826851	−2.19137
1.7	4.98711	7.192123	4.986877	−6.39655	−4.36839	−6.39606
1.8	4.180013	6.755284	4.179894	−9.51105	−9.01907	−9.51049
1.9	3.140064	5.853377	3.140088	−10.9789	−12.4152	−10.978
2.0	2.049895	4.611852	2.050076	−10.4869	−14.0171	−10.486

It is also noted that Euler's method gives very poor predictions of the variable y (which is $=dx/dt$) and this is the main reason for its inability to capture the correct solution. This situation can be corrected by using smaller step size in Euler's method. Thus taking $h = 0.02$, and following the same procedure as above the values of x and y obtained are at $t = 1.1 : (1.892052, 11.42531)$ and at $t = 1.2 : (3.06174, 11.74609)$. These are in much better agreement with the correct solution.

2.4.4 Boundary Value Problems

While solving an ordinary differential equation of order greater than 1, the values of the dependent variable and its derivatives must be known for some value of the independent variable. Such problems, called the initial value problems (IVP), can be solved by following any of the methods discussed above starting from the known values of variables. However, in case these values are known for different values of the independent variable, it is not possible to start the numerical solution. The most common method for solving such problems—called the boundary value problems—is the *shooting method*. Here we assume the values needed to convert the problem to IVP, determine its solution, and compare the computed value(s) at the "boundaries" with the prescribed values. Depending upon the nature of disagreement between these two, another "guess" for the initial values is made and the new IVP solved again. Such iterations are continued till convergence is obtained. For simple problems involving just 1 or 2 assumptions, one can adopt a heuristic approach to get new guesses, however, when a large number of equations are being solved, it is desirable to treat the task of selection of the initial values akin to the solution of nonlinear equations, and use the Warner's method (Section 2.2.1) to quickly arrive at the correct values of the variables being assumed. This is illustrated in Example 2.9.

Example 2.9

Solve the following second-order differential equation numerically.

$$\frac{d^2x}{dt^2} + \pi^2 x = 3(1 + \pi^2)e^{-t}$$

given the boundary conditions: at $t = 0, x = 3$ and at $t = 0.9, x = 0.601675$.

Solution

The given second-order equation can be converted into a system of two first-order differential equations by defining an additional variable y as follows.

$$\frac{dx}{dt} = y$$

$$\frac{dy}{dt} = 3(1 + \pi^2)e^{-t} - \pi^2 x$$

We are given at $t = 0$ $x = 3$, but the value of y is not known. Assuming $y = 0$, we solve the above system of equations using fourth-order Runge-Kutta method. The details are given in following table.

Detailed calculations for Example 2.9

t	x	y	k1x	k1y	k2x	k2y	k3x	k3y	k4x	k4y
0	3	0	0	3	0.15	1.4096	0.07048	1.3356	0.1336	−0.1727
0.1	3.00958	0.138631	.1386	−0.1976	0.1287	−1.7051	0.05338	−1.7002	−0.03139	−3.05816
0.2	3.01743	−0.02914	−0.02914	−3.08304	−0.1833	−4.3707	−0.2477	−4.2947	−0.4586	−5.3792
0.3	2.99494	−0.45902	−0.4590	−5.4017	−0.7291	−6.3533	−0.7767	−6.2200	−1.0810	−6.9340
0.4	2.91908	−1.08373	−1.0837	−6.9518	−1.4313	−7.4830	−1.4579	−7.3115	−1.8149	−7.5930
0.5	2.77446	−1.8193	−1.8193	−7.6046	−2.1995	−7.6714	−2.2029	−7.4838	−2.5677	−7.3126
0.6	2.5546	−2.57309	−2.5731	−7.3168	−2.9389	−6.9198	−2.9191	−6.7393	−3.2470	−6.1388
0.7	2.26233	−3.25265	−3.2527	−6.1352	−3.5594	−5.3199	−3.5186	−5.1685	−3.7695	−4.2035
0.8	1.90936	−3.77457	−3.7746	−4.1925	−3.9842	−3.0444	−3.9268	−2.9410	−4.06867	−1.7113
0.9	1.51494	−4.07249	−4.07249	−1.6941	−4.1572	−0.331	−4.0890	−0.2892	−4.1014	1.07999

It is seen that with initial assumption $y = 0$, we get the value of x at $t = 0.9$ as 1.51494 while the value given is 0.601675. Since actual value in less, we reduce y to -2 and solve the equation again. The results are given in following table.

Detailed calculations for Example 2.9... (contd)

t	x	y	k1x	k1y	k2x	k2y	k3x	k3y	k4x	k4y
0.0	3.0000	−2.0000	−2.0000	3.0000	−1.8500	2.3966	−1.8802	2.3226	−1.7677	1.7525
0.1	2.8129	−1.7635	−1.7635	1.7438	−1.6763	1.1750	−1.7047	1.1320	−1.6503	0.6185
0.2	2.6433	−1.6472	−1.6472	0.6098	−1.6167	0.1206	−1.6412	0.1056	−1.6367	−0.3110
0.3	2.4799	−1.6347	−1.6347	−0.3188	−1.6506	−0.6903	−1.6692	−0.6824	−1.7029	−0.9702
0.4	2.3137	−1.7019	−1.7019	−0.9765	−1.7508	−1.2027	−1.7621	−1.1786	−1.8198	−1.3175
0.5	2.1379	−1.8195	−1.8195	−1.3216	−1.8856	−1.3883	−1.8890	−1.3557	−1.9551	−1.3394
0.6	1.9491	−1.9554	−1.9554	−1.3411	−2.0224	−1.2489	−2.0178	−1.2159	−2.0769	−1.0526
0.7	1.7473	−2.0774	−2.0774	−1.0516	−2.1300	−0.8162	−2.1182	−0.7903	−2.1564	−0.5020
0.8	1.5351	−2.1569	−2.1569	−0.4986	−2.1818	−0.1488	−2.1643	−0.1365	−2.1705	0.2432
0.9	1.3181	−2.1706	−2.1706	0.2487	−2.1582	0.6733	−2.1370	0.6672	−2.1039	1.0962

Now, at $t = 0.9$, x is reduced to 1.31809, which is still much higher than the specified boundary condition. To estimate the value of y (at $t = 0$) which could satisfy the boundary condition, we use the information from the above two iterations as explained below.

For $y = 0$, the "error" at boundary $t = 0.9$ is $1.514938 - 0.601675 = 0.913263$ and for $y = -2$ the "error" is $1.318090 - 0.601675 = 0.716415$.

Assuming a linear relationship between these "errors" and the initially assumed value of y, we can get the value of y for which we can expect "error" = 0, from the following equation

$$\frac{y - 0}{0.913263} = \frac{y + 2}{0.716415} \quad \text{which gives } y = -9.2788$$

Using this as the starting assumption, the two equations have been solved again and the results are given in following table. It is seen the predicted value of x at $t = 0.9$ is 0.601678 which matches very well with the specified boundary condition of 0.601675. Detailed calculations for Example 2.9

t	x	y	k1x	k1y	k2x	k2y	k3x	k3y	k4x	k4y
0.0	3.0000	−9.2788	−9.2788	3.0000	−9.1288	5.9886	−8.9794	5.9145	−8.6873	8.7591
0.1	2.0970	−8.6860	−8.6860	8.8095	−8.2456	11.6569	−8.1032	11.4395	−7.5421	13.9992
0.2	1.2815	−7.5360	−7.5360	14.0496	−6.8335	16.4664	−6.7127	16.1198	−5.9240	18.1342
0.3	0.6057	−5.9134	−5.9134	18.1796	−5.0044	19.9196	−4.9174	19.4711	−3.9663	20.7341
0.4	0.1103	−3.9518	−3.9518	20.7701	−2.9133	21.6542	−2.8691	21.1417	−1.8377	21.5217
0.5	−0.1790	−1.8204	−1.8204	21.5447	−0.7432	21.4784	−0.7465	20.9468	0.2742	20.3993
0.6	−0.2544	0.2928	0.2928	20.4069	1.3132	19.3897	1.2623	18.8861	2.1814	17.4581
0.7	−0.1273	2.1997	2.1997	17.4496	3.0722	15.5744	2.9785	15.1438	3.7141	12.9690
0.8	0.1729	3.7307	3.7307	12.9453	4.3779	10.3897	4.2501	10.0703	4.7377	7.3562
0.9	0.601678	4.7510	4.7510	7.3194	5.1170	4.3283	4.9674	4.1477	5.1658	1.1551

A comparison of the intermediate values with the analytical solution (given in following table) shows a good agreement between the values. However, the correspondence at intermediate values is not as good as at the boundary. This can be attributed to truncation errors involved in numerical calculations.

t	x (Analytic)	x (Numerical)
0	3	3
0.1	2.096478	2.096958
0.2	1.280622	1.281531
0.3	0.604421	0.605655
0.4	0.108847	0.110265
0.5	−0.18041	−0.17898
0.6	−0.25568	−0.2544
0.7	−0.12828	−0.12732
0.8	0.172416	0.172937
0.9	0.601675	0.601678

Since the change in x-values in the step size $h = 0.1$ is quite appreciable and changes significantly at different values of t, it follows that for better accuracy smaller step size should be taken. Now if we solve these equations with a step of 0.05 the solution obtained with the initial assumption of $y = -9.2780$ gives at $t = 0.9, x = 0.602078$ which is slightly different from the earlier value of 0.601678. This suggests further reduction in the assumed value of y at $t = 0$. Thus taking $y = -9.283$, the solution obtained is

t	x (Analytic)	x (Numerical)
0	3	3
0.1	2.096478	2.096499
0.2	1.280622	1.280662
0.3	0.604421	0.604474
0.4	0.108847	0.108907
0.5	−0.18041	−0.18035
0.6	−0.25568	−0.25563
0.7	−0.12828	−0.12824
0.8	0.172416	0.17243
0.9	0.601675	0.601665

The agreement with the analytical values is now much better than earlier.

This points to the need for much greater accuracy of calculations in solving boundary value problems. If there are large truncation errors, the convergence to the correct boundary value may occur far away from the true solution. As an illustration of this we have also solved the problem using Euler's method. It is seen that convergence at the boundary point is obtained with an initial assumption of $y = -10.8956$, but the intermediate values are quite different from the analytical values:

t	x (Analytic)	x (Numerical Euler)
0	3	3
0.1	2.096478	1.91044
0.2	1.280622	0.85088
0.3	0.604421	−0.08705
0.4	0.108847	−0.8139
0.5	−0.18041	−1.25149
0.6	−0.25568	−1.34178
0.7	−0.12828	−1.05459
0.8	0.172416	−0.3934
0.9	0.601675	0.601668

This brings out rather strikingly the need for choosing proper method of solution and proper step size while solving boundary value problems.

2.5 LAPLACE TRANSFORMATION

Dynamic analysis of many thermal systems requires solution of differential equations in time (t). Rather than solving these equations directly it is helpful to use Laplace transformation which converts the differential equations in t to algebraic equations in Laplace transform variable s. This enables us to represent the input-output characteristic of any system in terms of an algebraic expression, called the transfer function. In systems having feedback control the transfer function helps analyze the stability of the system. In this section we shall study in brief these concepts.

The Laplace transform of any function $f(t)$ is indicated by symbol $Ł[f(t)]$ and defined by the equation:

$$Ł\{f(t)\} = \int_0^\infty f(t)e^{-st}dt = f(s) \tag{2.70}$$

Generally, the Laplace transform of a function $f(t)$ is indicated as $f(s)$. Let us obtain the transforms of a few simple functions *ab initio*.

Function unit step function $U(t) = f(t) = 0$ for $t < 0$ and $= 1$ for $t > 0$

$$Ł\{f(t)\} = \int_0^\infty f(t)e^{-st}dt = \int_{0+}^\infty 1\ e^{-st}dt = \left.\frac{e^{-st}}{-s}\right|_{0+}^\infty$$

$$\therefore f(s) = \frac{1}{s}$$

Ramp function $f(t) = t \cdot U(t)$

$$\alpha\{f(t)\} = \int_{0+}^\infty t \cdot e^{-st}dt = \left.\left| t \cdot \frac{e^{-st}}{-s}\right|\right._0^\infty - \int_0^\infty 1 \cdot \left\{\frac{e^{-st}}{-s}\right\}dt \quad \text{integrating by parts}$$

$$= 0 + \frac{1}{s}\int_0^\infty e^{-st}dt = \frac{1}{s^2}$$

Exponential function $f(t) = e^{-at}$

$$Ł\{f(t)\} = \int_o^\infty e^{-at} \cdot e^{-st}dt = \int_0^\infty e^{-(s+a)t}dt$$

$$= \left.\left|-\frac{e^{-(s+a)t}}{(s+a)}\right|\right._0^\infty = \frac{1}{s+a}$$

Proceeding in a similar manner the Laplace transform of any function can be found. Table 2.5 summarizes the transforms of commonly encountered functions.

Table 2.5 Laplace transforms of commonly encountered functions

Unit step function	$f(t) = U(t) = 0$	$t < 0$	$f(s) = \frac{1}{s}$
	$= 1$	$t > 0$	
Ramp function	$f(t) = t \cdot u(t)$		$f(s) = \frac{1}{s^2}$
Exponential function	$f(t) = e^{-at}$		$f(s) = \frac{1}{s+a}$
Sine function	$f(t) = \sin\ wt$		$f(s) = \frac{w}{s^2+w^2}$
Cosine function	$f(t) = \cos\ wt$		$f(s) = \frac{s}{s^2+w^2}$
Power function	$f(t) = t^n$		$f(s) = \frac{n!}{s^{n+1}}$
Unit Impulse function	$\delta(t) = \frac{d(U(t))}{dt}$		1

Starting from the basic definition of Laplace transforms, Eq. (2.70), we can also derive certain properties. For example

(a) *Linearity*: $Ł\{f_1(t) + f_2(t)\} = Ł\{f_1(t)\} + Ł\{f_2(t)\}$

(b) *Multiplication by a constant a*: $Ł\{af(t)\} = aŁ\{f(t)\}$

(c) *Scale change*:

$$Ł\{f(at)\} = \frac{1}{a}f(s/a) \quad \text{where } f(s) = Ł\{f(t)\}$$

$$Ł\{f(t/a)\} = af(as)$$

(d) *Differentiation of Laplace transform*:

$$\frac{d}{ds}f(s) = -Ł\{tf(t)\}$$

$$\frac{d^n}{ds^n}f(s) = Ł\{(-t)^n f(t)\} \quad n = 1, 2, 3, \ldots$$

(e) *Shifting theorems*:

$$Ł\{e^{at}f(t)\} = f(s-a)$$

$$Ł\{f(t-t_0)\} = e^{-st_0} f(s)$$

(f) *Transform of a derivative*:

$$Ł\left\{\frac{df(t)}{dt}\right\} = sf(s) - f(0)$$

where $f(0)$ is the value of $f(t)$ at $t = 0$

$$Ł\left\{\frac{d^2f}{dt^2}\right\} = s^2 f(s) - sf(0) - f'(0)$$

$$\text{where } f'(0) = \left\{\frac{df(t)}{dt}\right\}_{t=0}$$

(g) *Transform of integral*: $Ł\left\{\int_0^t f(t)dt\right\} = \frac{f(s)}{s}$

(h) *Final value theorem* :

$$\text{If } \lim_{\tau \to \infty} [f(\tau)] \text{ is finite}$$

$$\text{then } \lim_{\tau \to \infty} [f(\tau)] = \lim_{s \to \infty} [s f(s)]$$

By suitably combining the tabulated values of transforms with the properties mentioned above it should be possible to obtain the Laplace transforms for most functions encountered commonly.

Let us now see how Laplace transformation helps in solving a differential equation. We consider the basic equation encountered in the analysis of a pin-fin, namely.

$$\frac{d^2\theta}{dx^2} - m^2\theta = 0$$

with the known boundary conditions: at $x = 0$, $\theta = \theta_0$ and at

$$x = L, \quad -k\frac{d\theta}{dx} = h\theta$$

Applying Laplace transformation to the differential equation[3] we get

$$[s^2\theta(s) - s\theta_0 - \theta_0'] - m^2\theta(s) = 0$$

To proceed with the solution we assume $\theta_0' = C$, an unknown constant, whose value we would determine later by invoking the specified boundary condition at $x = L$. Rearranging the above equation we get

$$\theta(s) = \frac{s\,\theta_0 + C}{s^2 - m^2} = \frac{s}{s^2 - m^2}\theta_0 + \frac{C}{s^2 - m^2}$$

Now to get the solution in terms of $\theta(x)$, we need to find the functions of which the terms on the RHS are Laplace transforms. This process of inverse transformation is usually done by referring to the table of transforms and making use of the various properties of Laplace transforms enumerated above. Often we first need to subdivide the terms by making use of partial fractions. Thus here we can write the term $(s^2 - m^2)$ occurring in the denominator as a product of two terms $(s - m)(s + m)$ and expand the terms as follows, that is

$$\theta(s) = \frac{\theta_0}{2}\left[\frac{1}{s+m} + \frac{1}{s-m}\right] - \frac{C}{2m}\left[\frac{1}{s+m} - \frac{1}{s-m}\right]$$

Now, from the table of transforms we can directly write the inverse transforms as:

$$\theta(x) = \frac{\theta_0}{2}\left[e^{-mx} + e^{mx}\right] - \frac{C}{2m}\left[e^{-mx} - e^{+mx}\right]$$

which is the solution of the differential equation. To find the value of C, we use the specified boundary condition at $x = L$, that is

$$h\theta = -k\frac{d\theta}{dx}$$

---

[3] Here the independent variable is x.

This gives

$$h\left\{\frac{\theta_0}{2}\cdot\left(e^{-mL}+e^{mL}\right)-\frac{C}{2m}\left(e^{-mL}-e^{mL}\right)\right\}=-k\left\{\frac{\theta_0}{2}\cdot\left(-me^{-mL}+me^{mL}\right)\right\}$$
$$+k\left\{\frac{C}{2m}\left(-me^{-mL}-me^{mL}\right)\right\}$$

Simplifying we get

$$C=-\frac{m\theta_0\left[\frac{e^{mL}-e^{-mL}}{2}+N\frac{e^{mL}+e^{-mL}}{2}\right]}{\left[\frac{e^{mL}+e^{-mL}}{2}+N\frac{e^{mL}-e^{-mL}}{2}\right]}$$

where $N=\frac{h}{mk}$.

Substituting this value in the expression for θ, we get the complete solution as

$$\frac{\theta}{\theta_0}=\frac{\cosh m(L-x)+N\sinh m(L-x)}{\cosh mL+N\sinh mL}$$

Laplace transformation is also used to solve partial differential equations. By taking transform with respect to one of the two independent variables, say t, the partial differential equation gets transformed to an ordinary differential equation for the Laplace transform of the dependent variable as a function of the other variable. By solving this equation and taking inverse transform of the solution, we can get the solution of the original partial differential equation. This is illustrated in Example 2.10.

Example 2.10

Solve the partial differential equation (called wave equation)

$$\frac{\partial^2 Y}{\partial t^2}=v^2\frac{\partial^2 Y}{\partial x^2}$$

Subject to conditions

$$Y(0,t)=f(t)\quad\text{and}\quad\lim_{x\to\infty}Y(x,t)=0\quad\text{for}\quad t\geq 0$$

$$Y(x,0)=0\quad\left[\frac{\partial Y}{\partial t}\right]_{t=0}=0$$

Solution

Let us take a Laplace transform with respect t. This gives

$$s^2Y(s)-sY(x,0)-Y'(x,0)=v^2\frac{\partial^2 Y(s)}{\partial x^2}$$

Substituting initial conditions it simplifies to:

$$s^2 Y(s) = v^2 \frac{\partial^2 Y(s)}{\partial x^2}$$

$$\text{or} \quad \frac{\partial^2 Y}{\partial x^2} - \frac{s^2}{v^2} Y = 0$$

which is an ordinary differential equation for $Y(s, x)$ whose solution is

$$Y(x, s) = a\, e^{\frac{s}{v} x} + b\, e^{\frac{-s}{v} x}$$

To find the constants a and b we make use of the other specified conditions, after doing their Laplace transformation. Thus

$$Y(0, t) = f(t) \Rightarrow Y(0, s) = F(s)$$

$$\text{and} \quad \lim_{x \to \infty} (Y(x, t)) = 0 \Rightarrow \lim_{x \to \infty} (Y(x, s)) = 0$$

Substituting in the above expression for $Y(x, s)$ we get

$$F(s) = a + b; \quad a = 0; \quad \text{which gives } b = F(s)$$

Thus the solution for $Y(x, s)$ is

$$Y(x, s) = F(s)\, e^{-\frac{s}{v} x}$$

We have to take its inverse transform to get the solution for $Y(x, s)$. Referring to the table of transforms and the properties, we see its similarity with the second shifting theorem. Thus we can get

$$Y(x, t) = f\left(t - \frac{x}{v}\right)$$

as the solution to the partial differential equation.

2.5.1 Transfer Function

The concept of transfer function is of great help in simplifying and generalizing the dynamic response of thermal systems. Let us try to understand it in some detail through the simple example of a thermometer being used to measure the temperature of a fluid flowing in a pipe. Let us assume that, to begin with (i.e., at time $t = 0$), the thermometer and the fluid are in equilibrium. Then, the temperature of the fluid begins to change. Now, due to thermal inertia, the temperature as recorded by the thermometer will not be exactly equal to the fluid temperature at any instant. The relationship between the two will be primarily governed by the "thermal inertia" of the system which is characterized by the properties of the sensing device (like its mass, specific heat, volumetric expansion coefficient or the dependence of its electrical resistance on temperature, etc.) and its interaction with fluid (e.g., the heat-transfer coefficient between the sensor and the fluid, etc.). This can be modeled by using the basic principles of heat transfer, thermodynamics, and electrical engineering (if the sensor is an electrically sensitive device like thermistor or thermocouple). Usually such models result in ordinary differential equations which need to be solved to determine the system response (i.e., in our example, the temperature recorded by the thermometer) to the stimulus (i.e., the manner of variation of the fluid temperature). If we transform the governing equation using Laplace transformation, and define the variables suitably so that their values are set equal to zero at $t = 0$, then we can get a simple relationship between the stimulus (also called the input) and the response (also called the output), which is termed as the transfer function of the system.

Thus considering a system modeled by a second-order differential equation:

$$c_1 \frac{d^2\theta}{dt^2} + c_2 \frac{d\theta}{dt} + c_3 \, \theta = f(t) \tag{2.71}$$

where c_1, c_2, c_3 are given constants and $f(t)$ is a given function, we get by taking Laplace transform:

$$c_1 (s^2 \, \theta(s) - s \, \theta(0) - \theta'(0)) + c_2 (s \, \theta(s) - \theta(0)) + c_3 \, \theta(s) = F(s)$$

Now, if the initial conditions are $\theta(0) = 0$; $\theta'(0) = 0$, the above equation can be simplified to get

$$\frac{\theta(s)}{F(s)} = \frac{1}{s^2 c_1 + s c_2 + c_3} = G(s) \tag{2.72}$$

Here $G(s)$ is the transfer function of the system.

An important advantage of the concept of transfer function is that it can be used to predict the system response to any type of stimulus, by using the equation $\theta(s) = F(s) \cdot G(s)$. By inverting $\theta(s)$, the actual response with time, $\theta(t)$ can be obtained. This is usually represented by a block diagram of the kind shown in Fig. 2.6. The transfer function provides an insight into the nature of the system, and most commonly encountered thermal components are seen to be either first-order or second-order systems whose general behavior is well known.

For any complicated system if the transfer functions of the individuals' components are known, that of the whose system can be developed by combining these judiciously

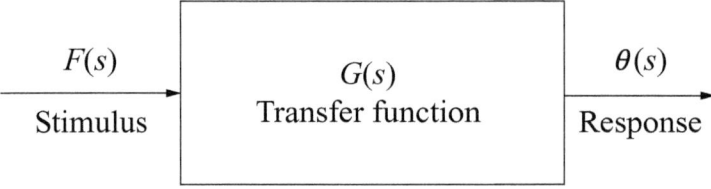

Fig. 2.6 Block diagram representing a transfer function.

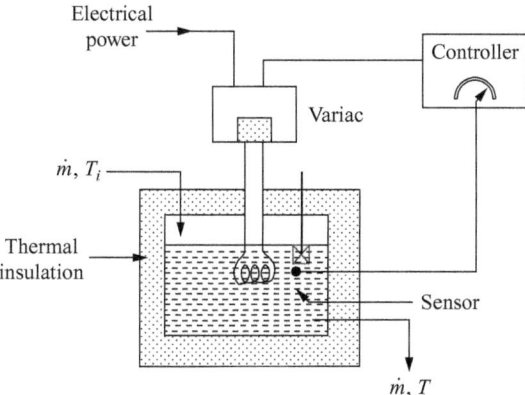

Fig. 2.7 Thermostatic bath.

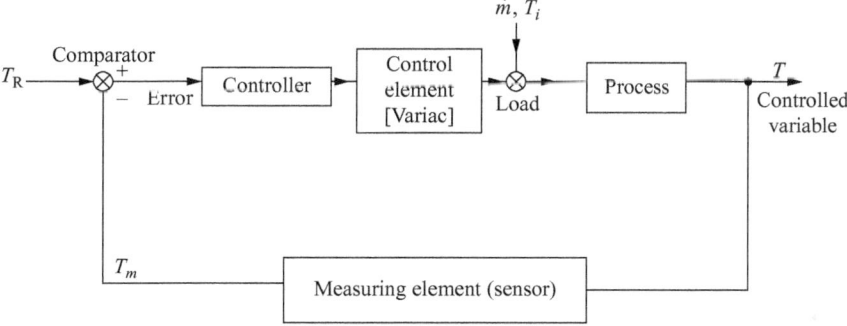

Fig. 2.8 Block diagram showing details of the control mechanism of a thermostatic bath.

taking due account of the transfer of "signals" from one component to another. As an example let us consider the simple control system of a thermostatic bath shown schematically in Fig. 2.7.

It consists of a well-insulated container in which heat-transfer oil can be maintained at any set point by providing heat input through an electrical heater. The heater voltage is set through the controller in response to the difference between the set point and the instantaneous temperature in the container as sensed by the sensor. This schematic diagram can be converted into a block diagram (Fig. 2.8) representing the nature of interaction between various components, as explained at length in Chapter 10.

Through detailed analysis of the processes occurring, the transfer functions relating the output of each block to the input can be obtained. By combining these individual transfer functions properly the dynamic analysis of the whole system can be carried out.

Fig. 2.9 indicates the rules for combining transfer functions. These can be easily derived from the basic definition.

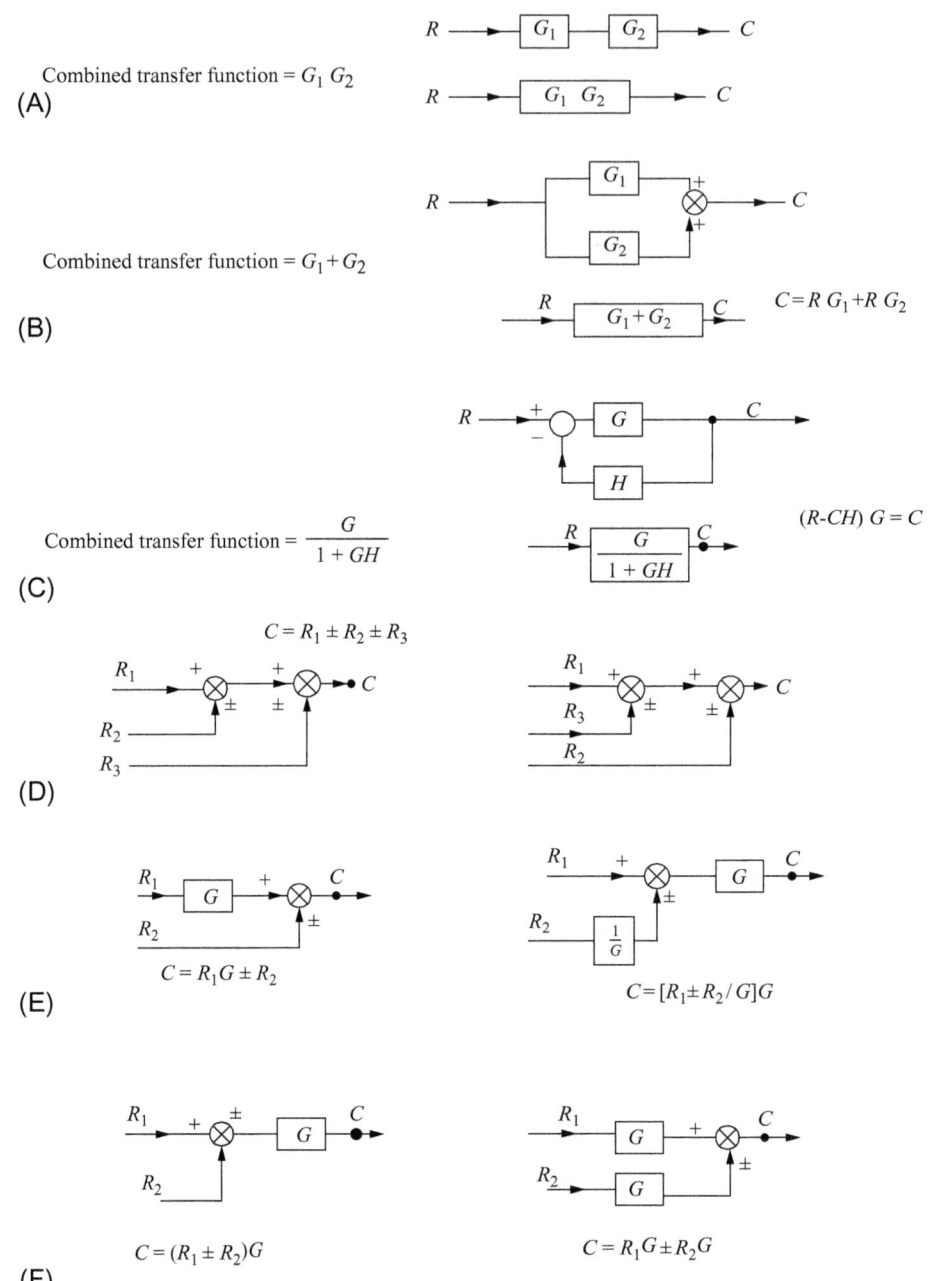

Fig. 2.9 Rules for combining transfer functions. (A) Systems in series. (B) Systems in parallel. (C) Feedback loop. (D) Rearranging summing junctions. (E) Moving a summing junction in front of a block. (F) Moving a summing junction beyond a block.

Example 2.11

Determine the overall transfer function for the block diagram shown in Fig. 2.10.

Solution

We first replace the inner loop, enclosed within dashed lines by its equivalent transfer function, that is, $G_a = \frac{G_{c2}\ G_{c3}}{1 + G_{c1}\ G_{c2}\ H_2}$.

The new block diagram obtained after this replacement, and its further rearrangements are shown in Fig. 2.11.

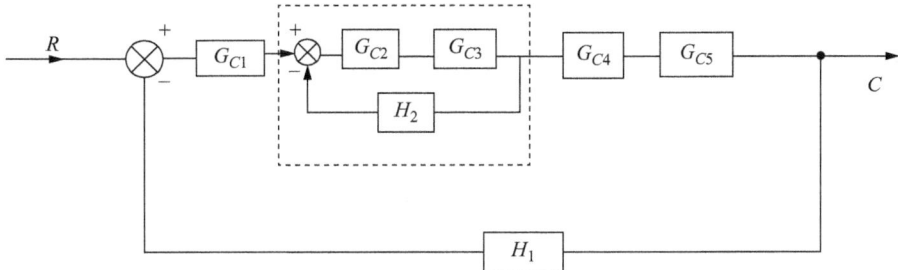

Fig. 2.10 Block diagram for Example 2.11.

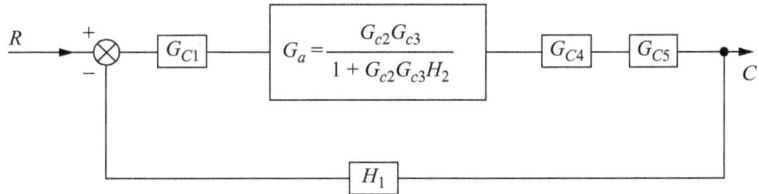

Combining all the blocks in series in the forward path this block diagram can be simplified to:

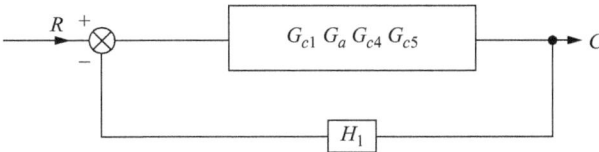

The feedback loop in this block can be replaced by its equivalent shown in the diagram below:

$$R \quad \boxed{\dfrac{G_{c1}\ G_a\ G_{c4}\ G_{c5}}{1 + H_1\ G_{c1}\ G_a\ G_{c4}\ G_{c5}}} \quad C$$

Thus the overall transfer function is : $\dfrac{G_{c1}\ G_a\ G_{c4}\ G_{c5}}{1 + H_1\ G_{c1}\ G_a\ G_{c4}\ G_{c5}}$

Fig. 2.11 Rearranging the block diagram of Example 2.11.

2.6 ANALYSIS OF UNCERTAINTY

The analysis of uncertainty of results has been an integral part of all scientific experimental studies. The uncertainty arises from many sources like accuracy of measuring instruments (like thermometers, thermocouples, thermistors, etc.), approximations made in the analysis of experimental results (like assuming 1-D heat transfer), uncontrolled changes in the environment (temperature or humidity) of the laboratory. To illustrate this, let us consider a simple example of determining the heat-transfer coefficient of boiling refrigerant flowing through a pipe. The simplest experimental set up would have a test section (say of dia. D) with thermocouples fixed on its surface to measure the surface temperature and one thermocouple immersed in the liquid to measure its temperature. Let us consider the case when the tube is heated by an electric heater wrapped over its outer surface, through which a uniform heat flux is imposed. The heat-transfer coefficient h would be calculated from various measurements as

$$h = \frac{Q/A}{(T_s - T_l)} \tag{2.73}$$

where Q is the heat input, A is the surface area, T_s the average of the various surface temperature values, and T_l is the temperature of the boiling refrigerant.

Now uncertainty in the h value arises from uncertainty in the measurement of all the terms on the RHS of Eq. (2.73), as also from various implicit approximations like neglecting axial conduction and the nonuniformity of imposed heat flux, as also the difference in temperatures between the external surface (where thermocouple are fixed) and the internal surface which is in contact with the liquid. By doing more rigorous calculations the influence of various approximations can be estimated, but the uncertainties in various primary measurements would remain. Usually these uncertainties are prescribed by the manufacturers of the measuring instruments. Using this information and assuming that these uncertainties are random and independent, and governed by normal (or Gaussian) distribution, the uncertainty of the result can be calculated from the following general formula for a quantity y which is a function of measured various $(x_1, x_2, \ldots, x_n)$;

$$\delta y = \sqrt{\left(\frac{\partial y}{\partial x_1} \cdot \delta x_1\right)^2 + \left(\frac{\partial y}{\partial x_2} \cdot \delta x_2\right)^2 + \cdots + \left(\frac{\partial y}{\partial x_n} \cdot \delta x_n\right)^2} \tag{2.74}$$

where δx_i represents the uncertainty in the value of the variable x_i. Example 2.12 illustrates this method.

Example 2.12

Determine the uncertainty in the value of heat-transfer coefficient h calculated using Eq. (2.73) from the following data.

$$Q = 200 \pm 1 \, \text{W} \quad D = 10 \pm 0.1 \, \text{mm} = 0.010 \pm 0.0001 \, \text{m} \quad L = 1 \, \text{m} \pm 0.5 \, \text{mm}$$

$$T_{surf} = 25 \pm 0.1°\text{C} \quad T_{liq} = 5 \pm 0.1°\text{C}$$

Solution

Since the primary variables measured are the tube diameter D and the length L, we first write Eq. (2.73) in terms of D and L (instead of surface area A) as

$$h = \frac{Q}{\pi DL(T_s - T_l)}$$

Various gradients in Eq. (2.74) can now be calculated as follows:

$$\frac{\partial h}{\partial Q} = \frac{1}{\pi \, DL(T_s - T_l)} = \frac{1}{\pi(0.010)(1)(25 - 5)} = 1.59155$$

$$\frac{\partial h}{\partial D} = \frac{Q}{(T_s - T_l)\pi L} \frac{(-1)}{D^2} = \frac{200}{\pi(25-5)(1)(0.01)^2} \frac{(-1)}{} = -31{,}830.99$$

$$\frac{\partial h}{\partial L} = \frac{Q}{(T_s - T_l)\pi D} \frac{(-1)}{L^2} = \frac{200}{\pi(25-5)(0.01)(1^2)} \frac{(-1)}{} = -318.31$$

$$\frac{\partial h}{\partial T_s} = \frac{Q}{\pi \, DL(T_s - T_l)^2} \frac{(-1)}{} = \frac{200}{\pi \times (0.01)(1)(25 - 5)^2} \frac{(-1)}{} = -15.9155$$

$$\frac{\partial h}{\partial T_l} = \frac{Q}{\pi D \, L(T_s - T_l)^2} = \frac{200}{\pi \times (0.01)(1)(25 - 5)^2} = 15.9155$$

The measured value of $h = \dfrac{200}{\Pi \times 0.01 \times 1(20)} = 318.31 \, \text{W/m}^2$

The uncertainty in h as given by Eq. (2.74) is

$$\delta h = \sqrt{(1.59155 \times 1)^2 + (31{,}830.99 \times 0.0001)^2 + (-318.31 \times 0.0005)^2 + (-15.9155 \times 0.1)^2}$$
$$+ (15.9155 \times 0.1)^2$$
$$= \sqrt{(2.5330 + 10.1321 + 0.02533 + 2.5330 + 2.5330)}$$
$$= 4.21 \, \text{W/m}^2 \, \text{K}$$

Thus the uncertainty in the value of the heat-transfer coefficient in 4.21 W/m^2 K.

Now the results of simulation, being a substitute for costly experimentation, would also be subject to similar uncertainties. Thus while estimating the heat-transfer coefficient of boiling refrigerant flowing through a tube during simulation, say of a refrigerant evaporator, we would have to proceed as follows.

(i) Get the pertinent data: inner diameter of the tube, mass flow rate of the refrigerant, its various thermo-physical properties, and the heat flux imposed on the tube, and an estimate of the uncertainty in the value of each variable.

(ii) Choose appropriate correlation applicable for the conditions in the tube (Section 3.3.3). This being the case of forced convection boiling, we could, for example, use Eq. (3.287). This in turn needs Eqs. (3.288)–(3.298), which demand iterative solution. Thus it is not possible to write an explicit equation (like Eq. 2.73 for h as a function of various known variables. The same is true of most system simulation problems. We therefore need alternative method to ascertain the uncertainty of the results of simulation.

The simplest approach to ascertain the effect of uncertainties is the factorial method. Here, after doing the initial simulation, all the uncertain parameter values are altered one by one, over the two extremes (i.e., the smallest and the largest values). If the number of uncertain parameters (N) is large, this method becomes unwieldy as the number of simulations needed becomes very high (i.e., 2^N). Further, if the simulation is done for mean parameter values also, the number of runs would be even greater, namely 3^N. This method is therefore suited only for small laboratory equipment, and can not be used for practical thermal systems with large number of uncertain variables.

The most common method for estimating the uncertainty in system simulation results is the statistical methodology of Monte Carlo analysis. It involves perturbing all the uncertain variables randomly between simulations. This is done in accordance with their probability distribution. If this is not known and only the lower and upper limiting values x_L and x_U of a variable are known, it is safe to assume a Gaussian normal distribution $N(\mu, \sigma)$ with mean $\mu = \frac{x_L + x_U}{2}$ and standard deviation $\sigma = \frac{\mu - x_L}{1.96}$ [5]. Then the simulation model is run for various randomly generated values of the uncertain variables. The model output values thus obtained also result in a probability distribution which, if the number of samples of the uncertain variable set is sufficiently large (typically about 100), approaches normal distribution. Thus a quantitative estimate of the uncertainty in simulation predictions can be obtained.

Application of Monte Carlo technique to the design of thermal equipment under uncertainties is discussed in Section 11.3.1.

2.7 ENGINEERING ECONOMICS

Like all other engineering products, the design of thermal systems is also governed by economic considerations. Thus, for example, while designing the various components of a thermal power plant we have to choose whether to design it for the lowest initial cost or for the lowest running cost (i.e., high efficiency) or for the lowest life cycle cost (i.e., a suitable combination of the initial cost and its lifetime running cost). It could also be designed for maximum net profit over its lifetime, which would include suitably accounting for the expenses and the returns (from selling power, for example). Since both the expenses and the returns from a project are spread over many years, and the "effective value" of money changes with time due to inflation, design engineers should have some basic knowledge about economics. In this section we shall review, in brief, some basic concepts of engineering economics.

2.7.1 Time Value of Money

For most engineering projects, a part of money needed is borrowed from the banks which lend it at certain interest rate. Usually this interest is compounded annually. Thus if an amount of money P is borrowed on loan, at an annual interest rate i, the amount F that will have to be returned in future, say after n years, would be

$$F = P(1 + i)^n \tag{2.75}$$

Thus Rs. 1000 borrowed today at an interest rate of 10% per annum, would have to be settled after 5 years with a payment of $F = 1000(1 + 0.1)^5 = 1610.51$ Rs.

This can also be interpreted alternatively as suggesting that the present worth of Rs. 1610.51 received after 5 years is Rs. 1000, that is

$$P = \frac{F}{(1 + i)^n} \tag{2.76}$$

Sometimes the compounding of the interest is done quarterly or half-yearly. In that case these formulae change to

$$F = P\left(1 + \frac{i}{m}\right)^{nm} \quad \text{or} \quad P = \frac{F}{(1 + \frac{i}{m})^{mn}} \tag{2.77}$$

where m is the number of compounding periods per year (=4 for quarterly, and =2 for half-yearly compounding). Thus in the above case of borrowing Rs. 1000 from bank at in interest rate of 10%, if the compounding is done quarterly, the amount to be paid after 5 years would be

$$F = 1000\left(1 + \frac{0.10}{4}\right)^{5 \times 4} = 1000(1 + 0.025)^{20} = 1638.6 \text{ Rs.}$$

While money invested in a bank (or any other security) would grow in accordance with the interest rate, inflation would eat into its actual worth. Thus an inflation rate of 5% would imply that what one could purchase in Rs. 100/- today, is likely to cost Rs. 105/- after a year. Often the interest rate used in above calculations are the net interest rates after accounting for inflation. In case these are gross interest rates, then inflation has to be accounted for by using the formula:

$$F = \frac{P(1 + i)^n}{(1 + f)^n} \quad \text{or} \quad P = \frac{F(1 + f)^n}{(1 + i)^n} \tag{2.78}$$

where f is the inflation rate. Alternatively, an effective interest rate accounting for inflation (i_{eff}) can be calculated as:

$$i_{eff} = \frac{i - f}{1 + f} \tag{2.79}$$

Example 2.13 explains these concepts in detail.

Example 2.13

A government bond purchased today for Rs. 10,000/- is guaranteed to give Rs. 20,000/- after 10 years. If the average inflation rate during this period is 5%, what is the effective rate of interest one will earn? What is the real worth of Rs. 20,000/- that one will get after 10 years?

Solution

To calculate the effective interest rate, we must first calculate the nominal interest rate using Eq. (2.75). This gives

$$20,000 = 10,000(1 + i)^8 \Rightarrow i = 0.07177$$

Therefore, effective interest rate accounting for inflation is (Eq. 2.79)

$$i_{eff} = \frac{0.07177 - 0.05}{1 + 0.05} = 0.020736 \quad \text{or} \quad 2.0736\%$$

Thus the real worth of the bond is $= 10,000(1.020736)^{10} = 12,278\,\text{Rs.}$

The same result could be obtained by deflating the sum of Rs. 20,000 in view of inflation:

$$\text{Real value} = \frac{20,000}{(1 + 0.05)^{10}} = 12,278.3\,\text{Rs.}$$

2.7.2 Present Worth Analysis

In this method all the projected expenses are brought forward to the present time and the net present worth assessed. If it is positive, the project meets the basic economic appraisal criterion.

We shall illustrate this method through an example of choosing the cooling tower for a thermal power plant.

Example 2.14

A thermal power plant designer has to choose between a traditional cooling tower and a newer design tower which costs about Rs. 45,000 more but is more efficient. This tower is expected to result in about 0.1% increase in the efficiency of the power plant which translates into an increase in annual income by Rs. 10,000. However, the operating cost of this new design tower is Rs. 3000 more than that of the traditional cooling tower, while its salvage value is Rs. 10,000 more. If the life of the cooling tower is 10 years, which design of the cooling tower should the designer choose? Given the interest rate is 9%.

Solution

The initial cost of the new-design tower is Rs. 45,000 more, but it will give an increase in income of Rs. $10,000 - 3000 =$ Rs. 7000 every year for next 10 years. The present worth of this additional income would be

$$\frac{7000}{(1+0.09)} + \frac{7000}{(1.09)^2} + \frac{7000}{(1.09)^3} + \cdots + \frac{7000}{(1.09)^{10}} = 44{,}923.6 \text{ Rs.}$$

Further the salvage value of the new-design tower, which will be obtained at the end of its life of 10 years is also more by Rs. 10,000. The present worth of that amount is

$$\frac{10{,}000}{(1+0.09)^2} = 4224.1 \text{ Rs.}$$

Thus the present worth of "benefits" accruing from the cooling tower of new design is $= 44{,}923 + 4224.1 = 49{,}147.7$ Rs., which is larger than the present worth of additional expense to be incurred in buying it, that is Rs. 45,000. So the designer should choose the cooling tower of new design.

In order to simplify the calculations of present/future worth of regular payments/receipts over many years, encountered in Example 2.14, simple formulae have been derived. Thus considering a series of regular (typically end of the year) payments of amount A, spread over n years, the total future value is

$$F = A(1+i)^{n-1} + A(1+i)^{n-2} + \cdots + A(1+i) + A$$
$$= \frac{A[(1+i^n) - 1]}{i} \qquad (2.80)$$

This equation can also be written as

$$A = F\left\{ \frac{i}{(1+i)^n - 1} \right\} \qquad (2.81)$$

which can be interpreted as the annual amount that should be put in a bank every year to receive an amount F at the end of n years. This term within braces is called the sinking fund factor—the term sinking fund referring to the amount that a company has to set aside (or sink) to buy say, a new machine (or replace an existing machine) at the end of stipulated period of n years.

The present worth of the amount F which would be received after n equal payments A at regular intervals can be obtained by combining Eqs. (2.76), (2.81). This gives us an equation to determine the equal installments of money (A) that will pay for a loan (P) taken today, over a certain period, n.

$$A = P \left\{ \frac{i(1 + i)^n}{(1 + i)^n - 1} \right\} \tag{2.82}$$

The term within braces on the right-hand side of Eq. (2.82) is known as the capital recovery factor.

Example 2.15

A company takes a loan of 15 lakhs from a bank to purchase a new machine, and agrees to repay the loan in equal monthly installment spread over 5 years. How much will be the monthly payment if the interest rate is 0.75% per month.

Solution

Since the payments are being made every month we use monthly interest rates, with $n = 60$ months. Substituting the values in Eq. (2.82) we get

$$A = (15 \times 10^5) \left\{ \frac{0.0075(1.0075)^{60}}{(1 + 0.0075)^{60} - 1} \right\}$$

$$= 31{,}137.5 \text{ Rs.}$$

2.7.3 Rate of Return Analysis

Another method of analyzing the viability of an engineering project is the rate of return analysis. Here all the revenue and expenses estimated to be incurred in the project are reduced to a single percentage number representing the rate of return on the investment. This can then be compared to the other investment returns, and the interest rates to arrive at a decision. The rate of return is defined as the compound interest rate at which the total sum of revenues received over a period of time brought to any point in time (usually the time of investment) equals the sum of costs at that point in time. This usually needs an iterative solution as shown in Example 2.16.

Example 2.16

Find the rate of return on investment in Example 2.14.

Solution

Bringing all revenue receipts to the present time, when the additional investment of Rs. 45,000 is to be made, we have to find the rate of return (r) such that the present worth of all receipts equals this amount. This demands that

$$\frac{7000}{(1 + r)} + \frac{7000}{(1 + r)^2} + \cdots + \frac{7000}{(1 + r)^{10}} + \frac{10{,}000}{(1 + r)^{10}} = 45{,}000$$

We have to find the value of r by trial and error, and it comes to $r = 0.1088\%$ or 10.88%. This being greater than the interest rate of 9% given in Example 2.14, the investment seems worthwhile from this perspective.

2.7.4 Life Cycle Costing

A popular method of assessing the economic viability of a project, especially of choosing between alternative designs/strategies, is the life cycle costing method. The term life cycle cost refers to the total cost incurred in a project during its entire life span. Thus for a typical air-conditioning plant the life cycle costs would include the initial capital investment needed to buy the equipment, the cost of electricity, the operating and maintenance costs likely to be incurred during its entire life span, less the net salvage value of the plant when it is decommissioned. Since the O&M costs are going to be spread over a long period, appropriate factors have to be incorporated to bring all the amounts to their present worth. Example 2.17 illustrates how this method can be employed to choose between alternate designs of equipment.

Example 2.17

Two designs of central plants are being evaluated for air conditioning of a building. Their key features are given in following table.

	Design A	Design B
Initial cost (lakhs)	35	28
Power consumption (kW)	50	55
Annual operation and maintenance cost (lakhs)	2.5	2.0
Salvage value (lakhs)	2.0	1.0

Taking the life of both the plants to be 10 years, and their actual operating time as 4500 h/year, determine the life cycle costs of both the alternatives. Take net interest rate as 5%, and assume the energy cost is Rs. 6/- per kWh in the first year, and would increase by 5% every year.

Solution

The life cycle cost, brought to the present time, would consist of
(a) initial capital cost
(b) cost of electricity over 10 years of operation
(c) O&M costs during 10 years
(d) minus the salvage value that would be obtained at the end of 10 years
Out of these constituents the terms (b)–(d) have to be brought to the present time.

Design A

Cost of electricity during the first year $= 50 \times 4500 \times 6 = 1.35 \times 10^6$ Rs.

Strictly speaking this would have to be paid bi-monthly or tri-monthly, depending on the service provider's agreement. However, we would assume, for simplicity, that the whole amount has to be paid at the end of year. This would escalate by 5% every year, and so the present worth of the energy expenses for 10 years would be

$$
\text{Energy costs} = \frac{1.35 \times 10^6}{1 + 0.05} + \frac{1.35 \times 10^6 \times 1.05}{(1 + 0.05)^2}
$$
$$
+ \frac{1.35 \times 10^6 \times 1.05^2}{1.05^3} + \cdots + \frac{(1.35 \times 10^6) \times 1.05^9}{(1.05)^{10}}
$$
$$
= \frac{1.35 \times 10^6 \times 10}{1.05}
$$
$$
= 1.2857 \times 10^7 \text{ Rs.}
$$

Similarly, the *O&M* costs can be brought to the present time as:

$$
\frac{2.5 \times 10^5}{1.05} + \frac{2.5 \times 10^5}{(1.05)^2} + \cdots + \frac{2.5 \times 10^5}{(1.05)^{10}} = \frac{2.5 \times 10^5 \{1.05^{10} - 1\}}{0.05(1.05)^{10}} = 19.30434 \times 10^5 \text{ Rs.}
$$

$$
\text{Present worth of salvage value} = \frac{2 \times 10^5}{1.05^{10}} = 1.2278 \times 10^5
$$

$$
\text{Thus for design A, LCC} = 35 \times 10^5 + 1.2857 \times 10^7 + 19.3043 \times 10^5 - 1.2278 \times 10^5
$$
$$
= 1.8164 \times 10^7 \text{ Rs.}
$$

Similarly we evaluate the life cycle cost for design B.

Cost of electricity consumed during the first year $= 55 \times 4500 \times 6 = 1.485 \times 10^6$ Rs. Following the same approach as for design A, the present worth of energy costs during 10 years of operation is

$$
\text{Energy costs} = \frac{1.485 \times 10^6 \times 10}{1.05} = 1.41429 \times 10^7 \text{ Rs.}
$$

$$
\text{Total O \& M costs (present worth)} = \frac{2 \times 10^5 \{1.05^{10} - 1\}}{0.05(1.05)^{10}}
$$
$$
= 1.54435 \times 10^6 \text{ Rs.}
$$

$$
\text{Present value of salvage value} = \frac{1.0 \times 10^5}{(1 + 0.05)^{10}} = 0.6139 \times 10^5 \text{ Rs.}
$$

$$
\text{Thus for design B, LCC} = 28 \times 10^5 + 1.41429 \times 10^7 + 1.54435 \times 10^6
$$
$$
- 0.6139 \times 10^5
$$
$$
= 1.8426 \times 10^7 \text{ Rs.}
$$

Thus we see that the life cycle cost of design A is lower in spite of the fact that its initial cost is much higher than of design B, and even its annual *O&M* cost is higher. This is clearly due to its lower electric power consumption.

There are also many other factors besides the time value of many that we have discussed in this section. These include consideration of various kinds of taxes, depreciation, the uncertainties inherent in estimation of future costs or receipts and risks associated with these. For a detailed exposition on these, standard texts on engineering economics should be consulted [6, 7].

REFERENCES

[1] S.C. Chapra, R.P. Canale, Numerical Methods for Engineers, sixth ed., McGraw Hill, Singapore, 2010.
[2] C.F. Gerald, P.O. Wheatley, Applied Numerical Analysis, seventh ed., Pearson Education, India, 2007.
[3] F.J. Warner, On solution of "Jury" problems with many degrees of freedom, J. Math. Tables Other Aids Comput. 11 (60) (1957) 268–271.
[4] www.bluebit.gr/matrix-calculator/.
[5] J. Schmidt, et al., Using Monte Carlo simulation to account for uncertainties in the spatial explicit modelling of biomass fired heat and power potentials in Austria, Discussion Paper, Department of Economics and Social Sciences, University of Natural Resources and Applied Life Sciences, Vienna, Austria, 2009.
[6] C.S. Park, G.P. Sharp-Bette, Advanced Engineering Economics, John Wiley and Sons Inc., New York, 1990.
[7] D. Newnan, J. Lavelle, T. Eschenbach, Engineering Economic Analysis, twelfth ed., Oxford University Press, New York, 2013.

CHAPTER 3

Review of Fundamentals

As the very name suggests, most thermal systems involve transfer of heat and work, often through the agency of fluids moving through them. Their simulation, therefore, demands a thorough knowledge of the principles of thermodynamics, heat transfer, mass transfer, and fluid flow. A brief review of the fundamental equations governing these phenomena is presented here. This is, of course, not sufficient to gain clarity of the basic concepts, but is only an aid to students who have already studied these subjects, but may have forgotten the detailed equations governing these phenomena which would be used in this book.

3.1 THERMODYNAMICS

The analysis of energy transfer is governed by the two laws of thermodynamics. The first law is essentially a statement of the law of conservation of energy. As applied to any object (or closed system) it can be mathematically expressed as [1]:

$$E_F - E_I = Q_{in} + W_{on} \tag{3.1}$$

which expresses the fact that the energy (E) of an object can be increased (subscript I referring to initial state, and F to the final state) in two manners, namely by giving it a heat input (Q_{in}) or by doing work on it (W_{on})

Work can be done on an object in many ways: by applying a mechanical force, fluid pressure, electrical voltage, magnetic field, etc. The general equation for calculating the work done on an object in a quasiequilibrium process (i.e., one where the external "force" and the internal resistive force of the object are only infinitesimally different from each other) is

$$W_{on} = \int_I^F f' \, dx' \tag{3.2}$$

Here f' is the internal resistive force of the object called the "generalized force" and x' is the generalized coordinate, that is, the property of the object which is directly influenced by the generalized force.

Thermal System Design and Simulation
http://dx.doi.org/10.1016/B978-0-12-809449-5.00003-6

Table 3.1 Generalized coordinates and forces

Object	Generalized coordinate	Generalized force
Spring weight (elastic object)	L, displacement (m)	Mechanical force, F (N)
Compressible fluid	V, Volume (m^3)	$-P$, Pressure (N/m^2)
Surface film	A, surface area (m^2)	σ, Surface tension (N/m^2)
Battery/capacitor	Q, Charge (Col)	V, voltage (V)
Magnetic induction	M, Magnetic moment (A/m^2)	H, Magnetic field strength (A/m)
Dielectric solid	Π, polarization (Col-m)	E, Electric intensity V/M
Multicomponent fluids	N_i, number of moles of component i (mols)	μ_i, Chemical potential of component i (J/mol)

Table 3.1 presents a summary of possible "generalized forces" and their corresponding "generalized coordinates."

Thus the work done on a capacitor can be calculated as

$$W_{capacitor} = \int_I^F V\, dQ \tag{3.3}$$

and the work done on a gas contained inside a cylinder is given by

$$W_{gas} = -\int_I^F P\, dV \tag{3.4}$$

Thus the work done on an ideal gas during its isothermal compression can be calculated as

$$W_{on} = -\int_I^F \frac{RT}{V} dV = -RT \ln\left(\frac{V_F}{V_I}\right)$$

$$= -RT \ln\left(\frac{P_I}{P_F}\right) = RT \ln\left(\frac{P_F}{P_I}\right) \tag{3.5}$$

Eq. (3.1) can also be applied to open systems, that is, objects through which fluid streams are entering and leaving, by visualizing the processes in these systems as a sequence of processes undergone by suitably defined closed systems whose boundaries deform in accordance with flowing streams [1, Chapter 6].

This gives the following general energy equation for open systems:

$$\frac{dE_{cv}}{dt} = \dot{Q}_{cv} + \dot{W}_{cv} + \sum \dot{m}_i \left(h_i + gz_i + \frac{Vel_i^2}{2} \right) - \sum \dot{m}_e \left(h_e + gz_e + \frac{Vel_e^2}{2} \right) \tag{3.6}$$

which essentially expresses the fact that

$$\begin{bmatrix} \text{Rate of increase of energy of} \\ \text{the control volume} \end{bmatrix} = \begin{bmatrix} \text{Rate of energy input as heat} \\ (\dot{Q}_{cv}) \text{ and work } (\dot{W}_{cv}) \end{bmatrix} +$$

$$\begin{bmatrix} \text{Rate at which energy is carried} \\ \text{into the control volume by the} \\ \text{entering streams} \end{bmatrix} - \begin{bmatrix} \text{Rate at which energy is} \\ \text{leaving the control volume} \\ \text{through exiting streams} \end{bmatrix}$$

The total "energy content" of a stream is quantified in terms of the sum total of its enthalpy (h_o), potential energy (gz), and the kinetic energy ($Vel^2/2$).

Most thermal systems are open systems, and so equation 3.6 is often needed in their analysis. A simplified version of this equation, applicable when these systems operate under steady state is the popular steady flow energy equation (SFEE):

$$\dot{Q}_{cv} + \dot{W}_{cv} + \sum \dot{m}_i \left(h_i + \frac{Vel_i^2}{2} + gz_i \right) = \sum \dot{m}_e \left(h_e + gz_e + \frac{Vel_e^2}{2} \right) \qquad (3.7)$$

A simpler version of the SFEE, for open systems with one stream entering and one stream leaving, is

$$\dot{Q}_{cv} + \dot{W}_{cv} = \dot{m} \left\{ (h_e - h_i) + (gz_e - gz_i) + \frac{Vel_e^2 - Vel_i^2}{2} \right\} \qquad (3.8)$$

$$= \Delta H + \Delta PE + \Delta KE$$

which expresses the fact that the changes in the enthalpy, the potential energy, and the kinetic energy of the flowing stream are caused by heat and work input to the control volume.

Further dividing both sides of this equation by the mass flow rate of the stream, we can express the first law for steady state, steady flow system in terms of specific properties as

$$\dot{q}_{cv} + \dot{w}_{cv} = (h_e - h_i) + g(z_e - z_i) + \frac{Vel_e^2 - Vel_i^2}{2} \qquad (3.9)$$

While deriving Eqs. (3.6)–(3.9), we make use of the mass conservation equation for the control volume

$$\frac{dM_{cv}}{dt} = \Sigma \dot{m}_i - \Sigma \dot{m}_e \qquad (3.10)$$

which expresses the fact that

$$\begin{Bmatrix} \text{Rate of increase of mass} \\ \text{within the control volume} \end{Bmatrix} = \begin{Bmatrix} \text{Sum of instantaneous} \\ \text{rates of mass inflow} \end{Bmatrix} - \begin{Bmatrix} \text{Sum of instantaneous} \\ \text{rates of mass out flow} \end{Bmatrix}$$

Traditionally, the work term in the various forms of the energy equation for open systems is written as $\dot{W}_{sh}$; that is, the shaft work output from the control volume. Now since $\dot{W}_{cv} = -\dot{W}_{sh}$, the above equations become:

General equation

$$\dot{Q}_{cv} = \dot{W}_{sh} + \frac{dE_{cv}}{dt} + \sum \dot{m}_e \left(h_e + gz_e + \frac{Vel_e^2}{2} \right) - \sum \dot{m}_i \left(h_i + gz_i + \frac{Vel_i^2}{2} \right) \quad (3.11)$$

SFEE

$$\dot{Q}_{cv} = \dot{W}_{sh} + \sum \dot{m}_e \left(h_e + gz_e + \frac{Vel_e^2}{2} \right) - \sum \dot{m}_i \left(h_i + gz_i + \frac{Vel_i^2}{2} \right) \quad (3.12)$$

$$\dot{Q}_{cv} = \dot{W}_{sh} + \Delta H + \Delta PE + \Delta KE \quad (3.13)$$

$$\dot{q}_{cv} = \dot{w}_{sh} + (h_e - h_i) + g(z_e - z_i) + \frac{Vel_e^2 - Vel_i^2}{2} \quad (3.14)$$

The general statement of the Second Law of Thermodynamics is usually expressed in the form of a mathematical inequality [1, Chapter 3]. The form which yields a quantitative relationship between the thermodynamic properties is obtained through the application of the general statement to a reversible process:

$$T\, dS = dU - \Sigma f'\, dx' \quad (3.15)$$

S being the total entropy and U the total internal energy of the object. When the mode of work transfer is through the deformation of a compressible fluid boundary—the most frequently encountered situation in the thermal systems—the above equation can be expressed as:

$$T\, dS = dU + PdV \quad (3.16)$$

This equation can also be expressed in terms of the specific thermodynamic properties indicated by small case letters (u, s, v) as

$$du = T\, ds - P\, dv \quad (3.17)$$

Eq. (3.17) is one of the fundamental relations of thermodynamics for compressible systems; the other relations that follow from it are

$$dh = T\, ds + v\, dP \quad (3.18)$$
$$df = -s\, dT - P\, dv \quad (3.19)$$
$$dg = -s\, dT + v\, dP \quad (3.20)$$

where f is the Helmholtz free energy, g the Gibbs function, and h the enthalpy, all per unit mass.

These equations can be seen as differential forms of the fundamental relations

$$u = u(s, v) \quad (3.21)$$
$$h = h(s, P) \quad (3.22)$$

$$f = f(T, v) \tag{3.23}$$

$$g = g(T, p) \tag{3.24}$$

Using these fundamental relations, equations relating various thermodynamic properties can be obtained. These enable us to calculate the property changes during various processes, using measurable properties like specific heats, compressibilities, etc. The most important of these relations are the so-called Maxwell's relations which arise from the consideration of the fact that thermodynamic properties like P, u, h, f, g, s, v, etc. are, by definition, state functions and therefore the differential terms du, dh, df, dg are "exact" differentials. Using the theorems of differential calculus this gives:

$$\text{from Eq. (3.17):} \quad \left(\frac{\partial T}{\partial v} \right)_s = - \left(\frac{\partial P}{\partial s} \right)_v \tag{3.25}$$

$$\text{from Eq. (3.18):} \quad \left(\frac{\partial T}{\partial P} \right)_s = \left(\frac{\partial v}{\partial s} \right)_P \tag{3.26}$$

$$\text{from Eq. (3.19):} \quad \left(\frac{\partial s}{\partial v} \right)_T = \left(\frac{\partial P}{\partial T} \right)_v \tag{3.27}$$

$$\text{from Eq. (3.20):} \quad \left(\frac{\partial s}{\partial P} \right)_s = - \left(\frac{\partial v}{\partial T} \right)_P \tag{3.28}$$

These equations are greatly helpful in finding changes in various thermodynamic properties in terms of measurable properties like the specific heats and the coefficient of volumetric expansion defined as:

$$C_p = \left(\frac{\partial h}{\partial T} \right)_P = T \left(\frac{\partial s}{\partial T} \right)_P$$

$$C_v = \left(\frac{\partial u}{\partial T} \right)_v = T \left(\frac{\partial s}{\partial T} \right)_v$$

$$\beta = \frac{1}{v} \left(\frac{\partial v}{\partial T} \right)_P$$

Thus, to express change in entropy in any process in terms of the changes in the temperature and pressure, we can write

$$s = s(T, P) \tag{3.29}$$

$$ds = \left(\frac{\partial s}{\partial T} \right)_p dT + \left(\frac{\partial s}{\partial P} \right)_T dP$$

$$= \frac{C_p}{T} dT - \left(\frac{\partial v}{\partial T} \right)_P dt$$

$$= \frac{C_p}{T} dT - \beta v \, dP \tag{3.30}$$

where we have used Eq. (3.28) to transform the second partial derivative on the right-hand side of the above equation. Thus change in a nonmeasurable property, s, can be expressed in terms of changes in measurable properties like pressure and temperature, C_p, and β.

Further using this result in Eq. (3.18), we can easily get

$$dh = C_p dT + (v - \beta v T) dP \tag{3.31}$$

Similarly change in the internal energy can be very conveniently expressed in terms of changes in the temperature and the volume

$$u = u(T, v) \tag{3.32}$$

$$du = \left(\frac{\partial u}{\partial T}\right)_v dT + \left(\frac{\partial u}{\partial v}\right)_T dv$$

$$= C_v dT + \left\{ T\left(\frac{\partial s}{\partial v}\right)_T - P \right\} dv \tag{3.33}$$

by using Eq. (3.17).

Further, using the Maxwell's relation (3.27) in this equation, we get

$$du = C_v dT + \left\{ T\left(\frac{\partial P}{\partial T}\right)_v - P \right\} dv \tag{3.34}$$

which can be used to estimate change in the internal energy in any process from known changes in the temperature and volume.

Using these equations, along with the equations of state representing the $P - v - T$ data, the thermodynamic properties of various substances have been computed and tabulated for a wide range of pressures and temperatures. Most common among these are the *steam tables* giving the properties of water, and the *refrigerant property tables* giving properties of commonly used refrigerants. For ease of refrigeration system calculations, the thermodynamic data of refrigerants have also been plotted in the form of *Pressure-Enthalpy* or $p - h$ charts. The main features of this chart are indicated in Fig. 3.1, adopted from Dhar [1]. Similarly the calculations of the steam cycles of thermal power plants are greatly facilitated by *Enthalpy-Entropy* or $h - s$ chart (commonly known as the Mollier Chart) shown in Fig. 3.2, also adopted from Dhar [1]. Many softwares have also been developed which enable computation of all thermodynamic properties for any thermodynamic state of a variety of working substances. REFPROP is one such versatile proprietary software developed by the National Institute of Standards and Technology (NIST), United States for predicting the thermodynamic and thermophysical properties of variety of fluids including refrigerants and refrigerant mixtures. The software

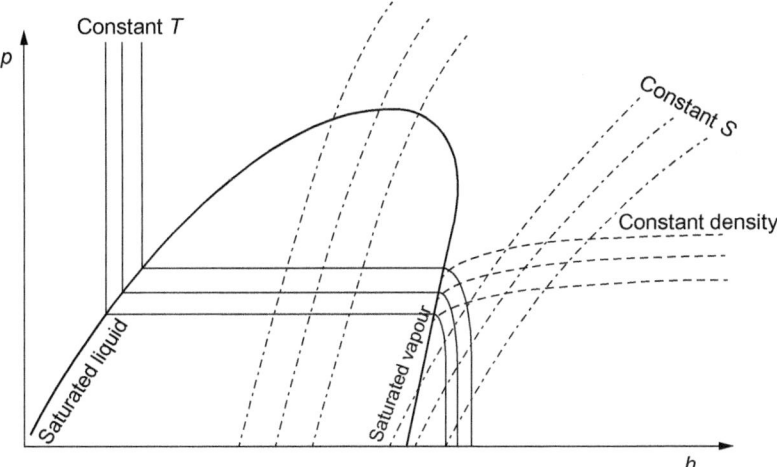

Fig. 3.1 Pressure-enthalpy ($p - h$) chart for a refrigerant.

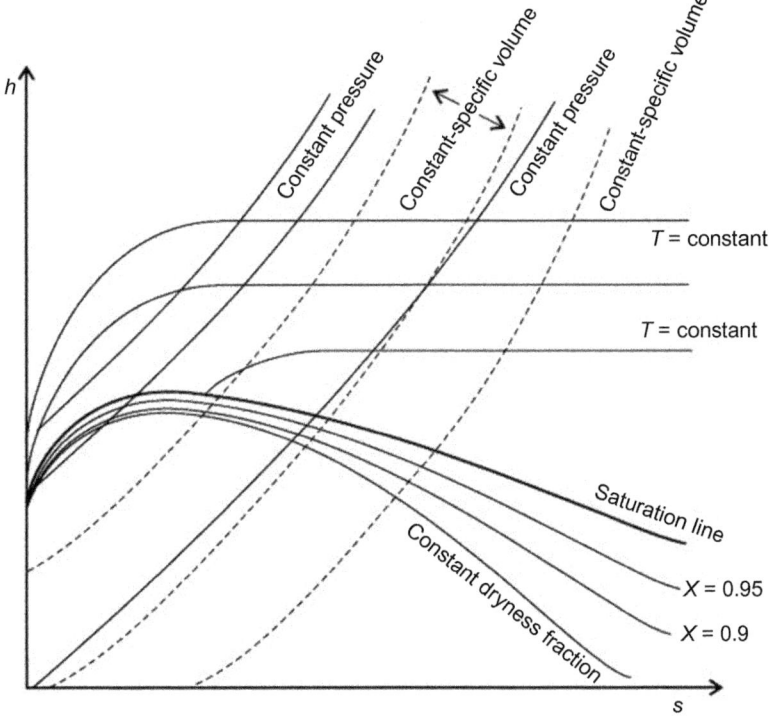

Fig. 3.2 Enthalpy-entropy ($h - s$) chart for steam.

predicts all the thermodynamic and thermophysical properties of the substance when its two independent properties are specified and can also plot various thermodynamic diagrams like $p - h$, $T - s$, $p - T$, $p - v$, etc. Another software which can predict only thermodynamic properties of a limited number of substances is CATT or computer-aided thermodynamic tables. It is a user friendly freely downloadable software which can be used as a substitute for traditional steam and refrigerant property tables.

As an illustration of the direct application of the above equations in analysis of various thermodynamic processes, let us consider an ideal gas undergoing isentropic compression. We can use the thermodynamic relations to find out a relationship between the pressure, the temperature, and the volume of the gas before and after compression.

We start from Eq. (3.30). For an isentropic process, $ds = 0$, and we get

$$\frac{C_p}{T} dT = \beta v \, dp \tag{3.35}$$

Further, for an ideal gas:

$$\beta = \frac{1}{v} \left(\frac{\partial v}{\partial T} \right)_P dt = \frac{1}{v} \left(\frac{\partial}{\partial T} \left\{ \frac{RT}{P} \right\} \right)_P$$
$$= \frac{1}{v} \cdot \frac{R}{P} = \frac{1}{T} \tag{3.36}$$

which, when substituted in Eq. (3.35), gives

$$dT = \frac{v}{C_p} dP = \frac{RT}{C_p} \frac{dP}{P}$$
$$\text{or} \quad \frac{dT}{T} = \frac{R}{C_p} \cdot \frac{dP}{P} \tag{3.37}$$

Assuming C_p to be constant the above equation can be integrated over the process states to get

$$\frac{T_2}{T_1} = \left(\frac{P_2}{P_1} \right)^{R/C_p} = \left(\frac{P_2}{P_1} \right)^{\frac{\gamma-1}{\gamma}} \tag{3.38}$$

where we have used the fact that for an ideal gas, $C_p - C_v = R$; and γ is the ratio of the two specific heats, that is $\gamma = C_p/C_v$.

Eq. (3.38) can be combined with the ideal gas equation of state, $P_1 v_1 = RT_1$ to get the well-known relationship for isentropic process:

$$P_1 v_1^{\gamma} = P_2 v_2^{\gamma} \tag{3.39}$$

Often the nonisentropic processes are represented by a similar equation, with an index n replacing γ. Such a general process is termed as polytropic process and its end states are related by equation

$$P_1 v_1^n = P_2 v_2^n \tag{3.40}$$

Using this relationship in Eq. (3.4), an expression for work done in a polytropic process can be obtained as

$$W_{in} = \frac{P_2 V_2 - P_1 V_1}{n - 1} = \frac{P_1 V_1}{n - 1} \left\{ \left(\frac{P_2}{P_1} \right)^{\frac{n-1}{n}} - 1 \right\} \tag{3.41}$$

The above equation is for a closed system like gas trapped inside a piston-cylinder arrangement. In engineering practice we come across, more frequently, open systems like compressors and turbines where a steady stream of the fluid enters at a certain pressure/temperature and leaves at a different pressure/temperature. It can be easily proved that in any open system undergoing reversible steady flow process, the work input per unit mass flow rate, is given by the expression

$$w_{in} = \int_1^2 v dP + g(z_2 - z_1) + \frac{Vel_2^2 - Vel_1^2}{2} \tag{3.42}$$

where subscript 1 refers to the state of the fluid at the inlet to the control volume and 2 refers to the state at the exit.

In most practical cases, changes in the potential and kinetic energies are insignificant, and if the process end states follow Eq. (3.40), we get

$$w_{in} = \int_1^2 v dP = \frac{n}{n - 1} (P_2 v_2 - P_1 v_1) = \frac{n}{n - 1} P_1 v_1 \left\{ \left(\frac{P_2}{P_1} \right)^{\frac{n-1}{n}} - 1 \right\} \tag{3.43}$$

Further if the working fluid can be treated as an ideal gas, above equation can be transformed into

$$w_{in} = \frac{n}{n - 1} R(T_2 - T_1) = \frac{n}{n - 1} RT_1 \left\{ \left(\frac{P_2}{P_1} \right)^{\frac{n-1}{n}} - 1 \right\} \tag{3.44}$$

If the compression process is isentropic, $n = \gamma$ and $\frac{\gamma}{\gamma - 1} R = C_p$, and we get

$$w_{in} = C_p(T_2 - T_1) = C_p T_1 \left\{ \left(\frac{P_2}{P_1} \right)^{\frac{\gamma-1}{\gamma}} - 1 \right\} \tag{3.45}$$

If the mass flow rate through the device is $\dot{m}$, the total power input to the control volume is

$$\dot{W}_{in} = \dot{m} w_{in} = \dot{m} C_p T_1 \left\{ \left(\frac{P_2}{P_1} \right)^{\frac{\gamma-1}{\gamma}} - 1 \right\} \tag{3.46}$$

Eqs. (3.44)–(3.46) are often used in the analysis of compressors and turbines (cf. Section 6.2.1).

3.1.1 Thermodynamics of Multicomponent Systems

Many thermal systems encounter chemical reactions, for example, most power-producing systems have combustors-furnaces where fuel is burnt to produce "heat." Thermodynamic analysis of chemical reactions is therefore needed to estimate the state of the products of combustion, from the given state of the reactants. Since the chemical compositions of the reactants and the products of a chemical reaction differ, we need to first understand the method of calculating the thermodynamic properties of multicomponent systems [1, Chapter 8].

The fundamental thermodynamic relations for multicomponent systems can be obtained from Eq. (3.15) by incorporating an additional generalized force term, that is, μ_i, the chemical potential of compound i, and its corresponding generalizes coordinate, N_i, the number of moles of component i. Thus the fundamental equation for a compressible multicomponent system becomes (cf. Eq. 3.17)

$$dU = T\,dS - P\,dV + \sum_{i=1}^{r} \mu_i\,dN_i \tag{3.47}$$

where we have used total properties U, S, V, and N_i, unlike Eq. (3.17) which was in terms of the specific properties.

Eq. (3.47) can be visualized as the differential from of a fundamental relation of the form

$$U = U(S, V, N_1, N_2, \ldots, N_r) \tag{3.48}$$

and from Eqs. (3.47), (3.48) it follows that:

$$\left(\frac{\partial U}{\partial S}\right)_{V,N_i} = T \quad \left(\frac{\partial U}{\partial V}\right)_{S,N_i} = -P \quad \left(\frac{\partial U}{\partial N_i}\right)_{S,V,N_{j\neq i}} = \mu_i \tag{3.49}$$

Further, as in the case of single-component systems, we can also define other fundamental relations as:

$$H = H(S, P, N_1, N_2, \ldots, N_r) : dH = TdS + VdP + \sum_{i=1}^{r} \mu_i\,dN_i \tag{3.50}$$

$$F = F(T, V, N_1, N_2, \ldots, N_r) : dF = -SdT - PdV + \sum_{i=1}^{r} \mu_i\,dN_i \tag{3.51}$$

$$G = G(T, P, N_1, N_2, \ldots, N_r) : dG = -SdT + VdP + \sum_{i=1}^{r} \mu_i\,dN_i \tag{3.52}$$

It can be seen that if the composition of a multicomponent system remains constant, $dN_i = 0$, the above equations reduce to the corresponding functional relations for a single-component compressible system.

Of great utility in multicomponent analysis is the concept of partial molar property, $\bar{M}_i$, defined as

$$\bar{M}_i = \left(\frac{\partial M}{\partial N_i}\right)_{T,P,N_{j\neq i}} \tag{3.53}$$

Its utility lies in the fact that any molar property M of the mixture can be simply related to the component partial molar properties $\bar{M}_i$ as

$$M = \sum_{i=1}^{r} N_i \bar{M}_i \tag{3.54}$$

Eq. (3.53) also suggests a physical interpretation of the partial property. It is the rate at which the property value for the entire multicomponent system changes with the number of moles of component i in the substance when T, P and the extent of other components remains fixed. For properties like volume, which can be easily measured, this approach could be directly employed to determine the partial molar volume. For binary mixtures there exists an even simpler method to determine any partial molar property at any composition from a plot of specific molar property as a function of molar composition of the binary mixture. This is indicated in Fig. 3.3.

At any composition x_1, the specific property for the mixture is given by the ordinate AB. We draw a tangent to the $M - x_1$, plot at B and locate the intersections of this tangent at the ordinates corresponding to mole-fractions $x_1 = 0$ and $x_1 = 1$. It can

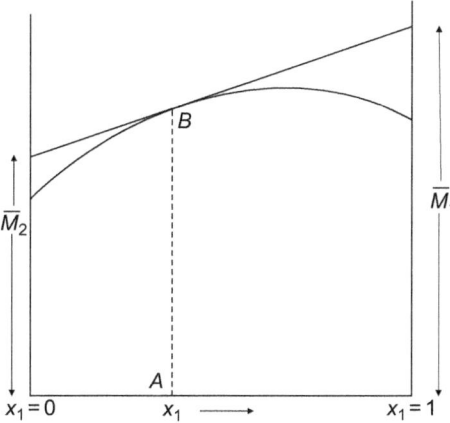

Fig. 3.3 Determining partial molar properties.

be easily proved that the lengths of these ordinates correspond to the two partial molar properties, as shown in Fig. 3.3.

Mathematically, it amounts to equations

$$\bar{M}_1 = M + (1 - x_1)\frac{dM}{dx_1} \tag{3.55}$$

$$\bar{M}_2 = M - x_1\frac{dM}{dx_1} \tag{3.56}$$

Two of the important partial molar properties, namely the partial molar volumes and the partial molar enthalpies of binary mixtures can also be determined from experimental measurement of volume changes on mixing and the heat of mixing. These quantities can be easily found experimentally by isothermal, isobaric mixing of the two constituents of a mixture, and would be related to the partial molar properties by following equations:

$$\Delta V_{mix} = (N_1 \bar{V}_1 + N_2 \bar{V}_2) - (N_1 v_1^0 + N_2 v_2^0) \tag{3.57}$$

$$\Delta H_{mix} = (N_1 \bar{H}_1 + N_2 \bar{H}_2) - (N_1 h_1^0 + N_2 h_2^0) \tag{3.58}$$

where the terms with superscript 0 refer to the properties of the pure constituents. By a little mathematical manipulation it can be established that

$$(\bar{V}_1 - v_1^0) = \Delta v_{mix} + (1 - x_1)\frac{d\Delta v_{mix}}{dx_1} \tag{3.59}$$

$$(\bar{V}_2 - v_2^0) = \Delta v_{mix} - x_1\frac{d\Delta v_{mix}}{dx_1} \tag{3.60}$$

$$(\bar{H}_1 - h_1^0) = \Delta h_{mix} + (1 - x_1)\frac{d\Delta h_{mix}}{dx_1} \tag{3.61}$$

$$(\bar{H}_2 - h_2^0) = \Delta h_{mix} - x_1\frac{d\Delta h_{mix}}{dx_1} \tag{3.62}$$

$$\text{where} \quad \Delta v_{mix} = \frac{\Delta V_{mix}}{N_1 + N_2} = x_1(\bar{V}_1 - v_1^0) + x_2(\bar{V}_2 - v_2^0) \tag{3.63}$$

$$\Delta h_{mix} = \frac{\Delta H_{mix}}{N_1 + N_2} = x_1(\bar{H}_1 - h_1^0) + x_2(\bar{H}_2 - h_2^0) \tag{3.64}$$

Eqs. (3.57), (3.58) can also be written as:

$$V = N_1 v_1^0 + N_2 v_2^0 + \Delta V_{mix} \tag{3.65}$$

$$H = N_1 h_1^0 + N_2 h_2^0 + \Delta H_{mix} \tag{3.66}$$

which gives us the procedure to calculate the mixture properties from those of the pure components by using the data on volume change in mixing and the heat of mixing. To determine other properties like the internal energy U, the entropy S, etc., we need to take recourse to the basic thermodynamic relationships and the $p-v-T-x$ data, which are usually modeled in the form of suitable equation of state for the mixture. As in the case of single-component fluids, there exist numerous equations of state for mixtures. The simplest of all is the Dalton's model of "ideal-gas-mixtures" where it is presumed that each component behaves as an ideal gas at the temperature and the volume of the mixture. It therefore follows that each component, i, will not be at the mixture pressure, P, but exert a smaller pressure, called the partial pressure, P_i, whose magnitude can be determined using the ideal gas equation of state as

$$P_i V = N_i \bar{R} T \tag{3.67}$$

where $\bar{R}$ is the universal gas constant (=8.3143 kJ/kmol K). It is also postulated that a mixture of ideal gases itself would behaves like an ideal gas. It therefore follows that

$$PV = N\bar{R}T \tag{3.68}$$

$$\text{where} \quad N = \Sigma N_i$$

Dividing the above equation, we get

$$\frac{P_i}{P} = \frac{N_i}{N} = x_i, \quad \text{the mole fraction} \tag{3.69}$$

and since $\sum x_i = 1$, we also get

$$\sum P_i = P \tag{3.70}$$

Now, from the ideal gas model which postulates the absence of any intermolecular interaction, it also follows that any extensive property of the mixture equals the sum of the values of that extensive property for each of the components as it exists in the mixture, that is, at the mixture temperature T and the partial pressure P_i. We can thus write

$$U = \sum U_i \text{ or } \quad U = Nu = \sum N_i u_i^0 \quad \text{or}$$

$$u(T, P) = \sum u_i^0(T, P_i) x_i \tag{3.71}$$

$$H = \sum H_i \text{ or } \quad H = Nh = \sum N_i h_i^0 \quad \text{or}$$

$$h(T, P) = \sum h_i^0(T, P_i) x_i \tag{3.72}$$

$$S = \sum S_i \text{ or } \quad S = Ns = \sum N_i s_i^0 \quad \text{or}$$

$$s(T, P) = \sum s_i^0(T, P_i) x_i \tag{3.73}$$

An alternative approach is the Amagat's model which presumes that each component exists at the temperature and the pressure of the mixture, occupying only a part of the total volume, called the partial volume V_i. When the components can be treated as ideal gases, we get an expression for V_i as

$$PV_i = N_i \bar{R} T \tag{3.74}$$

which on summation gives

$$P \sum V_i = \left(\sum N_i \right) \bar{R} T \tag{3.75}$$

Further since the mixture is also an ideal gas, we have

$$PV = NRT \tag{3.76}$$

This gives

$$V = \Sigma V_i \tag{3.77}$$

and on dividing the earlier two equations, we get

$$\frac{V_i}{V} = \frac{N_i}{N} = x_i \tag{3.78}$$

Dalton's model of ideal gas mixtures is often employed to analyze mixtures of gases and vapor like air–water vapor mixtures by making a few simplifying assumptions. In most of these cases the amount of water vapor present is much smaller than the gaseous components. Its partial pressure in the mixture is therefore quite low permitting us to treat it as an ideal gas. However, if the temperature of the mixture is such that the saturation pressure of the water vapor at that temperature is lower than its partial pressure in the mixture, condensation of the water vapor takes place. This can be understood by referring to the $T - s$ diagram of water vapor in a gas–water vapor mixture shown in Fig. 3.4 [1, Chapter 8].

The state point 1 shows the initial condition of water vapor present in the gas–water vapor mixture at temperature T_1, the partial pressure of the water vapor being P_{v1}.

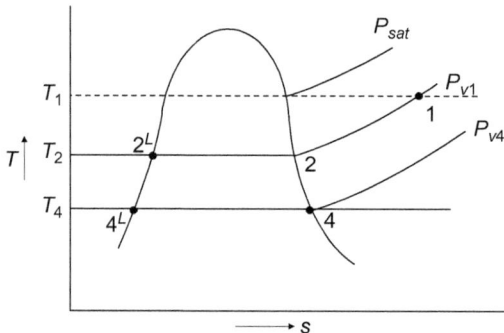

Fig. 3.4 Cooling a gas-vapor mixture.

As the mixture is cooled, its composition remains unaltered, so the partial pressure of the water vapor also remains constant. This isobaric cooling of the water vapor will continue till its reaches temperature T_2, which is the saturation temperature of water vapor at pressure P_{v1}. So the condensation of water vapor begins and a few droplets of water vapor (at state 2^L) are formed. This temperature T_2 is therefore called the *Dew point temperature*. If the temperature of the air is further reduced, it is postulated that the partial pressure of water vapor will fall in accordance with the $p - T$ relationship for saturated water. Thus, at temperature T_4 (Fig. 3.4), the partial pressure P_{v4} would equal the saturation pressure at this temperature. The mole fraction of the water vapor in the mixture would therefore be lower than at 1 (or 2) since the excess water would have condensed out from air. The enthalpy of the condensate, and the water vapor would correspond to that of the saturated liquid water and the saturated water vapor at that temperature, respectively.

In many thermal systems we encounter equipment through which moist air flows and exchanges both heat and moisture. The analysis of these equipment, like cooling towers, cooling and dehumidifying coils, desiccant dehumidifiers, etc., needs the thermodynamic properties of moist air, and these are usually found by making the aforementioned assumptions, namely the moist air is an ideal mixture of dry air (an ideal gas) and water vapor, also considered as an ideal gas in view of its low partial pressure. We shall briefly discuss here the formulae used for the calculation of these properties.

The amount of water vapor present in the moist air is specified in many ways, two of the popular measures being the relative humidity, ϕ, and the humidity ratio, w. These are defined as follows:

$$\phi = \frac{\text{Mole fraction of water vapor in air}}{\substack{\text{Mole fraction of water vapor in moist air saturated at} \\ \text{the same temperature and pressure}}} \tag{3.79}$$

$$\phi = \frac{\text{Mass of water vapor in a certain volume of the mixture}}{\substack{\text{Mass of remaining gases (i.e., dry air, in air water vapor} \\ \text{moisture)}}} \tag{3.80}$$

We can derive mathematical formulae for both ϕ and w using the laws of ideal gas mixtures discussed earlier.

Thus, using Eq. (3.69) in Eq. (3.79), we get

$$\phi = \frac{P_v/P}{P_{sat}/P} = \frac{P_v}{P_{sat}} \tag{3.81}$$

where P_v is the actual partial pressure of water vapor in the mixture and P_{sat} is the saturation pressure of water vapor at the mixture temperature (see Fig. 3.4).

Similarly, for air–water vapor mixture, we can write Eq. (3.80) as

$$w = \frac{m_v}{m_a} = \frac{(P_v V)/(R_v T)}{(P_a V)/(R_a T)} = \frac{P_v}{P_a} \cdot \frac{R_a}{R_v} = \frac{M_v}{M_a} \cdot \frac{P_v}{P_a} \tag{3.82}$$

where m_v is the mass of water-vapor and m_a the mass of dry air in a certain volume V of the moist air at temperature T; R_v and R_a being the corresponding gas constants, and M_v and M_a the corresponding molecular mass. Further using Eq. (3.70) and substituting the values of M_v and M_a, we get

$$w = 0.622 \frac{P_v}{P - P_v} \tag{3.83}$$

The other thermodynamic properties can be calculated using Eqs. (3.71)–(3.73). However, in air-conditioning practice, a peculiar convention is generally followed. The moist air-specific properties are defined per kilogram of dry air (and *not* per kilogram of moist air). This is so because during various air-conditioning processes, the moisture content w changes and so does the total mass flow rate of the moist air, but the mass flow rate of the dry air does not change. Defining various thermodynamic properties on per kilogram dry air basis thus greatly simplifies the calculations. Eqs. (3.71)–(3.73) have therefore to be suitably modified to account for this convention, and the fact that the moisture content w is defined on mass basis, rather than on mole basis. Keeping this in view, we get following expressions for moist air enthalpy and entropy

$$h = h_{da} + w h_v \text{ kJ/kg da} \tag{3.84}$$

$$s = s_{da} + w s_v \text{ (kJ/K)/kg da} \tag{3.85}$$

where the subscript *da* refers to the dry air and *v* refers to the water vapor. Further since water vapor usually exists at very low pressures in the air, its enthalpy is not influenced much by the pressure, and often an approximation can be made that $h_v \approx h_{sat}$, the enthalpy of saturated vapor at the prevailing temperature. We shall illustrate the use of these formulae with the help of an example taken from Dhar [1, Example 8.16].

Example 3.1

Moist air at 35°C and a specific humidity of 15 g/kg da passes over the cooling coil of an air-conditioner maintained at 10°C. The air leaves the cooling coil at 12°C with a relative humidity of 90%, and the condensate drips down at 10°C. The volume flow rate of the air leaving the coil is 20 m³/min. Determine the rate at which water is condensed and the rate at which heat is removed from the cooling coil. Given atmospheric pressure is 100 kPa and the temperature is 27°C. Take specific heat of the dry air, C_{po} as 1.0 kJ/kg K.

Solution

We first find the state of air at the coil exit.

Given $\phi = 90\%$, temperature = 12°C

At 12°C, the saturation pressure is (from tables) = 1.42 kPa

$$\therefore P_v = 0.9 \times 1.42 = 1.278 \text{ kPa (Eq. 3.81)}$$

Substituting it in Eq. (3.83) we get

$$w = \frac{0.622 \times 1.278}{100 - 1.278} = 8.052 \times 10^{-3} \text{ kg/kg da}$$

The volume flow rate of air at exit = 20 m³/min

$$\text{Density of dry air at exit} = \frac{P_{da}}{R_a T}$$

$$\therefore \quad \text{Mass flow rate of dry air at exit} = \frac{20 \times (100 - 1.278) \times 10^3}{(8314.3/29)(285)}$$

$$= 24.16 \text{ kg/min}$$

$$\text{The moisture condensation rate} = 24.16(w_{inlet} - w_{exit})$$

$$= 24.16(15 - 8.052) \times 10^{-3} \text{ kg/min}$$

$$= 167.86 \text{ g/min}$$

To find the heat removal rate from the cooling coil, we apply the SFEE to the coil. This gives

$$\dot{m}_{da}h_{inlet} = \dot{m}_{da}h_{exit} + \dot{m}_{cond.}h_{water} + \dot{Q}$$

The enthalpy of air at the inlet and the exit of the cooling coil can be found using Eq. (3.84), while the enthalpy of the condensate (water) at 10°C is found from the tables as 42.01 kJ/kg. Thus we get

$$\dot{Q} = \dot{m}_{da}(h_{inlet} - h_{exit}) - \dot{m}_{cond.}h_{water}$$

$$= 24.16[1.0(35 - 12) + 15 \times 10^{-3} \times 2565.3 - 8.052 \times 10^{-3} \times 2523.4]$$

$$- 167.86 \times 10^{-3} \times 42.01$$

$$= 987.4 \text{ kJ/min} = 16.457 \text{ kW}$$

In the above calculations h_v values at the inlet and the exit of the coil are taken from tables as the saturated water vapor enthalpies at 35 and 12°C, respectively.

Processes involving moist air are encountered in a variety of systems like the air-conditioning plants, cooling towers, engines, power plants, etc. To simplify their analysis, properties of moist air have been calculated over a wide range of temperatures and moisture content, and plotted in the form of a chart with Dry Bulb Temperature (DBT) on the abscissa and the humidity ratio w, and the partial pressure of water vapors, P_v as the ordinates. Fig. 3.5 shows such a chart—known as the *psychrometric chart*—and a few typical processes are also indicted therein.

For mixing of the two streams, the final condition is found using the SFEE and the moisture conservation equations as:

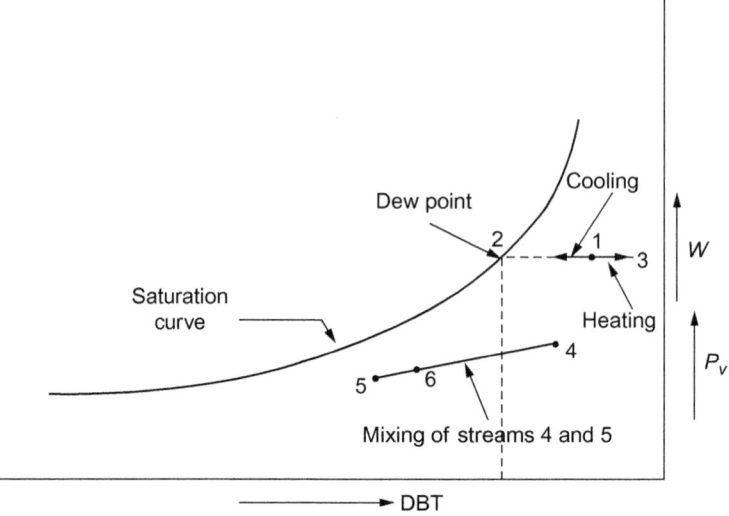

Fig. 3.5 Typical processes depicted on a psychrometric chart.

$$m_4 w_4 + m_5 w_5 = m_6 w_6 \tag{3.86}$$

$$m_4 h_4 + m_5 h_5 = m_6 h_6 \tag{3.87}$$

Continuing with our discussion on the Dalton's model of Ideal Gas Mixtures, combining the model Eqs. (3.71)–(3.73) with the definition of partial molar property (Eq. 3.53), it can be easily established that:

$$\bar{H}_i(T, P) = \left(\frac{\partial H}{\partial N_i}\right)_{T,P,N_{j\neq i}} = h_i^0(T, P_i) = h_i^0(T, P) \tag{3.88}$$

$$\bar{U}_i(T, P) = \left(\frac{\partial U}{\partial N_i}\right)_{T,P,N_{j\neq i}} = u_i^0(T, P_i) = u_i^0(T, P) \tag{3.89}$$

where we have used the fact that for ideal gases u and h are functions of T alone, and not dependent on the pressure.

Similarly from Eqs. (3.74), (3.76), (3.53) it follows that

$$\bar{V}_i(T, P) = \left(\frac{\partial V}{\partial N_i}\right)_{T,P,N_{j\neq i}} = \frac{\bar{R}T}{P} = \frac{V_i^0(T, P)}{N_i} = v_i^0(T, P) \tag{3.90}$$

Eqs. (3.88), (3.90) imply that the partial molar properties ($\bar{U}_i, \bar{H}_i, \bar{V}_i$) of an ideal gas mixture are equal to the properties of the pure component at the mixture temperature and pressure. It thus follows from Eqs. (3.57), (3.58) that the heat of mixing and the volume change on mixing of the components of an ideal gas is zero. However, there would be a change in the entropy during such a mixing process which can be estimated using Eq. (3.73) as follows:

$$\Delta s = s(T,P) - \sum x_i s_i^0(T,P)$$

$$= \sum s_i^0(T,P_i)x_i - \sum x_i s_i^0(T,P)$$

$$= \sum x_i[s_i^0(T,P_i) - s_i^0(T,P)]$$

$$= \sum x_i \left[\int_P^{P_i} ds_i^0 \right]_T = \sum x_i \left[\int_P^{P_i} \frac{-\bar{R}}{P} dP \right]$$

$$= -\bar{R} \sum x_i \ln(P_i/P) = -\bar{R} \sum x_i \ln x_i \qquad (3.91)$$

where we have used Eq. (3.18), at constant temperature to estimate the change in entropy in terms of the change in the pressure of an ideal gas.

Since x_i is < 0, Δs is always positive, indicating the inherent irreversibility of such mixing.

Dalton's model, and the law of partial pressures, is also sometimes extended to mixtures of real gases. Here, the partial pressures of various components are calculated at the mixture temperature and volume using any of the real gas equation of state like the Vender-Waals equation, Redlich-Kwong equation, or Martin-Hou equation. The total pressure of the mixture is then estimated, in accordance with Eq. (3.70), as the sum of these partial pressures.

More commonly used methods for analysis of multicomponent systems have been developed based on the concepts of chemical potential, μ_i, defined above and its surrogate quantity, fugacity defined below. From Eq. (3.52) it is clear that

$$\bar{G}_i = \left(\frac{\partial G}{\partial N_i} \right)_{T,P,N_{j\neq i}} = \mu_i \qquad (3.92)$$

So, to find the chemical potential of a component in the multicomponent mixture, we could start from an equation for the Gibb's function. Now, since $G = H - TS$, we could start from Eqs. (3.72), (3.73), and get an expression for Gibb's function for an ideal-gas mixture, and from there obtain an expression for μ_i using Eq. (3.92). The calculations finally give

$$\mu_i = \bar{G}_i = g_i + \bar{R}T \ln x_i \qquad (3.93)$$

This gives, at constant T

$$d\mu_i = dg_i + \bar{R}T \, d(\ln x_i)$$
$$= v_i dp + \bar{R}T \, d(\ln x_i) \quad \{\text{using the basic Eq. (3.20)}\}$$

$$= \frac{\bar{R}T}{P} dP + \bar{R}T \, d(\ln x_i)$$

$$= \bar{R}T \, d(\ln Px_i) \qquad (3.94)$$

Eqs. (3.93), (3.94) indicate the mathematical difficulties involved in calculation of μ_i. It is clear that as $x_i \to 0$, $\ln x_i \to -\infty$ and Eq. (3.93) would become unstable. Again in Eq. (3.94) we note similar situation with $P \to 0$, even when $x_i \neq 0$.

In order to resolve these difficulties, Lewis introduced the concept of fugacity, a new thermodynamic property based on generalization of Eq. (3.94) obtained above for a mixture of ideal gases. The fugacity f_i of any component i is defined as the property of a multicomponent mixture which is related to the chemical potential change in any isothermal process by the equation

$$\text{for constant } T: \quad d\mu_i = \bar{R}T \, d(\ln f_i) \qquad (3.95)$$

From Eqs. (3.94), (3.95) it follows that for ideal gas mixtures:

$$f_i = Px_i \qquad (3.96)$$

To bring in consistency of Eq. (3.95) with the above result, a limiting relationship is associated with Eq. (3.95), namely

$$\lim_{P \to 0} \left(\frac{f_i}{Px_i} \right) = 1 \qquad (3.97)$$

since all mixtures of gases exhibit ideal gas behavior at very low pressures.

Now, from Eq. (3.96), it follows that for ideal gas mixtures $f_i = Px_i = P_i$, the partial pressure of component i in the mixture; the fugacity of a component in a mixture can therefore be considered as a pseudo-partial pressure.

The thermodynamics of multicomponent systems is largely based on these concepts of the chemical potential μ_i and the fugacity f_i. Thermodynamic relations have been developed using which the fugacities of various components can be determined from experimental $p-v-T$ data, and all other thermodynamic properties from the fugacities. However, the data required, and the extent of numerical calculations often become prohibitive. Therefore, simple models for estimation of fugacity have been postulated using which a reasonable prediction of the thermodynamic properties can be made. One such popular model, called *ideal solution model*, postulates that the fugacity of component i in the mixture is equal to the product of its mole fraction and the fugacity of the pure component at the same pressure and temperature:

$$f_i = f_i^0 x_i \qquad (3.98)$$

This relationship, known as Lewis-Randall rule, has been found to be applicable to many gaseous mixtures, liquid and even solid solutions.

Many interesting conclusions emerge from using Eq. (3.98) in various thermodynamic relations. It can be shown that various thermodynamic properties of an ideal solution can be obtained from those of its components by using following equations.

$$v^{id} = \sum x_i v_i^0 \tag{3.99}$$

$$h^{id} = \sum x_i\, h_i^0 \tag{3.100}$$

$$s^{id} = \sum x_i s_i^0 - \bar{R} \sum x_i \ln x_i \tag{3.101}$$

$$g^{id} = \sum x_i g_i^0 + \bar{R}T \sum x_i \ln x_i \tag{3.102}$$

$$\mu^{id} = g_i^0 + \bar{R}T \ln x_i \tag{3.103}$$

where all pure component properties are at the mixture temperature and pressure, and the superscript *id* indicates ideal solution model.

In multicomponent systems with various phases in equilibrium with each other, besides equality of the pressure and the temperature, the fugacity and the chemical potential of various components in all the phases should be equal. From this result very important relationships for liquid-vapor equilibrium can be derived, which help in the analysis of stability of multicomponent systems. One such relation between the equilibrium mole fractions of the vapor and the liquid phases of an ideal solution, called the Raoult's law states:

$$P_i^{sat} x_i = P y_i \tag{3.104}$$

where x_i/y_i denote the mole fraction of component i in the liquid/vapor phase; P_i^{sat} is the saturation pressure of component i at the mixture temperature T, and P is the mixture pressure. Using this relationship, the $P - x - y$ and $T - x - y$ diagrams for binary mixtures can be easily drawn.

3.1.2 Thermodynamics of Reactive Mixtures

While discussing the thermodynamics of multicomponent systems above, we made an implicit assumption that the total number of moles of various constituents remains constant. This assumption is clearly not valid if there occurs a chemical reaction between various components. Not only would the number of moles of the initial components (the reactants) change, new compounds are also likely to be formed. Though the number of moles would change, the total mass of the system would always be conserved (in the absence of nuclear reactions). Therefore, the total mass of reactants would always equal to the total mass of the products of any reaction. Considering, for example, the well-known equation for the combustion of propane gas:

$$C_3H_8 + 5O_2 \rightarrow 3CO_2 + 4H_2O$$

We note that the number of moles is not conserved but if we find the masses of the reactants and the products, we note that 44 kg of propane and 160 kg of oxygen combine to form 132 kg of carbon dioxide and 72 kg of water. Thus the total mass of the reactants and the products is the same (i.e., 204 kg). This implies that in any general chemical reaction, with reactants R_1 and R_2 and the products P_3 and P_4 written as

$$\nu_1 R_1 + \nu_2 R_2 \rightarrow \nu_3 P_3 + \nu_4 P_4 \qquad (3.105)$$

The coefficients ν_1, ν_2, ν_3, and ν_4 have to be chosen in such a manner that the total number of atoms of each element which is present in the reactants, namely $R_1, R_2, \ldots$ is equal to the number of atoms of that element in the products $P_3, P_4, \ldots$. We shall illustrate this with the help of an example involving combustion of ammonia gas, where the number of modes of NO in the products is known to be double the number of moles of NO_2:

$$NH_3 + aO_2 \rightarrow 2bNO + bNO_2 + cH_2O$$

Balancing the atoms of various elements, we get the following equations:

$$N \text{ balance } 1 = 2b + b \Rightarrow b = 1/3$$
$$H \text{ balance } 3 = 2c \Rightarrow c = 3/2$$
$$O \text{ balance } 2a = 2b + b \times 2 + c \Rightarrow a = 17/12$$

The chemical equation is therefore

$$NH_3 + \frac{17}{12}O_2 \rightarrow \frac{2}{3}NO + \frac{1}{3}NO_2 + \frac{3}{2}H_2O$$

Most fossil fuels are hydrocarbons, and if during combustion all the carbon is converted to carbon dioxide and hydrogen to water, the combustion is said to be complete. Accordingly, for the complete combustion of a general hydrocarbon, by balancing C, H, and N of the products and the reactants, we get the chemical equation:

$$C_xH_y + \left(x + \frac{y}{4}\right)(O_2 + 3.76N_2) \rightarrow xCO_2 + \frac{y}{2}H_2O + 3.76\left(x + \frac{y}{4}\right)N_2 \quad (3.106)$$

where it has been presumed that the combustion air is dry and consists of a mixture of only oxygen and nitrogen in the molar ratio 1 : 3.76.

If the actual amount of air used in τ times the theoretical air needed for complete combustion as per the above equation, and $\tau > 1$, then we still have complete combustion of the fuel and the excess oxygen leaves with the products. The chemical equation is

$$C_xH_y + \tau \left(x + \frac{y}{4}\right)(O_2 + 3.76N_2) \rightarrow xCO_2 + \frac{y}{2}H_2O + (\tau - 1)\left(x + \frac{y}{4}\right)O_2$$
$$+ 3.76\tau\left(x + \frac{y}{4}\right)N_2 \qquad (3.107)$$

On the other hand, if $\tau < 1$, incomplete combustion occurs with CO and CO_2 present in the products. If we assume that all the hydrogen is converted into water in spite of deficiency of oxygen, we get the following general equation for incomplete combustion:

$$C_xH_y + \tau\left(x + \frac{y}{4}\right)(O_2 + 3.76N_2) \rightarrow (x - \alpha)CO_2 + \alpha CO + \frac{y}{2}H_2O$$
$$+ 3.76\tau\left(x + \frac{y}{4}\right)N_2 \qquad (3.108)$$

where

$$\alpha = 2(1 - \tau)\left(x + \frac{y}{4}\right)$$

The first law analysis of a chemical reaction would depend on whether the reaction is being carried out in a *closed vessel* or in an *open system* with reactants flowing in and the products flowing out. The second situation is more frequently encountered and can be analyzed by applying SFEE (Eq. 3.12). Indicating by suffix 1 the state of the reactants and by suffix 2 that of the products, we get (cf. Eq 3.13)

$$Q_w = W_{sh} + \sum(N_{Pi}\bar{H}_{Pi})_2 - \sum(N_{Ri}\bar{H}_{Ri})_1 \qquad (3.109)$$

where we have neglected the changes in the KE and PE, which are generally very small in comparison to other terms. This equation is greatly simplified by assuming that the reacting mixtures are ideal solutions, so that we can replace the partial molar enthalpies by the pure component enthalpies, to get (cf. Eq 3.100)

$$Q_{cv} = W_{sh} + \sum(N_{Pi}h_{Pi}^0)_2 - \sum(N_{Ri}h_{Ri}^0)_1 \qquad (3.110)$$

Comparing Eq. (3.110) with the SFEE for a single-component Eq. (3.12), we note an important difference. In Eq. (3.12) the enthalpy values at the exit and inlet to the control volume are for the same substance, and therefore we can refer to any tables for thermodynamic properties to get these values. In Eq. (3.110) the enthalpy differences are between different substances (namely the products and the reactants), and therefore we need the absolute values of their enthalpies. This difficulty is overcome by defining the enthalpy of formation, Δh_f of compound, as the enthalpy change which occurs during the formation of the compound from its stable elements at a standard reference state, that is

$$\Delta h_f = h_{compound} - \sum(\nu_i h_i)_{element} \qquad (3.111)$$

where ν_i is the stoichiometric coefficient of the ith element in forming a single mole of the compound. The values of Δh_f for a large number of chemical compounds have been tabulated, determined using statistical thermodynamics, by consideration of the bond energies involved in forming the compound, as also experimentally by measuring the heat transfer during the chemical reaction in which 1 kmol of the compound is formed from the stable elements at the standard reference state.

It follows from the definition that the enthalpy of formulation of stable elements (like H_2, O_2, N_2) is zero at the reference pressure and temperature. We therefore get from Eq. (3.111) a simple expression for the enthalpy of compounds:

$$h(T_{ref}, P_{ref}) = \Delta h_f \tag{3.112}$$

Thus a common datum is obtained from all compounds/elements participating in a chemical reaction.

The Basic SFEE for a chemical reaction can now be transformed as follows.

$$Q_{cv} = W_{sh} + \sum N_{Pi}\{h^0_{Pi}(T_2, P_2) - h^0_{Pi}(T_{ref}, P_{ref})\}$$
$$- \sum N_{Ri}\{h^0_{Ri}(T_1, P_1) - h^0_{Ri}(T_{ref}, P_{ref})\}$$
$$+ \left\{\sum N_{Pi}h^0_{Pi}(T_{ref}, P_{ref}) - \sum N_{Ri}h^0_{Ri}(T_{ref}, P_{ref})\right\} \tag{3.113}$$

Clearly, out of the three terms on the RHS of above equation involving enthalpies, the first two can be evaluated using the standard thermodynamic tables for the substances since these involve enthalpy change for a substance due to change in the temperature and the pressure. The third term, called the enthalpy of reaction ΔH_r at the reference state, can be evaluated using Eq. (3.112)

$$\Delta H_r = \sum N_{Pi} \cdot h^0_{Pi}(T_{ref}, P_{ref}) - \sum N_{Ri} \cdot h^0_{Ri}(T_{ref}, P_{ref})$$
$$= \sum N_{Pi} \cdot \Delta h_{fi} - \sum N_{Ri} \cdot \Delta h_{fi} \tag{3.114}$$

Combining Eqs. (3.113), (3.114), the first law equation for a chemical reaction under steady flow conditions can be written as

$$Q_{cv} = W_{sh} + \Sigma N_{Pi}(h^*_i)_p - \Sigma N_{Ri}(h^*_i)_R \tag{3.115}$$
$$\text{and} \quad h^*_i = h_i(T, P) - h_i(25°C, 1\ atm) + \Delta h_{fi} \tag{3.116}$$

where h^*_i is the total molar enthalpy of compound i involved in the reaction. The reference temperature has been taken as 25°C and the reference pressure as 1 atm.

During the combustion of a fuel heat is liberated and Q_{cv} value in Eq. (3.115) is negative. Traditionally, the heating value of a fuel (earlier termed as the calorific value) HV is defined as the energy obtained in combustion of the fuel, when both the reactants and the products are at reference conditions. Thus putting it in Eq. (3.115) we get

$$HV = -Q_{cv} = \sum N_{Ri}\Delta h_{fi} - \sum N_{Pi}\Delta h_{fi}$$
$$= -\Delta H_R \tag{3.117}$$

Most common fuels contain hydrogen which, on combustion, is converted into water. If water is present in the products in the liquid state, we get more heat output than when it is in the vapor state. Accordingly two heating values are usually defined for any fuel, namely the higher heating value (*HHV*), with water as liquid and the lower heating value (*LHV*), when water is present in the products of combustion in vapor form. Clearly

$$HHV = LHV + \text{Energy released during condensation of water}$$
$$\text{vapor formed during combustion}$$

We shall illustrate these calculations with a simple example.

Example 3.2
CNG, the ecofriendly fuel for automobiles can be considered to contain pure methane. Determine its *LHV* and *HHV*.

Solution
The stoichiometric equation for methane (CH_4) can be written using Eq. (3.106) as

$$CH_4 + 2(O_2 + 3.76N_2) \rightarrow CO_2 + 2H_2O + 2 \times 3.76N_2$$

The heating value can be found using Eq. (3.117). This gives

$$
\begin{aligned}
HHV &= (\Delta h_f)_{CH_4} + 0 - (\Delta h_f)_{CO_2} - 2(\Delta h_f)_{H_2O(l)} - 0 \\
&= (-74,850) - (-393,520) - 2(-285,830) \\
&= 890330 \text{ kJ}
\end{aligned}
$$

$$
\begin{aligned}
\text{and} \quad LHV &= (\Delta h_f)_{CH_4} - (\Delta h_f)_{CO_2} - (\Delta h_f)_{H_2O(v)} \\
&= (-74,850) - (-393,520) - 2(-241,820) \\
&= 802,310 \text{ kJ}
\end{aligned}
$$

The first law analysis presented above is for open system where the reactants flow in and the products flow out. In case the reaction occurs inside a closed vessel, we revert back to the basic first law (i.e., Eq. 3.1). Expressing it in the terminology of multicomponent reacting systems (with $E \equiv U$ and $W_{cv} = -W_{sh}$), this equation can be written as

$$Q_{in} = W_{sh} + \Delta U = W_{sh} + U_p - U_R = W_{sh} + \sum N_{Pi}(\bar{U}_i)_P - \sum (N_{Ri})\bar{U}_{iR} \quad (3.118)$$

where $\bar{U}_i$ represents the partial molar internal energy of component *i* in the mixture. Again assuming that the mixtures behave as ideal solutions we can write

$$\bar{U}_i = u_i^0 = h_i^0 - Pv_i^0 \quad (3.119)$$

Then following the same approach as with open systems to bring the enthalpy values of different substances to a common datum, we can get the energy balance equation as:

$$Q = W_{sh} + \sum N_{Pi}(h_i^* - Pv_i)_P - \sum N_{Ri}(h_i^* - Pv_i)_R \qquad (3.120)$$

where h_i^* (given by Eq. 3.116) and v_i are calculated for each species at the corresponding pressure and temperature. If the reactants and the products both can be modeled as ideal gases, we can substitute $pv = \bar{R}T$, and the above equation becomes

$$Q = W_{sh} + \sum N_{Pi}(h_i^*)_P - \sum N_{Ri}(h_i^*)_R - \bar{R}T_p \sum N_{Pi} + \bar{R}T_r \sum N_{Ri} \qquad (3.121)$$

This equation can be used to estimate the maximum temperature reached in sudden combustion of fuel in spark ignition engines, as illustrated in Example 3.3.

Example 3.3

Determine the maximum temperature reached in a spark ignition engine when a stoichiometric mixture of CNG (i.e., methane) is burnt suddenly (i.e., adiabatically), given the initial temperature of the reactant mixture is 600°K.

Solution

During the process of sudden combustion it can be assumed that there is no heat or work transfer. Eq. (3.121) thus becomes

$$\sum N_{Pi}(h_i^*)_P - \bar{R}T_p \sum N_{Pi} = \sum N_{Ri}(h_i^*)_R - \bar{R}T_R \sum N_{Ri}$$

To use this general equation for the present problem, we first write the chemical equation for combustion of methane

$$CH_4 + 2(O_2 + 3.76N_2) \rightarrow CO_2 + 2H_2O + 2 \times 3.76N_2$$

The energy equation, as applied to the combustion of methane can now be written as

$$[1 \cdot (h_i^*)_{CO_2}] + [2(h_i^*)_{H_2O} + 7.52(h_i^*)_{N_2}] - 8.31143\,T_p(1 + 2 + 7.52)$$
$$= [1 \cdot (h_i^*)_{CH_4}] + [2(h_i^*)_{O_2} + 7.52(h_i^*)_{N_2}]_{600K} - 8.3143 \times 600(1 + 2 + 7.52)$$

The values of the terms on the RHS can be found from the thermodynamic property tables since the temperature is known. This gives (taking C_p for methane = 2.2537 kJ/kg)

$$RHS = (-74{,}850 + 2.2537 \times 16 \times (600 - 298)) + 2(9247)$$
$$+ 7.52(8891) - 8.3143 \times 600 \times 10.52 \text{ kJ}$$
$$= -31{,}085.6 \text{ kJ}$$

We need to find, by trial and error, the value of the product temperature T_p such that the total value of the LHS of the above equation equals $-31{,}085.6$.

Taking $T_p = 2500$ K, as the first approximation, we find the values of various terms on the LHS from the thermodynamic tables and substitute in the expression. This gives

$$LHS = (121{,}926 - 393{,}520) + 2(98{,}964 - 241{,}820) + 7.52 \times 74{,}312$$
$$- 8.3143 \times 2500 \times 10.52$$
$$= -217{,}145.8$$

Since LHS is much smaller than the RHS, actual T_p is much higher than 2500 K. By trial and error, it can be found that for $T_p \approx 3015$ K the LHS is approximately equal to RHS.

3.2 FLUID FLOW

In thermal systems we come across both confined flows—for example, liquids flowing inside the tubes of a heat exchanger, air flowing inside the ducts of an air-conditioning system—and external flows, for example, air flowing over the heat sink of a thermo-electric cooler, flue gases flowing over the finned tubes of water-tube boilers, etc. The flow fields in these practical situations are so complex that we have to take recourse to empirical correlations to predict the parameters like pressure drop needed in the analysis of thermal systems. We shall review these correlations in this section after a brief recapitulation of some basic concepts.

A basic assumption which is made in the analysis of fluid flows in engineering systems is the "no-slip" boundary condition which presumes that the fluid "sticks" to the solid surface with which it is in contact. Thus the fluid layer immediately in touch with a solid surface acquires its velocity and temperature, that is

$$V_{fluid} = V_{solid} = 0 \quad \text{(when solid surface is at rest)}$$
$$T_{fluid} = T_{solid}$$

Since bulk of the fluid is moving, there is a velocity gradient near the solid surface, and in this region (where 99% of the velocity change occurs) the viscous effects are significant. It has been shown experimentally that this region—termed the boundary layer—is usually very thin in most fluids with small to moderate viscosity. Fig. 3.6 shows how boundary layer would develop in external and internal flows. While in external flows the boundary layer goes on thickening, in confined flows the boundary layers merge after some distance (Fig. 3.6B) and thereafter the velocity profile stabilizes.

This region is termed as the fully developed flow region, and in most thermal system analysis confined flows are considered to be "fully developed."

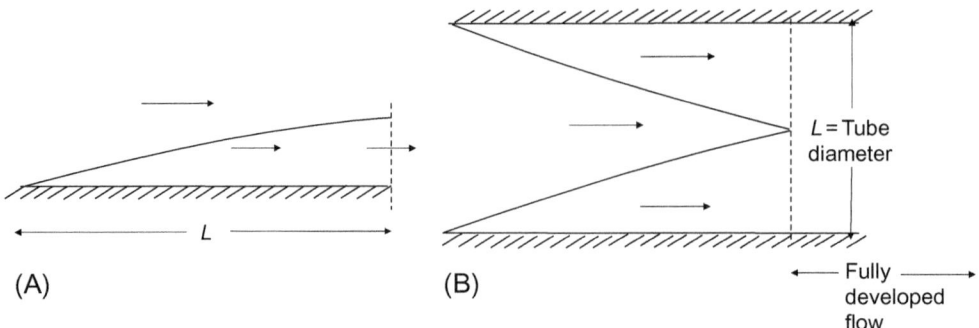

Fig. 3.6 Boundary layer development: (A) external flow and (B) internal flow.

The nature of flow is often characterized into two categories—laminar or smooth and turbulent or microscopically fluctuating. This categorization is usually done on the basis of Reynolds number Re defined as:

$$Re = \frac{\rho VL}{\mu} \tag{3.122}$$

where ρ is the fluid density, V the velocity, μ the dynamic viscosity, and L the characteristic length (see Fig. 3.6). For external flows, transition from laminar to turbulent flows occurs at $Re \approx 2 \times 10^5$ whereas for confined flows the limiting Reynolds number is 2300.

For laminar flows theoretical analysis predicts that the wall shear stress τ_w is given by the equation:

$$\tau_w = \frac{1}{2}\rho \bar{U}^2 \cdot C_f \tag{3.123}$$

where $\bar{U}$ is the mean velocity and C_f the fanning friction factor (or skin-friction coefficient) is given as

$$C_f = \frac{16}{Re_D} \quad \text{for flow in circular pipe of diameter } D \tag{3.124}$$

$$= \frac{0.664}{Re_x} \quad \text{for flow over a flat plate, at position } x \tag{3.125}$$

For noncircular pipes, we replace D by hydraulic diameter, D_H defined as:

$$D_H = \frac{4 * \text{Cross sectional area}}{\text{Wetted perimeter}} \tag{3.126}$$

For pipe flow, the wall shear stress is constant, while for flow over a flat plate, it changes along the flow direction.

For circular pipe flow, the wall shear stress can be related to the pressure drop as:

$$\tau_w = \frac{r}{2}\left(-\frac{dp}{dx}\right) \tag{3.127}$$

Substituting Eq. (3.123) in the above equation, we get

$$\left(-\frac{dp}{dx}\right) = \frac{4C_f}{D}\left(\frac{1}{2}\rho\bar{U}^2\right) = \frac{f}{D}\left(\frac{1}{2}\rho\bar{U}^2\right) \tag{3.128}$$

where $f = 4C_f$ is called the Darcy friction factor. For laminar flow, using C_f from Eq. (3.124) we get

$$f = \frac{64}{Re_D} \tag{3.129}$$

For turbulent flow, a simple formula, given first by Blasius, is

$$f = \frac{0.3164}{Re_D^{-0.25}} \quad (4000 < Re_D < 10^5) \tag{3.130}$$

More accurate results can be obtained from the implicit equation given by Prandtl (for $Re_D > 4000$):

$$\frac{1}{\sqrt{f}} = 2.0\log(Re_D f^{1/2}) - 0.8 \tag{3.131}$$

An explicit formula, which reproduces the results of above formula quite accurately in the region $10^4 < Re_D < 5 \times 10^6$, has been given by Petukhov [2]:

$$f = (0.79\ln Re_D - 1.64)^{-2} \tag{3.132}$$

These equations are for hydraulically smooth pipes. For rough pipes, like the commercial CI or cement pipes, the following formula suggested by Colebrook, based on the experimental data of Moody, is recommended:

$$\frac{1}{\sqrt{f}} = -2\log\left[\frac{\epsilon/D}{3.7} + \frac{2.51}{Re_D\sqrt{f}}\right] \tag{3.133}$$

where ϵ is the equivalent roughness of pipe surface. An explicit formula, which reproduces the results from the above implicit formula within ±2%, proposed by Haaland [3], is

$$\frac{1}{\sqrt{f}} \approx -1.8\log\left[\frac{6.9}{Re_D} + \left(\frac{\epsilon/D}{3.7}\right)^{1.11}\right] \tag{3.134}$$

Churchill [4] has developed a relationship which, though a bit complex, is valid for all ranges of Reynolds numbers:

$$f = 8 \left[\left(\frac{8}{Re_D} \right)^{12} + \frac{1}{(A+B)^{1.5}} \right]^{1/12}$$

$$A = \left[2.457 \ln \left\{ \frac{1}{(7/Re_D)^{0.9} + (0.27\epsilon/D)} \right\} \right]^{16} \tag{3.135}$$

$$B = \left\{ \frac{37,530}{Re_D} \right\}^{16}$$

For turbulent flow over a flat plate, an approximate formula for skin friction coefficient, suggested by White [5] is

$$C_f = \frac{0.455}{[\ln(0.06 Re_x)]^2} \tag{3.136}$$

The external flow configuration most commonly encountered in thermal systems is that of cross-flow over a banks of tubes (bare or finned). Fig. 3.7 shows the two possible arrangements of the tubes, such as (a) in-line tube configuration and (b) the staggered tube configuration. The flow field inside the tube bank is very complex as there is boundary layer separation over each tube and the resulting wakes of various tubes interact with each other in a manner which depends on the tube bank configuration, the tube pitches P_L and P_T, and the tube diameter D (Fig. 3.7).

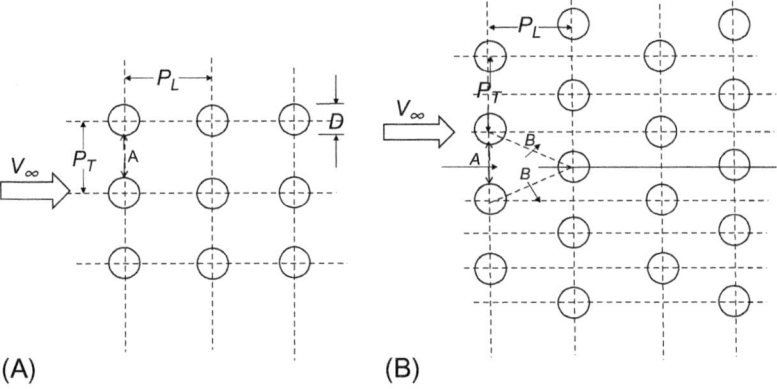

(A)   (B)

Fig. 3.7 Flow across tube banks.

If V_∞ is the velocity of the fluid in the free stream before it enters the tube bank, it is evident from Fig. 3.7, that the velocity of the fluid will increase to V_{max}, given by

$$V_{max} = V_\infty \cdot \frac{P_T}{P_T - D} \tag{3.137}$$

at the throat area A between adjacent tubes. This is the location of the maximum fluid velocity in the in-line tube banks. In staggered tube banks, this velocity may not be the maximum velocity. If the area of flow through the two diagonal rows (i.e., 2(B) in Fig. 3.7) is lesser than (A), the fluid velocity will be higher here than at (A). Its magnitude is

$$(V_{max})_{diag} = V_\infty \cdot \frac{P_T}{2\{[P_L^2 + (P_T/2)^2]^{1/2} - D\}} \tag{3.138}$$

The pressure drop in tube banks is generally correlated with the maximum fluid velocity. A commonly used correlation, given by Zukauskas [6] is

$$\Delta P = N\chi \left(\frac{\rho V_{max}^2}{2} \right) f \tag{3.139}$$

where the friction factor f and correction factor χ are functions of $Re_{D,max}$ and the tube bank configuration. Their values have to be obtained from the graphs given by Zukauskas [6].

An earlier correlation given by Robinson and Briggs [7] and applicable for banks of finned tubes is

$$\Delta P = 18.93 \frac{G^2}{\rho} N_r Re^{-0.316} \left(\frac{P_T}{D_b} \right)^{-0.927} \left(\frac{P_T}{P_d} \right)^{0.515} \tag{3.140}$$

where D_a and D_b are the outer and the inner diameters of the fins, respectively; P_d is the diagonal pitch ($= [(P_T/2)^2 + P_L^2]$), N_r the number of transverse rows, and Re the Reynolds number is based on the fin inner diameter D_b, that is, $Re = \frac{D_b G}{\mu}$, G being the mass velocity of the fluid.

Example 3.4

Estimate the friction factor for water flowing through a copper tube of 10 mm diameter for Reynolds number ranging from 3000 to 20,000, using various formulae and compare the results.

Solution

We first compute friction factor using explicit formulae based on Re_D. These calculations can be easily done using an Excel sheet. The value of tube roughness is taken from ASHRAE Handbook of Fundamentals as 1.52 μm.

Re_D	$f = \dfrac{0.3164}{Re_D^{0.25}}$ (Blasius)	$f = (0.79 \ln Re_D - 1.64)^{-2}$ (Petukhov)	$f^{-1/2} = -1.8 \log \left[\dfrac{6.9}{Re_D} + \left(\dfrac{\epsilon/D}{307}\right)^{1.11}\right]$ (Haaland)
3000	0.042752	0.045559	0.044343
5000	0.037627	0.038619	0.037731
7000	0.034591	0.03488	0.034152
9000	0.032484	0.032431	0.031801
12,000	0.03023	0.02993	0.029397
15,000	0.02859	0.028185	0.027714
20,000	0.026606	0.026151	0.025751

$B = \left(\dfrac{37,530}{Re_D}\right)$	$A = \left[2.457 \ln \left\{\dfrac{1}{(7/Re_D)^{0.9}+(0.27\epsilon/D)}\right\}\right]^{16}$	$f = 8\left[\left(\dfrac{8}{Re_D}\right)^{12} + \dfrac{1}{(A+B)^{1.5}}\right]^{1/12}$ (Churchill)
3.59846×10^{17}	1.05262×10^{18}	0.043087
1.01517×10^{14}	3.79336×10^{18}	0.038081
4.6611×10^{11}	8.33604×10^{18}	0.034512
8,359,434,122	1.46068×10^{19}	0.032175
83,783,228.97	2.70194×10^{19}	0.029794
2,358,288.24	4.27151×10^{19}	0.028136
23,636.17	7.52751×10^{19}	0.026213

To illustrate the procedure for calculation using the implicit equation, the values of friction factor have been calculated using the implicit Eq. (3.131) given by Prandtl. The calculations have been done using Excel software. A good estimate of f is taken from one of the values calculated using the explicit formulae, and it is substituted in Eq. (3.131). The LHS and RHS of the equation differ considerably. Then f values are then altered slightly and by trial and error the value of f giving very small difference between the two sides is obtained. Result obtained are given in following table:

Re_D	f	Error: $(LHS - RHS)$ of Eq. (3.131)
3000	0.04353	-5.1×10^{-5}
5000	0.0374	6.53×10^{-5}
7000	0.034018	-7.8×10^{-5}
9000	0.031768	7.03×10^{-5}
12,000	0.029448	-5.5×10^{-5}
15,000	0.027811	1.61×10^{-5}
20,000	0.025888	-2.1×10^{-5}

It can be seen that the values predicted using Haaland equation are quite close to those obtained using the implicit equation of Prandtl over the entire range of Reynolds. Students are advised to check the values obtained by implicit Eq. (3.133).

3.2.1 Compressible Flow

In the above analysis we have implicitly assumed that the density of the fluid remains constant during the flow, or in other words, the fluid is incompressible. This is generally true of most liquids; and for gases and vapors too unless the velocities are very high (>20–40 m/s) or the conduit is extremely long. For analysis of such compressible flows, we need to start from the basic SFEE. Considering the specific case of flow of a gas through an orifice/nozzle, the heat transfer, the work transfer, and the potential energy terms in Eq. (3.14) can be neglected. This gives

$$h_e - h_i + \frac{V_e^2 - V_i^2}{2} = 0$$

$$\text{or} \quad h + \frac{V^2}{2} = h^0 = \text{constant} \tag{3.141}$$

The term h^0 is called the total or stagnation enthalpy of a flowing stream. For an ideal gas, this also leads to the definition of the total temperature, T^0:

$$C_p T + \frac{V^2}{2} = C_p T^0 \tag{3.142}$$

$$\text{or} \quad T^0 = T + \frac{V^2}{2 C_p} \tag{3.143}$$

The specific heat C_p can also be expressed in terms of the specific heat ratio γ as:

$$\frac{C_p}{C_v} = \gamma; \quad C_p - C_v = R; \quad C_p = R \cdot \frac{\gamma}{\gamma - 1} \tag{3.144}$$

Substituting this in Eq. (3.143) gives

$$\frac{T^0}{T} = 1 + \frac{V^2}{\frac{2\gamma \cdot RT}{\gamma - 1}} \tag{3.145}$$

The velocity of sound c in an ideal gas at temperature T is given by equation:

$$c = \sqrt{\gamma RT} \tag{3.146}$$

Using this in Eq. (3.145) and rearranging terms we get

$$\frac{T^0}{T} = 1 + \frac{\gamma - 1}{2} \cdot M^2 \tag{3.147}$$

where M is the Mach number $= \frac{V}{c}$.

Like total (stagnation) enthalpy and total (stagnation) temperature, in compressible flow analysis we also define total or stagnation pressure, P^0 as the pressure obtained in an isentropic process of decelerating the flow to zero velocity. The relationship between P^0 and P can be obtained using Eq. (3.38) which relates isentropic processes:

$$\frac{P^0}{P} = \left(\frac{T^0}{T}\right)^{\frac{\gamma}{\gamma-1}} \tag{3.148}$$

Using Eq. (3.147) we can also write

$$\frac{P^0}{P} = \left(1 + \frac{\gamma-1}{2}M^2\right)^{\frac{\gamma}{\gamma-1}} \tag{3.149}$$

The analysis of a nozzle can now be done using Eq. (3.141). Indicating by subscripts 1 and 2 the conditions at the inlet and the exit of the nozzle, Eq. (3.141) gives

$$h_1^0 = h_2^0$$
$$\text{or} \quad T_1^0 = T_2^0 \tag{3.150}$$

Further, the velocity at the entrance to the nozzle is usually quite low. Therefore, we can write $T^0 = T_1$. Using Eq. (3.145) we can now write

$$\frac{T_2^0}{T_2} = 1 + \frac{V_2^2}{\frac{2\gamma \cdot RT_2}{\gamma-1}} \tag{3.151}$$

Substituting for T_2^0 from Eq. (3.151) in Eq. (3.150), with approximation $T_1^0 = T_1$ we get

$$T_1 = T_2^0 = T_2 + \frac{V_2^2}{\frac{2\gamma \cdot R}{\gamma-1}}$$
$$\Rightarrow \quad V_2^2 = \frac{2\gamma}{\gamma-1} \cdot R(T_1 - T_2) \tag{3.152}$$

$$\text{or} \quad V_2^2 = \frac{2\gamma RT_1}{\gamma-1} \cdot \left(1 - \frac{T_2}{T_1}\right) \tag{3.153}$$

Substituting for T_2/T_1 from Eq. (3.38), and expressing the temperature T_1 in terms of the pressure and the density at state 1, we get

$$V_2 = \sqrt{\frac{2\gamma}{\gamma-1} \cdot \left(\frac{P_1}{\rho_1}\right)\left(1 - \left(\frac{P_2}{P_1}\right)^{\frac{\gamma-1}{\gamma}}\right)} \tag{3.154}$$

Using this the mass flow rate can be calculated as:

$$\dot{m} = V_2 A_2 \rho_2$$

$$= A_2 \sqrt{\frac{2\gamma}{\gamma - 1}(P_1 \rho_1)\left[\left(\frac{P_2}{P_1}\right)^{2/\gamma} - \left(\frac{P_2}{P_1}\right)^{\frac{\gamma+1}{\gamma}}\right]} \qquad (3.155)$$

which is considerably different from the expression for incompressible flow:

$$\dot{m} = A_2 \cdot \rho \cdot \sqrt{\frac{2\Delta P}{\rho}} = A_2 \sqrt{2\rho(P_1 - P_2)} \qquad (3.156)$$

The above expression for the mass flow rate for compressible flows shows that as (P_2/P_1) decreases, the mass flow rate increases, till the ratio (P_2/P_1) falls to a critical value corresponding to a maxima in $\dot{m}$. This situation called choked flow occurs when

$$\left(\frac{P_2}{P_1}\right) = \left(\frac{2}{\gamma + 1}\right)^{\frac{\gamma}{\gamma-1}} \qquad (3.157)$$

Any further decreases in P_2 beyond this value does not increase the mass flow rate, and the velocity V_2 reaches local acoustic value.

3.2.2 Two-Phase Flow

We come across two–phase flow in numerous thermal systems like power plants and refrigeration systems. In the condensers of both these systems the working fluid changes state from vapor to liquid phase. In the evaporator of a refrigeration system, as also the boiler of a power plant, the working substance boils and converts from liquid to vapor phase. Since there is a large difference in the density of the two phases, phase change results in considerable changes in the velocity of mixture during the flow. As a result numerous flow patterns are encountered in two–phase flows depending on the thermophysical properties and the mass velocities of the two phases, and the orientation of the tubes through which the flow is occurring. The prediction of these flow patterns remains an uphill task even now and numerous flow pattern maps have been prepared to help in this. We need to consult handbooks and specialist books on this topic for a detailed understanding of the phenomena as also the methods for prediction of flow patterns and pressure drops. Here only a very brief introduction to this topic is presented along with a few important correlations for analyzing two–phase flows.

We begin by defining some basic terms used in the description of two–phase flows. Vapor fraction (or quality) x is the ratio of mass flow rate of the vapor phase to the total mass flow rate, that is

$$x = \frac{\dot{m}_v}{\dot{m}_v + \dot{m}_l} \qquad (3.158)$$

where $\dot{m}_v$ is the vapor mass flow rate and $\dot{m}_l$ the mass flow rate of the liquid phase.

Superficial velocity j is the volumetric flow rate divided by the tube cross-sectional area (A_c), that is

$$j_v = \frac{\dot{m}_v}{A_c \rho_v}; \quad j_e = \frac{\dot{m}_l}{A_c \rho_l}; \quad j = j_v + j_e \tag{3.159}$$

Cross-sectional void fraction, ϵ_{cs}, is defined as the ratio of the tube cross-sectional area occupied by vapor phase to the total cross-sectional area, that is

$$\epsilon_{cs} = \frac{A_v}{A_c} \tag{3.160}$$

The volumetric void fraction, ϵ_{vol} is defined in terms of the volume flow rates $(\dot{V})$ of the two phases as:

$$\epsilon_{vol} = \frac{\dot{V}_v}{\dot{V}_v + \dot{V}_l} \tag{3.161}$$

The cross-sectional void fraction ϵ_{cs} is the most widely used term, and often referred to as the void fraction, ϵ. Using this void fraction, the mean velocity of liquid and vapor phases can be found as:

$$U_v = \frac{\dot{m}_v/\rho_v}{A\epsilon} = \frac{\dot{m}(x)/\rho_v}{A\epsilon} = \frac{\dot{m}x}{A\rho_v\epsilon} \tag{3.162}$$

$$U_l = \frac{\dot{m}_l/\rho_l}{A(1-\epsilon)} = \frac{\dot{m}(1-x)/\rho_l}{A(1-\epsilon)} = \frac{\dot{m}(1-x)}{A\rho_l(1-\epsilon)} \tag{3.163}$$

where $\quad \dot{m} = \dot{m}_v + \dot{m}_l$

Combining these we get the relationship:

$$\frac{x}{1-x} = \frac{\rho_v}{\rho_l} \cdot \frac{U_v}{U_l} \cdot \frac{\epsilon}{1-\epsilon} \tag{3.164}$$

If the velocities of the two phases are equal, the flow is said to be homogeneous, and we get a simple relationship of void fraction (ϵ) to the vapor fraction (x):

$$\epsilon = \frac{x/\rho_v}{\frac{1-x}{\rho_l} + \frac{x}{\rho_v}} \quad \text{for homogeneous flow}$$

$$= \frac{1}{1 + \frac{1-x}{x} \cdot \frac{\rho_v}{\rho_l}} \tag{3.165}$$

The ratio of the velocities of the two phases is called the slip ratio $S = U_v/U_l$. Introducing S in Eq. (3.164) and rearranging we can get a general expression for ϵ in terms of vapor fraction x as:

$$\epsilon = \frac{1}{1 + \frac{1-x}{x} \cdot \frac{\rho_v}{\rho_l} \cdot S} \tag{3.166}$$

for most flows $S > 1$, and the actual void fraction is smaller than that calculated on the basis of assumption of homogeneous flow.

The prediction of void fraction in two-phase flows is of considerable importance since it influences the pressure gradient and the heat-transfer coefficient. It is also strongly linked to the flow pattern encountered during the flow which, as mentioned earlier, is influenced by a number of parameters, and thus rather difficult to predict with good accuracy. Over the years a number of correlations, mostly empirical or at best semiempirical, have been developed and validated against limited experimental data. One of the earliest models was given by Zivi, who presumed that the total kinetic energy of the two phases will tend to approach a minimum. On this basis he developed a correlation for the velocity ratio S:

$$S = \frac{U_v}{U_l} = \left(\frac{\rho_l}{\rho_v}\right)^{1/3} \tag{3.167}$$

Putting it in Eq. (3.166), we get Zivi's void fraction expression as:

$$\epsilon = \frac{1}{1 + \frac{1-x}{x}\left(\frac{\rho_v}{\rho_l}\right)^{2/3}} \tag{3.168}$$

More accurate predictions have been reported with the so-called drift-flux models, which attempt to correlate the drift velocities defined as

$$U_{vu} = U_v - j; \quad U_{lu} = U_l - j \tag{3.169}$$

to the various parameters governing the flow. The final result of this model is the following void fraction equation:

$$\epsilon = \frac{x}{\rho_v}\left[Co\left(\frac{x}{\rho_v} + \frac{1-x}{\rho_l}\right) + \frac{\bar{U}_{vu}}{G}\right]^{(-1)} \tag{3.170}$$

Here $\bar{U}_{vu}$ is the weighted mean drift velocity, Co is the distribution parameter whose values are empirically correlated, and G is the mass velocity of the total flow. A special feature of this equation is that it captures the effect of mass velocity G on the void fraction.

One of the earliest drift-flux correlations was given by Rouhani and Axelsson [8] for vertical channels (inner diameter d_i):

$$\bar{U}_{vu} = 1.18\left[\frac{g\sigma(\rho_L - \rho_v)}{\rho_l^2}\right]^{1/4} \tag{3.171}$$

$$\text{with} \quad Co = 1 + 0.2(1 - x)\left(\frac{gd_i\rho_l^2}{G^2}\right)^{1/4} \tag{3.172}$$

For horizontal tubes, Steiner [9] recommends

$$\bar{U}_{vu} = 1.18(1 - x)\left[\frac{g\sigma(\rho_L - \rho_v)}{\rho_l^2}\right]^{1/4} \tag{3.173}$$

$$\text{with} \quad Co = 1 + 0.12(1 - x) \tag{3.174}$$

In a recent review Ghajar and Tang [10] have recommended the following correlation for estimating void fraction:

$$\epsilon = \frac{j_v}{Co(j_v + j_l) + U_{vm}} \tag{3.175}$$

$$\text{where} \quad Co = \frac{j_v}{j_v + j_l}[1 + (j_l/j_v)^{zz}] \tag{3.176}$$

$$zz = \left(\frac{\rho_v}{\rho_l}\right)^{0.1} \tag{3.177}$$

$$\text{and} \quad U_{vm} = 2.9(1.22 + 1.22\sin\theta)^{P_{atm}/P_{sys}}$$
$$\cdot \left[\frac{gd_i\sigma(1 + \cos\theta)(\rho_l - \rho_v)}{\rho_l^2}\right]^{1/4} \tag{3.178}$$

where θ is the angle of inclination of the tube, P_{atm} the atmospheric pressure, and P_{sys} the system pressure. This equation is to be used with parameters in SI units.

Example 3.5

Calculate the void fraction along the length of a refrigeration evaporator tube at positions where the vapor fraction is 0.2, 0.4, 0.6, 0.8, 0.9, and 0.95 using various correlations. Given mass flow rate of refrigerant-22 = 0.0184 kg/s, tube diameter = 10 mm, refrigerant temperature = 6°C.

Solution

We first find from the refrigerant property tables various thermophysical properties:

$$\rho_l = 1260.8 \text{ kg/m}^3 \quad \rho_v = \frac{1}{0.03913} = 25.556 \text{ kg/m}^3 \quad \sigma = 10.81 \times 10^{-3} \text{ N/m}$$

We then set up an Excel sheet to calculate the void fraction using Eq. (3.165) (homogeneous flow model), Eq. (3.168) (Zivi's correlation), Eq. (3.170) along with Eqs. (3.173), (3.174) (Steiners' correlation), and Eq. (3.175) along with Eqs. (3.176), (3.178) (Ghajar and Tang model). The results are summarized in the following table

x	Homog.	Zivi	Steiner	Ghajar and Tang
0.2	0.92502	0.770793	0.810434	0.805788
0.4	0.970493	0.899675	0.890483	0.890314
0.6	0.986667	0.952779	0.934286	0.93131
0.8	0.994958	0.981754	0.968753	0.959181
0.9	0.997753	0.991808	0.984599	0.97169
0.95	0.998934	0.996103	0.992341	0.978426

It can be seen that the values predicted by Steiner's and Ghajar and Tang's correlations are near each other.

The prediction of pressure drop in two-phase flows encountered in equipment like the condensers, the evaporators, and the boilers is a fundamental requirement in simulating their performance. In view of the dependence of the pressure drop on flow pattern and the void fraction changes inside these equipment—about both of which there is considerable uncertainty—accurate prediction of two-phase flow pressure drop is rather difficult. We shall present here only a brief introduction to this complex topic.

The total pressure drop in two-phase flow is usually attributed to three reasons namely friction at the tube walls, change in elevation, and change in momentum due to change in vapor quality during condensation/boiling. Thus

$$\left(-\frac{dp}{dz}\right) = \left(-\frac{dp}{dz}\right)_F + \left(-\frac{dp}{dz}\right)_G + \left(-\frac{dp}{dz}\right)_M \qquad (3.179)$$

where subscripts F, G, and M refer to pressure gradients due to wall friction, change in gravitational head, and the change in momentum, respectively.

The simplest approach to estimate these terms is through the assumption of homogeneous flow, which is usually applicable at high velocities and low x values in vertical tubes. The two-phase flow is considered as equivalent to a single-phase flow with the density ρ_H and viscosity μ_H given by the following equations:

$$\frac{1}{\rho_H} = \frac{x}{\rho_v} + \frac{1-x}{\rho_l} \qquad (3.180)$$

and

$$\frac{1}{\mu_H} = \frac{x}{\mu_v} + \frac{1-x}{\mu_l} \qquad (3.181)$$

The friction pressure gradient is now calculated using the standard single-phase Eq. (3.128):

$$\left(-\frac{dp}{dz}\right)_F = \frac{f}{D}\frac{G^2}{2\rho_H} \tag{3.182}$$

where G is the mass velocity and the friction factor f is calculated using any of the single-phase correlations with the Reynolds number defined as:

$$Re_H = \frac{GD}{\mu_H} \tag{3.183}$$

The pressure drop due to elevation is simply calculated as:

$$\left(-\frac{dp}{dz}\right)_G = -\rho_H g \sin\theta \tag{3.184}$$

where θ is the angle of inclination measured from the horizontal.

The pressure drop due to momentum change is

$$\left(-\frac{dp}{dz}\right)_M = \frac{d}{dz}\left(\frac{G^2}{\rho_H}\right) = -G^2\frac{d}{dz}\left(\frac{1}{\rho_H}\right) = \left(\frac{G}{\rho_H}\right)^2\frac{d\rho_H}{dz} \tag{3.185}$$

It thus depends on the rate of change of vapor quality x along the flow direction (since ρ_H is a function of x), which is governed by the rate at which heat transfer is taking place.

This homogeneous flow model gives reasonable results only in limited situations and for better estimation of the two-phase flow pressure drop the separated flow models need to be used. These models consider the two phases as if these were two separate streams flowing within the tube. Clearly, the cross-sectional area occupied by the vapor stream is $A_c\epsilon$, where ϵ is the void fraction defined earlier (Eq. 3.160). The average density of the two-phase flow ρ_S is now defined as:

$$\rho_S = \rho_l(1 - \epsilon) + \rho_v\epsilon \tag{3.186}$$

and the pressure drop due to gravitational head is

$$\left(-\frac{dp}{dz}\right)_G = -\rho_S g \sin\theta \tag{3.187}$$

The momentum pressure drop term has to account for separate velocities of the two phases, and on simplification becomes

$$\left(-\frac{dp}{dz}\right)_M = G^2\left\{\left[\frac{(1-x)^2}{\rho_L(1-\epsilon)} + \frac{x^2}{\rho_v\epsilon}\right]_2 - \left[\frac{(1-x)^2}{\rho_L(1-\epsilon)} + \frac{x^2}{\rho_v\epsilon}\right]_1\right\} \tag{3.188}$$

The frictional pressure drop term is usually calculated indirectly through the correlations for the ratio of the two-phase frictional pressure gradient to the single-phase gradient for the liquid flowing alone in the pipe, that is $(dp_F/dz)_L$; or the vapor flowing alone in the pipe, that is $(dp_F/dz)_V$; or the frictional pressure gradient for the total flow in the channel with liquid-phase properties, $(dp/dz)_{LO}$. These pressure gradient ratios are defined as:

$$\phi_L^2 = \frac{(dp/dz)_F}{(dp_F/dz)_L} \tag{3.189}$$

$$\phi_V^2 = \frac{(dp/dz)_F}{(dp_F/dz)_V} \tag{3.190}$$

$$\phi_{LO}^2 = \frac{(dp/dz)_F}{(dp_F/dz)_{LO}} \tag{3.191}$$

These ratio terms are then correlated with flow parameters. Over the years, numerous correlations have been developed using this approach. One of widely recommended methods, especially suited for two-phase flow of refrigerants, is that of Grönnerud [11, Chapter 5]:

$$\phi_{LO}^2 = 1 + \left(\frac{dp}{dz}\right)_{Fr} \left[\frac{\rho_L/\rho_G}{(\mu_L/\mu_G)^{0.25}} - 1\right] \tag{3.192}$$

where
$$\left(\frac{dp}{dz}\right)_{Fr} = f_{FR}[x + 4(x^{1.8} - x^{10} f_{FR}^{0.5})] \tag{3.193}$$

Further $f_{FR} = 1.0$ for $Fr_L \geq 1$

$$= F_{rL}^{0.3} + 0.0055(\ln(1/F_{rL}))^2 \text{ for } Fr_L < 1 \tag{3.194}$$

where $$Fr_L = \frac{G^2}{g d_i \rho_L^2} \tag{3.195}$$

Here $\left(\frac{dp_F}{dz}\right)_{LO}$ is calculated as follows:

$$\left(\frac{dp_F}{dz}\right)_{LO} = f_l \cdot \frac{G^2}{2\rho_l D} \tag{3.196}$$

$$f_l = \frac{0.3164}{Re^{0.25}} \tag{3.197}$$

$$Re = \frac{GD}{\mu_l} \tag{3.198}$$

Müller-Steinhagen and Heck [12, Chapter 5] have proposed a simple, and purely empirical correlation which has been reported to give even better prediction of the pressure drop in a variety of two-phase flows. This correlation is essentially an empirical interpolation between the pressure gradients in all-liquid and all-vapor flows.

$$\frac{dp}{dz} = A(1-x)^{1/3} + Bx^3 \tag{3.199}$$

$$\text{where} \quad A = \left(\frac{dp}{dz}\right)_{LO} + 2\left[\left(\frac{dp}{dz}\right)_{VO} - \left(\frac{dp}{dz}\right)_{LO}\right]x \tag{3.200}$$

$$\text{and} \quad B = \left(\frac{dp}{dz}\right)_{VO} = f_v\frac{G^2}{2D\rho_v} \tag{3.201}$$

$$\left(\frac{dp}{dz}\right)_{LO} = f_l\frac{G^2}{2D\rho_l} \tag{3.202}$$

where friction factors f_l, f_v are calculated using the appropriate Reynolds number in the single-phase Eqs. (3.197), (3.198). In a recent review [13] these two correlations have been reported to give the best results for a variety of working substances.

Example 3.6

Calculate the frictional pressure gradient for refrigerant flowing through a tube at the same conditions as in Example 3.5, using the correlations of Grönnerud and Muller et al.

Solution

In addition to the thermophysical properties indicated in Example 3.5, following properties are also needed for these calculations.

$$\mu_l = 204.4 \times 10^{-6} \text{ Pa s} \quad \mu_v = 11.77 \times 10^{-6} \text{ Pa s}$$

Grönnerud Correlation

$$G = \frac{0.184}{\left(\frac{\Pi}{4} \times 0.01^2\right)} = 234.3292 \text{ kg/ms}^2$$

$$Re = \frac{GD}{\mu_l} = \frac{234.3292 \times 0.01}{204.4 \times 10^{-6}} = 11{,}464.25$$

$$f_l = \frac{0.3164}{Re^{0.25}} = \frac{0.3164}{(11{,}464.25)^{0.25}} = 0.030577$$

$$Fr_L = \frac{G^2}{gd_i\rho_l^2} = \frac{(234.3292)^2}{9.81 \times 0.01 \times (1260.8)^2} = 0.352121$$

Since $Fr_L < 1$, the value of f_{FR} is calculated using Eq. (3.194) as

$$f_{FR} = (0.352121)^{0.3} + 0.0055(\ln(1/0.352121))^2 = 0.737144$$

$$\left(\frac{dp}{dz}\right)_{LO} = f_l \cdot \frac{G^2}{2\rho_l D} = \frac{0.030577 \times (234.3292)^2}{2 \times 1260.8 \times 0.01} = 66.585 \text{ N/m}^2$$

The remaining calculations dependent on x have been done through an Excel sheet, and the results are given in following table.

x	$(dp/dz)_{Fr}$	ϕ_{LO}	$(dp/dz)_F$
0.2	0.310118	8.184591	544.76
0.4	0.861139	20.95023	1394.42
0.6	1.602443	38.12419	2537.5
0.8	2.29083	54.07223	3598.9
0.9	2.219709	54.42456	3489.3

Müller et al. Correlation

$$Re_v = \frac{GD}{\mu_v} = \frac{234.3292 \times 0.01}{11.77 \times 10^{-6}} = 199{,}090.3 \tag{3.203}$$

$$f_v = \frac{0.3164}{Re_v^{0.25}} = \frac{0.3164}{(199{,}090.3)^{0.25}} = 0.014979 \tag{3.204}$$

$$\left(\frac{dp}{dz}\right)_{VO} = \frac{f_v G^2}{2D\rho_v} = \frac{0.014979 \times (234.3292)^2}{2 \times 0.01 \times 25.556} = 1609.18 \tag{3.205}$$

Rest of the calculations involving x, as done on the Excel software are given in following table

x	A	$(dp/dz)^F$
0.2	683.3518	647.236
0.4	1300.145	1199.53
0.6	1916.938	1759.86
0.8	2533.731	2305.31
0.9	2842.128	2491.82

It can be seen that the pressure drops predicted by the two correlations are substantially different, with Grönnerud correlation mostly predicting higher values. This is also reported in the literature.

3.3 HEAT TRANSFER

The transfer of energy between two objects at different temperatures is generally termed as "heat transfer." If the two objects are in contact, the mechanism of heat transfer is called *conduction*; if they are separated from each other by an evacuated space (to prevent any indirect contact also) the heat transfer is said to occur by *radiation*. If one of the two objects in contact is a fluid, which is in motion—either externally forced or induced due to temperature gradient itself—then the heat transfer is termed as *convection*. The microscopic analysis of the mechanisms of these three modes of heat transfer is quite intricate, and from engineering perspective, it suffices to know the phenomenological laws which enable us to predict its magnitude. These laws are enunciated below.

Conduction

Fourier's Law of Heat Conduction

In a homogeneous substance, the local heat flux (q) is proportioned to the local temperature gradient (dT/dx):

$$\frac{\dot{Q}}{A} = q = -k\frac{dT}{dx} \tag{3.206}$$

where k, the constant of proportionality, is called the thermal conductivity of the substance. The magnitude k varies widely, from 0.027 W/mk for air at 27°C to 386 W/m for copper. It is also dependent on the temperature.

Radiation

Stefan-Boltzmann Law

The emissive power (radiative energy emitted per unit time per square meter of surface area) of a black body (i.e., a surface which absorbs all radiations incident on it) is proportional to the fourth power of its absolute temperature

$$E_b = \sigma T^4 \tag{3.207}$$

where σ, the constant of proportionality, is the Stefan Boltzmann constant ($= 5.67 \times 10^{-8}$ W/m^2 K^4).

Convection

Newton's Law of Cooling

The rate of heat transfer from a surface at temperature T_s to a (colder) fluid at temperature T_∞ is proportional to the difference of their temperatures:

$$\frac{Q}{A} = q = h(T_s - T_\infty) \tag{3.208}$$

where the "constant" of proportionality h is called the (convective) heat-transfer coefficient. Unfortunately, this h is not really a constant and depends very strongly on fluid properties and its velocity, and in some cases, even on the temperature difference $(T_s - T_\infty)$. In fact this equation should be seen more as a definition of h rather than as a physical law of heat transfer.

From a practical perspective, the study of conductive and radiative heat transfer under different conditions essentially involves judicious use of Eqs. (3.206), (3.207), while in the field of convective heat transfer the focus is on developing equations to predict the heat-transfer coefficient in different situations encountered in practice. We shall briefly review these in the following sections.

3.3.1 Conductive Heat Transfer

Starting from the basic Fourier's law, Eq. (3.206) and combining it with the first law of thermodynamics, we can derive the general equation for three-dimensional heat conduction in any solid as:

$$\rho c \frac{\partial T}{\partial \tau} = \frac{\partial}{\partial x}\left(k\frac{\partial T}{\partial x}\right) + \frac{\partial}{\partial y}\left(k\frac{\partial T}{\partial y}\right) + \frac{\partial}{\partial z}\left(k\frac{\partial T}{\partial z}\right) + \dot{Q}_v''' \tag{3.209}$$

where $\dot{Q}_v'''$ is the volumetric heat generation rate within the solid. If the influence of temperature on the thermal conductivity of the material is insignificant, the above equation can be simplified to:

$$\rho c \frac{\partial T}{\partial \tau} = k\left(\frac{\partial^2 T}{\partial x^2} + \frac{\partial^2 T}{\partial y^2} + \frac{\partial^2 T}{\partial z^2}\right) + \dot{Q}_v''' \tag{3.210}$$

If there is no internal heat generation, $\dot{Q}_v''' = 0$, the above equation can be written as:

$$\frac{\partial T}{\partial \tau} = \alpha\left(\frac{\partial^2 T}{\partial x^2} + \frac{\partial^2 T}{\partial y^2} + \frac{\partial^2 T}{\partial z^2}\right) \tag{3.211}$$

where $\alpha = k/\rho c$ (in m^2/s) is a thermophysical property of the material called the thermal diffusivity. Its significance is obvious from Eq. (3.211). When there is no internal heat generation, and the thermal conductivity of the material does not change much within the solid, thermal diffusivity is the only property which influences the temperature distribution in the material.

Further, under steady-state conditions, when there is no change in the temperature with time, the temperature distribution is governed by the Poisson's (partial differential) equation:

$$\frac{\partial^2 T}{\partial x^2} + \frac{\partial^2 T}{\partial y^2} + \frac{\partial^2 T}{\partial z^2} = -\frac{\dot{Q}_v'''}{k} \tag{3.212}$$

Further if, in addition, there is no internal heat generation, we get

$$\frac{\partial^2 T}{\partial x^2} + \frac{\partial^2 T}{\partial y^2} + \frac{\partial^2 T}{\partial z^2} = 0 \tag{3.213}$$

which is the well-known Laplace (partial differential) equation.

The above equations in Cartesian coordinates can also be transformed into other coordinate systems as indicated below.

$$\nabla^2 = \frac{\partial^2}{\partial x^2} + \frac{\partial^2}{\partial y^2} + \frac{\partial^2}{\partial z^2}$$

$$= \frac{1}{r}\frac{\partial}{\partial r}\left(r\frac{\partial}{\partial r}\right) + \frac{1}{r^2}\frac{\partial^2}{\partial \phi^2} + \frac{\partial^2}{\partial z^2} \tag{3.214}$$

in cylindrical coordinates r, ϕ, z

$$= \frac{1}{r^2}\frac{\partial}{\partial r}\left(r^2\frac{\partial}{\partial r}\right) + \frac{1}{r^2 \sin\theta} \cdot \frac{\partial}{\partial \theta}\left(\sin\theta\frac{\partial}{\partial \theta}\right)$$

$$+ \frac{1}{r^2 \sin^2\theta} \cdot \frac{\partial^2}{\partial \phi^2} \tag{3.215}$$

in spherical coordinates r, θ, ϕ

The operator ∇^2 is called the Laplacian operator.

Most heat conduction problems thus involve solution of these partial differential equations, which often reduce to ordinary differential equations in many situations, as illustrated below.

Considering the simplest case of one-dimensional steady heat conduction in a plane wall, with no internal heat generation, Eq. (3.213) reduces to

$$\frac{d^2 T}{dx^2} = 0 \tag{3.216}$$

If T_1 and T_2 are the temperatures at the two ends of the wall (Fig. 3.8), the solution of the above equation is

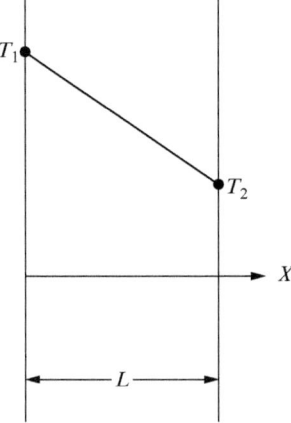

Fig. 3.8 Steady one-dimensional heat conduction in a plane wall.

$$T = \frac{T_2 - T_1}{L}x + T_1 \tag{3.217}$$

The basic Fourier's law, Eq. (3.206), then gives

$$\frac{\dot{Q}}{A} = -k\left(\frac{T_2 - T_1}{L}\right)$$

$$\text{or} \quad \dot{Q} = \frac{T_1 - T_2}{L/kA} \tag{3.218}$$

This equation is usually viewed, by analogy with Ohm's law for current flow, as indicating the flow of heat (cf. current) taking place due to the temperature difference (cf. potential difference) $T_1 - T_2$, across a thermal resistance L/kA (cf. electrical resistance). Using this analogy, many complicated situations, like a composite wall of Fig. 3.9, can be very easily analyzed to get

$$\dot{Q} = \frac{T_1 - T_2}{L_1/k_1 A} = \frac{T_2 - T_3}{L_2/k_2 A} = \frac{T_3 - T_4}{L_3/k_3 A}$$

$$= \frac{T_1 - T_4}{[(L_1/k_1 A) + (L_2/k_2 A) + (L_3/k_3 A)]} \tag{3.219}$$

A frequently encountered situation is that of heat conduction across the walls of a tube (Fig. 3.10). This can be analyzed starting from Laplace equation written in cylindrical coordinates (r, ϕ, z) (Eq. 3.214). Often the heat conduction is predominantly in the radial direction. This equation can therefore be simplified to:

$$\frac{1}{r}\frac{\partial}{\partial r}\left(r\frac{\partial T}{\partial r}\right) = 0 \tag{3.220}$$

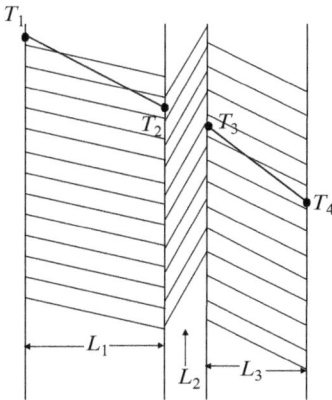

Fig. 3.9 Heat conduction through a composite wall.

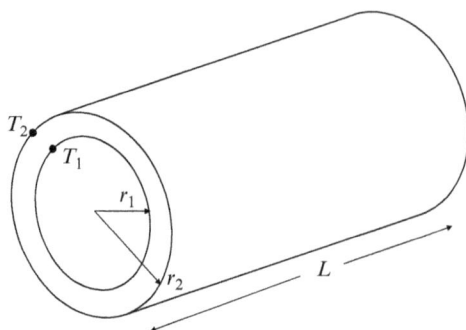

Fig. 3.10 Steady heat conduction across a tube.

which when integrated under boundary conditions $T = T_1$ at $r = r_1$ and $T = T_2$ at $r = r_2$ gives the temperature distribution as:

$$\frac{T_1 - T}{T_1 - T_2} = \frac{\ln(r/r_1)}{\ln(r_2/r_1)} \tag{3.221}$$

and the heat-transfer rate as:

$$\dot{Q} = \frac{T_1 - T_2}{[\ln(r_2/r_1)/(2\pi kL)]} \tag{3.222}$$

Eq. (3.222) can again be interpreted akin to Ohm's law for electric current, giving the thermal resistance for tube wall as

$$R = \frac{\ln(r_2/r_1)}{2\pi kL} \tag{3.223}$$

Using this approach of electrical analogy we can also analyze more complex problems like, for example, heat gained by an insulated pipe carrying cold water to an air-conditioning plant (Fig. 3.11). We now introduce in addition a change in the boundary conditions. Rather than the wall temperatures, let the known cold fluid temperature T_i and the environment temperature T_o, form the boundary conditions. At both the boundaries heat transfer between the surface and the fluid occurs through convection, and we make use of Eq. (3.208) to predict its magnitude.

Under steady state, the rate of heat transfer across all surfaces must be equal, and this gives us the following equation (see Fig. 3.11):

$$\dot{Q} = \frac{T_{io} - T_{wo}}{\left\{\frac{\ln(r_3/r_2)}{2\pi k_i L}\right\}} = \frac{T_{wo} - T_{wi}}{\left\{\frac{\ln(r_2/r_1)}{2\pi k_w L}\right\}} = \frac{T_{wi} - T_i}{1/h_i A_{wi}} = \frac{T_o - T_{io}}{1/h_o A_{io}} \tag{3.224}$$

$$= \frac{(T_o - T_i)}{\left[\left\{\frac{\ln(r_3/r_2)}{2\pi k_i L}\right\} + \left\{\frac{\ln(r_2/r_1)}{2\pi k_w L}\right\} + \frac{1}{h_i A_{wi}} + \frac{1}{h_o A_{io}}\right]} \tag{3.225}$$

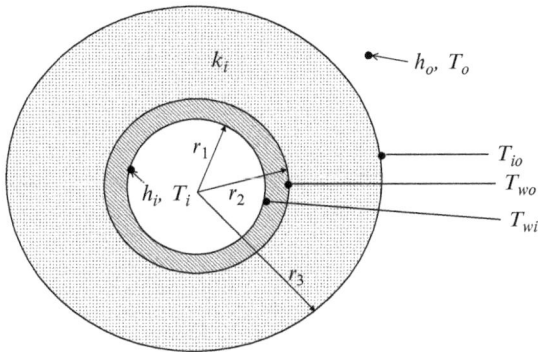

Fig. 3.11 Heat transfer to cold water flowing through a insulated pipe.

Here A_{io} is the external surface area of the insulated surface, k_i the thermal conductivity of the insulating material, A_{wi} the inside surface area of the tube walls, k_w the thermal conductivity of the tube material, h_o the heat-transfer coefficient at the insulated surface, and h_i the heat-transfer coefficient at the inside surface of the tube wall.

Above equation is used to define the overall heat-transfer coefficient U as:

$$\frac{1}{UA} = \frac{1}{h_i A_{wi}} + \frac{\ln(r_2/r_1)}{2\pi k_w L} + \frac{\ln(r_3/r_2)}{2\pi k_i L} + \frac{1}{h_o A_{io}} \tag{3.226}$$

so that we can be write Eq. (3.225) as

$$\dot{Q} = UA(T_o - T_i) \tag{3.227}$$

The area A in the above equations could be any of the two surface areas such as A_{wi} or A_{io}, and the U value would then correspond to that particular area.

In many practical situations involving heat transfer across the walls of an uninsulated pipe, we often encounter situations where one of the heat-transfer coefficients (usually of the surfaces exposed to air) is much lower than the other. In such equipment, it is possible to increase the value of U, and hence the rate of heat transfer, substantially by increasing the effective area of the surface where the convective heat-transfer coefficient is low by providing fins. Fig. 3.12 shows one of the numerous designs of fins being employed in thermal equipment these days.

While the fins increase surface area, temperature all over the finned surface can not be the same as at the base surface since temperature gradient is needed to conduct heat across them. The temperature over the finned surface progressively decreases (for the case of heat loss from the surface) as we move away from the base surface. This reduces the effectiveness of the fins, and this effect is quantified by fin efficiency, η_f, which is defined as

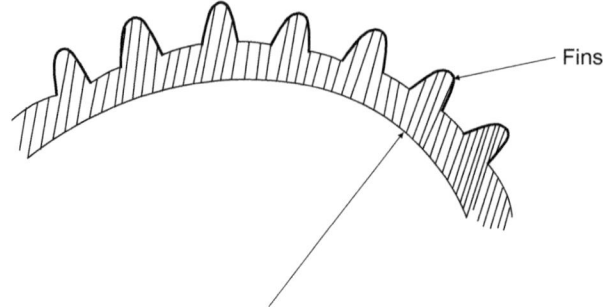

Fig. 3.12 Fins over a tube.

$$\eta_f = \frac{\text{Actual heat transfer from the fin}}{\text{Heat transfer from the fin had its entire surface been at the base temperature}} \qquad (3.228)$$

The expressions for fin efficiency for various shapes of fins have been obtained by solution of the basic heat conduction equation, and can be seen in standard heat-transfer texts and handbooks [14, 15]. Typical practical values range from 0.8 to 0.95.

In a similar manner the overall efficiency of a surface with fins, η_t can also be defined as follows:

$$A \cdot \eta_t = (A - A_f)1 + \eta_f A_f$$

$$\text{or} \quad \eta_t = 1 - \frac{A_f}{A}(1 - \eta_f) \qquad (3.229)$$

where A is the total surface area of the finned surface, A_f being the surface area of the fins only.

Thus for a heat exchanger using finned tubes to transfer heat between two fluids, the overall heat-transfer coefficient U can be calculated as:

$$\frac{1}{UA} = \frac{1}{(\eta_t h A)_c} + \frac{R_{fc}}{(\eta_t A)_c} + \frac{\ln(r_o/r_i)}{2\pi k_w L} + \frac{R_{fh}}{(\eta_t A)_h} + \frac{1}{(\eta_t h\, A)_h} \qquad (3.230)$$

where the subscripts c and h refer to the cold surface and the hot surface, respectively; and the terms R_{fc} and R_{fh} quantify the thermal resistance arising from the fouling of the two surfaces of the heat exchanger. The values of these resistances depend upon the nature of the fluid/surface in contact, and recommended values can be seen from standard handbooks. Typical values range from 0.00009 m^2 K/W for distilled water in contact with a copper surface to 0.00176 m^2 K/W for engine exhaust.

3.3.2 Radiative Heat Transfer

Stefan-Boltzmann law, Eq. (3.207), is the basic equation for radiative heat-transfer calculations. Most of the engineering equipment can not be treated as "black surfaces" as these reflect and some, like glass, even transmit a portion of the radiation incident on them. Accordingly we define three properties of a surface, namely the absorptivity α, the reflectivity (ρ), and the transmittivity (τ) as the fraction of energy incident which is absorbed, reflected, or transmitted by it. Clearly

$$\alpha + \rho + \tau = 1 \tag{3.231}$$

Only glass, and few plastics, can transmit radiation. Most objects are opaque and for these:

$$\alpha + \rho = 1 \tag{3.232}$$

For surfaces which are not black, we also define emissivity, ϵ, as the ratio of its emissive power (E) to that of a black body (E_b) at the same temperature. This gives

$$E = \epsilon E_b = \epsilon \sigma T^4 \tag{3.233}$$

It can be shown that the absorptivity of a surface equals its emissivity at the same temperature, that is

$$\alpha = \epsilon \tag{3.234}$$

Both α and ϵ are weakly dependent on the temperature and so, to simplify the calculations, most engineering surfaces are modeled as *gray surfaces*, that is, those for which α and ϵ are equal and constant over the entire operating range of temperatures.

To calculate radiative exchange between two surfaces 1 and 2, we also need to define the shape (or view) factor, F_{12}. This gives the fraction of energy leaving surface 1, that is intercepted by surface 2. Correspondingly, F_{21} gives the fraction of the energy leaving surface 2, that reaches surface 1. It can be shown that these shape factors are related by the equation:

$$A_1 F_{12} = A_2 F_{21} \tag{3.235}$$

where A_1 is the area of surface 1 and A_2 is the area of surface 2.

These shape factors depend primarily on the size, shape, and orientation of the surfaces, and their values for typical configuration can be seen from handbooks/textbooks on heat transfer.

Using these concepts, the radiative heat transfer between two gray surfaces can be determined. The final result is usually expressed in the following form:

$$\dot{Q}_{12} = \frac{\sigma T_1^4 - \sigma T_2^4}{\frac{1-\epsilon_1}{\epsilon_1 A_1} + \frac{1}{A_1 F_{12}} + \frac{1-\epsilon_2}{\epsilon_2 A_2}} \tag{3.236}$$

where subscript 1 indicates properties of surface 1 and subscript 2 those of surface 2.

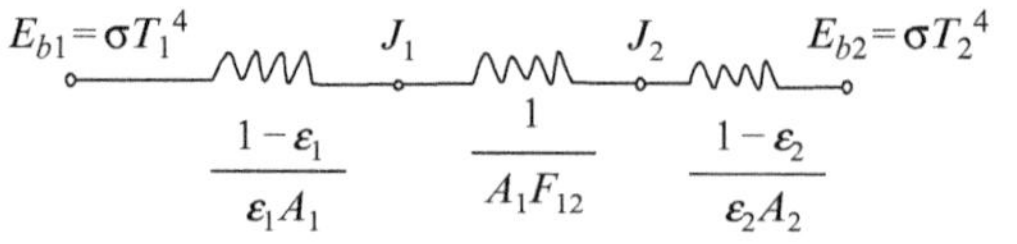

$$E_{b1}=\sigma T_1^{\,4} \quad\text{\textasciitilde}\quad J_1 \quad\text{\textasciitilde}\quad J_2 \quad\text{\textasciitilde}\quad E_{b2}=\sigma T_2^{\,4}$$

$$\frac{1-\varepsilon_1}{\varepsilon_1 A_1} \qquad \frac{1}{A_1 F_{12}} \qquad \frac{1-\varepsilon_2}{\varepsilon_2 A_2}$$

Fig. 3.13 Equivalent electrical circuit for radiative heat transfer between two surfaces.

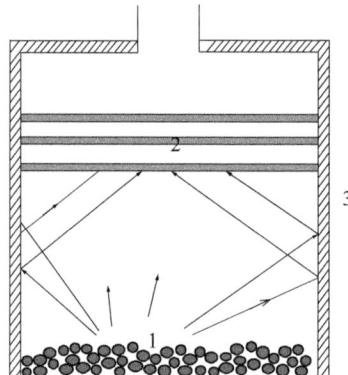

Fig. 3.14 Reradiating enclosure of a boiler furnace.

This result can also be interpreted to define an equivalent electrical circuit for radiation heat transfer as shown in Fig. 3.13. Here the terms J_1 and J_2 stand for the radiosity of the surfaces, which is the sum total of all the radiation (whether emitted or reflected) that leaves any surface.

This concept of electrical analogy can be easily extended to more complex configurations like three surface enclosures, surfaces with radiation shield interposed between them, etc. As an illustration of the approach, we consider the example of a furnace in which a hot bed of fuel, is exchanging heat with a bundle of water tubes of a boiler, (2), Fig. 3.14, both directly as also via the surrounding enclosure of insulating bricks, (3), assumed for the sake of simplicity, to be isothermal. Fig. 3.15 shows the electrical circuit analog for this enclosure on the basis of which the calculations for the net heat transfer to the water tubes can be done.

If the surrounding enclosure is perfectly insulating, there is no heat transfer across surface 3, and therefore $E_{b3} = J_3$ and the circuit reduces to that given in Fig. 3.16A. The three resistances in between J_1 and J_2 can now be replaced by the equivalent resistance R_Δ, using the basic rules for resistances in series and parallel to get the electric circuit of Fig. 3.16B, from which the heat-transfer rate from the furnace to the boiler tubes can be immediately written as

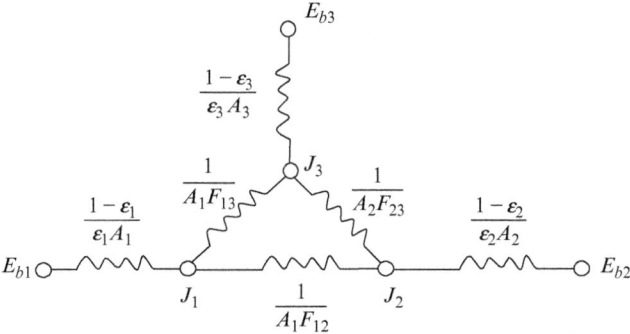

Fig. 3.15 Electric circuit analog for a reradiating enclosure.

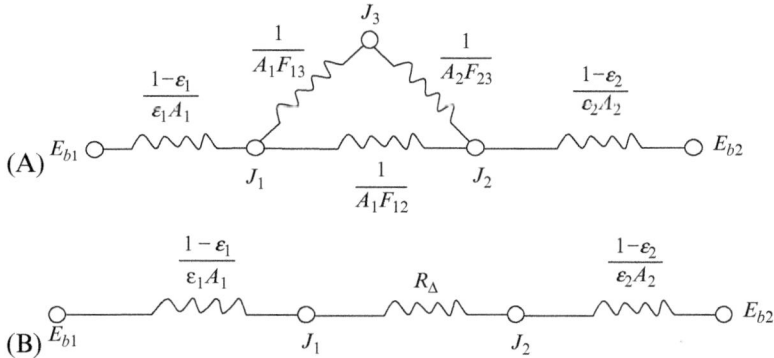

Fig. 3.16 Simplifying the electric circuit analog of reradiating enclosure.

$$Q_{12} = \frac{\sigma(T_1^4 - T_2^4)}{\frac{1-\epsilon_1}{\epsilon_1 A_1} + R_\Delta + \frac{1-\epsilon_2}{\epsilon_2 A_2}} \tag{3.237}$$

$$\text{where} \quad R_\Delta = \frac{R_{sum}(1/A_1 F_{12})}{R_{sum} + (1/A_1 F_{12})} \tag{3.238}$$

$$\text{with} \quad R_{sum} = \frac{1}{A_1 F_{13}} + \frac{1}{A_2 F_{23}} \tag{3.239}$$

Throughout above analysis, it has been implicitly assumed that the radiative heat transfer between the surfaces is not influenced by the presence of the intervening medium. However, this assumption is not valid if between the surfaces we have gases which can absorb and emit radiation. In such cases, the analysis must account for absorption, transmission, and emission of radiation.

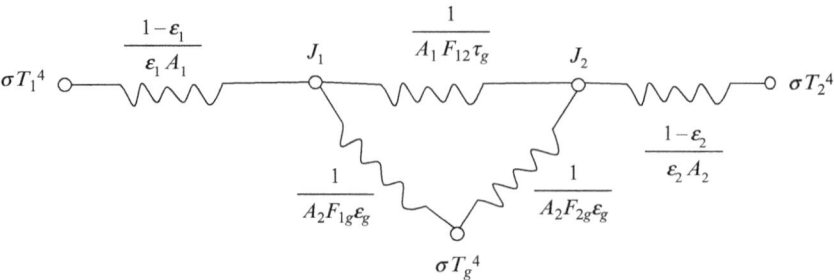

Fig. 3.17 Electric circuit analog for radiative exchange between two gray surfaces forming an enclosure filled with a participating medium (gray gas).

To illustrate the simplified approach of analyzing such situation we again consider the case of heat transfer between two gray surfaces forming an enclosure, but now in the presence of a participating medium, say a gas of overall emissivity ϵ_g, at a temperature T_g. The manner in which the equivalent circuit of Fig. 3.13 would be modified can be appreciated by visualizing how the heat transfer from/to the surfaces would now occur. The total radiation leaving surface 1, in the process of arriving at surface 2 would get attenuated due to absorptive power of the intervening gas. Thus the net amount that would reach surface 2 would be $J_1 A_1 F_{12}(\tau_g)$. Similar factor would apply to the total radiation leaving surface 2 and arriving at surface 1. Therefore, the net radiative exchange would be $(J_1 - J_2) A_1 F_{12} \tau_g$, which implies that the thermal resistance between radiosities J_1 and J_2 (Fig. 3.13) would now be $1/A_1 F_{12} \tau_g$.

Similarly considering the radiation exchange between the surface 1 and the gas, we can easily see that its value is $A_1 F_1 \epsilon_g (J_1 - \sigma T_g^4)$; while that between the surface 2 and the gas would be $A_2 F_{2g} \epsilon_g (J_2 - \sigma T_g^4)$. These can be incorporated in Fig. 3.13, by inclusion of additional resistances between nodes J_1 and J_2 and the gas node σT_g^4. This results in a new electric analog for this case, shown in Fig. 3.17. From this diagram we can calculate the net heat transfer between the surfaces and the gas as was done above using Fig. 3.16.

3.3.3 Convective Heat Transfer

To determine the performance of thermal equipment encountering heat transfer by convection the main task is to estimate the value of the heat-transfer coefficient, defined in Eq. (3.208). Theoretical estimation of this heat-transfer coefficient is rarely possible for most situations encountered in practice; we have therefore to rely on various empirical/semiempirical correlations. These correlations attempt to capture the variables which influence the heat-transfer coefficient through a few dimensionless numbers. Since the mechanism of convective heat transfer depends strongly on the nature of fluid motion and its interaction with the heat-transfer surface, the correlations also vary greatly depending on whether the flow is external or internal (confined); natural convection or forced convection, laminar or turbulent, and single-phase or two-phase flow. In this section we shall present the commonly employed correlations for each of these cases.

Forced Convection

In most thermal systems the fluid flow is caused by external agents—pumps, fans, or blowers, and their characteristics govern the velocity of the flow. This is termed as forced convection. In contrast, in natural convection, the fluid flow is induced by buoyancy forces arising from temperature difference in the neighborhood of a surface. As discussed in Section 3.2 the nature of flow in forced convection depends greatly on whether the flow is external to a surface or within a confined space like a tube, pipe, or duct (Fig. 3.6). The correlations for heat-transfer coefficient are therefore quite different in these two cases.

External Flow

As discussed earlier in Section 3.2, external flows are characterized by a slowly growing hydrodynamic boundary layer. Akin to it is the thermal boundary layer, the region near the surface where bulk of the temperature change occurs. This also grows in thickness, slowly as the fluid flows over the hot (or cold) surface. The nondimensional numbers that characterize these boundary layers are the Reynolds number and the Prandtl number, defined as:

$$Re_x = \frac{\rho U_\infty x}{\mu} \quad Pr = \frac{\mu C_p}{k} = \frac{\nu}{\alpha} \tag{3.240}$$

where x the characteristic length, and all other symbols have their usual meanings.

The heat-transfer coefficient is correlated through the dimensionless number, called the Nusselt number and defined as:

$$Nu_x = \frac{hx}{k} \tag{3.241}$$

The general form of heat-transfer correlations for external flow is

$$Nu_x = f(Re_x, Pr) \tag{3.242}$$

with the function f depending upon the geometric configuration as also whether the flow is laminar or turbulent (Section 3.2). Typical equations are enumerated below.

Isothermal Flat Plate Laminar flow ($Re_x < 5 \times 10^5$)

$$\text{for } Pr \approx 1 : Nu_x = 0.332 \quad Re_x^{1/2} Pr^{1/3} \tag{3.243}$$

$$\text{for } Pr \ll 1 : Nu_x = 0.565 \quad Re_x^{1/2} Pr^{1/2} \tag{3.244}$$

$$\text{for } Pr \gg 1 : Nu_x = 0.339 \quad Re_x^{1/2} Pr^{1/3} \tag{3.245}$$

Turbulent flow (from the leading edge itself)

$$Pr \approx 1: Nu_x = 0.0296\ Re_x^{0.8} \tag{3.246}$$

Cross Flow Over a Long Isothermal Cylinder for $Re_D\ Pr > 0.2$

$$(Nu_D)_{avg} = \frac{\bar{h}D}{k} = 0.3 + \frac{0.62Re_D^{1/2}Pr^{1/3}}{\left[1 + \left(\frac{0.4}{Pr}\right)^{2/3}\right]^{1/4}} \left\{1 + \left(\frac{Re_D}{282,000}\right)^{5/8}\right\}^{4/5} \tag{3.247}$$

where Re_D is the Reynolds number based on the diameter of the cylinder.

Cross Flow Across a Bank of Isothermal Cylinders

$$(Nu_D)_{avg} = C \cdot Re_{D,\max}^m Pr^{0.36} \left(\frac{Pr}{Pr_s}\right)^{1/4} \tag{3.248}$$

where values of C (varying between 0.03–1.04) and m (varying between 0.4–0.8) depend upon Reynolds number, and the tube bank configuration (i.e., aligned or staggered). These values can be obtained from the Handbook of Heat Transfer [15]. The Prandtl number term accounts for the influence of property change due to temperature difference between the tube surface and the free stream.

Internal Flow

In internal flows the two boundary layers, growing from the leading edge onward, are confined by the walls of the tube/duct, and soon the whole flow becomes boundary layer flow, Fig. 3.6B. Therefore in most practical situations bulk of the length of the tube/duct encounters this fully developed flow (Section 3.2). The characteristic length for these flows in the so-called hydraulic diameter defined by Eq. (3.126), and the commonly used equations to predict the heat-transfer coefficient are

Fully Developed Laminar Flow

$$Nu = 3.66 \quad \text{isothermal wall} \tag{3.249}$$
$$= 4.364 \quad \text{uniform wall heat flux} \tag{3.250}$$

Fully Developed Turbulent Flow Gnielinski correlation:

$$Nu_D = \frac{(f/8)(Re_D - 10^3)Pr}{1 + 12.7(f/8)^{1/2}(Pr^{2/3} - 1)} \tag{3.251}$$

where the friction factor f is obtained from the Moody's charts, or from Eq. (3.133) based on this data, or from Eq. (3.132). This equation gives accurate results ($\pm10\%$) in the range $0.5 \leq Pr \leq 10^6$ and $2300 \leq Re_D \leq 5 \times 10^6$.

Simpler alternatives suggested by Gnielinski are

$$Nu_D = 0.0214(Re_D^{0.8} - 100)Pr^{0.4} \tag{3.252}$$

$$\text{for} \quad 0.5 \leq Pr \leq 1.5 \quad 10^4 \leq Re_D \leq 5 \times 10^6$$

$$\text{and} \quad Nu_D = 0.012(Re_D^{0.87} - 280)Pr^{0.4} \tag{3.253}$$

$$\text{for} \quad 1.5 \leq Pr \leq 500 \quad 3 \times 10^3 \leq Re_D \leq 10^6$$

Natural (or Free) Convection

As mentioned above, in natural convection, the fluid flow is induced by buoyancy forces arising from temperature difference in the neighborhood of a surface. The direction of flow depends on whether the surface is hotter or colder than the surrounding fluid and its velocity will depend on the magnitude of the temperature difference. The dimensionless number, which replaces Reynolds number in natural convection, is the Grashof's number defined as

$$Gr_x = \frac{g\beta\Delta T x^3}{\nu^2} \tag{3.254}$$

where β is the coefficient of thermal expansion of the fluid; ν is dynamic viscosity; ΔT the temperature difference between the surface and the quiescent fluid; and x the characteristic length, which is measured from the leading edge in the direction of flow. The commonly used correlations for typical configurations are given below.

Vertical Flat Surfaces
Churchill and Chu Correlation [22]

$$(Nu)_{avg} = \left\{ 0.825 + \frac{0.387 Ra^{1/6}}{[1 + (0.492/Pr)^{9/16}]^{8/27}} \right\}^2 \tag{3.255}$$

$$Ra = Gr \cdot Pr \tag{3.256}$$

Both Nu and Gr are based on the height of the flat surface. This equation is for turbulent flow. The equation recommended for both laminar and turbulent regions is identical to Eq. (3.255) except that the coefficient in the denominator term is 0.437 instead of 0.492.

Simplified Correlations

$$(Nu)_{avg} = 0.59 Ra^{1/4} \quad 10^4 < Ra < 10^9 \quad \text{(laminar)} \tag{3.257}$$

$$= 0.10 Ra^{1/3} \quad 10^9 < Ra < 10^{13} \quad \text{(turbulent)} \tag{3.258}$$

Horizontal Surfaces

Here the heat-transfer coefficient is also strongly influenced by the direction and temperature of the surface, that is, whether it is hotter or cooler than the surroundings, and is facing up or down. This is because the presence of the plate obstructs flow, if the plate is hot and facing downward, or if it is cold and facing upward as shown in Fig. 3.18.

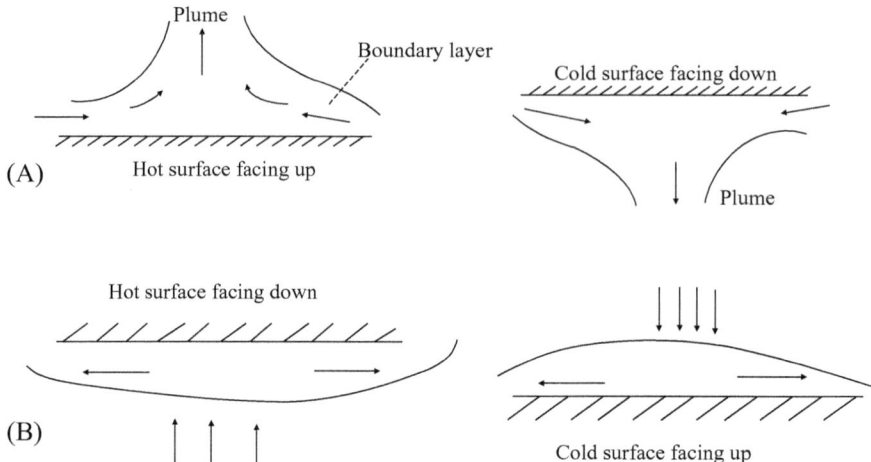

Fig. 3.18 Natural convection around horizontal surfaces.

For these situations, the classical correlations given by McAdams give reasonable prediction.

For the case (A), Fig. 3.18

$$(Nu)_{avg} = 0.54 Ra^{1/4} \quad \text{for} \quad 10^4 \leq Ra \leq 10^7 \tag{3.259}$$

$$= 0.15 Ra^{1/3} \quad \text{for} \quad 10^7 \leq Ra \leq 10^{11} \tag{3.260}$$

For the case (B), Fig. 3.18

$$(Nu)_{avg} = 0.27 Ra^{1/4} \quad 10^5 \leq Ra \leq 10^{10} \tag{3.261}$$

The characteristic length L in these correlations is defined as

$$L = \frac{\text{Area of the plane surface}}{\text{Perimeter of the surface}} \tag{3.262}$$

Thus, for a circular disc of diameter $D, L = D/4$.

The natural convection around a horizontal cylinder resembles that on a vertical surface, the only difference being that here the wall is curved, with cylinder diameter D being the height of the wall. The correlation of Churchill and Chu, for heat transfer from an isothermal horizontal cylinder is therefore quite similar to that for a vertical surface (i.e., Eq. 3.255)

$$(Nu)_{avg} = \left\{ 0.6 + \frac{0.387 Ra_D^{1/6}}{[1 + (0.559/Pr)^{9/16}]^{8/27}} \right\}^2 \tag{3.263}$$

$$\text{for} \quad 10^{-5} < Ra_D < 10^{12}$$

The classical simplified correlations of McAdams, also give reasonable prediction in their range of validity

$$(Nu_D)_{avg} = 0.53Ra^{1/4} \quad \text{for} \quad 10^4 < Ra < 10^9 \tag{3.264}$$

$$= 0.13Ra^{1/3} \quad \text{for} \quad 10^9 < Ra < 10^{12} \tag{3.265}$$

Heat Transfer With Phase Change

In many thermal equipment the fluid also undergoes change of phase due to heat transfer, for example, in boilers the liquid is converted to vapor phase and in condensers the vapor is condensed to liquid phase. As discussed in Section 3.2.2, phase change introduces considerable complexity in the flow, resulting in numerous flow patterns depending on the flow direction, the velocity, and the direction of heat transfer. We shall briefly discuss the equations governing heat transfer in both condensation and boiling.

Heat Transfer in Condensation

Condensation occurs when hot vapors come in contact with a surface at a temperature lower than the saturation temperature at the prevailing pressure. As condensation occurs, drops of condensate usually coalesce to form a thin film which flows down due to gravity. The rate of heat transfer is strongly influenced by the film thickness which, in turn, depends on the height and the shape of the surface, that is, whether it is a plane vertical surface, or the cylindrical external surface of a tube. For condensation inside tubes, the phenomenon is even more complex due to the differing velocities of the liquid and vapor phases which can give rise to complex flow patterns (as discussed in Section 3.2.2)

The average heat-transfer coefficient during film condensation on a vertical surface of height L is given by the Nusselt's equation:

$$h = 0.943 \left[\frac{\rho_l g (\rho_l - \rho_v) k_l^3 h_{fg}}{\mu_l (T_{sat} - T_w) L} \right]^{1/4} \tag{3.266}$$

where T_w is the wall temperature and T_{sat} the saturation temperature of the vapors condensing on it, all other symbols have their usual meanings.

This equation is recommended for laminar, nonwavy liquid film, which is characterized by Reynolds number (defined as follows) being less than 30.

$$Re_l = \frac{4\Gamma}{\mu_l}$$

where $\Gamma = \dot{m}_l/b$ = mass flow rate of condensate per unit breadth of the surface.

For $Re_l > 30$, the film becomes wavy and for $Re_l > 1800$, the film is turbulent. The average heat-transfer coefficient for a film (total length L) that may have wavy and turbulent regions can be calculated from the following correlation given by Chen et al. [16]:

$$\frac{h_{avg}}{k_l}\left(\frac{v_l^2}{g}\right)^{1/3} = [Re_l^{-0.44} + (5.82 \times 10^{-6})Re_l^{0.8}Pr_l^{1/3}]^{1/2} \tag{3.267}$$

For a more frequently encountered situation of condensation over a horizontal tube of diameter D, the average heat-transfer coefficient is given by a similar equation:

$$h = 0.729\left[\frac{\rho_l g(\rho_l - \rho_v)k_l^3 h_{fg}}{\mu_l(T_w - T_{sat})L}\right]^{1/4} \tag{3.268}$$

For condensation inside tubes, the flow pattern exerts a strong influence on the mechanism of heat transfer and therefore the heat-transfer coefficient changes with vapor fraction. For low flow rates, stratified flow occurs in horizontal tubes, and Chato's correlation (quite similar to the above correlation) gives a reasonable prediction:

$$(Nu)_{avg} = \frac{\bar{h}D}{k_l} = 0.555\left[\frac{\rho_l g(\rho_l - \rho_v)D^3 h_{fg}}{k_l \mu_l(T_{sat} - T_w)}\right]^{1/4} \tag{3.269}$$

for $Re_{vo} = \frac{GD}{\mu_v} < 35{,}000$, where G is the mass velocity of flow. When the flow rates are high enough to cause formation of a nearly symmetric annular liquid film with vapor core, the heat-transfer coefficient can be predicted using Shah's [17] correlation:

$$Nu = 0.023Re_l^{0.8}Pr_l^{0.4}\left[1 + \frac{3.8}{p_r^{0.38}}\left(\frac{x}{1-x}\right)^{0.76}\right] \tag{3.270}$$

$$\text{where}\quad Re_l = \frac{GD(1-x)}{\mu_l} \tag{3.271}$$

This correlation gives good results in the range:

$$0.002 < p_r\left(= \frac{P}{P_{crit}}\right) < 0.44 \quad 0 < x < 1$$

$$Re_{lo} = \frac{GD}{\mu_l} > 350 \quad Pr_l > 0.5$$

$$10.8\,\text{kg/s m}^2 < G < 1599\,\text{kg/s m}^2$$

Here the heat-transfer coefficient varies significantly as the condensation proceeds and x changes from a high value around 1, corresponding to dry vapor, to low values around 0 (for complete condensation $x = 0$).

Example 3.7

Refrigerant-22 is condensing inside a 10 mm diameter tube at a temperature of 40°C. Estimate the heat-transfer coefficient at positions where the vapor fraction is 0.2, 0.4, 0.6, 0.8, 0.9, and 0.95 using Shah's correlation. Given the mass flow rate is 0.0184 kg/s.

Solution

We first find the thermophysical properties of R-22 at 40°C

$$\rho_l = 1128.5 \text{ kg/m}^3; \quad \rho_v = \frac{1}{0.01511} \text{ kg/m}^3; \quad \text{Pressure} = 1.5336 \text{ M Pa}$$

$$\mu_l = 139.4 \times 10^{-6} \text{ Pa s}; \quad k_l = 76.9 \times 10^{-3} \text{ W/m K}; \quad C_{pl} = 1.339 \text{ kJ/kg K}$$

$$Pr_l = \frac{139.4 \times 10^{-6} \times 1.339 \times 10^3}{76.9 \times 10^{-3}} = 2.42726$$

$$\text{Reduced pressure} \quad p_r = \frac{1.5336}{4.99} = 0.307335$$

$$G = \frac{0.184}{\frac{\pi}{4} \times (0.01)^2} = 234.276 \text{ kg/s m}^2$$

The heat-transfer coefficient values at different positions along the condenser can now be calculated using above values in Eq. (3.270), in the Excel sheet. The results are tabulated below:

x	Re_e	Nu	h (W/m² K)
0.2	13,444.8	202.49	1557.1
0.4	10,083.6	281.05	2161.3
0.6	6722.4	344.10	2646.2
0.8	3361.2	392.43	3017.8
0.9	1680.6	406.81	3128.4
0.95	840.3	406.80	3128.3

Heat Transfer in Boiling

The mechanism of heat transfer in boiling is quite different from that in condensation due to the formation of bubbles. These bubbles break the boundary layer near the heating surface and can cause a local micro-pumping action, pushing the hot fluid in the proximity of the heating surface into the bulk of the liquid. In quiescent fluid conditions, as for example, in a fire tube boiler drum, these bubbling vapors govern the rate of heat transfer. In flow boiling, the contribution of bubbling can be overwhelmed by the high velocity of flow. Accordingly we have separate correlations to predict the heat transfer in these distinctly different situations.

In *pool boiling*, that is, where the fluid is not moving we may have different local flow conditions around the heating surface depending on the rate of bubble formation. In most engineering equipment, we encounter nucleate boiling where bubbles nucleate, grow, and eventually depart from the hot surface and induce local movement of fluid by natural convection.

Over the years numerous correlations have been developed to predict the heat-transfer coefficient in nucleate pool boiling. One of the simplest, yet giving reasonable results, is the correlation based on reduced pressure, given by Mostinski [18]:

$$h_{nb} = 0.00417 \, q^{0.7} \, p_{crit}^{0.69} F_p \tag{3.272}$$

$$\text{where} \quad F_p = 1.8 \, p_r^{0.17} + 4p_r^{1.2} + 10p_r^{10} \tag{3.273}$$

Here the heat transfer h_{nb} is in W/m^2 K, the heat flux q in W/m^2, and the critical pressure, p_{crit} in kPa.

A more recent correlation is that given by Gorenflo [19] wherein the ratio of actual heat-transfer coefficient to that under a reference condition is correlated to the ratios of heat flux, of surface roughness, and the reduced pressure:

$$\frac{h_{nb}}{h_{ref}} = F_{PF} \cdot (q/q_{ref})^n (Rp/Rp_{ref})^{0.133} \tag{3.274}$$

where

$$F_{PF} = 1.2p_r^{0.27} + 2.5p_r + \frac{p_r}{1 - p_r} \tag{3.275}$$

$$n = 0.9 - 0.3p_r^{0.3} \tag{3.276}$$

Here Rp is the surface roughness in μm; the reference surface roughness, Rp_{ref} being 0.4 μm (when Rp not known, its value is set equal to Rp_{ref}). The other reference conditions are taken as $Pr_{ref} = 0.1$, and $q_{ref} = 20{,}000$ W/m^2. The values of h_{ref} in W/m^2 K have been tabulated for a wide range of fluids in handbooks (see, e.g., [15, p. 657])—typical values being 5600 W/m^2 K for water, 4500 W/m^2 K for refrigerant 134a and 7000 W/m^2 K for ammonia.

For water, the equations for F_{PF} and n are different

$$F_{PF} = 1.73p_r^{0.27} + \left(6.1 + \frac{0.68}{1 - p_r}\right) p_r^2 \tag{3.277}$$

$$n = 0.9 - 0.3p_r^{0.15} \tag{3.278}$$

This method has been found to be accurate over a wide range of values of heat flux and pressure [$0.0005 \leq p_r \leq 0.95$].

Example 3.8

Determine the heat-transfer coefficient in pool boiling of water at atmospheric pressure as a function of heat flux using the Mostinski and the Gorenflo correlations.

Solution

Taking atmospheric pressure as 1.013 bar, and the critical pressure as $P_{crit} = 220.64$ bars we get

$$p_r = \frac{1.013}{220.64} = 0.0045912$$

while the Prandtl number is $Pr = \dfrac{\mu C_p}{k} = \dfrac{281.7 \times 10^{-6} \times 4216}{0.6791} = 1.749$

Mostinski correlation

$$F_p = 1.8p_r^{0.17} + 4p_r^{1.2} + 10p_r^{10}$$
$$= 0.720775 + 0.006257 + 0$$
$$= 0.727032$$

Eq. (3.272) reduces to

$$h_{nb} = 0.00417q^{0.7}(220.64 \times 100)^{0.69}(0.727032) \quad \text{here } p_{cr} \text{ has to be taken in kPa}$$
$$= 3.01184 \, q^{0.7}$$

Gorenflo correlation (taking $R_p \approx R_{pref}$)

$$\frac{h_{nb}}{h_{ref}} = F_{pf} \left(\frac{q}{q_{ref}} \right)^n \quad (1)$$

for water $h_{ref} = 5600 \, \text{W/m}^2 \quad q_{ref} = 20,000 \, \text{W/m}^2$

$$n = 0.9 - 0.3p_r^{0.15} = 0.766214$$

$$F_{pf} = 1.73p_r^{0.27} + \left(6.1 + \frac{0.68}{1 - p_r} \right) p_r^2$$
$$= 0.404358 + 0.000142983$$
$$= 0.4045$$

Substituting the values of various terms, the correlation simplifies to:

$$h_{nb} = 5600(0.4045)(q/20,000)^{766,214} \quad\quad (3.279)$$
$$= 1.14709q^{0.766214}$$

The heat-transfer coefficient values as predicted by the two correlations for a few values of q are tabulated below

q (W/m²)	h_{nb} (W/m²k)	
	Mostiski	Gorenflo
5000	1169.8	769.8
10,000	1900.3	1307.5
15,000	2524.0	1782.5
20,000	3087.1	2220.8
25,000	3609.0	2633.7
30,000	4100.3	3027.4

It can be seen that there is considerable difference between the values predicted by the two correlations, Gorenflo's correlation predicting values about 30% lower than the values obtained by Mostinski correlation.

In *forced convection boiling* as, for example, would occur inside the tubes of a water-tube boiler, the rate of heat transfer is influenced both by the nucleate boiling (NB) as also the forced convection vaporization (C). The correlations to predict the heat-transfer coefficient have to incorporate both these influences judiciously. The earliest correlations presumed that the contributions of both these mechanisms are additive. Thus

$$h_{TP} = h_C + h_{NB} \tag{3.280}$$

The convective boiling coefficient h_c is calculated using the traditional single-phase flow Dittus-Boelter correlation considering only the liquid portion of the two-phase stream:

$$h_C = 0.023 \left[\frac{G(1-x)D}{\mu} \right]^{0.8} Pr_l^{0.4} \; \frac{k_l}{D} F \tag{3.281}$$

where F is an empirically determined function of the Martinelli's parameter X_{tt} defined as:

$$X_{tt}^{-1} = \left(\frac{x}{1-x} \right)^{0.9} \left(\frac{\rho_l}{\rho_g} \right)^{0.5} \left(\frac{\mu_g}{\mu_l} \right)^{0.1} \tag{3.282}$$

The nucleate boiling component h_{NB} is determined from the equation:

$$h_{NB} = 0.00122 \left[\frac{k_l^{0.79} C_{pl}^{.45} \rho_l^{0.49}}{\sigma^{0.5} \mu_l^{0.29} h_{fg}^{0.24} \rho_g^{0.24}} \right] \Delta T_{sat}^{0.24} \Delta P_{sat}^{0.75} S \tag{3.283}$$

Chen gives curves relating factors F and S to X_{tt} and Re_{TP}, the two-phase Reynolds number. These curves are usually approximated by following equations [20]:

$$F = 1 \qquad\qquad\qquad \text{for } X_{tt}^{-1} \leq 0.1$$
$$\quad = 2.35\left(0.213 + X_{tt}^{-1}\right)^{0.736} \quad \text{for } X_{tt}^{-1} > 0.1 \tag{3.284}$$

$$S = \frac{1}{1 + 2.53 \times 10^{-6} Re_{TP}^{1.17}} \tag{3.285}$$

where $\quad Re_{TP} = Re_l F^{1.25}$ $\qquad\qquad\qquad\qquad\qquad\qquad$ (3.286)

$$Re_l = \left[\frac{G(1-x)D}{\mu_l}\right]; \Delta T_{sat} = \text{wall temperature} - \text{saturation temperature;}$$

$\Delta P_{sat} = $ saturation pressure at wall temperature $-$ pressure

However, newer correlations generally compare the values of heat-transfer coefficient as calculated by formulae for nucleate boiling and forced convective vaporization and choose the larger of the two values. One such general correlation, based on a wide range of data, and applicable for both horizontal and vertical flows has been given by Shah [21]:

Choose h as the largest of the values given by following equations:

$$h/h_f = 230 \, Bo^{0.5} \qquad \text{for } Bo > 3 \times 10^{-5} \tag{3.287}$$
$$\quad = 1 + 46 \, Bo^{0.5} \quad \text{for } Bo < 3 \times 10^{-5} \tag{3.288}$$
$$h/h_f = 1.8 \, [Co(0.38 \, Fr_l^{-0.3})^n]^{-0.8} \tag{3.289}$$
$$h/h_f = F \exp\{2.47[Co(0.38 \, Fr_l^{-0.3})^n]^{-0.15}\} \tag{3.290}$$
$$h/h_f = F \exp\{2.74[Co(0.38 \, Fr_l^{-0.3})^n]^{-0.1}\} \tag{3.291}$$

where h_f is calculated using Gnielinski correlation, Eq. (3.251) and

$$F = 14.7 \, Bo^{0.5} \quad \text{if } Bo > 0.0011 \tag{3.292}$$
$$\quad = 15.43 \, Bo^{0.5} \quad \text{if } Bo < 0.0011 \tag{3.293}$$

$n = 0$, for vertical tube and

$$n = 0 \quad \text{for horizontal tube if } Fr_l > 0.04 \tag{3.294}$$
$$\quad = 1 \quad \text{otherwise } \{\text{i.e., for } Fr_l < 0.04\} \tag{3.295}$$
$$Co = \left(\frac{1-x}{x}\right)^{0.8} \left(\frac{\rho_v}{\rho_l}\right)^{0.5} \tag{3.296}$$
$$Bo = \frac{q}{Gh_{fg}} \tag{3.297}$$
$$Fr_l = G^2/(\rho_l^2 g d_i) \tag{3.298}$$

Another more recent correlation is that given by Kandilkar [23]. Here too, the heat-transfer coefficient is taken as the larger of the values given by following equations:

1. *Nucleate-boiling dominated region*

$$\frac{h}{h_f} = 0.6683 \left(\frac{\rho_l}{\rho_v}\right)^{0.1} \left(\frac{x}{1-x}\right)^{0.16} f_2(Fr_l) + 1058 \, Bo^{0.7} F_{fl} \tag{3.299}$$

2. *Convective boiling dominated region*

$$\frac{h}{h_f} = 1.136 \left(\frac{\rho_l}{\rho_v}\right)^{0.45} \left(\frac{x}{1-x}\right)^{0.72} f_2(Fr_l) + 667.2 Bo^{0.7} F_{fl} \tag{3.300}$$

where h_f is calculated using modified Gnielinski correlation, that is

$$h_f = \frac{Re_l Pr_l (f/8)(k_l/D)}{1.07 + 12.7(Pr_l^{2/3} - 1)(f/8)^{0.5}} \tag{3.301}$$

$$\text{for} \quad 10^4 \leq Re_e \leq 5 \times 10^6$$

$$\text{and} \quad = \frac{(Re_l - 1000) Pr_l (f/8)(k_l/D)}{1 + 12.7(Pr_l^{2/3} - 1)(f/8)^{0.5}} \tag{3.302}$$

$$\text{for} \quad 2000 \leq Re_e \leq 10^4$$

$$\text{where} \quad f = [0.79 \ln(Re_l) - 1.64]^{-2} \tag{3.303}$$

$$Re_l = \frac{G(1-x)D}{\mu_l} \tag{3.304}$$

$$\text{and} \quad f_2(Fr_l) = (25 \, Fr_l)^{0.3} \quad \text{when } Fr_l < 0.04 \text{ in horizontal tube}$$

$$= 1, \qquad \text{otherwise} \tag{3.305}$$

F_{fl} is a fluid and tube material-specific coefficient whose values can be seen from a handbook. Typical values are

$$F_{fl} = 1.0 \quad \text{for water boiling in the copper tube, and}$$

$$\text{for all fluids with stainless steel}$$

$$= 2.2 \quad \text{for refrigerant R-22 in copper tubes}$$

$$= 1.63 \quad \text{for refrigerant R-134a in copper tubes}$$

In forced convective boiling the heat-transfer coefficient varies very significantly with vapor function x. Therefore boilers and refrigerant evaporators need to be analyzed as heat exchangers with the overall transfer coefficient varying along the length (Section 4.1.4). Further, since the local heat flux values (and hence Bo values) are generally not known *a priori*, but depend upon the values of the heat-transfer coefficients on the two sides of the heat-transfer surface, iterations are necessary to determine the local boiling heat-transfer coefficient in these equipment.

Example 3.9

Determine the heat-transfer coefficient at various positions along the vertical tubes of a boiler under following conditions.

> Tube material: AISI 304 SS
> Tube ID: 45 mm
> Tube thickness: 2.5 mm
> Water flow rate at entrance: 0.4 kg/s
> Pressure: 12 bar
> Temperature difference between the wall and water: 15°C

Calculate the values of the heat-transfer coefficient for x varying from 0.05 to 0.5 by different methods.

Solution

At 12 bar pressure the saturation temperature of water is 188°C. At this temperature various thermophysical properties as obtained from ASHRAE Handbook of Fundamentals are

$$\rho_l = 878.3 \, \text{kg/m}^3 \qquad C_{pl} = 4438 \, \text{J/kg K} \qquad \mu_l = 143.4 \times 10^{-6} \, \text{Pa s}$$
$$k_l = 669.7 \times 10^{-3} \, \text{W/m K} \quad \sigma = 41 \times 10^{-3} \, \text{N/m} \quad \mu_v = 15.3 \times 10^{-6} \, \text{Pa s}$$
$$\rho_v = 6.13 \, \text{kg/m}^3 \qquad h_{fg} = 1985.2 \, \text{kJ/kg} \qquad T_{sat} = 461 \, \text{K}$$

It is given that $\Delta T = 15°C$. This implies that the wall temperature is $= 461 + 15 = 476°K$. The corresponding ΔP_{sat} is therefore

$$\Delta P_{sat} = [\text{Saturation pressure at } 476°K - 12 \text{ bar}] = 16.565 - 12 = 4.565 \text{ bar}$$

Using this data in Eqs. (3.280)–(3.305) we can determine the heat-transfer coefficient for different values of x as predicted by the correlations given by Chen, Kandilkar, and Shah. The calculations are conveniently done using Excel software and the results are summarized in following table.

Chen

x	Re_l	$1/X_{tt}$	F	h_c	Re_{tp}	S	h_{nb}	h_{TP}
0.05	74,977.7	0.676	2.155	5741.0	195,800.1	0.2028	9419.9	15,160.9
0.1	71,031.5	1.325	3.225	8227.8	307,032.0	0.1306	6068.6	14,296.4
0.2	63,139.1	2.748	5.225	12,129.6	498,753.8	0.0785	3646.4	15,776
0.3	55,246.7	4.464	7.314	15,259.8	664,527.8	0.0574	2666.1	17,925.9
0.4	47,354.3	6.644	9.693	17,876.6	809,901.7	0.0461	2140.6	20,017.2
0.5	39,461.9	9.570	12.591	20,069.3	935,926.4	0.0392	1820.4	21,889.7

Kandilkar

q	$Bo \times 10^3$	$Fr_l \times 10^2$	f	h_f	h/hf_{nb}	htp_{nb}
227,413.8	0.455	0.19	1.914	2418.2	5.53	13,377.0
214,446.0	0.430	0.19	1.937	2318.5	5.43	12,574.9
236,639.9	0.474	0.19	1.988	2115.9	5.86	12,404.8
268,888.8	0.539	0.19	2.049	1908.3	6.41	12,228.6
300,257.5	0.601	0.19	2.122	1694.9	6.92	11,721.5
328,345.6	0.658	0.19	2.214	1474.2	7.37	10,858.0

h/hf_c	htp	Final h_{TP}
4.3296	10,469.8	13,377
5.1139	11,856.6	12,574.9
7.0525	14,922.3	14,922.3
9.2004	17,557.2	17,557.2
11.6351	19,720.0	19,720.0
14.5609	21,466.1	21,466.1

Shah

Co	htp_a	htp_b	F	htp_c	htp_d	Final h_{TP}
0.88087	11,870.05	4817.64	0.32931	9872.0	12,771.7	12,771.68
0.48451	11,051.5	7451.43	0.31978	11,639.5	14,107.6	14,107.6
0.25325	10,594.72	11,426.66	0.33592	14,785.2	16,476.8	16,476.8
0.16455	10,185.66	14,550.90	0.35808	17,409.9	18,193.9	18,193.9
0.11555	9559.48	17,146.73	0.37839	19,489.9	19,215.5	19,489.9
0.08354	8695.30	19,333.54	0.39569	21,017.8	19,551.6	21,017.8

It should be noted that while Chen's correlation directly gives the value of the two-phase heat-transfer coefficient, in other correlations the final value is selected as the largest of the values predicted by correlations applicable in different regimes.

On comparing the values predicted by these three correlations, it is seen that the prediction of the correlations given by Kandilkar and Shah are in reasonable agreement with each other, but differ considerably from those of Chen's correlation at the lowest value of the dryness fraction.

3.4 MASS TRANSFER

In many thermal systems we encounter transfer of mass along with transfer of heat. Typical examples are the cooling and dehumidifying coils of air-conditioning plants, the cooling towers, solid and liquid desiccant-based air-conditioning systems, etc. To carry out the analysis of such equipment involving simultaneous heat and mass transfer, a basic understanding of the phenomenon of mass transfer is needed. The transport of chemical species between a stream and a surface occurs when the concentration of the species in the stream is different from that on the surface. Thus, just as temperature in the driving potential of heat transfer, concentration of a species is the driving potential for mass transfer. Further, akin to conduction and convection heat transfer, we have diffusive and convective mass transfer. Even their governing equations are quite similar.

Thus the *diffusive mass transfer* is governed by the following equation:

$$\text{Diffusive mass transfer} \quad j_1 = -D_{12} \frac{dC_1}{dx} \tag{3.306}$$

where j_1 is the diffusive mass flux in $(kg/m^2 \ s)$ of species 1, C_1 the concentration in (kg/m^3), and D_{12} is diffusivity in (m^2/s), both of species 1, diffusing through a mixture of species 1 and 2.

The similarity of this governing equation, called the Fick's law of diffusion, with Fourier's law of heat conduction, Eq. (3.206), is quite obvious. The above equation is sometimes written in the form of mass fraction of species, m_{fi}, as

$$j_1 = -\rho D_{12} \frac{dm_{f1}}{dx} \tag{3.307}$$

$$\text{where} \quad m_{f1} = \frac{m_1}{m} \quad \text{and} \quad \rho = \frac{m}{V}$$

m being total mass of species 1 and 2 in volume V, their individual masses being m_1 and m_2, respectively; ρ is the local density of the mixture.

The *Convective mass transfer* is governed by the equation (Fig. 3.19):

$$j_{1,s} = \rho h_m (m_{f1,s} - m_{f1,\infty}) \tag{3.308}$$

where $j_{1,s}$ is the mass flux of species 1 leaving the surface; $m_{f1,s}$ and $m_{f1,\infty}$ are the mass fractions of species 1 at the surface and in the free stream, respectively; and ρ is the density of the mixture. The constant h_m is called the mass transfer coefficient.

The similarity of this equation with the basic equation for convective heat transfer, Eq. (3.208) is quite evident. This similarity is the basis of analogy between heat and mass transfer which permits conversion of all the convective heat-transfer coefficient correlations into correlations for convective mass transfer by substituting the Nusselt and

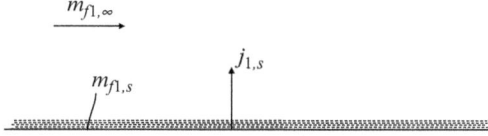

Fig. 3.19 Convective mass transfer.

Prandtl numbers by Sherwood and Schmidt numbers (Sh and Sc), respectively, which are also defined by analogy as follows:

$$Nu = \frac{hL}{k} \quad Sh = \frac{h_m L}{D} \tag{3.309}$$

$$Pr = \frac{\nu}{\alpha} \quad Sc = \frac{\nu}{D} \tag{3.310}$$

The Reynolds number remains unaltered while the Grashof's number is modified as follow.

$$Gr = \frac{g\beta \Delta T L^3}{\nu^2} \quad \text{(heat transfer)} \tag{3.311}$$

$$Gr = \frac{g(\Delta\rho/\rho)L^3}{\nu^2} \quad \text{(mass transfer)} \tag{3.312}$$

Here $\Delta\rho$ represents the density difference, which in mass transfer is due to concentration difference. In natural convective heat transfer, $\beta\Delta T$ term represents $(\Delta\rho/\rho)$, where the change in density is caused by the temperatures difference, ΔT.

We shall illustrate this analogy with a simple example.

Example 3.10

Determine the rate of evaporation of water from a thin film of water left on a roof tile. The pertinent data are

Tile size : 25 cm × 15 cm   Air velocity : 15 kmph
Temperature of air and water film : 25°C   Relative humidity of air : 50%

Solution

As the tiles are usually laid over each other as shown in Fig. 3.20, we assume the flow of air over a tile akin to the idealized situation of laminar flow over a flat plate. The local convective heat-transfer coefficient is given by Eq. (3.243), that is

$$Nu_x = 0.332Re_x^{1/2}Pr^{1/3}$$

and the average heat-transfer coefficient over a length L can be obtained by integration as:

$$(Nu)_{avg} = 0.664Re_L^{1/2}Pr^{1/3}$$

Using the analogy between heat and mass transfer, the equation for calculating the average mass transfer coefficient can be written as:

$$(Sh)_{avg} = 0.664Re_L^{1/2}Sc^{1/3}$$

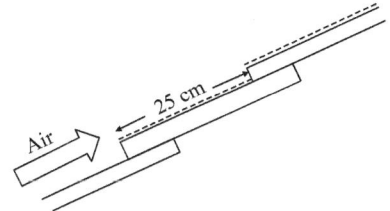

Fig. 3.20 Air flowing over a wet tile.

Now the properties of dry air at 25°C (as taken from the tables) are

$$\rho = 1.185 \, \text{kg/m}^3 \quad \nu = 1.55 \times 10^{-5} \, \text{m}^2/\text{s}$$

while for the water vapor at 25°C

$$D = 2.88 \times 10^{-5} \, \text{m}^2/\text{s} \quad \text{and} \quad Sc = 0.6$$

The Reynolds number can now be calculated as

$$Re_L = \frac{(15 \times 1000/3600) \times 0.25}{1.55 \times 10^{-5}} = 67{,}204$$

which being less than 5×10^5 confirms the initial assumption of laminar flow. Substituting this in the above equation, average value of the mass transfer coefficient can be found:

$$(Sh)_{avg} = 0.664(67{,}204)^{0.5}(0.6)^{1/3}$$

$$= 145.18$$

$$(h_m)_{avg} = \frac{145.18 \times 2.88 \times 10^{-5}}{0.25} = 0.0167 \, \text{m/s}$$

The rate of evaporation of water can now be determined using Eq. (3.308), as

$$\text{Evaporation rate} = j \times (\text{Area of the tile})$$

$$j = 1.185 \times 0.0167 \times (m_{f1,s} - m_{f1,\infty})$$

The two mass fractions can be related to the humidity ratio w, using Eq. (3.80) as:

$$m_{f1,s} = \frac{w_s}{1 + w_s} \quad m_{f1,\infty} = \frac{w_\infty}{w_\infty + 1}$$

and the w's can be found using Eq. (3.83).

The partial pressure of water vapor in the air, just touching the water film is equal to the saturation pressure of water vapor at 25°C, which is 3169 N/m². Taking atmospheric pressure as 1.0133 bar or 101,330 N/m², we can get

$$w_s = 0.622 \times \frac{3169}{101{,}330 - 3169} = 0.02008 \, \text{kg/kg da}$$

and therefore $m_{f1,s} = 0.019685$ kg/kg moist air.

The partial pressure of water vapor in air can be found using Eq. (3.81) as

$$P_{v,\infty} = 0.5 \times 3169 = 1584.5\,\text{N/m}^2$$

This gives $w_\infty = 0.622 \times \dfrac{1584.5}{101{,}330 - 1584.5} = 0.00988\,\text{kg/kg da}$

and therefore $m_{f1,\infty} = 0.009784\,\text{kg/kg moist air}$

Substituting values we get

$$\text{Evaporation rate} = \{1.185 \times 0.0167 \times (0.019685 - 0.009784)\} \times \{0.15 \times 0.25\}$$
$$= 7.348 \times 10^{-6}\,\text{kg/s}$$
$$= (0.02645\,\text{kg/h})$$

Hence, the rate of evaporation of water from each tile is $0.02645\,\text{kg/h}$.

The other correlations for convective heat transfer given in Section 3.3.3 can also be similarly transformed into correlations for convective mass transfer.

3.4.1 Simultaneous Heat and Mass Transfer

A situation of considerable practical importance, often encountered in thermal systems, is that of air flowing over a wet surface at a temperature different from air temperature. Here, besides the convective heat transfer, convective mass transfer also occurs. The latent heat of evaporation, needed to convert the liquid water into water vapor is taken from the liquid film. Under steady-state conditions, the total heat transfer from the surface over which the liquid film is formed, is thus given by

Total heat transfer = Sensible heat transfer due to convection +
Latent heat transfer required to evaporate
water

Thus considering air flowing over a wetted surface, Fig. 3.21, the total heat transfer dq_T through an infinitesimal area dA can be written as

$$dq_T = h_c\,dA(T_i - T_{a\infty}) + h_{fg} \cdot h_m \rho_{a\infty} dA(w_i - w_{a\infty}) \tag{3.313}$$

where the mass fraction terms in Eq. (3.308) have been replaced by humidity ratio whose values are almost equal to these (see Example 3.10). The terms with subscript ∞ indicate the properties of free air stream, and those with subscript i indicate the properties of saturated air at the liquid-film-air interface.

The relationship between the heat and mass transfer coefficients is given by the dimensionless number, Lewis number, defined as:

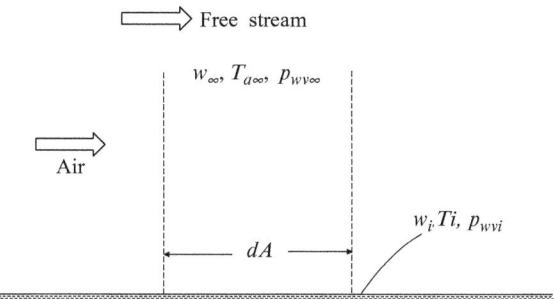

Fig. 3.21 Simultaneous heat and mass transfer.

$$Le = \frac{h_c}{h_m \rho_{a\infty} C_{pa\infty}} \tag{3.314}$$

Now, for air-water vapor mixture, under usual conditions encountered in practice, it is seen that $Le \approx 1$. This gives

$$h_m = \frac{h_c}{\rho_{a\infty} C_{pa\infty}} \tag{3.315}$$

Substituting this in Eq. (3.313) and simplifying, we get

$$dq_T = \frac{h_c \, dA}{C_{pa\infty}} \left[C_{pa\infty}(T_i - T_{a\infty}) + h_{fg}(w_i - w_{a\infty}) \right] \tag{3.316}$$

It can be shown, by rearranging terms and introducing the definition of moist air enthalpy, Eq. (3.84), that the term within the square brackets above equals the difference in the enthalpies of the saturated air at the interface and the free stream air, that is

$$dq_T = \frac{h_c \, dA}{C_{pa\infty}} (h_i - h_{a\infty}) \tag{3.317}$$

The quantity $(h_i - h_{a\infty})$ is called the *Enthalpy Potential*. It determines the direction of overall heat transfer between air and the wet surface, with which it is in contact.

Eq. (3.317) is the basic equation used in the analysis of cooling towers and cooling and dehumidifying coils.

Often, to calculate the convective mass transfer rate of water evaporating from a wet surface, Eq. (3.315) is combined with the basic Eq. (3.308) to get an expression for mass flux in terms of the surface heat-transfer coefficient:

$$j_{1,s} = \frac{h_c}{C_{pa\infty}} (m_{f1,s} - m_{f1,\infty}) \tag{3.318}$$

This equation is used to find the change of moisture content of air as it passes through equipment having simultaneous heat and mass transfer (see Example 4.11).

REFERENCES

[1] P.L. Dhar, Engineering Thermodynamics: A Generalized Approach, Elsevier, New Delhi, 2008 (revised reprint 2014).

[2] B.S. Petukhov, Heat transfer and friction in turbulent pipe flow with variable physical properties, in: J.P. Hartnett, T.F. Irvine (Eds.), Advances in Heat Transfer, vol. 6, Academic Press, New York, 1970.

[3] S.E. Haaland, Simple and explicit formulas for the friction factor in turbulent pipe flow, J. Fluids Eng. 105 (1983) 89–90.

[4] S.W. Churcuill, Friction-factor equation spans all fluid flow regimes, Chem. Eng. 84 (24) (1977) 91–92.

[5] F.M. White, A new integral method for analyzing the turbulent boundary layer with arbitrary pressure gradients, J. Basic Eng. 91 (1969) 371–378.

[6] A. Zukauskas, R. Ulinskas, Efficiency parameters for heat transfer in the tube banks, Heat Transfer Eng. 6 (2) (1985) 19–25.

[7] K.K. Robinson, D.E. Brigss, Pressure drop of air flowing across triangular pitch banks of finned tubes, Chem. Eng. Prog. Symp. Ser. 62 (64) (1966) 177–184.

[8] S.Z. Rouhani, Z. Axelsson, Calculation of void volume fraction in the subcooled and quality boiling regions, Int. J. Heat Mass Transfer 13 (1970) 383–393.

[9] D. Steiner, Heat Transfer to Boiling Saturated Liquids. VDIWarmeatlas (VDI Heat Atlas), Verein Deutscher Ingenieure, VDI-Gesellschaft Verfahrenstechnik und Chemieingenieurwesen (GCV), Dusseldorf, Germany, 1993.

[10] A.J. Ghajar, C.C. Tang, Advances in void fraction, flow pattern maps and non-boiling heat transfer two-phase flow in pipes with various inclinations, Adv. Multiphase Flow Heat Transfer 1 (2009) 1–52.

[11] R. Grönnerud, Investigation of liquid holdup, flow resistance and heat transfer in circulation type evaporators, part IV: two phase flow resistance in boiling refrigerants, Annexe 1972-1, Bull. Inst. Froid (1979).

[12] H. Müller-Steinhagen, K. Heck, A simple friction pressure drop correlation for two phase flow in pipes, Chem. Eng. Prog. 20 (1986) 297–308.

[13] M.B.O. Didi, N. Kattan, J.R. Thome, Prediction of two phase pressure gradients of refrigerants in horizontal tubes, Int. J. Refrig. 25 (2002) 935–947.

[14] A.F. Mills, Heat and Mass Transfer, Irwin, Chicago, 1995.

[15] A. Bejan, A.D. Kraus, Heat Transfer Handbook, John Wiley, NY, 2003.

[16] S.L. Chen, F.M. Gerner, C.L. Tien, General film condensation correlations, Exp. Heat Transfer 1 (1987) 93–107.

[17] M.M. Shah, A general correlation for heat transfer during film condensation inside pipes, Int. J. Heat Mass Transfer 22 (1979) 547–556.

[18] I.L. Mostinski, Application of the rule of corresponding states for calculation of heat transfer and critical heat flux, Teploenergetika 4 (1963) 66.

[19] D. Gorenflo, Pool boiling, in: VDI Heat Atlas, VDI-Verlag, Düsseldorf, Germany, 1993.

[20] https://www.thermalfluidscentral.org/encyclopedia/index.php/   Heat_Transfer_Predictions_for_Forced_Convective_Boiling.

[21] M.M. Shah, A new correlation for saturated boiling heat transfer: equations and further study, ASHRAE Trans. 88 (1) (1982) 185–196.

[22] S.W. Churchill, H.H.S. Chu, Correlating equations for laminar and turbulent free convection from a vertical plate, Int. J. Heat Mass Transfer, 18 (11) (1975) 1323–1329.

[23] S.G. Kandilkar ed, Handbook of phase change: Boiling and condensation, Taylor and Francis, Philadelphia, 1999.

CHAPTER 4

Modeling of Thermal Equipment

In this chapter we shall discuss the methodology of developing detailed models for simulating the performance of typical thermal equipments like heat exchangers; cooling and dehumidifying coils; cooling towers; absorbers and regenerators for desiccant solutions; reciprocating and rotating equipments like compressors and internal combustion engines, etc. Most of these equipments involve fluids exchanging heat and mass with each other, sometimes also receiving or absorbing some work. Therefore, the main objective of simulation is to predict the rates of heat transfer, mass transfer, pressure drop, and work transfer. The conservation equations like the steady flow energy equation (3.7) and the mass conservation equation, usually form the starting point of model development. When the fluids are separated from each other by a metallic surface the rate of heat transfer between the fluids is calculated using the combined conduction and convection heat-transfer equation (3.225). The mass transfer is usually by convection; it is therefore estimated using Eq. (3.308). In equipment involving simultaneous heat and mass transfer, the concept of enthalpy potential (Section 3.4.1) is invoked to determine the rate of total heat transfer. Since the temperature and the species concentration in the fluids change along the direction of flow, we write the basic conservation equations for an infinitesimal section of the equipment, and then integrate these equations, either analytically or numerically, to find the overall heat and mass transfer rates. As discussed in Section 3.2, the pressure drop has many components: the pressure drop due to sudden area change, acceleration of flow, and friction. The largest contribution usually arises from friction in the pipes, which is calculated using the Darcy equation (3.128). We shall also illustrate the application of these basic principles to the development of models for other applications involving heat and mass transfer like cooling of electronic equipment, manufacturing processes, heat treatment, dehydration of foods, etc.

4.1 HEAT EXCHANGERS

A heat exchanger enables transfer of heat between two fluids without allowing them to come in direct contact with each other. This is an ubiquitous component of thermal systems which operate over a wide range of operating conditions with a variety of working fluids ranging from the gaseous products of combustion leaving the furnace of a

Thermal System Design and Simulation
http://dx.doi.org/10.1016/B978-0-12-809449-5.00004-8

147

thermal power plant at 1000°C, to liquid helium at 4 K being produced in a refrigeration plant. The specific designs of heat exchangers, therefore, vary greatly. However, from the perspective of the methodology of calculation of heat-transfer rate, these are usually classified into three generic categories based on the relative directions of flow of the two fluids, that is, parallel-flow, counter-flow, and cross-flow configuration (see Fig. 4.1).

In most industrial heat exchangers the flow configuration is usually much more complex; it can be visualized as a mixture of *a* and *c*, or of *b* and *c*, or even a combination of *a*, *b*, and *c*. In the configurations shown in Fig. 4.1, both the fluids move across the heat exchanger only once. However, in industrial equipment, the fluids may move forward and backward many times. This is termed as multipass configuration. In the heat exchanger of Fig. 4.2, for example, one of the fluids is flowing in single pass, as before, while the other fluid has two passes. As a result of this multipassing, the fluids have different relative flow directions in different passes; thus in Fig. 4.2 in the lower pass the two fluids are flowing in the same direction, while in the upper pass these are flowing in opposite directions. We shall first develop the basic procedure for modeling of heat transfer in single-pass parallel and counter-flow heat exchangers and then extend these to more complex configurations.

4.1.1 Single-Pass Parallel and Counter-Flow Heat Exchangers

Fig. 4.1A and B shows two of the possible modes of operation of a single-pass heat exchanger. The corresponding variation in the temperatures of the two fluids along the

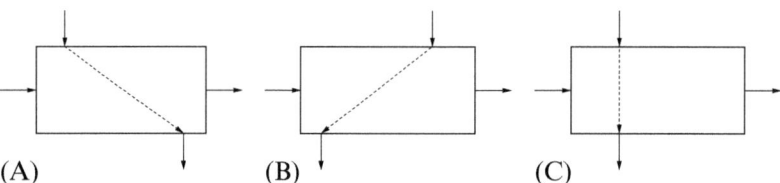

(A)   (B)   (C)

Fig. 4.1 Heat exchanger configurations. (A) Parallel flow. (B) Counter flow. (C) Cross flow.

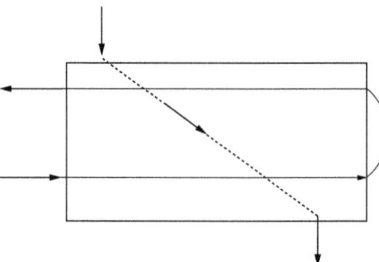

Fig. 4.2 Multipass configuration.

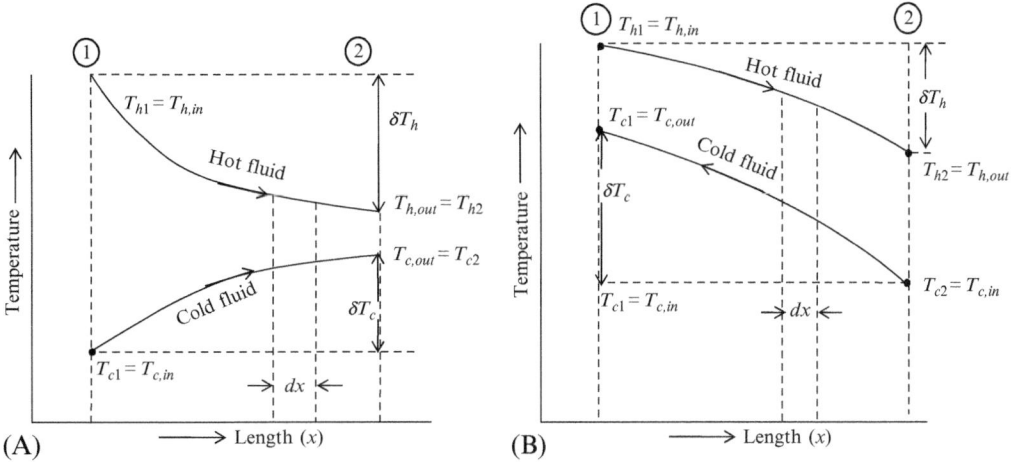

Fig. 4.3 Temperature variation along the heat exchanger length. (A) Parallel flow heat exchanger. (B) Counter flow heat exchanger (CFHE).

length of the heat exchangers is shown in Fig. 4.3A and B. It is evident from these figures that the temperature difference between the two fluids varies along the heat exchanger length.

To find the total rate of heat transfer we would therefore need to first write the equation for heat transfer through an infinitesimal length, and then integrate the expression suitably. If the surface area across which heat transfer takes place in the infinitesimal length of the heat exchanger, dx, is represented by dA, we can use Eq. (3.227), to get an expression for dQ, the rate of heat transfer across the area:

$$dQ = U dA (T_h - T_c) \tag{4.1}$$

where T_h and T_c are the local temperatures of the hot and the cold fluids, respectively.

In accordance with the energy conservation principle this heat transfer would reduce the temperature of the hot fluid and increase the temperature of the cold fluid. The magnitudes of these temperature changes are given by following equations.

$$|dT_h| = \frac{dQ}{C_h} \tag{4.2}$$

$$|dT_c| = \frac{dQ}{C_c} \tag{4.3}$$

with

$$C_h = \dot{m}_h c_h \quad \text{and} \quad C_c = \dot{m}_c c_c \tag{4.4}$$

where $\dot{m}$ represents the mass flow rate, and c, the specific heat of the fluid; subscript h refers to the properties of the hot fluid and subscript c to the properties of the cold fluid.

Now, as is clear from Fig. 4.3A and B the contribution of these changes in fluid temperatures to the change in the temperature difference $(T_h - T_c)$ depends on their relative directions of flow. Taking the direction of flow of hot fluid as the reference direction with increasing heat-transfer area (and increasing heat exchanger length), it can be easily seen from Fig. 4.3A and B that

$$d(T_h - T_c) = -\frac{dQ}{C_h} - \frac{dQ}{C_c}, \quad \text{for PFHE} \tag{4.5}$$

$$d(T_h - T_c) = -\frac{dQ}{C_h} + \frac{dQ}{C_c}, \quad \text{for CFHE} \tag{4.6}$$

Eqs. (4.5), (4.6) could be unified as

$$d(T_h - T_c) = -dQ\left\{\frac{1}{C_h} \pm \frac{1}{C_c}\right\} \tag{4.7}$$

With the +ve sign being used with PFHE and −ve sign for CFHE.

Combining Eqs. (4.1), (4.7) we get

$$\frac{d(T_h - T_c)}{T_h - T_c} = -UdA\left\{\frac{1}{C_h} \pm \frac{1}{C_c}\right\} \tag{4.8}$$

Now, if we assume that the overall heat-transfer coefficient U, and the heat capacities of the fluids, C_h and C_c, are constant and do not change along the heat exchanger length, the above equation can be integrated over the complete heat exchanger (location 1 to location 2 in Fig. 4.3), to get

$$\ln\left\{\frac{T_{h2} - T_{c2}}{T_{h1} - T_{c1}}\right\} = -U \cdot A\left\{\frac{1}{C_h} \pm \frac{1}{C_c}\right\} \tag{4.9}$$

or

$$\frac{T_{h2} - T_{c2}}{T_{h1} - T_{c1}} = \exp\left[-\frac{UA}{C_h}\left\{1 \pm \frac{C_h}{C_c}\right\}\right] \tag{4.10}$$

Eq. (4.10) relates the temperature difference between hot and cold fluids at the two ends of the heat exchanger to its important design parameters, viz. U, A, C_h, and C_c. As is evident from Fig. 4.3A, this equation also predicts that for PFHE, the temperature difference between the two fluids at the exit will always be smaller than that at the entrance since the exponential term on the RHS of Eq. (4.10) will always be less than 1. In CFHE, however, the relationship between $T_{h2} - T_{c2}$ and $T_{h1} - T_{c1}$ would depend upon the relative values of C_h and C_c. Thus we have

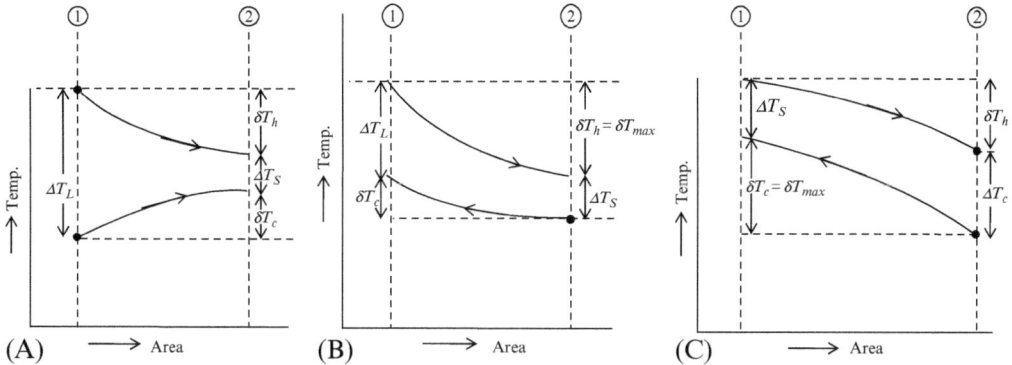

Fig. 4.4 Temperature distribution in PFHE and CFHE. (A) PFHE. (B) CFHE with $C_h < C_c$. (C) CFHE with $C_h > C_c$.

$$\text{for } C_h < C_c : \quad (T_{h2} - T_{c2}) < T_{h1} - T_{c1}$$
$$C_h > C_c : \quad (T_{h2} - T_{c2}) > T_{h1} - T_{c1}$$
$$C_h = C_c : \quad (T_{h2} - T_{c2}) = T_{h1} - T_{c1}$$

These situations are shown in Fig. 4.4A–C, with symbol ΔT representing the temperature difference $(T_h - T_c)$ and δT the temperature change of each fluid.

Indicating by term C_{min}, the smaller of C_h and C_c, Eq. (4.10) can be written in a generalized form as

$$\frac{\Delta T_S}{\Delta T_L} = \exp\left[-\frac{UA}{C_{min}}\left\{1 \pm \frac{C_{min}}{C_{max}}\right\}\right] \tag{4.11}$$

where

C_{min} = smaller of C_c and C_h;
C_{max} = larger of C_c and C_h;
ΔT_S = smaller of $(T_{h2} - T_{c2})$ and $(T_{h1} - T_{c1})$; and
ΔT_L = larger of $(T_{h2} - T_{c2})$ and $(T_{h1} - T_{c1})$.

Eq. (4.11) is the fundamental equation governing the performance of single-pass parallel-flow and counter-flow heat exchangers. It can be used to solve both the simulation problems, where the aim is to predict the total heat transfer and the exit temperatures of two fluids entering a given heat exchanger at known temperatures; as also the design problem where the objective is to find out the area of the heat exchanger required for a certain amount of heat transfer to occur between two fluids entering at prescribed temperatures. This is illustrated in Examples 4.1 and 4.2, taken from Dhar [1].

Example 4.1

Determine the surface area of (a) CFHE, (b) PFHE which would transfer heat at a rate of 2 kW between a hot fluid available at 110°C and a cold fluid available at 40°C. The heat capacities of the two fluids are $C_h = 100$ W/K and $C_c = 50$ W/K. Take the overall heat-transfer coefficient for the heat exchangers as 400 W/m² K.

Solution

The required heat-transfer rate = 2000 W.

$$C_{max} = C_h = 100 \text{ W/K} \quad \therefore \quad \delta T_{min} = \delta T_h = 2000/100 = 20°C$$
$$C_{min} = C_c = 50 \text{ W/K} \quad \therefore \quad \delta T_{max} = \delta T_c = 2000/50 = 40°C$$

Using this information we can draw the temperature profile for both the CFHE and PFHE as shown in Fig. 4.5. We can thus calculate the areas of the two heat exchangers as follows:

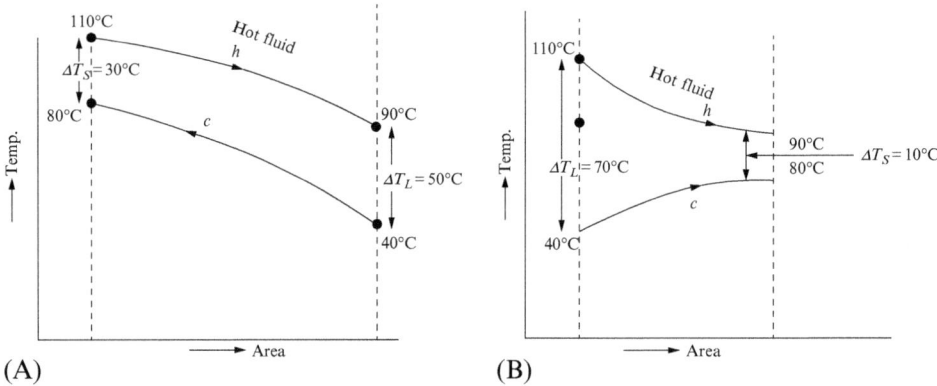

Fig. 4.5 Temperature profiles of heat exchangers of Example 4.1. (A) CFHE. (B) PFHE.

(a) *CFHE* from Fig. 4.5A

$$\Delta T_L = 50°C \quad \Delta T_S = 30°C$$

Substituting in Eq. (4.11) we get

$$\frac{30}{50} = \exp\left[-\frac{400 \cdot A}{50}\left(1 - \frac{50}{100}\right) \right]$$

which gives $A = 0.1277$ m².

(b) *PFHE* from Fig. 4.5B

$$\Delta T_L = 70°C \quad \Delta T_S = 10°C$$

Substituting in Eq. (4.11) we get

$$\frac{10}{70} = \exp\left[-\frac{400 \cdot A}{50}\left(1 + \frac{50}{100}\right) \right]$$

which gives $A = 0.16216$ m².

Example 4.2

Consider the problem of simulation of two heat exchangers, one CFHE with $A = 0.1277$ m^2 and another PFHE with $A = 0.16216$ m^2. Determine the rate of heat transfer in each for the following data:

1. Inlet temperature of hot fluid $= 110°$C
2. Inlet temperature of cold fluid $= 40°$C
3. Heat capacity of hot fluid $= 100$ W/K
4. Heat capacity of cold fluid $= 50$ W/K
5. Overall heat-transfer coefficient of each of the heat exchangers $= 400$ W/m^2 K

Solution

Using Eq. (4.11) and substituting the data in its RHS gives for CFHE:

$$\frac{\Delta T_S}{\Delta T_L} = \exp\left[-\frac{400 \times 0.1277}{50}\left(1 - \frac{50}{100}\right)\right] = 0.6$$

Now from Fig. 4.4B and C, it is evident that if δT_{max} represents the larger of δT_h and δT_c and δT_{min} represents the smaller of the two, then we can write

$$\Delta T_S + \delta T_{max} = \Delta T_L + \delta T_{min} = 110 - 40 = 70$$

Further

$$\frac{\delta T_{max}}{\delta T_{min}} = \frac{C_{max}}{C_{min}} = \frac{100}{50} = 2$$

Combining the above three equations we get

$$0.6 = \frac{\Delta T_S}{\Delta T_L} = \frac{70 - \delta T_{max}}{70 - \delta T_{min}} = \frac{70 - 2 \cdot \delta T_{min}}{70 - \delta T_{min}}$$

which gives $\delta T_{min} = 20°$C, $\delta T_{max} = 40°$C

$$\therefore \quad Q = C_{max}\delta T_{min} = 100 \times 20 = 2000 \text{ W}$$

Similarly for PFHE we get

$$\frac{\Delta T_S}{\Delta T_L} = \exp\left[-\frac{400 \times 0.16216}{50}\left(1 + \frac{50}{100}\right)\right] = 0.142856$$

Here, $\Delta T_L = 110 - 40 = 70°$C. Substituting above we get $\Delta T_s = 10°$C for PFHE, Fig. 4.4A gives $\delta T_{max} + \delta T_{min} = \Delta T_L - \Delta T_S = 70 - 10 = 60°$C.

Further

$$\frac{\delta T_{max}}{\delta T_{min}} = \frac{C_{max}}{C_{min}} = 2 \quad \therefore \ \delta T_{min} = 20°\text{C}$$

and

$$Q = C_{max}\delta T_{min} = 100 \times 20 = 2000 \text{ W}$$

It is evident from above examples that Eq. (4.11) can be very conveniently used to solve the both design and simulation problems without any iteration. This makes it more versatile than the conventional methods [2, Chapter 11], where, to avoid iterations, we have to use different procedures for the design and simulation problems. Further, for simulation problems we have to use different expressions for the effectiveness of CFHE and PFHE while here just one equation suffices.

Of course, the conventional equation for design problems, and the expressions for heat exchanger effectiveness used to solve simulation problems, can also be derived from Eq. (4.11). Thus, referring to Figs. 4.3 and 4.4, we can write

$$\delta T_h = T_{h1} - T_{h2} = \frac{Q}{C_h} \tag{4.12}$$

$$\delta T_c = \pm(T_{c2} - T_{c1}) = \frac{Q}{C_c} \tag{4.13}$$

Defining δT_{min} as the smaller of δT_c and δT_h and δT_{max} as the larger of the two, as in Example 4.2, we can write

$$\delta T_{max} + \Delta T_S = \delta T_{min} + \Delta T_L = T_{h,in} - T_{c,in} \quad \text{for CFHE} \tag{4.14}$$

and

$$\Delta T_L = \delta T_{max} + \delta T_{min} + \Delta T_S \quad \text{for PFHE} \tag{4.15}$$

Transposing terms in Eqs. (4.14), (4.15) we get

$$\Delta T_L - \Delta T_S = \delta T_{max} \pm \delta T_{min} \tag{4.16}$$

using $+$ for PFHE and $-$ for CFHE.

Eqs. (4.12), (4.13) can be easily expressed in terms of δT_{max} and δT_{min} defined above, as:

$$\delta T_{min} = Q/C_{max} \tag{4.17}$$
$$\delta T_{max} = Q/C_{min} \tag{4.18}$$

Combining Eq. (4.11) with Eqs. (4.16)–(4.18), we get

$$
\begin{aligned}
\ln\left(\frac{\Delta T_L}{\Delta T_S}\right) &= \frac{UA}{C_{min}}\left(1 \pm \frac{C_{min}}{C_{max}}\right) \\
&= \frac{UA}{Q}\left\{\frac{Q}{C_{min}} \pm \frac{Q}{C_{max}}\right\} \\
&= \frac{UA}{Q}(\delta T_{max} \pm \delta T_{min}) \\
&= \frac{UA}{Q} \cdot (\Delta T_L - \Delta T_S)
\end{aligned}
\tag{4.19}
$$

which gives

$$Q = UA \cdot LMTD \tag{4.20}$$

where

$$LMTD = \frac{\Delta T_L - \Delta T_S}{\ln(\Delta T_L / \Delta T_S)} \tag{4.21}$$

is the *Logarithmic Mean Temperature Difference* between the hot and the cold fluids.

Eq. (4.20) is the equation traditionally recommended for heat exchanger design calculations and, as is evident from the above derivation, it is applicable to both parallel- and counter-flow heat exchangers.

We shall now derive the expressions for effectiveness of PFHE and CFHE, traditionally used for predicting the performance of these heat exchangers, from Eq. (4.11).

The effectiveness ϵ is defined as:

$$\epsilon = \frac{\text{Actual heat transfer}}{\substack{\text{Maximum possible heat transfer in any heat exchanger for the same} \\ \text{fluid inlet temperatures}}}$$

Referring to Fig. 4.4, it should be evident that the maximum possible heat transfer, for the given fluid inlet temperatures would occur in a counter-flow heat exchanger of infinite area, with the fluid stream having lower heat capacity experiencing the maximum possible temperature change of $T_{h,in} - T_{c,in}$. Accordingly

$$\epsilon = \frac{Q}{C_{min}(T_{h,in} - T_{c,in})} \tag{4.22}$$

$$= \frac{C_{min}\delta T_{max}}{C_{min}(T_{h,in} - T_{c,in})} = \frac{\delta T_{max}}{T_{h,in} - T_{c,in}} \tag{4.23}$$

Further, Eq. (4.16) gives

$$\Delta T_L - \Delta T_S = \delta T_{max} \pm \delta T_{min}$$

$$= \delta T_{max} \pm \delta T_{max} \cdot \frac{C_{min}}{C_{max}} \tag{4.24}$$

Defining heat capacity ratio C_R as:

$$C_R = \frac{C_{min}}{C_{max}} \tag{4.25}$$

We get from Eq. (4.24):

$$\delta T_{max} = \frac{\Delta T_L - \Delta T_S}{1 \pm C_R} \tag{4.26}$$

Substituting Eq. (4.26) in Eq. (4.23) and using Eq. (4.11) we get

1. For PFHE

$$\epsilon = \frac{\Delta T_L - \Delta T_S}{1 + C_R} \cdot \frac{1}{\Delta T_L}$$

$$= \frac{1 - \exp\left[-\frac{UA}{C_{min}}(1 + C_R)\right]}{1 + C_R} \qquad (4.27)$$

2. For CFHE using Eq. (4.14), we get

$$\epsilon = \frac{\delta T_{max}}{\Delta T_S + \delta T_{max}}$$

Substituting for δT_{max} from Eq. (4.26) we get

$$\epsilon = \frac{\frac{\Delta T_L - \Delta T_S}{1 - C_R}}{\Delta T_S + \frac{\Delta T_L - \Delta T_S}{1 - C_R}} = \frac{\Delta T_L - \Delta T_S}{\Delta T_L - C_R \Delta T_S} = \frac{1 - \Delta T_S / \Delta T_L}{1 - C_R \cdot \Delta T_S / \Delta T_L}$$

Substituting for $\Delta T_S / \Delta T_L$ from Eq. (4.11) we get the expression for effectiveness of CFHE as

$$\epsilon = \frac{1 - \exp\left(-\frac{UA}{C_{min}}(1 - C_R)\right)}{1 - C_R \exp\left(-\frac{UA}{C_{min}}(1 - C_R)\right)} \qquad (4.28)$$

Eqs. (4.27), (4.28) are the usual expressions for effectiveness of PFHE and CFHE [2]. The quantity $\frac{UA}{C_{min}}$ is termed as the *number of transfer units* (NTU) and is indicative of the thermal size of the heat exchanger. Larger the value of NTU, larger would be the heat exchanger effectiveness, and hence the total heat transfer, other conditions remaining unaltered. Of course larger NTU does not necessarily imply large physical size of the heat exchanger since it depends not only upon the surface A, but also on the values of the overall heat-transfer coefficient U and the minimal heat capacity rate C_{min}.

Example 4.3
Solve Example 4.2 using effectiveness approach.

Solution
PFHE
The effectiveness of the heat exchanger can be found using Eq. (4.27) as:

$$\epsilon = \frac{1 - \exp\left[-\frac{400 \times 0.16216}{50}\left(1 + \frac{50}{100}\right)\right]}{1 + \frac{50}{100}}$$

$$= 0.57143$$

$$\therefore \quad Q = \epsilon C_{min}(T_{h,in} - T_{c,in}) = 0.57143 \times 50 \times (110 - 40)$$

$$= 2000 \text{ W}$$

CFHE

$$\epsilon = \frac{1 - \exp\left(-\frac{400 \times 0.1277}{50}\left(1 - \frac{50}{100}\right)\right)}{1 - \left(\frac{50}{100}\right)\exp\left(-\frac{400 \times 0.1277}{50}\left(1 - \frac{50}{100}\right)\right)}$$

$$= \frac{1 - 0.6}{1 - 0.5 \times 0.6} = 0.57143$$

$$\therefore Q = \epsilon C_{min}(T_{h,in} - T_{c,in}) = 0.57143 \times 50 \times (110 - 40)$$

$$= 2000 \text{ W}$$

Besides determining the overall heat-transfer rate in a heat exchanger, we may also be interested in finding the temperature variation of the fluids as these flow through the heat exchanger. This can be done by integrating Eq. (4.8) from the hot fluid entry station to a location x, within the heat exchanger (with $x \leq L$). This gives

$$\ln\left(\frac{T_h - T_c}{T_{h1} - T_{c1}}\right) = -U \cdot \left(\frac{Ax}{L}\right)\left\{\frac{1}{C_h} \pm \frac{1}{C_c}\right\}$$

$$\text{or} \quad \frac{T_h - T_c}{T_{h1} - T_{c1}} = \exp\left[-\left(\frac{x}{L}\right) \cdot \frac{UA}{C_h} \cdot \left\{1 \pm \frac{C_h}{C_c}\right\}\right] \tag{4.29}$$

where T_h and T_c are the temperatures of the hot and the cold fluids, respectively, at the location x along the heat exchanger. Since this equation involves two unknowns T_h and T_c, it is necessary to express one of them in terms of the other to get separate equations for the temperature distribution of the hot and the cold fluids. This can be done by using energy balance over the heat exchanger length $0 - x$, which gives

$$C_h(T_{h1} - T_h) = \pm C_c(T_c - T_{c1}) \tag{4.30}$$

Rearranging terms, we get

$$T_c = T_{c1} \pm \frac{C_h}{C_c}(T_{h1} - T_h) \tag{4.31}$$

with +ve sign for PFHE and −ve sign for CFHE.

Combining Eqs. (4.29), (4.31), and using Eq. (4.30) we get

$$(T_h - T_c) = (T_{h1} - T_{c1})\exp\left[-\frac{x}{L} \cdot \frac{UA}{C_h} \cdot \left\{1 \pm \frac{C_h}{C_c}\right\}\right]$$

$$T_h - T_{c1} = \pm\frac{C_h}{C_c}(T_{h1} - T_h) + (T_{h1} - T_{c1})\exp\left[-\frac{x}{L} \cdot \frac{UA}{C_h} \cdot \left\{1 \pm \frac{C_h}{C_c}\right\}\right]$$

$$T_h\left\{1 \pm \frac{C_h}{C_c}\right\} = T_{c1} \pm \frac{C_h}{C_c} \cdot T_{h1} + (T_{h1} - T_{c1})\exp\left[-\frac{x}{L} \cdot \frac{UA}{C_h} \cdot \left\{1 \pm \frac{C_h}{C_c}\right\}\right]$$

This equation can be rearranged to give an expression for nondimensional temperature of the hot fluid as:

$$\frac{T_h - T_{c1}}{T_{h1} - T_{c1}} = \frac{\pm\frac{C_h}{C_c} + \exp\left[-\frac{x}{L} \cdot \frac{UA}{C_h} \cdot \left\{1 \pm \frac{C_h}{C_c}\right\}\right]}{1 \pm \frac{C_h}{C_c}} \tag{4.32}$$

This equation is useful only for PFHE where both the temperatures T_{h1} and T_{c1}, the inlet temperatures of the hot and the cold fluids, respectively, are known. For PFHE, thus we can calculate the temperature of the hot fluid at any location x from this equation. The local cold fluid temperature can then be determined using energy balance.

For CFHE, the temperature T_{c1} is also unknown, and we therefore need to transform Eq. (4.32) by expressing T_{c1} in terms of T_{c2}, the cold fluid inlet temperature and other heat exchanger design parameters. To do so, we use Eq. (4.32) to find the temperature of the hot fluid at heat exchanger outlet (i.e., $x/L = 1$). This gives

$$\frac{T_{h2} - T_{c1}}{T_{h1} - T_{c1}} = \frac{\exp\left[-\frac{UA}{C_h} \cdot \left\{1 - \frac{C_h}{C_c}\right\}\right] - \frac{C_h}{C_c}}{1 - \frac{C_h}{C_c}} \tag{4.33}$$

Subtracting from 1 both sides of the above equation, and rearranging terms we get

$$\frac{T_{h1} - T_{h2}}{T_{h1} - T_{c1}} = \frac{1 - \exp\left[-\frac{UA}{C_h}\left\{1 - \frac{C_h}{C_c}\right\}\right]}{1 - \frac{C_h}{C_c}} \tag{4.34}$$

Further, since the total heat lost by hot fluid equals that gained by the cold fluid:

$$C_h(T_{h1} - T_{h2}) = C_c(T_{c1} - T_{c2}) \tag{4.35}$$

Using this in the above equation we get

$$\frac{T_{c1} - T_{c2}}{T_{h1} - T_{c1}} = \frac{(C_h/C_c)\left(1 - \exp\left[-\frac{UA}{C_h}\left\{1 - \frac{C_h}{C_c}\right\}\right]\right)}{1 - C_h/C_c} \tag{4.36}$$

Adding 1 to both sides gives

$$\frac{T_{h1} - T_{c2}}{T_{h1} - T_{c1}} = \frac{1 - \frac{C_h}{C_c} \cdot \exp\left[-\frac{UA}{C_h}\left\{1 - \frac{C_h}{C_c}\right\}\right]}{1 - \frac{C_h}{C_c}} \tag{4.37}$$

Dividing Eq. (4.32) by Eq. (4.37) we get

$$\frac{T_h - T_{c1}}{T_{h1} - T_{c2}} = \frac{\exp\left[-\frac{x}{L} \cdot \frac{UA}{C_h}\left\{1 - \frac{C_h}{C_c}\right\}\right] - \frac{C_h}{C_c}}{1 - \frac{C_h}{C_c} \cdot \exp\left[-\frac{UA}{C_h}\left\{1 - \frac{C_h}{C_c}\right\}\right]} \tag{4.38}$$

Since at $x = 0$, $T_h = T_{h1}$, it follows from Eq. (4.38) that:

$$\frac{T_{h1} - T_{c1}}{T_{h1} - T_{c2}} = \frac{1 - \frac{C_h}{C_c}}{1 - \frac{C_h}{C_c} \exp\left[-\frac{UA}{C_h}\left\{1 - \frac{C_h}{C_c}\right\}\right]} \tag{4.39}$$

Subtracting Eq. (4.39) from Eq. (4.38) we get

$$\frac{T_h - T_{h1}}{T_{h1} - T_{c2}} = \frac{\exp\left[-\frac{x}{L}\cdot\frac{UA}{C_h}\left\{1 - \frac{C_h}{C_c}\right\}\right] - 1}{1 - \frac{C_h}{C_c}\cdot\exp\left[-\frac{UA}{C_h}\left\{1 - \frac{C_h}{C_c}\right\}\right]} \tag{4.40}$$

Adding 1 to both sides we get the expression for nondimensional temperature of the hot fluid at any location x as:

$$\frac{T_h - T_{c2}}{T_{h1} - T_{c2}} = \frac{\exp\left[-\frac{x}{L}\cdot\frac{UA}{C_h}\left\{1 - \frac{C_h}{C_c}\right\}\right] - \frac{C_h}{C_c}\cdot\exp\left[-\frac{UA}{C_h}\left\{1 - \frac{C_h}{C_c}\right\}\right]}{1 - \frac{C_h}{C_c}\cdot\exp\left[-\frac{UA}{C_h}\left\{1 - \frac{C_h}{C_c}\right\}\right]} \tag{4.41}$$

Since both T_{h1} and T_{c2} are known, being the inlet temperatures of the hot and the cold fluids, respectively; Eq. (4.41) can be used to predict the temperature of the hot fluid at any position in the heat exchanger. The corresponding cold fluid temperature can then be determined using energy balance.

The expressions for effectiveness and temperature distribution in counter-flow heat exchanger derived above (Eqs. 4.28, 4.41) become indeterminate (i.e., 0/0) if $C_h = C_c$. We can get valid expressions for this case either mathematically—that is, by writing series expansion for the exponential terms and taking limits—or by physical reasoning. It is instructive to follow the physical reasoning, coupling it with the basic Eq. (4.11) which shows that when $C_h = C_c$, for CFHE $\Delta T_s = \Delta T_L$, that is, $T_{h2} - T_{c2} = T_{h1} - T_{c1}$. Further, it is evident from Eq. (4.29) that for any value of (x/L) the RHS of the equation is equal to 1 if $C_h = C_c$. This gives

$$T_h - T_c = T_{h1} - T_{c1} = T_{h2} - T_{c2} = \text{Constant} \tag{4.42}$$

This implies that the temperature difference between the hot and the cold fluids is constant throughout this heat exchanger. It therefore follows from Eq. (4.1) that the rate of heat transfer per unit area is the same at every section. $Q(x)$, the total heat transfer up to any location x can then be expressed as:

$$Q(x) = \frac{x}{L}\cdot Q \tag{4.43}$$

where Q is the total heat transfer in the heat exchanger. The temperature of hot fluid at any location x can then be written, using Eqs. (4.2), (4.43) as:

$$T_h = T_{h1} - \frac{x}{L}\frac{Q}{C_h} \tag{4.44}$$

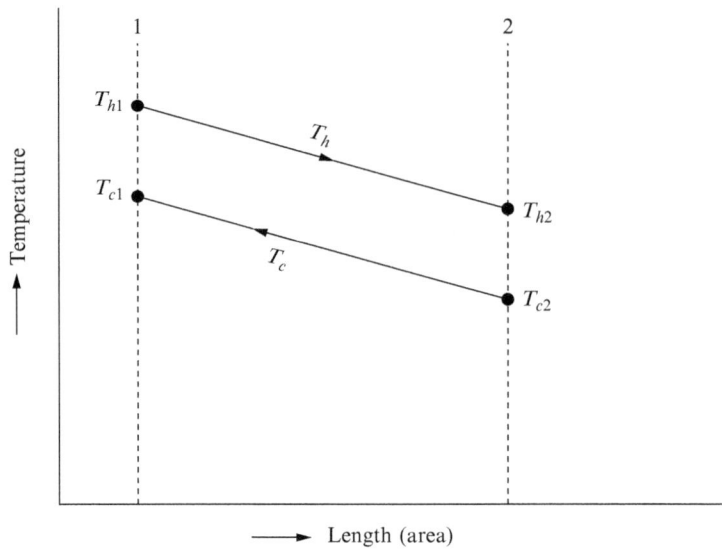

Fig. 4.6 Temperature distribution in a CFHE with $C_h = C_c$.

which shows that the temperature varies linearly with x. Since the temperature difference $(T_h - T_c)$ is constant, it follows that the temperature T_c also varies linearly along the heat exchanger, as shown in Fig. 4.6.

The effectiveness of ϵ of such a heat exchanger can now be determined by transforming Eq. (4.23) using Eqs. (4.42), (4.44) as indicated below.

$$\epsilon = \frac{T_{h1} - T_{h2}}{T_{h1} - T_{c2}} = \frac{T_{h1} - T_{h2}}{(T_{h1} - T_{c1}) + (T_{c1} - T_{c2})} = \frac{1}{\frac{T_{h1} - T_{c1}}{T_{h1} - T_{h2}} + 1}$$

$$T_{h1} - T_{h2} = \frac{Q}{C_h} = \frac{UA(T_{h1} - T_{c1})}{C_h}$$

$$\therefore \quad \epsilon = \frac{1}{1 + C_h/UA} = \frac{NTU}{NTU + 1} \tag{4.45}$$

since $NTU = \frac{UA}{C_h}$.

The total heat transfer Q can now be calculated from the following equation:

$$Q = \epsilon \cdot C_h(T_{h1} - T_{c2}) \tag{4.46}$$

The temperature distribution is linear and given by Eq. (4.44).

Another interesting limiting case, often encountered in practice is where one of the fluids undergoes a phase change at a constant temperature. This can be captured in the analysis by taking $C_{max} = \infty$. This gives $C_R = 0$, and the expressions for effectiveness, both for parallel- and counter-flow heat exchangers, become

$$\epsilon = 1 - \exp(-NTU) \tag{4.47}$$

4.1.2 Single-Pass Cross-Flow Heat Exchanger

Fig. 4.7 shows the schematic diagram of a typical parallel plate heat exchanger where the two streams flow at right angles to each other through alternate spaces between parallel plates. Often, to increase the effective area of contact between the fluids and the plates, these spaces contain crimped thin sheets of metal which act like "fins" for the primary heat-transfer surface provided by the parallel plates, and also as separate passages for the flow of the two fluids. The fluid flowing in a passage thus does not mix with that flowing in adjacent passages, and will undergo temperature change due to heat transfer across the primary surface area and the passage surface area pertaining to that individual passage. The temperature of the primary surface would thus vary in both the orthogonal directions as shown in Fig. 4.8.

Assuming the overall heat-transfer coefficient U to be constant over the entire plate area, and referring it to the base surface area, we can write the following equation for calculating the heat transfer in the elemental area $[dx\ dy]$ of Fig. 4.8 as:

$$dQ = U \cdot (dx\ dy)(T_h - T_c) \tag{4.48}$$

This heat transfer would cause a drop in the hot fluid temperature, and a rise in the cold fluid temperature, which can be estimated by writing the energy balance for the two fluids. For this we have to first estimate the mass flowing through each passage.

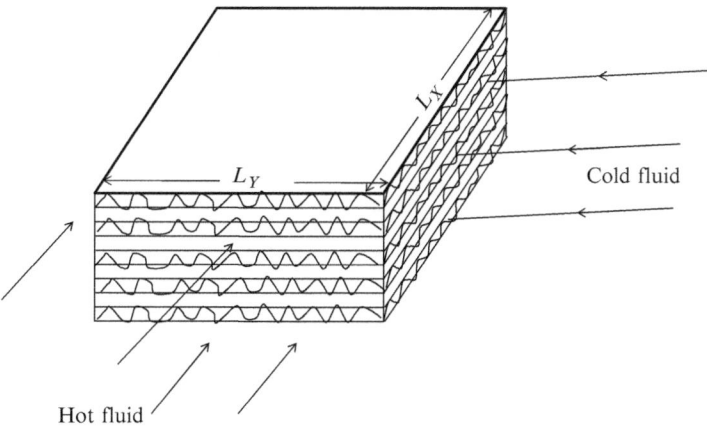

Fig. 4.7 Cross-flow heat exchanger with both streams unmixed.

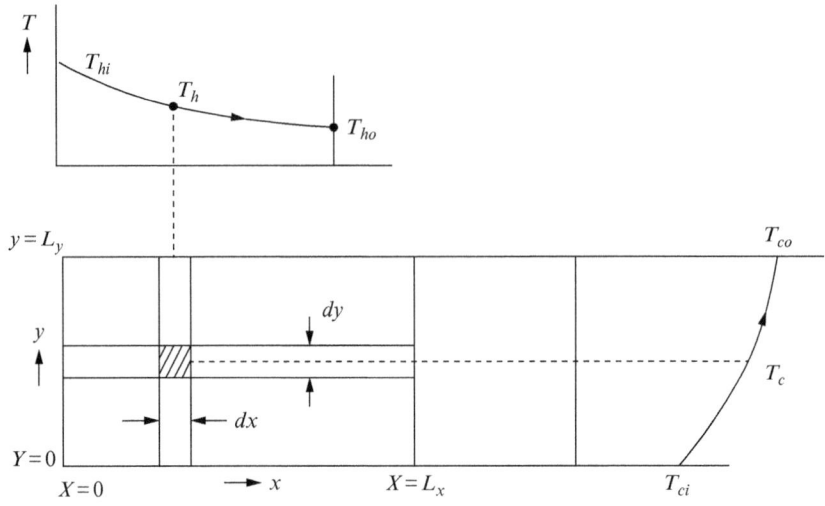

Fig. 4.8 Temperature variation of fluids in a cross-flow heat exchangers.

Assuming that the mass flowing through a single plate is distributed uniformly across the plate width, we can write for the two fluids (Fig. 4.8):

$$d\dot{m}_h = \frac{\dot{m}_h}{L_y} \cdot dy \tag{4.49}$$

$$d\dot{m}_c = \frac{\dot{m}_c}{L_x} \cdot dx \tag{4.50}$$

The heat balances of the two fluids, accounting for the heat transferred through the elemental area $[dx\, dy]$, can now be written as:

$$dQ = -\left(\frac{\dot{m}_h}{L_y}\right) dy \cdot c_h \cdot \left(\frac{\partial T_h}{\partial x}\right) \cdot dx$$

$$= \frac{\dot{m}_c}{L_x} \cdot dx \cdot c_c \left(\frac{\partial T_c}{\partial y}\right) dy$$

where c stands for the specific heat of the fluid. Substituting for dQ from Eq. (4.48) we get

$$U(T_h - T_c) = -\frac{\dot{m}_h c_h}{Ly} \cdot \left(\frac{\partial T_h}{\partial x}\right) = \frac{\dot{m}_c c_c}{L_x}\left(\frac{\partial T_c}{\partial y}\right) \tag{4.51}$$

We define the nondimensional temperatures θ_h and θ_c as:

$$\theta_h = \frac{T_h - T_{ci}}{T_{hi} - T_{ci}} \quad \text{and} \quad \theta_c = \frac{T_c - T_{ci}}{T_{hi} - T_{ci}} \tag{4.52}$$

where additional subscript i refers to the conditions at entry to the heat exchanger. Further we define NTU for the heat exchanger, assuming C_{min} to correspond to the hot fluid, that is

$$NTU = \frac{U \cdot L_x L_y}{\dot{m}_h c_h} \tag{4.53}$$

We also define nondimensional distances along the heat exchanger as:

$$x^* = \frac{x}{L_x} \quad \text{and} \quad y^* = \frac{y}{L_y} \tag{4.54}$$

Using Eqs. (4.52)–(4.54) we can transform Eq. (4.51) to:

$$\frac{1}{NTU} \cdot \frac{\partial \theta_h}{\partial x^*} + \theta_h = \theta_c \tag{4.55}$$

$$\frac{1}{NTU \cdot C_R} \cdot \frac{\partial \theta_c}{\partial y^*} + \theta_c = \theta_h \tag{4.56}$$

where $C_R = \frac{\dot{m}_h c_h}{\dot{m}_c c_c}$ is the ratio of the heat capacities of the two fluids.

The boundary conditions for solving these two partial differential equations for the temperature distributions in the heat exchanger are

$$\theta_h(0, y^*) = 1; \quad \theta_c(x^*, 0) = 0 \tag{4.57}$$

It is not possible to get a simple analytical solution for these two simultaneous partial differential equations. A series solution was obtained by Nusselt, based on which the following expression for the effectiveness of the heat exchanger can be obtained [3, p. 128]:

$$\epsilon = 1 - \exp(-NTU) - \exp[-(1 + C_R)NTU] \sum_{n=1}^{\infty} C_R^n P_n(NTU)$$

$$\text{where } P_n(y) = \frac{1}{(n+1)!} \sum_{j=1}^{n} \frac{(n+1-j)}{j!} y^{n+j} \tag{4.58}$$

Proceeding in a similar manner we can derive expressions for the effectiveness of a cross-flow heat exchanger with one of the fluids mixed and the other unmixed, as also for the (rarely encountered) case when both the fluids are mixed. To illustrate the approach of handling these cases, we shall analyze a heat exchanger in which the hot fluid is mixed and the cold fluid is unmixed.

Referring to Fig. 4.7, we can see that in this case T_h varies only along x-direction. Due to mixing of the hot fluid its temperature does not vary along the transverse y direction. Since the cold fluid is unmixed, the mass of cold fluid flowing through a passage, say at position x, exchanges heat with the hot fluid at constant temperature,

$T_h(x)$. The rate of variation of the cold fluid temperature along the y-direction, at a particular position x, can be written using Eq. (4.51) as:

$$\left(\frac{\partial T_c}{\partial y}\right)_x = (T_h - T_c)\frac{UL_x}{\dot{m}_c C_c}$$

$$= (T_h - T_c)\frac{(NTU \cdot C_R)}{L_y} \tag{4.59}$$

Since at any specified location along x-axis, T_h is fixed and T_c is a function of y only, we can integrate above equation for limits $y = 0$ to $y = L_y$ to get

$$\int_{T_c(x,0)}^{T_c(x,Ly)} \frac{dT_c}{T_h - T_c} = \frac{(NTU \cdot C_R)}{L_y} \int_0^{Ly} dy$$

$$\Rightarrow -\ln\left\{\frac{T_h(x) - T_c(x, L_y)}{T_h(x) - T_c(x, 0)}\right\} = NTU \cdot C_R$$

which gives $\qquad \dfrac{T_h(x) - T_c(x, L_y)}{T_h(x) - T_c(x, 0)} = \exp(-C_R \cdot NTU) \tag{4.60}$

The temperature change of the cold fluid from the inlet to the outlet can therefore be written as:

$$\left\{\frac{T_c(x, L_y) - T_c(x, 0)}{T_h(x) - T_c(x, 0)}\right\} = 1 - \exp(-NTU \cdot C_R)$$

$$= \Gamma \quad \text{(say)} \tag{4.61}$$

The total heat transfer through the passage situated along the y-axis at position x is therefore:

$$dq = d\dot{m}_c \cdot c_c(T(x, Ly) - T_c(x, 0)) \tag{4.62}$$

This heat transfer would cause a drop in the temperature of the hot fluid, whose magnitude can be found from the energy balance equation:

$$dq = -\dot{m}_h c_h dT_h(x) \tag{4.63}$$

Combining Eqs. (4.61)–(4.63) we get

$$dq = d\dot{m}_c \cdot c_c \Gamma (T_h(x) - T_c(x, 0)) = -\dot{m}_h \cdot c_h \cdot dT_h(x) \tag{4.64}$$

this gives

$$\frac{dT_h(x)}{T_h(x) - T_c(x, 0)} = -\Gamma \frac{c_c d\dot{m}_c}{\dot{m}_h c_h}$$

Integrating this over the length of the plate in x-direction (i.e., L_x), we get

$$\int_{T_h(0)}^{T_h(Lx)} \frac{dT_h(x)}{T_h(x) - T_c(x,0)} = \frac{-\Gamma c_c}{\dot{m}_h c_h} \int_{x=0}^{x=Lx} d\dot{m}_c$$

which gives on integration:

$$\ln\left\{ \frac{T_h(L_x) - T_c(L_x,0)}{T_h(0) - T_c(L_x,0)} \right\} = \frac{-\Gamma c_c \dot{m}_c}{\dot{m}_h c_h} = \frac{-\Gamma}{C_R}$$

$$\text{or} \quad \frac{T_h(L_x) - T_c(L_x,0)}{T_h(0) - T_c(L_x,0)} = \exp(-\Gamma/C_R) \tag{4.65}$$

Now, clearly

$$T_c(L_x,0) = T_{ci} \quad \text{The cold fluid inlet temperature}$$
$$T_h(0) = T_{hi} \quad \text{The hot fluid inlet temperature}$$
$$T_h(L_x) = T_{ho} \quad \text{The hot fluid outlet temperature}$$

Eq. (4.65) can therefore be written as

$$\frac{(T_{ho} - T_{ci})}{(T_{hi} - T_{ci})} = \exp(-\Gamma/C_R)$$

from which an expression for effectiveness ϵ can be easily obtained, keeping in view our assumption that the hot fluid has lower heat capacity, as

$$\epsilon = \frac{T_{hi} - T_{ho}}{T_{hi} - T_{ci}} = 1 - \exp(-\Gamma/C_R)$$

$$= 1 - \exp[-(1 - \exp(-NTU \cdot C_R))/C_R] \tag{4.66}$$

Eq. (4.66) is the desired expression for effectiveness of a cross-flow heat exchanger with fluid having lower heat capacity mixed, and the other fluid unmixed.

Proceeding similarly, following expressions can be derived for other possible cases of cross-flow heat exchangers:

1. C_{min} unmixed, C_{max} mixed

$$\epsilon = \frac{1}{C_R}[1 - \exp\{-C_R[1 - \exp(-NTU)]\}] \tag{4.67}$$

2. Both fluids mixed

$$\epsilon = \frac{1}{\frac{1}{1-\exp(-NTU)} + \frac{C_R}{1-\exp(-NTU \cdot C_R)} - \frac{1}{NTU}} \tag{4.68}$$

Shah and Sekulic [3] give charts showing ϵ values of various types of heat exchangers as a function of NTU and C_R.

Example 4.4

Determine the surface area of a cross-flow heat exchanger which would enable the transfer of heat between two fluids as stipulated in Example 4.1. Solve the problem for following three cases:

1. both fluids unmixed;
2. hot fluid mixed, cold unmixed; and
3. cold fluid mixed, hot unmixed.

Solution

We rewrite the following data/results from Example 4.1:

$$\dot{m}_h c_h = 100 \text{ W/K} \quad \text{Temperature range: } 110-90°\text{C}$$
$$\dot{m}_c c_c = 50 \text{ W/K} \quad \text{Temperature range: } 40-80°\text{C}$$
$$U = 400 \text{ W/m}^2 \text{ K}$$

Its effectiveness is (since C_{min} is C_{cold}):

$$\epsilon = \frac{T_{co} - T_{ci}}{T_{hi} - T_{ci}} = \frac{80 - 40}{110 - 40} = 0.5714$$

Further $C_R = \frac{50}{100} = 0.5$.

Here, both ϵ and C_R are given and we can find NTU from the expressions for effectiveness of the three cases (Eqs. 4.58, 4.66, 4.67).

Both Fluids Unmixed

The expression for ϵ, Eq. (4.58), does not permit its rearrangement to express NTU as a function of C_R and ϵ. Hence the solution has to be found by iteration. We expect the required surface area to be in between that for the parallel-flow and the counter-flow heat exchangers. Therefore, to start the iterations we assume $A = 0.15$ m^2.

This gives

$$NTU = \frac{400 \times 0.15}{50} = 1.2$$

Putting this value in Eq. (4.58) we get

$$\epsilon = 1 - e^{-1.2} - e^{-1.5 \times 1.2} \left\{ \sum_{n=1}^{\infty} 0.5^n \frac{1}{(n+1)!} \sum_{j=1}^{n} \frac{(n+1-j)}{j!} (1.2)^{n+j} \right\}$$

$$= 1 - 0.301194 - 0.165299 \left\{ \frac{0.5 (1+1-1)}{2!} \frac{}{1!} (1.2)^2 \right.$$

$$+ \frac{0.5^2}{3!} \left[\frac{(2+1-1)}{1!} (1.2)^3 + \frac{(2+1-2)}{2!} (1.2)^4 \right]$$

$$\left. + \frac{0.5^3}{4!} \left[\frac{(3+1-1)}{1!} (1.2)^4 + \frac{(3+1-2)}{2!} (1.2)^5 + \frac{(3+1-3)}{3!} (1.2)^6 \right] + \cdots \right\}$$

$$= 0.698806 - 0.165299(0.36 + 0.1872 + 0.0479 + 0.00823 + \cdots)$$
$$= 0.599$$

Since the required effectiveness is 0.5714, a heat exchanger with slightly lesser surface area would be needed. We assume the area as 0.14 m^2 which gives $NTU = \frac{400 \times 1.4}{50} = 1.12$.

Substituting this value in Eq. (4.58) we get, following the same steps of calculation as above,

$$\epsilon = 1 - e^{-1.12} - e^{-1.12 \times 1.5} \left\{ \sum_{n=1}^{\infty} 0.5^n \frac{1}{(n+1)!} \sum_{j=1}^{n} \frac{(n+1-j)}{j!} (1.12)^{n+j} \right\}$$
$$= 1 - 0.32628 - 0.186374\{0.3136 + 0.149859 + 0.035478 + 0.00565 + \cdots\}$$
$$= 0.5797$$

which is approximately (within 1.5%) of the desired value. Thus the required surface area for unmixed-unmixed cross-flow heat exchanger is slightly less than 0.14 m^2. By extrapolation from the above two results the required surface area can be estimated as 0.137 m^2.

Hot Fluid Mixed, Cold Fluid Unmixed

Since the hot fluid has larger heat capacity, the effectiveness for this heat exchanger is given by Eq. (4.67). Assuming area 0.15 m^2 gives $NTU = 1.2$ (as before) and putting this in Eq. (4.67) gives

$$\epsilon = \frac{1}{0.5}[1 - \exp\{-0.5(1 - e^{-1.2})\}]$$
$$= 2[1 - \exp\{-0.3494\}] = 0.58978$$

which is larger than the required value of 0.5714.

Taking the next assumption for the area as 0.14 m^2, with $NTU = 1.12$ (as above), we get

$$\epsilon = 2[1 - \exp(-0.5[1 - e^{-1.12}])]$$
$$= 2[1 - \exp(-0.33686)] = 0.5719$$

which matches almost exactly with the required value. Thus the required surface area is 0.14 m^2.

Hot Fluid Unmixed, Cold Fluid Mixed

This corresponds to the case when the fluid with lower heat capacity (i.e., C_{min}) is mixed and the other fluid is unmixed. The effectiveness ϵ is given by Eq. (4.66).

Starting with the assumption of area $A = 0.14 \text{ m}^2$, $NTU = 1.12$, we get

$$\epsilon = 1 - \exp\left[-\frac{\{1 - \exp(-1.12 \times 0.5)\}}{0.5} \right]$$
$$= 1 - \exp[-0.8575818] = 0.57581$$

Since this is larger than the desired value of effectiveness of the heat exchanger, we choose as next assumption an area of 0.135 m^2, for which $NTU = 1.08$, and the effectiveness is

$$\epsilon = 1 - \exp\left[-\frac{\{1 - \exp(-1.08 \times 0.5)\}}{0.5}\right]$$

$$= 1 - \exp[-0.8345035] = 0.56591$$

which is lesser than the desired value of ϵ. Taking $A = 0.138$ m^2, $NTU = 1.104$, and ϵ is

$$\epsilon = 1 - \exp\left[-\frac{\{1 - \exp(-1.104 \times 0.5)\}}{0.5}\right]$$

$$= 1 - \exp[-0.84840587] = 0.5719$$

which is nearly the same as the desired value of 0.5714. Hence the required area is $A = 0.138$ m^2.

Comment
The heat exchanger surface areas required for the three cases are
1. both fluids unmixed: $A = 0.137$ m^2
2. hot fluids mixed, cold unmixed: $A = 0.140$ m^2
3. hot fluids unmixed, cold mixed: $A = 0.138$ m^2
In Example 4.1 we had found the areas required for parallel-flow and counter-flow configurations as 0.16216 and 0.1277 m^2, respectively. The areas required by various cross-flow configurations lie in between these limiting values, as expected.

4.1.3 Multipass Heat Exchangers

In industrial applications involving large magnitudes of heat transfer, use of basic parallel- or counter-flow configuration would demand very long lengths of the tubes. To get compact designs with modest foot print we go in for multipass heat exchangers, where the fluids traverse the heat exchanger length more than once.

Fig. 4.9 shows a few typical multipass configurations in shell and tube heat exchangers. Fig. 4.9A and B shows the shell-side fluid traversing the heat exchanger length only once, while the tube-side fluid traverses it twice (in Fig. 4.9A) and thrice (in Fig. 4.9B), respectively. We call these as one-shell pass and two/three tube pass heat exchangers. Since one-shell pass is most commonly encountered, it is treated as the default configuration and these heat exchangers are termed as two-pass/three-pass heat exchangers. In contrast, in the heat exchangers shown in Fig. 4.9C and D, the shell side fluid also traverses the heat exchangers twice and these are designated as two–shell pass, two-tube pass and two-shell pass, four-tube pass heat exchangers, respectively.

It is intuitively evident that the effectiveness of these multipass heat exchangers would be better than those of purely parallel-flow heat exchangers but lesser than those of pure counter-flow heat exchangers. To determine the value of this effectiveness we can proceed, as in Section 4.1.1, by writing the basic equations governing the transfer of heat between the two fluids, and then solve these, taking care of the specific conditions

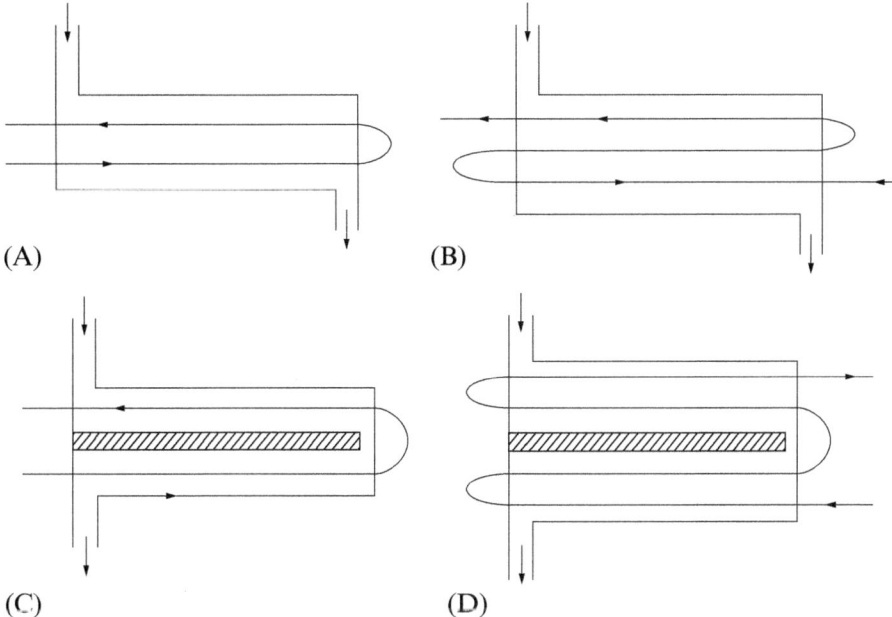

Fig. 4.9 Typical multipass heat exchanger configurations.

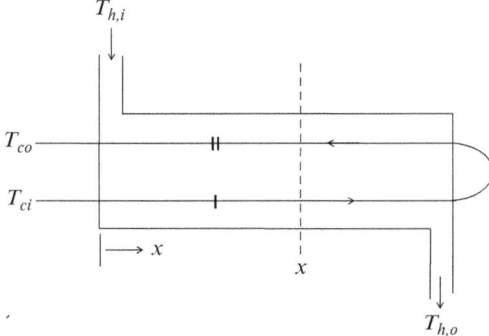

Fig. 4.10 Two-pass heat exchanger.

of the fluids. We shall illustrate this procedure by taking the case of a two-pass heat exchanger, shown in Fig. 4.10.

We assume, as before, that the hot fluid, flowing in the shell, has lower heat capacity. We distinguish between the temperatures of the cold fluid flowing in the two passes through the tubes by additional subscripts I and II, signifying the pass number, as indicated in the figure. Considering the energy balance of the tube-side fluid in the two passes, and of the shell-side fluid (which is assumed to be thoroughly mixed in the transverse direction), we can easily get the following equations for the rate of change of temperature along the heat exchanger length, in the direction of flow of the hot fluid (cf. Eqs. 4.1–4.3):

$$C_c \frac{dT_{CI}}{dx} = UA'(T_h - T_{CI}) \tag{4.69}$$

$$C_c \frac{dT_{CII}}{dx} = -UA'(T_h - T_{CII}) \tag{4.70}$$

$$C_h \frac{dT_h}{dx} = -UA'(2T_h - T_{CI} - T_{CII}) \tag{4.71}$$

where value of A', the heat-transfer area per unit length, is assumed to be equal in both the passes.

Differentiating Eq. (4.71) we get

$$C_h \frac{d^2 T_h}{dx^2} = -UA' \left(2 \frac{dT_h}{dx} - \frac{dT_{CI}}{dx} - \frac{dT_{cII}}{dx} \right)$$

Using Eqs. (4.69), (4.70) in the above equation we get

$$C_h \frac{d^2 T_h}{dx^2} = -UA' \left(\frac{2dT_h}{dx} - \frac{UA'}{C_c}(T_h - T_{CI}) + \frac{UA'}{C_c}(T_h - T_{CII}) \right)$$

This can be simplified to get a differential equation for the hot fluid temperature as:

$$\frac{d^2 T_h}{dx^2} = -\frac{UA'}{C_h} \left\{ \frac{2dT_h}{dx} - \frac{UA'}{C_c}(T_{CII} - T_{CI}) \right\} \tag{4.72}$$

Considering the heat balance on the right-hand side of the heat exchanger, ahead of the position x (Fig. 4.10), we get

$$C_h(T_h - T_{ho}) = C_c(T_{CII} - T_{CI}) \tag{4.73}$$

Using Eq. (4.73) in Eq. (4.72) we get

$$\frac{d^2 T_h}{dx^2} = -\frac{UA'}{C_h} \left\{ \frac{2dT_h}{dx} - \frac{UA'}{C_c} \cdot \frac{C_h}{C_c} \cdot (T_h - T_{ho}) \right\}$$

$$= -\frac{UA'}{C_h} \cdot 2 \cdot \frac{dT_h}{dx} + \left(\frac{UA'}{C_c} \right)^2 (T_h - T_{ho})$$

This can be simplified further, by defining $z = T_h - T_{ho}$, to get a second-order differential equation for z:

$$\frac{d^2 z}{dx^2} + \frac{2UA'}{C_h} \cdot \frac{dz}{dx} - \left(\frac{UA'}{C_c} \right)^2 z = 0 \tag{4.74}$$

This is the standard form of a second-order differential equation whose solution is

$$z = C_1 e^{m_1 x} + C_2 e^{m_2 x} \tag{4.75}$$

where m_1 and m_2 are the roots of the characteristic equation

$$m^2 + \frac{2UA'}{C_h} \cdot m - \left(\frac{UA'}{C_c}\right)^2 = 0$$

Clearly $\quad m_1 = -\frac{UA'}{C_h}\left[1 - \sqrt{1 + \left(\frac{C_h}{C_c}\right)^2}\right]$ \hfill (4.76)

$$m_2 = -\frac{UA'}{C_h}\left[1 + \sqrt{1 + \left(\frac{C_h}{C_c}\right)^2}\right] \tag{4.77}$$

To determine the constants C_1 and C_2, we use the prescribed initial conditions, such as

(i)   at $x = 0 \quad T_h = T_{hi} \quad$ and $\quad z = T_{hi} - T_{ho}$

which gives $\quad C_1 + C_2 = T_{hi} - T_{ho}$ \hfill (4.78)

(ii)   at $x = 0 \quad C_h\left(\dfrac{dT_h}{dx}\right)_{x=0} = -[UA'(T_{hi} - T_{ci}) + UA'(T_{hi} - T_{co})]$

which gives $\left(\dfrac{dz}{dx}\right)_{x=0} = -\dfrac{UA'}{C_h}[2T_{hi} - T_{ci} - T_{co}] = C_1 m_1 + C_2 m_2$ \hfill (4.79)

Solving Eqs. (4.78), (4.79), and setting $C_h/C_c = C_R$, the heat capacity ratio, we get

$$C_1 = \frac{-[T_{hi} + T_{ho} - T_{ci} - T_{co} - (T_{hi} - T_{ho})\sqrt{1 + C_R^2}]}{2\sqrt{1 + C_R^2}} \tag{4.80}$$

$$C_2 = \frac{T_{hi} + T_{ho} - T_{ci} - T_{co} + (T_{hi} - T_{ho})\sqrt{1 + C_R^2}}{2\sqrt{1 + C_R^2}} \tag{4.81}$$

We thus get the temperature profile of the hot fluid along the heat exchanger as

$$T_h = T_{ho} + C_1\, e^{m_1 x} + C_2\, e^{m_2 x} \tag{4.82}$$

where $C_1, C_2, m_1,$ and m_2 are given by Eqs. (4.80), (4.81), (4.76), and (4.77), respectively.

The temperature of the hot fluid at the end of the heat exchanger T_{ho} can also be found from this equation by substituting $x = L$, the heat exchanger length. This gives

$$T_{ho} = T_{ho} + C_1\, e^{m_1 L} + C_2\, e^{m_2 L}$$

from which we get

$$\exp\{(m_1 - m_2)L\} = \left(-\frac{C_2}{C_1}\right)$$

Substituting the values of m_1, m_2, C_1, and C_2, obtained above, we get

$$\exp\left\{\frac{2UA'L}{C_h}\sqrt{1+C_R^2}\right\} = \frac{T_{hi} + T_{ho} - T_{ci} - T_{co} + (T_{hi} - T_{ho})\sqrt{1+C_R^2}}{T_{hi} + T_{ho} - T_{ci} - T_{co} - (T_{hi} - T_{ho})\sqrt{1+C_R^2}} \qquad (4.83)$$

The term $\frac{U(2A'L)}{C_h}$ on the left-hand side of the above equation is clearly the NTU of the heat exchanger.

The effectiveness ϵ is defined in terms of temperature differences (Eq. 4.23), which can be written here as:

$$\epsilon = \frac{T_{hi} - T_{ho}}{T_{hi} - T_{ci}} \qquad (4.84)$$

Using this equation the terms on the right-hand side of Eq. (4.83) can be expressed as a function of ϵ. Simplifying, we finally get an expression for ϵ in terms of NTU and C_R as:

$$\epsilon = 2\left[1 + C_R + \sqrt{1+C_R^2}\frac{1 + \exp\left\{-NTU\sqrt{1+C_R^2}\right\}}{1 - \exp\left\{-NTU\sqrt{1+C_R^2}\right\}}\right]^{-1} \qquad (4.85)$$

Following a similar procedure, expressions for the effectiveness of multipass heat exchangers of other configurations can be obtained. It is seen that above equation is valid for heat exchangers with even number of passes, greater than 2 also. Expressions for effectiveness of heat exchangers with odd number of passes and more than 1 shall pass, can be found in the book on heat exchanger design by Shah and Sekulic.

Example 4.5

Determine the surface area of a two-pass heat exchanger which would enable transfer of heat between two fluids as stipulated in Example 4.1.

Solution

The given data regarding the two fluids are
1. Hot fluid:

$$\dot{m}_h c_h = 100 \text{ W/K} \quad \text{Temperature range: } 110-90°C$$

2. Cold fluid:

$$\dot{m}_c c_c = 50 \text{ W/K} \quad \text{Temperature range: } 40-80°C \quad \text{and} \quad U = 400 \text{ W/m}^2 \text{ K}$$

Clearly $C_R = \frac{50}{100} = 0.5$, and $\epsilon = 0.5714$ from Example 4.1.

Using this data in Eq. (4.85), we can find the value of NTU by trial and error; or we could first rearrange this equation to express NTU as a function of C_R and ϵ. This could be done by rewriting Eq. (4.85) as

$$\frac{2}{\epsilon} = 1 + C_R + \sqrt{1 + C_R^2} \cdot \frac{1 + \exp\left\{-NTU\sqrt{1 + C_R^2}\right\}}{1 - \exp\left\{-NTU\sqrt{1 + C_R^2}\right\}}$$

Transposing terms we get

$$\frac{\frac{2}{\epsilon} - (1 + C_R)}{\sqrt{1 + C_R^2}} = \frac{1 + \exp\left\{-NTU\sqrt{1 + C_R^2}\right\}}{1 - \exp\left\{-NTU\sqrt{1 + C_R^2}\right\}}$$

Calling the expression on LHS, involving known parameters ϵ and C_R, as E, we can write from above equation

$$E = \frac{\frac{2}{\epsilon} - (1 + C_R)}{\sqrt{1 + C_R^2}} = \frac{1 + \exp\left\{-NTU\sqrt{1 + C_R^2}\right\}}{1 - \exp\left\{-NTU\sqrt{1 + C_R^2}\right\}}$$

and on rearrangement, we get

$$\frac{E - 1}{E + 1} = \exp\left\{-NTU\sqrt{1 + C_R^2}\right\}$$

which gives

$$NTU = -(1 + C_R^2)^{-1/2}\ln\left\{\frac{E - 1}{E + 1}\right\} \tag{4.86}$$

where

$$E = \frac{\frac{2}{\epsilon} - (1 + C_R)}{\sqrt{1 + C_R^2}} \tag{4.87}$$

Substituting the given data in Eqs. (4.86), (4.87) we can directly get the NTU for this heat exchanger as follows:

$$E = \frac{2/0.5714 - (1 + 0.5)}{\sqrt{1 + 0.25}} = 1.789$$

$$NTU = -(1.25)^{-1/2}\ln\left\{\frac{0.789}{2.789}\right\} = 1.129368$$

This gives the total surface area required in the heat exchanger, A, as

$$A = \frac{NTU\, C_{min}}{U} = \frac{1.129368 \times 50}{400} = 0.14117 \text{ m}^2$$

It can be seen on comparing with the solution to Example 4.1 that this area is less than that needed for purely parallel-flow heat exchanger, but more than that for a counter-flow heat exchanger performing the same duty.

Many multipass configurations, like those in cross-flow heat exchangers, can be analyzed by considering these as a train of heat exchangers. Fig. 4.11 shows a typical arrangement of three-pass cross-flow heat exchanger with the fluids mixed between passes, the overall flow configuration being like that in a counter-flow heat exchanger. Similar flow configuration is also encountered in a typical shell and tube heat exchanger with baffles in the shell-side directing the shell-side fluid up and down across the tube. We shall derive an equation for the overall effectiveness of the heat exchanger in terms of the effectiveness of each pass (see Fig. 4.11).

Considering the case when the hot fluid has lower heat-capacity than the cold fluid, we can write following equations for each pass

$$\epsilon_{12} = \frac{T_{h1} - T_{h2}}{T_{h1} - T_{c2}}; \quad \epsilon_{23} = \frac{T_{h2} - T_{h3}}{T_{h2} - T_{c3}}; \quad \epsilon_{34} = \frac{T_{h3} - T_{h4}}{T_{h3} - T_{c4}}$$

From the first equation for ϵ_{12} we can easily get

$$1 - \epsilon_{12} = \frac{T_{h2} - T_{c2}}{T_{h1} - T_{c2}}$$

$$\text{and} \quad 1 - \epsilon_{12} C_R = 1 - \frac{T_{h1} - T_{h2}}{T_{h1} - T_{c2}} \cdot \left(\frac{T_{c1} - T_{c2}}{T_{h1} - T_{h2}} \right) = \frac{T_{h1} - T_{c1}}{T_{h1} - T_{c2}}$$

Dividing above two equations we get

$$\frac{1 - \epsilon_{12}}{1 - \epsilon_{12} C_R} = \frac{T_{h2} - T_{c2}}{T_{h1} - T_{c1}} \tag{4.88}$$

Following similar procedure we can get expressions for the ratio of temperature differences at the two ends of each of these "component" heat exchangers (or the "passes" of a multipass heat exchanger) as:

$$\frac{1 - \epsilon_{23}}{1 - \epsilon_{23} C_R} = \frac{T_{h3} - T_{c3}}{T_{h2} - T_{c2}} \tag{4.89}$$

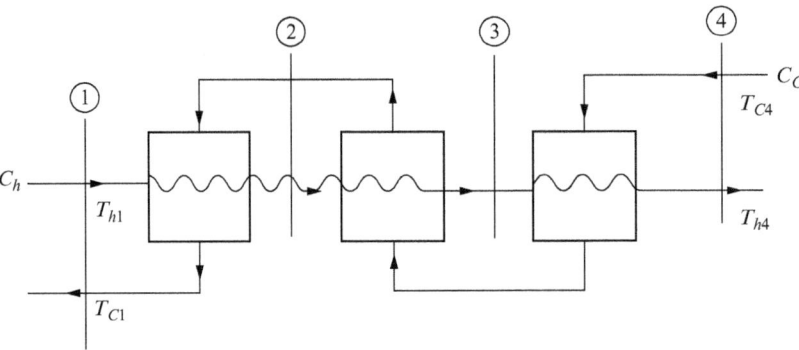

Fig. 4.11 Three-pass overall counter-flow heat exchanger.

$$\frac{1 - \epsilon_{34}}{1 - \epsilon_{34} C_R} = \frac{T_{h4} - T_{c4}}{T_{h3} - T_{c3}} \qquad (4.90)$$

Proceeding similarly, we can derive the following equation for the overall effectiveness ϵ of the heat exchanger:

$$\frac{1 - \epsilon}{1 - \epsilon C_R} = \frac{T_{h4} - T_{c4}}{T_{h1} - T_{c1}} \qquad (4.91)$$

From Eqs. (4.88) to (4.91) it immediately follows that:

$$\left(\frac{1 - \epsilon}{1 - \epsilon C_R}\right) = \left(\frac{1 - \epsilon_{12}}{1 - \epsilon_{12} C_R}\right) \left(\frac{1 - \epsilon_{23}}{1 - \epsilon_{23} C_R}\right) \left(\frac{1 - \epsilon_{34}}{1 - \epsilon_{34} C_R}\right)$$

$$= Z \quad \text{say} \qquad (4.92)$$

which gives

$$\epsilon = \frac{Z - 1}{Z C_R - 1} \qquad (4.93)$$

Clearly, if $\epsilon_{12} = \epsilon_{23} = \epsilon_{34} = \epsilon_p$ then,

$$Z = \left(\frac{1 - \epsilon_p}{1 - \epsilon_p C_R}\right)^3$$

This expression can easily be generalized for the case of a n-pass heat exchanger as:

$$Z = \left(\frac{1 - \epsilon_p}{1 - \epsilon_p C_R}\right)^n \qquad (4.94)$$

while Eq. (4.93) remains unaltered.

The above results have been derived for the case when the hot fluid has lower heat capacity. Following similar steps, it can be easily shown that we get the same expression for ϵ, given by Eqs. (4.92), (4.93), even for the case when the cold fluid has lesser heat capacity than the hot fluid.

Example 4.6

Determine the surface area required in a two-pass cross-flow heat exchanger to transfer the stipulated amount of heat between the two fluids of Example 4.1. Assume that the hot fluid is mixed and the cold fluid unmixed, and the effectiveness of each pass is the same.

Solution

The given data regarding the two fluids are

1. Hot fluid:

$$\dot{m}_h c_h = 100 \text{ W/K} \quad \text{Temperature range: } 110-90°\text{C}$$

2. Cold fluid:

$$m_c c_c = 50 \text{ W/K} \quad \text{Temperature range: } 40-80°\text{C}$$

and $U = 400 \text{ W/m}^2 \text{ K}$.

Clearly $C_R = \frac{50}{100} = 0.5$, and $\epsilon = 0.5714$ from Example 4.1.

Substituting these data in Eq. (4.94), we get

$$Z = \left(\frac{1 - 0.5714}{1 - 0.5714 \times 0.5} \right) = 0.6$$

$$= \left(\frac{1 - \epsilon_p}{1 - \epsilon_p \times 0.5} \right)^2$$

$$\Rightarrow \frac{1 - \epsilon_p}{1 - 0.5\epsilon_p} = 0.774597$$

which gives $\epsilon_p = 0.36788$.

The effectiveness of each pass ϵ_p, can be related to the NTU by Eq. (4.67), since here the hot fluid with larger heat capacity is mixed. This gives

$$0.36788 = \frac{1}{0.5}[1 - \exp\{-0.5[1 - \exp(-NTU)]\}]$$

solving this equation we get

$$NTU = 0.521776$$

which gives $A = \dfrac{NTU \; C_{min}}{U} = \dfrac{0.521776 \times 50}{400} = 0.06522 \text{ m}^2$

$\therefore$ Total surface area of the heat exchanger $= 0.13044 \text{ m}^2$

This is lesser than the area obtained for the corresponding single-pass configuration in Example 4.4. It is even lesser than the area requirement for the two-pass heat exchanger of Example 4.5.

Eqs. (4.93), (4.94) can be employed to determine the effectiveness of a variety of heat exchanger configurations by judiciously using the expressions for effectiveness of basic simple configurations. Details can be seen in Shah and Sekulic [3] and Kays and London [4].

4.1.4 Heat Exchangers With Varying Heat-Transfer Coefficients

In the analysis of heat exchangers presented above, it has been presumed that the overall heat-transfer coefficient, U, governing the rate of heat transfer between the two fluids remains constant throughout the heat exchanger length. This assumption is generally valid in most commonly encountered situations involving transfer of heat between fluids in single phase. However, in applications where either (or both) of the fluids undergo a phase change—as, for example, in boilers, condensers, or refrigeration evaporators—the overall heat-transfer coefficient can change very significantly along the heat exchanger. During phase change, heat transfer can occur without any change in the temperature and therefore the "heat capacity" defined by Eq. (4.4) becomes infinite. The concept of NTU of the whole heat exchanger is clearly inapplicable, and therefore we need to determine the heat transfer in such heat exchangers from the first principles, by writing the energy and mass balance for the fluids. We shall illustrate the procedure for a few typical cases in this section.

(a) Shell and Tube Heat Exchanger With Phase Change in the Tube-Side Fluid

We first consider a shell and tube heat exchanger where the tube side fluid undergoes phase change. Typical practical examples are the waste heat recovery boiler where water is converted into steam by hot flue gases surrounding the tubes, and the so-called DX chillers used in central air-conditioning plants where water flowing in the shell is cooled by refrigerant liquid-vapor mixture flowing through the tubes. Usually these are multipass heat exchangers, but we shall first develop the equations for a single-pass configuration and then extend these to multipass configuration. Fig. 4.12 shows the schematic diagram of such a heat exchanger. We indicate by symbol θ the temperature of the shell-side fluid at any location along the heat exchanger, and by symbol θ_i, the temperature of the tube-side fluid, the subscript i indicating the pass number (thus for a single-pass heat exchanger, i would be equal to 1). Tube wall temperature is indicated by ϕ_i, with additional subscripts s and t distinguishing the shell-side tube wall temperature ϕ_{si}, and the tube inner wall temperature ϕ_{ti}. Since the tube-side fluid is undergoing phase change, its heat-transfer coefficient changes significantly along the

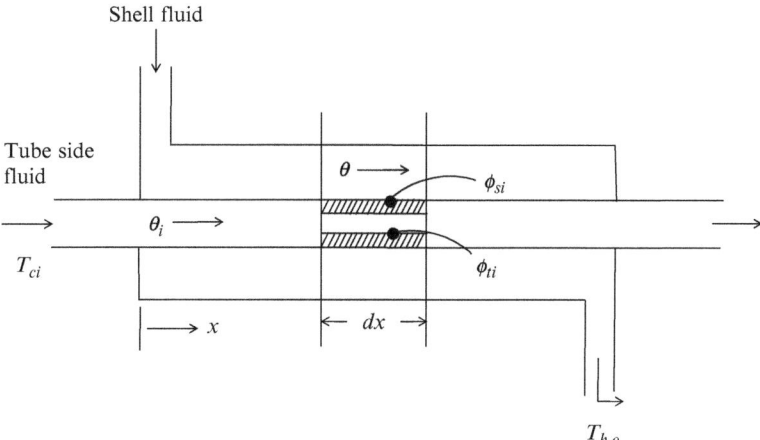

Fig. 4.12 Shell and tube heat exchanger with phase change in tube-side fluid.

flow direction. Further, if the pressure drop during the flow is significant, as is usually the case, the temperature of the tube side fluid also changes in accordance with the saturation pressure-temperature relationship of the fluid. The heat-transfer and the fluid-flow phenomena thus get coupled and we need to solve the energy-balance and the momentum-balance equations simultaneously.

To write the energy balance equations, we consider a small elemental length dx of the heat exchanger, as shown in Fig. 4.12, and equate the heat gained by the evaporating tube-side fluid to the heat lost by the shell-side fluid. This gives

$$dQ = \dot{m}_i dh_i = -\dot{m}c \, d\theta \tag{4.95}$$

where $\dot{m}$ represents the mass flow rate of shell-side stream, $\dot{m}_i$ that of the tube-side fluid, and h_i its enthalpy at location x. Further since this heat is transferred through the tubes, we can equate it to the convective heat-transfer rate at the two surfaces of the tube, and the conduction heat-transfer rate through the tube. This gives

$$dQ = \alpha(dx \, A'_s)(\theta - \phi_{si}) = \frac{\phi_{si} - \phi_{ti}}{(R'_{tw}/dx)} = \alpha_i(dx \, A'_t)(\phi_{ti} - \theta_i) \tag{4.96}$$

where α is the convective heat-transfer coefficient on the shell side of the tube, α_i that of the tube-side fluid; A'_s and A'_t are the external and the internal tube surface areas per unit length (these may differ significantly because of fins); and R'_{tw} is the thermal resistance of the tube wall per unit length. This equals $\frac{\ln(r_o/r_i)}{2\pi k}$, where r_o and r_i are the outer and the inner radii of the tube and k the thermal conductivity of the tube material.

Combining Eqs. (4.95), (4.96), we get differential equations for the tube fluid enthalpy h_i and the shell-side fluid temperature θ, and two algebraic equations relating various temperatures.

$$\dot{m}_i \frac{dh_i}{dx} = A'_t \alpha_i (\phi_{ti} - \theta_i) \tag{4.97}$$

$$-\dot{m}c \frac{d\theta}{dx} = A'_s \alpha (\theta - \phi_{si}) \tag{4.98}$$

$$(\theta - \phi_{si})\alpha A'_s + (\theta_i - \phi_{ti})\alpha_i A'_t = 0 \tag{4.99}$$

$$(\phi_{si} - \phi_{ti}) = \frac{\theta - \theta_i}{1 + \frac{1}{\alpha A'_s R'_{tw}} + \frac{1}{\alpha_i A'_t R'_{tw}}} \tag{4.100}$$

Written in this form, these equations can be easily extended to heat exchangers with more than one tube-side pass, usually encountered in practice (Fig. 4.13).

Thus, for the general case of a n-pass heat exchanger, we can obtain the set of equations representing micro- and macro-level energy balances as follows.

(a) Eq. (4.97) represents the fact that the enthalpy of the tube-side fluid increase due to heat transfer from the tube wall. This would clearly be directly applicable for each of the passes in which the fluid flows in the same direction as the shell fluid

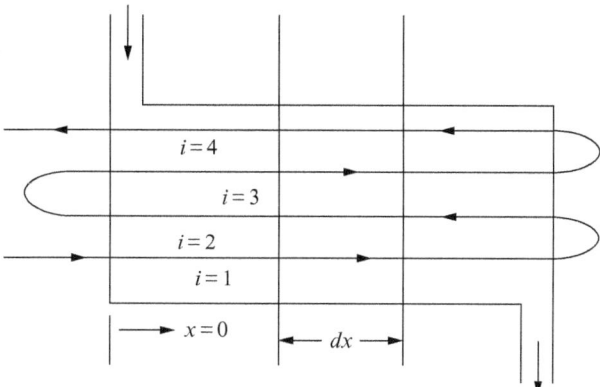

Fig. 4.13 Multipass heat exchanger with phase change in the tube-side fluid.

(i.e., x-direction). For other passes, where the tube fluid is flowing in the opposite direction, a negative sign would have to be introduced since enthalpy would be reducing in the x-direction. Thus for a n-pass heat exchanger we will have

$$\dot{m}_i \frac{dh_i}{dx} = \pm A_t' \alpha_i (\phi_{ti} - \theta_i) \quad \text{for } i = 1, 2, \ldots, n \tag{4.101}$$

$+$ sign is applicable for tube passes with tube-side fluid flowing parallel to the shell-side fluid; $-$ sign for tube passes where the tube-side fluid is flowing in counter flow to the shell-side fluid.

(b) Eq. (4.98) relates the heat transfer from the shell-side fluid to the drop in its temperature. Since in a multipass heat exchanger, the shell-side fluid when passing over an elemental length dx (see Fig. 4.13) transfers heat to each of the passes, Eq. (4.98) would be modified to

$$- \dot{m}_c \frac{d\theta}{dt} = \sum_{i=1}^{n} A_s' \alpha (\theta - \phi_{si}) \tag{4.102}$$

(c) Eq. (4.99) represents the fact that at each section of the tube, under steady-state conditions, the convective heat transfer from the shell-side fluid to the tube outer surface equals the convective heat transfer to the tube-side fluid from its inner surface. Clearly this remains valid for each section along any of the tubes. This gives

$$(\theta - \phi_{si}) \alpha A_s' = (\phi_{ti} - \theta_i) \alpha_i A_t' \quad \text{for } i = 1, 2, \ldots, n \tag{4.103}$$

(d) Eq. (4.100) relates the temperature difference across the tube wall, at any section, to the overall temperature difference between the shell-side and the tube-side fluids. This would be directly applicable to any section along any of the tubes of the multipass heat exchanger, giving

$$(\phi_{si} - \phi_{ti}) = \frac{\theta - \theta_i}{1 + \frac{1}{\alpha A_s' R_{tw}'} + \frac{1}{\alpha_i A_t' R_{tw}'}} \quad \text{for } i = 1, 2, \ldots, n \qquad (4.104)$$

Besides the above equations, we also need to have equations for pressure drop of the tube-side (evaporating) fluid since its temperature at any position along the heat exchanger, θ_i, is dependent on the pressure, p_i. Accordingly, we write for each tube pass

$$\frac{dp_i}{dx} = f_p(h_i, \ldots, \dot{m}_i) \quad \text{for } i = 1, 2, \ldots, n \qquad (4.105)$$

where function f_p, on the right-hand side of above equations is usually one of the empirical equations, mentioned in Chapter 3, for prediction of pressure gradient in two-phase flows.

This pressure gradient is a strong function of the fluid mass flow rate and vapor fraction, which in turn is directly related to the fluid enthalpy h_i, thus coupling the flow and heat-transfer equations for the tube-side fluid. The pressure drop on the shell side where there is no phase change is, on the other hand, usually not influenced significantly by the change in the shell-fluid temperature, and can therefore be calculated independently using the correlation suitable for the shell-side configuration.

Thus the simulation problem reduces to solution of $2n + 1$ ordinary differential equations, (4.101), (4.102), (4.105), and $2n$ algebraic equations, (4.103), (4.104). The total number of variables involved is $6n + 1$, that is, $\phi_{si}, \phi_{ti}, \alpha_i, h_i, p_i, \theta_i$, and θ. However, not all of these are independent. Thus θ_i and p_i are related to each other by the saturation pressure-temperature relationship of the tube-side fluid; and α_i, the local heat-transfer coefficient of the tube-side fluid, is dependant on the mass flow rate, the local dryness fraction, and the heat flux. Thus we can write

$$\theta_i = f_\theta(p_i) \qquad (4.106)$$
$$\alpha_i = f_\alpha(\dot{m}_c, h_i, \phi_{ti}, \theta_i) \qquad (4.107)$$

where function f_θ is the saturation pressure-temperature relationship of the tube-side fluid, and function f_α is the empirical correlation for determination of local heat-transfer coefficient of a fluid during boiling (Chapter 3), suitable for the configuration under consideration.

Through these relationships, the number of independent variables gets reduced to $4n + 1$, that is, $\phi_{si}, \phi_{ti}, h_i, p_i$, and θ. Further, by using $2n$ algebraic equations, (4.103), (4.104), it is possible to express ϕ_{si} and ϕ_{ti} as functions of rest of the variables. Thus finally the simulation problem is reduced to solving $2n + 1$ ordinary differential equations (i.e., Eqs. 4.101, 4.102, 4.105), in $2n + 1$ variable, that is, h_i, p_i, and θ.

To solve these equations completely, we also need an equal number of initial/boundary conditions. These can be obtained by considering the compatibility conditions at the two ends of the heat exchanger, depicted for a typical case of four-pass heat exchanger in Fig. 4.14. Thus for the two possible configurations of the four-pass heat exchanger, the $4 \times 2 + 1$ conditions needed to solve the nine ordinary differential equations are

1. Configuration (A, Fig. 4.14) *first tube pass parallel to shell-fluid flow*

$$\text{At } x = 0 \quad \theta, p_1, h_1, \text{ are known} \quad \text{(i, ii, iii)}$$
$$p_2 = p_3 \quad \text{(iv)}$$
$$h_2 = h_3 \quad \text{(v)}$$
$$\text{At } x = L \quad p_1 = p_2 \quad \text{(vi)}$$
$$h_1 = h_2 \quad \text{(vii)}$$
$$p_4 = p_3 \quad \text{(viii)}$$
$$h_4 = h_3 \quad \text{(ix)}$$

2. Configuration (B, Fig. 4.14) *first tube pass fluid in-counter flow to the shell-side fluid*

$$\text{At } x = 0 \quad \theta \text{ is known} \quad \text{(i)}$$
$$p_3 = p_4 \quad \text{(ii)}$$
$$h_3 = h_4 \quad \text{(iii)}$$
$$p_1 = p_2 \quad \text{(iv)}$$
$$h_1 = h_2 \quad \text{(v)}$$
$$\text{At } x = L \quad p_1, h_1 \text{ are known} \quad \text{(vi, vii)}$$
$$p_2 = p_3 \quad \text{(viii)}$$
$$h_2 = h_3 \quad \text{(ix)}$$

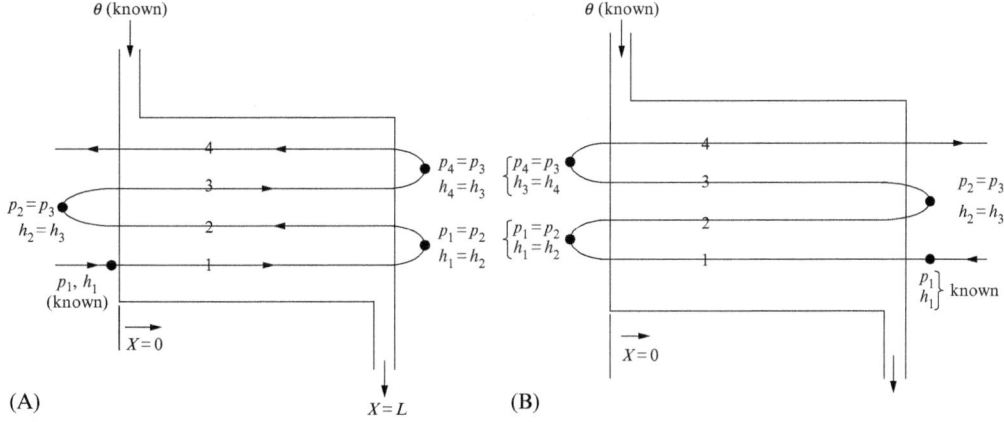

Fig. 4.14 Initial/boundary conditions for heat exchanger.

Clearly, in both the cases we have mixed initial and boundary value problems, which can only be solved iteratively. The values of the variables not known at $x = 0$ are assumed, and the differential equations are solved as initial value problems to find the values of the variables at $x = L$.

Thus, for example, for configuration (A) of the four-pass exchanger of Fig. 4.14, we assume the values of $p_2, h_2, p_4,$ and h_4 at $x = 0$, and solve the differential equations as initial value problems to obtain the values of all the variables at $x = L$. We then check for the correctness of the assumptions through the compatibility conditions, that is, Eqs. (vi)–(ix). Depending upon the extent of the errors in satisfaction of these equations, new assumptions are made for the four variables and iterations carried out till the convergence is reached. The above procedure can be easily generalized for simulation of any n-pass heat exchanger.

This method has been successfully applied by Dhar and Arora [5] for simulation of a eight-pass 40-TR DX-chiller manufactured by a prominent Indian company. The predicted heat-transfer rate is 40.4TR, which differs from the cited value by 1%. The predicted refrigerant side pressure drop is 2.5 psi (17.2 kPa) as against the approximate value of 2 psi (13.8 kPa) indicated by the company.

(b) Shell and Tube Heat Exchanger With Phase Change in the Shell-Side Fluid

Typical examples of such heat exchangers are the condensers and the flooded chillers used in large refrigeration plants, and the fire-tube boilers used for generating steam at moderate pressures. We shall illustrate the approach of simulating such systems by considering the case of a two-pass flooded chiller shown in Fig. 4.15.

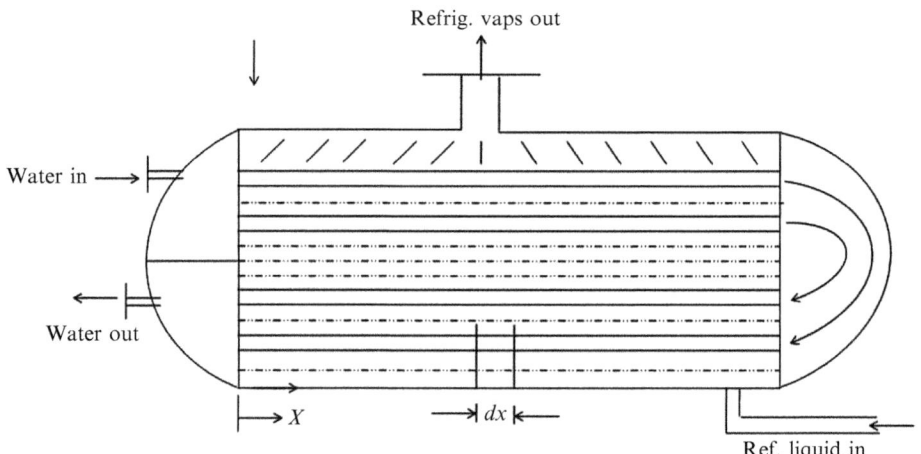

Fig. 4.15 Flooded liquid chiller.

Here the tubes, through which the water to be chilled is circulated, are submerged in a pool of refrigerant, contained in the shell. As the refrigerant boils, the heat-transfer coefficient outside the tubes does not remain constant, and we need to develop a procedure to predict the total heat transfer taking place in the heat exchanger. Since the pool of refrigerant boils at a constant temperature, the situation here is much simpler than in the DX-chiller considered above, and as indicated by Dhar and Jain [6], it is possible to integrate the differential equation governing the temperature change of the coolant along the heat exchanger length. This equation can be directly obtained by considering the energy balance of a small control volume of the coolant as it flows through a tube, see Fig. 4.15. This gives

$$\dot{m}c\frac{d\theta}{dx} = -UA'_s(\theta - \theta_r) \tag{4.108}$$

where θ is the coolant temperature at any location x; θ_r the refrigerant temperature; U the local overall heat-transfer coefficient; $\dot{m}$ the mass flow rate of the coolant and c its specific heat; and A'_s is the surface area per unit length. Usually this surface area is taken as the tube outside surface area (often finned) and U is also defined with respect to this area. The overall heat-transfer coefficient U incorporates the effects of the convective heat-transfer resistances of the boiling refrigerant, and water flowing through the tubes, the tube metal wall resistance, and various fouling resistances. Out of these, only the boiling heat-transfer coefficient varies along the tube length, and all others can be assumed to remain constant. Thus we can write the following simplified expression for U:

$$\frac{1}{U} = C_1 + \frac{1}{C_2\alpha_r} \tag{4.109}$$

where α_r is the heat-transfer coefficient of the boiling refrigerant and C_1 and C_2 are constants depending on the tube geometry and other thermal resistances enumerated above, whose values can found following procedures explained in Chapter 3. Now, as is well known, (see Chapter 3) the heat-transfer coefficient in pool boiling depends on the temperature difference between the hot surface and the boiling liquid (i.e., $(\phi_{si} - \theta_r)$). Conventionally this dependence is indicated by expressing α_r as a function of the heat flux, q, as

$$\alpha_r = Bq^n \tag{4.110}$$

where B and n are constants depending upon the thermophysical properties of the refrigerant and the tube geometry (Chapter 3). Combining Eqs. (4.109), (4.110), we can write

$$\frac{1}{U} = \frac{\theta - \theta_r}{q} = C_1 + \frac{1}{BC_2q^n}$$

$$\text{or} \quad D_1 q^{1-n} + q - \frac{(\theta - \theta_r)}{C_1} = 0 \tag{4.111}$$

$$\text{where } D_1 = \frac{1}{BC_1 C_2}$$

Since θ_r is constant, Eq. (4.111) provides us with a method to compute the heat flux q at any position along the tubes as a function of the coolant temperature. Thus in the first pass, where the coolant entry temperature is known, q at the entry point can be evaluated by substituting $\theta = \theta_i$ in Eq. (4.111). Now, the variation of coolant temperature along the heat exchanger length is governed by Eq. (4.108), and to relate it to the variation in local heat flux, we combine it with Eq. (4.111) to get

$$\dot{m} c \frac{d\theta}{dx} = -UA'_s(\theta - \theta_r) = -A'_s \cdot q \tag{4.112}$$

Further, by differentiating Eq. (4.111) we get

$$\frac{d\theta}{dx} = C_1 \cdot \left[D_1 \cdot \frac{1-n}{q^n} + 1 \right] \frac{dq}{dx} \tag{4.113}$$

Combining Eqs. (4.112), (4.113) we get

$$\frac{\dot{m} c}{q} \frac{d\theta}{} = -(dx A'_s) = \dot{m} c \left[\frac{C_1 D_1 (1-n)}{q^{n+1}} + \frac{C_1}{q} \right] dq \tag{4.114}$$

Eq. (4.114) can be easily integrated to express the local heat flux q, at any position along the tube, as a function of the heat flux at entry, that is, (q_i, at $x = 0$) and other design parameters as:

$$- A'_s x = \dot{m} c \left[-\frac{C_1 D_1 (1-n)}{n} \left\{ \frac{1}{q^n} - \frac{1}{q_i^n} \right\} + C_1 \ln \left(\frac{q}{q_i} \right) \right]$$

$$\text{or} \quad k_1 \left(\frac{1}{q^n} - \frac{1}{q_i^n} \right) + k_2 \ln \left(\frac{q}{q_i} \right) + x = 0 \tag{4.115}$$

where $k_1 = \frac{C_1 D_1 (n-1) \dot{m} c}{n A'_s}$ and $k_2 = \frac{\dot{m} c C_1}{A'_s}$. Eq. (4.115) enables us to determine the local heat flux at any position x along the tube, using the value of q_i obtained by substituting the known value of fluid inlet temperature in Eq. (4.111). By substituting $x = L$, in Eq. (4.115), we can obtain the heat flux at the tube outlet, q_L. Substituting this value of q_L in Eq. (4.111), we can determine the coolant outlet temperature, θ_L as:

$$\theta_L = \theta_r + C_1 D_1 q_L^{1-n} + C_1 q_L \tag{4.116}$$

Thus both the coolant inlet and outlet temperatures are known and the total heat transfer in that pass can be easily computed. The same procedure can be applied successively to

Table 4.1 Rated and predicated performance of the flooded chiller

Section no.	$\dot{m}$ (kg/s)	θ_i (°C)	θ_r (°C)	TR Pred.	TR Rated	TR % Diff	PD Pred.	PD Rated	PD % Diff.
1	18.93	11.67	2.22	115.3	111.25	3.64	0.113	0.135	−16.3
2	18.93	14.44	−0.95	203.5	200.0	1.75	0.113	0.135	−16.3
3	22.71	11.67	2.54	124.56	120.0	3.8	0.157	0.185	−15.24
4	22.71	15.56	2.88	186.30	180.0	3.5	0.157	0.185	−15.24
5	25.24	11.67	3.24	120.61	115.64	4.3	0.189	0.224	−15.38
6	28.39	11.67	1.91	154.23	150.0	2.82	0.234	0.276	−15.63
7	28.39	15.56	1.82	235.05	225.0	4.47	0.234	0.276	−15.63

other passes, using the outlet temperature of a pass as the inlet temperature for the next pass and so on. The total heat transfer in the chiller can thus be computed accurately.

This simulation procedure has been successfully used to predict the performance of a four-pass 120TR flooded chiller with integrally finned tubes being manufactured by a prominent Indian company. As Table 4.1 (Dhar and Saraf [7]) shows at a glance, the predicted and the rated values of chiller capacity are quite close to each other.

It can be seen that the chiller capacity is predicted very well, while there are larger deviations in pressure drop prediction. The reasons for the same have been discussed by Dhar and Saraf [7].

Example 4.7

Determine the rate of heat transfer in a two pass fire-tube type waste heat recovery boiler of the following specifications:

1. First pass tube diameter = 20 cm, length = 4 m;   Second pass has 10 tubes of 6 cm ID.
2. Water inlet temperature = 50°C;   Steam generation temperature = 150°C.
3. Flue gas inlet temperature = 400°C;   Flow rate = 0.3 kg/s.

Solution

Fig. 4.16 shows the schematic diagram of the boiler producing steam at 150°C. Since the feed water is entering at 50°C, the temperature of water in the vicinity of entry point will be lesser than the saturation temperature of 150°C. However, since the mass of water in the shell is very large, we assume that the water surrounding the tubes is also at ~150°C.

Since water is boiling in a pool without any significant motion, we use the correlation given by Gorenflo [8] Eqs. (3.274), (3.277), (3.278) to determine the heat-transfer coefficient h_{nb} of boiling water.

The surface roughness correction term $(Rp/Rpref)^{0.133}$ is taken as 1. We find the value of pressure from our problem data:

$$\text{Boiler pressure} = \text{Saturation pressure at } 150°C = 4.758 \text{ bar}$$

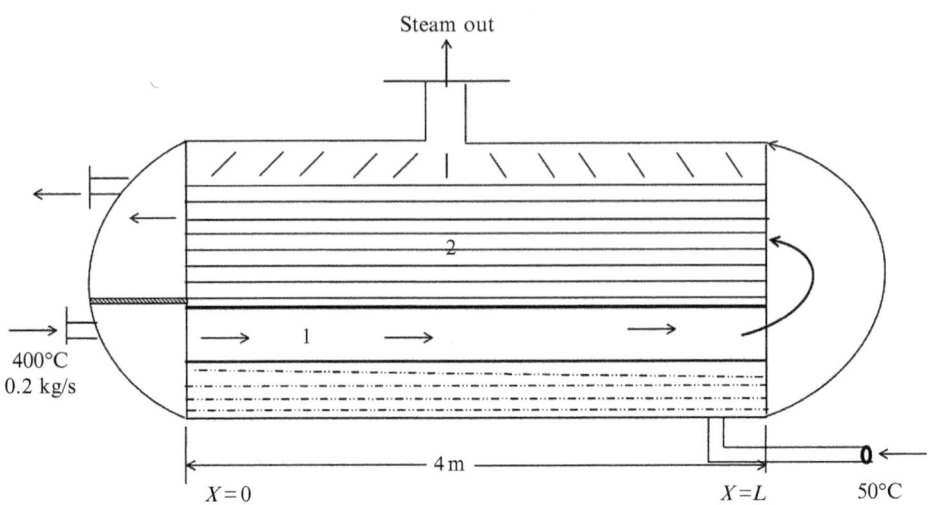

Fig. 4.16 Waste heat recovery boiler of Example 4.7.

This gives $p_r = \frac{4.758}{220.6} = 0.0216$, where the critical pressure of water is taken as 220.6 bar. Further using Eqs. (3.277), (3.278) we get

$$n = 0.9 - 0.3p_r^{0.15} = 0.9 - 0.3(0.0216)^{0.15} = 0.731$$

$$F_{PF} = 1.73(0.0216)^{0.27} + 6.1(0.0216)^2 + \frac{0.68(0.0216)^2}{1 - 0.0216} = 0.6174$$

Substituting these values in Eq. (3.274) we get: $\frac{h_{nb}}{h_{ref}} = 0.6174 \left(\frac{q}{20,000}\right)^{0.731}$.

Now for water $h_{ref} = 5600 \text{ W/m}^2 \text{ K}$ which gives $h_{nb} = \frac{5600 \times 0.6174}{(20,000)^{0.731}} \cdot q^{0.731} = 2.4814q^{0.731}$. Thus comparing with Eq. (4.110), used in the above formulation, here

$$\alpha_r = h_{nb} = 2.4814q^{0.731}; \quad \text{which gives } B = 2.4814 \text{ and } n = 0.731.$$

We now determine the heat-transfer coefficient for hot gases flowing inside the big tube constituting the first pass of the boiler. We assume the thermophysical properties of the gas to be the same as those of air. At an estimated mean temperature of 350°C, these are [2]

$$\rho = 0.558 \text{ kg/m}^3 \quad c_p = 1.057 \text{ kJ/kg K} \quad Pr = 0.687$$

$$\mu = 314 \times 10^{-7} \text{ Ns/m}^2 \quad k = 48.3 \times 10^{-3} \text{ W/m K}$$

To determine the heat-transfer coefficient, we use simplified Gnielinski correlation, Eq. (3.252) viz.

$$Nu = 0.0214(Re_D^{0.8} - 100)Pr^{0.4}$$

which is mentioned to be valid in the range:

$$0.5 \leq Pr \leq 1.5 \quad 10^4 \leq Re_D \leq 5 \times 10^6$$

Here

$$Re_D = \frac{\rho VD}{\mu} = \frac{GD}{\mu} = \frac{\dot{m}}{\frac{\pi}{4}D^2} \cdot \frac{D}{\mu} = \frac{4\dot{m}}{\pi \mu D}$$

$$= \frac{4 \times 0.3}{\pi \times (314 \times 10^{-7}) \times 0.2} = 6.0824 \times 10^4$$

Thus both Pr and Re_D are within the range of validity of this equation.

Substituting these values in the above equation we get

$$Nu = 0.0214((6.0824 \times 10^4)^{0.8} - 100)(0.687)^{0.4}$$

$$= 121.88$$

This gives

$$\alpha_i = \frac{121.88 \times 48.3 \times 10^{-3}}{0.2} = 29.43 \, \text{W/m}^2 \, \text{K}$$

The overall heat-transfer coefficient at any section, Eq. (4.109), can be written as:

$$\frac{1}{U} = \frac{1}{\alpha_i} + \text{Metal wall resistance} + \text{Fouling resistance} + \frac{1}{\alpha_r}$$

where we have neglected the small difference in the inner and outer surface areas/unit length of the tube. We take the value of metal wall resistance plus the fouling resistance as $10^{-3} \, \text{m}^2 \, \text{K/W}$. This gives

$$\frac{1}{U} = \frac{1}{29.43} + 0.001 + \frac{1}{\alpha_r} = 0.03498 + \frac{1}{\alpha_r}$$

Thus comparing with Eq. (4.109), we have

$$C_1 = 0.03498 \quad \text{and} \quad C_2 = 1$$

We can now initiate the calculation of heat flux at various positions along the central tube of the boiler. To determine the heat flux at the flue gas entry point $q_{1,0}$, where the gas temperature is known ($=400°C$), we use Eq. (4.111). Substituting the values of various constants, as obtained above, we get

$$D_1 = \frac{1}{BC_1 C_2} = \frac{1}{2.4814 \times 0.03498 \times 1} = 11.52$$

and Eq. (4.111) becomes

$$11.52q^{0.269} + q - \frac{(400 - 150)}{0.3498} = 0$$

$$\text{or} \quad 11.52q^{0.269} + q - 7146.94 = 0$$

We solve this equation by trial and error, to get

$$q = 7020 \, \text{W/m}^2 \, \text{K} = q_{1,0}$$

Next, we find the heat flux at the end of this pass by using Eq. (4.115). The constants k_1 and k_2 are

$$k_1 = \frac{C_1 D_1 (n-1) \dot{m} c}{n A'_s} = \frac{0.03498 \times 11.52(0.731-1)(0.3 \times 1057)}{0.731 \times (\pi \times 0.2)} = -74.838$$

$$k_2 = \frac{\dot{m} c C_1}{A'_s} = \frac{0.3 \times 1057 \times 0.03498}{\pi \times 0.2} = 17.65$$

Substituting these values, Eq. (4.115) becomes, for $x = L = 4$ m:

$$-74.838 \left(\frac{1}{q^{0.731}} - \frac{1}{0.7020^{0.731}} \right) + 17.65 \ln \left(\frac{q}{7020} \right) + 4 = 0$$

This equation again can be solved only by trial and error, to get

$$q = 5600 \, \text{W/m}^2 \, \text{K} \quad \text{at } x = L$$

$$= q_{1L}$$

We can now find the temperature of the flue gases at the end of the first pass by substituting q_{1L} value obtained above in Eq. (4.111). This gives

$$11.52(5600)^{0.269} + 5600 = \frac{(\theta - 150)}{0.03498}$$

$$\Rightarrow \theta = 350^\circ \text{C}$$

The total heat transfer in the first pass is therefore

$$Q_1 = 0.3 \times 1.057 \times (400 - 350) = 15.855 \, \text{kW}$$

The mean temperature of the hot gases in the first pass turns out to be 375°C, which is different from the assumed value of 350°C, at which various thermophysical properties have been evaluated. Since the variation in these properties for this difference in average temperature is not going to significantly influence the results, we need not carry out another iteration of calculations and can use the same values for calculations of the second pass.

To find the heat transfer through the second pass, we first find the heat-transfer coefficient of gases in each tube, assuming that the flow is equally distributed in all the 10 tubes. To do so, we first determine the Reynolds number of flow:

$$Re = \frac{4 \dot{m}}{\pi \mu D} = \frac{4 \times (0.3/10)}{\pi \times (314 \times 10^{-7}) \times 0.06} = 20,275$$

Substituting it in the simplified Gnielinski correlation, we get

$$Nu = 0.0214(20,275^{0.8} - 100)(0.687)^{0.4}$$

$$= 49.535$$

$$\text{which gives} \quad \alpha_i = \frac{49.535 \times 48.3 \times 10^{-3}}{0.06} = 39.88 \text{ W/m}^2 \text{ K}$$

$$\Rightarrow \frac{1}{U} = \frac{1}{\alpha_i} + 0.001 + \frac{1}{\alpha_r} = 0.02608 + \frac{1}{\alpha_r}$$

which gives the values of the coefficients corresponding to Eq. (4.109) as

$$C_1 = 0.02608 \quad \text{and} \quad C_2 = 1$$

$$\therefore \quad D_1 = \frac{1}{BC_1C_2} = \frac{1}{2.4814 \times 0.02608 \times 1} = 15.452$$

Using this value of D_1 in Eq. (4.111), with $\theta = 350°C$, the heat flux at the entrance to the second pass can be determined from the equation

$$15.452q^{0.269} + q - \frac{(350 - 150)}{0.02608} = 0$$

By trial and error we get $q \simeq 7500 \text{ W/m}^2 = q_{2L}$.

To find the heat flux at the end of the second pass, we can use Eq. (4.115), appropriately accounting for the change in the direction of flow. First, the new values of coefficients k_1 and k_2 are to be found.

$$k_1 = \frac{0.02608 \times 15.452(0.731 - 1)(0.03)(1057)}{0.731 \times (\pi \times 0.06)} = -24.947$$

$$k_2 = \frac{0.03 \times 1057 \times 0.02608}{\pi \times 0.06} = 4.387$$

Eq. (4.115) thus gives

$$-24.947 \left(\frac{1}{q^{0.731}} - \frac{1}{7500^{0.731}} \right) + 4.387 \ln \left(\frac{q}{7500} \right) + 4 = 0$$

whose solution (by trial and error) is

$$q = 3040 \text{ W/m}^2$$

The exit temperature of hot gases can now be found by substituting this value of q in Eq. (4.111). This gives

$$15.452(3040)^{0.269} + 3040 - \frac{\theta - 150}{0.02608} = 0$$

$$\text{or} \quad 133.622 + 3040 = \frac{\theta - 150}{0.02608}$$

$$\text{which gives} \quad \theta = 232.8°C$$

Total heat transfer in the second pass is therefore

$$Q_2 = (0.3)(1.057)(350 - 232.8) = 37.16 \text{ kW}$$

Thus the total heat-transfer rate in the complete boiler is

$$Q = Q_1 + Q_2 = 15.855 + 37.16 = 53.015 \, kW$$

Note: The mean temperature of the gas in the second pass based on above calculations is $(350 + 232.8)/2 = 291.4°C$. Thermophysical property values at this temperature differ from those at $350°C$ (corresponding to which the values have been used in the above calculations) by about 5% to 10%. The students are advised to do another iteration of calculations using the properties evaluated at $290°C$ and find out the influence of thermophysical properties on the predicted rate of heat transfer.

4.1.5 Pressure Drop

While the main purpose of a heat exchanger is to transfer heat between two fluids, it is also important that this be achieved with minimal consumption of power for circulating the fluids through the heat exchanger. The pumping power is directly governed by the drop in the pressure of the fluids flowing through the heat exchanger. Therefore, determination of this pressure drop is also an important objective of heat exchanger simulation.

The pressure drop in a heat exchanger can be considered to have two major constituents, that is (a) the pressure drop inside the core of the heat exchanger (i.e., in the tubes, the shell, the plates, the finned passages, etc.) where the heat transfer takes place and (b) the pressure drop in the flow distribution devices such as the inlet and the outlet headers, pipe bends, etc. Ideally, we would like the second component to be as small as possible, but in modern compact heat exchangers like the plate heat exchangers, this is usually not the case. We therefore need to understand the procedure for calculation of both these constituents of the pressure drop. Since the pressure drop depends very strongly on the geometry of the heat exchanger, we need to develop these procedures separately for each configuration.

(a) Plate-Fin Heat Exchangers

Fig. 4.17 shows the typical configuration of a plate-fin heat exchanger. As the flow enters the passages in the plates, there is a sudden reduction in the free-flow area from that in the incoming passage. Due to sudden contraction, the velocity increases in the *vena contracta* and then stabilizes to the steady value in the passage. The net pressure drop due to this entrance phenomenon is indicated as Δp_{12} in Fig. 4.17.

The bulk of the pressure drop, of course, occurs within the length of heat exchanger passages mainly due to skin friction. This core pressure drop is indicated as Δp_{23} in Fig. 4.17. If there is an appreciable rise in the temperature of a fluid while passing through the heat exchanger, the resulting fall in density may cause a significant acceleration of the flow which would also contribute to the core pressure drop. On

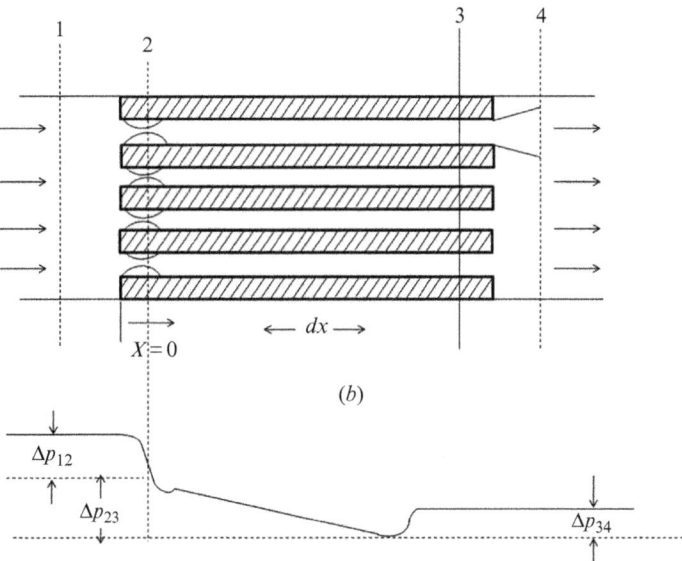

Fig. 4.17 Pressure drop in plate fin heat exchangers.

the other hand, a significant rise in density due to drop in temperature would reduce the total pressure drop in the core.

As the fluid leaves the heat exchanger and enters the outlet duct, there is a velocity drop due to sudden increase in the flow area. This would cause a small pressure recovery, indicated as Δp_{34} in Fig. 4.17. The total pressure drop in the plate-fin heat exchanger is therefore given by the equation:

$$\Delta p = \Delta p_{12} + \Delta p_{23} - \Delta p_{34} \tag{4.117}$$

We shall see below how each of these components can be evaluated.

The first of these components, i.e., Δp_{12}, the pressure drop at entry into the heat exchanger core can be visualized as arising due to two reasons, that is, Δp_{area}, the pressure drop due to reduction in the flow area, from A_1 to A_2 (Fig. 4.17); and $\Delta p_{free\ exp.}$ arising due to the irreversibilities associated with free expansion following sudden contraction. Since the fluid density change during flow from Section 1 to 2 is very insignificant, we can apply Bernoulli's equation to estimate the pressure drop due to area change alone as

$$\Delta p_{area} = \rho_1 \left(\frac{U_2^2}{2} - \frac{U_1^2}{2} \right) \tag{4.118}$$

where ρ_1 is the fluid density at Section 1, U_1 the velocity at that section, and U_2 the velocity inside the heat exchanger passages. The velocity change from U_1 to U_2 can be related to the change in the free flow area by the continuity equation:

$$\rho_1 A_1 U_1 = \rho_2 A_2 U_2$$

Introducing σ as the ratio of core free flow area (A_2) to the frontal area (A_1), we get

$$U_1 = \frac{A_2}{A_1} U_2 = \sigma U_2 \tag{4.119}$$

since $\rho_1 \approx \rho_2$. We can now write Eq. (4.118) as

$$\begin{aligned}
\Delta p_{area} &= \frac{\rho_1 U_2^2}{2} \cdot (1 - \sigma^2) \\
&\approx \frac{(\rho_2 U_2)^2}{2\rho_i}(1 - \sigma^2) \\
&= \frac{G^2}{2\rho_i}(1 - \sigma^2) \tag{4.120}
\end{aligned}$$

where G is the mass velocity inside the heat exchanger core and ρ_i is the fluid density at core inlet ($\rho_i \approx \rho_1 \approx \rho_2$).

The second component of the entry pressure loss, namely the pressure drop due to irreversible free expansion downstream of the *vena contracta* is usually estimated with the help of an empirical factor, termed the contraction loss coefficient k_c which, when multiplied with the dynamic velocity head at the inlet to the heat exchanger core, gives the magnitude of this pressure drop:

$$\Delta p_{free\ exp.} = k_c \frac{\rho_i U_2^2}{2} = k_c \frac{G^2}{2\rho_i} \tag{4.121}$$

The value of k_c depends on the contraction ratio σ, the Reynolds number, and the flow geometry. For most practical situations involving turbulent flow its value lies between 0.2 and 0.4, the larger value being for lower value σ (for more accurate values refer to Kays and London [4, Chapter 5]).

The total pressure drop at the entrance to the heat exchanger core can now be written, by combining Eqs. (4.120), (4.121), as:

$$\Delta p_{12} = \Delta p_{area} + \Delta p_{free\ exp.} = \frac{G^2}{2\rho_i}(1 - \sigma^2 + k_c) \tag{4.122}$$

Following a similar approach, we can analyze the impact of fluid exit from the heat exchanger core on the pressure drop. Since there will be a deacceleration of flow due to area increase, this would result in a pressure rise which can be calculated by an expression similar to Eq. (4.120). There would also be a *pressure drop* due to irreversible free expansion and momentum changes associated with sudden increase in area, which can again be estimated using an expression similar to Eq. (4.121) with the contraction coefficient k_c, replaced by an expansion coefficient k_e. The net *pressure rise* at the exit of the heat exchanger can therefore be estimated as:

$$\Delta p_{34} = \frac{G^2}{2\rho_o}(1 - \sigma^2 - k_e) \tag{4.123}$$

where ρ_o represents the fluid density at heat exchanger exit.

The main contribution to the pressure drop in a heat exchanger, however, arises from the flow through the heat exchanger core. This can further be subdivided into two constituents: the friction pressure drop and that due to fluid acceleration caused by change in the temperature due to heat transfer. The friction pressure drop is usually estimated from the standard Darcy equation (Chapter 3, Eq. 3.128):

$$\Delta p_{fric} = \frac{4fL}{D}\frac{\rho_m V_m^2}{2} = \frac{4fL}{D_h}\frac{G^2}{2\rho_m} \tag{4.124}$$

where ρ_m represents the mean fluid density within the heat exchanger, V_m the mean velocity, f the friction factor, D_h the hydraulic diameter, and L the heat exchanger length.

Similarly the acceleration pressure drop, due to drop in density associated with heating of the fluid within the heat exchanger, can be estimated from momentum conservation equation (3.185) as:

$$\Delta p_{acc} = \frac{(GA)(V_o - V_i)}{A} = G^2\left(\frac{1}{\rho_o} - \frac{1}{\rho_i}\right) \tag{4.125}$$

where V_o, V_i are the fluid velocities at the outlet and the inlet to the heat exchanger.

The total core pressure drop is thus

$$\Delta p_{23} = \Delta p_{fric} + \Delta p_{acc} = \frac{4fL}{D_h}\frac{G^2}{2\rho_m} + G^2\left(\frac{1}{\rho_o} - \frac{1}{\rho_i}\right) \tag{4.126}$$

The total pressure drop in a plate-fin heat exchanger can thus be calculated as:

$$\Delta p = \Delta p_{12} + \Delta p_{23} - \Delta p_{34} = \frac{G^2}{2\rho_i}\left[(1 - \sigma^2 + k_c) + \frac{4fL}{D_h}\cdot\frac{\rho_i}{\rho_m} + 2\left(\frac{\rho_i}{\rho_o} - 1\right)\right.$$
$$\left. -(1 - \sigma^2 - k_e)\frac{\rho_i}{\rho_o}\right] \tag{4.127}$$

(b) Tube Fin Heat Exchangers

This is the most popular heat exchanger configuration for transferring heat between a liquid, which flows inside the tubes, and a gas, which flows outside the tubes. Since the heat-transfer coefficients for gas flows are much lesser than those for liquid flows, fins are usually provided outside the tubes. For determining the pressure drop inside the tubes, we follow the same procedure as for plate-fin surfaces, using appropriate values of k_c, k_e, and f depending on the tube cross section (circular, oval, or flat end) and other parameters (such as Reynolds number and the entrance/exit area ratios). While determining the pressure drop of the fluid flowing outside the tubes, we need to distinguish between two design configurations. For flow over bank of tubes (whether bare or finned) (see Fig. 3.7), there is expansion and contraction of the flow in each tube row. Thus the behavior of the first and the last rows is not much different from those in the interior rows. Therefore, it is customary to combine these expansion and contraction losses with the core friction losses. The friction factor f (generally obtained experimentally) thus incorporates the entrance and the exit effects which need not be accounted for separately. The total pressure drop is thus given by

$$\Delta p = \frac{G^2}{2\rho_i} \left[\frac{4fL}{D_h} \cdot \frac{\rho_i}{\rho_m} + 2\left(\frac{\rho_i}{\rho_o} - 1 \right) \right] \tag{4.128}$$

On the other hand for heat exchangers having flat fins on an array of tubes (very common in small-sized heat exchangers like the condenser and the evaporator coils of window air conditioners), all the components of the total pressure drop indicated in Eq. (4.117) are relevant. However, in these heat exchangers, the free flow area near the entrance and the exit of the heat exchanger, are different from that in the core due to obstruction by the tubes. This needs to be accounted for while calculating the entrance, exit, and the core losses. Introducing G', the mass velocity at the leading edge and σ', the corresponding area ratio, we can use continuity equation to get

$$G'\sigma' = G \cdot \sigma = \text{Mass flow rate}$$

$$\text{where} \quad \sigma' = \frac{A_{o,leading\ edge}}{A_{fr}} \quad \text{and} \quad \sigma = \frac{A_{o,core}}{A_{fr}} \tag{4.129}$$

The values of k_c and k_e could be found using σ' in the charts given by Kays and London [4]. The expression for the total pressure drop can therefore be written as:

$$\Delta p = \frac{G^2}{2\rho_i} \left[\frac{4fL}{D_h} \cdot \frac{\rho_i}{\rho_m} + 2\left(\frac{\rho_i}{\rho_o} - 1 \right) \right] + \frac{G'^2}{2\rho_i} \left[(1 - \sigma'^2 + k_c) - (1 - \sigma'^2 - k_e) \frac{\rho_i}{\rho_o} \right] \tag{4.130}$$

(c) Shell and Tube Heat Exchangers

The tube-side pressure drop in shell and tube heat exchangers can be calculated from Eq. (4.127) by using appropriate values of the factors k_c, k_e, and f. In single-pass heat exchangers, the entrance and exit losses are usually insignificant, and often neglected. Eq. (4.128) can then be used for estimating the pressure drop. In multipass heat exchangers (see Fig. 4.13), especially those using U-tubes or hairpin bends, the pressure drop due to these bends needs to be accounted for. This is usually done using appropriate coefficients, determined experimentally. Shah and Sekulic [3] give detailed procedures for a variety of bends. The estimation of shell-side pressure drop is extremely complicated due to complex flow field caused by the presence of baffles. These baffles repeatedly change the direction of flow of the shell fluid. Besides, we also have leakage through the clearance space around the tubes in the baffles. Semiempirical methods have therefore been devised for determination of the total pressure drop, details of which can be found in handbooks like Shah and Sekulic [3].

(d) Plate Heat Exchangers

The total pressure drop in plate heat exchangers can also be visualized as consisting of three components, that is, the pressure drop at the inlet and the outlet manifolds and ports, the pressure drop in the plate passages, and that due to elevation changes (for vertical flow heat exchangers). The pressure drop in the inlet/outlet manifolds and ports is generally put at 1.5 times the velocity head per pass. The pressure drop inside the plate passages, can be determined using Eq. (4.126), taking appropriate value of friction factor f estimated on the basis of the Reynolds number for flow through the plates. Accordingly the total pressure drop on one fluid side in a plate heat exchanger is given by

$$\Delta p = \frac{1.5 G_p^2 n_p}{2\rho_i} + \frac{4fL}{D_h}\frac{G^2}{2\rho_m} + G^2\left(\frac{1}{\rho_o} - \frac{1}{\rho_i}\right) + \rho_m g L \tag{4.131}$$

$$G_p = \frac{\dot{m}}{\frac{\pi}{4}D_p^2} \tag{4.132}$$

where D_p is the port diameter, G_p the mass velocity in the port, n_p the number of passes on the given fluid side, D_h the equivalent diameter of plate passage (usually taken as twice the plate spacing), and L the height of the outlet port above the inlet port. The friction factor f for flow through the channel can be calculated using the empirical relation

$$f = \frac{K}{Re^z} \tag{4.133}$$

where values of constants K and z, which depend upon the plate geometry and the Reynolds number, can be seen from the tables given in Bejan and Kraus [9].

Example 4.8

Estimate the pressure drop on the air side of a air-to-flue gas cross-flow plate fin heat exchanger from the following data:

$$\text{Air flow rate at inlet} = 1\,\text{m}^3/\text{s}$$
$$\text{Air inlet temperature} = 30°\text{C}$$
$$\text{Air inlet pressure} = 101\,\text{kPa}$$
$$\text{Air outlet temperature} = 130°\text{C}$$
$$\text{Minimum free flow area} = 0.15\,\text{m}^2$$
$$\text{Free flow area/frontal area} = 0.6$$
$$\text{Length of plate in the direction of air flow} = 0.8\,\text{m}$$

$$\text{Hydraulic diameter} = 0.004\,\text{m}$$

$$\text{Friction factor} = 0.05$$

Solution

To use Eq. (4.127) to determine the pressure drop, we first find the air density at the inlet and the outlet of the heat exchanger using ideal gas equation of state. This gives

$$\rho_i = \left(\frac{P}{RT}\right)_i = \frac{101 \times 1000}{(8314.3/29)303} = 1.1626\,\text{kg/m}^3$$

$$\text{and} \quad \rho_o = \left(\frac{P}{RT}\right)_o = \frac{101 \times 1000}{(8314.3/29)403} = 0.8742\,\text{kg/m}^3$$

where we have taken, as a first approximation, the exit pressure $P_o \approx P_i = 101\,\text{kPa}$.

The mean density ρ_m is taken as the harmonic mean of these values [10, Section 4.8.1]:

$$\frac{1}{\rho_m} = \frac{1}{2}\left(\frac{1}{\rho_i} + \frac{1}{\rho_o}\right) = \frac{0.86014 + 1.1439}{2} = 1.002\,\text{kg/m}^3$$

$$\rightarrow \quad \rho_m = 0.998\,\text{kg/m}^3$$

The mass velocity can now be calculated using value of ρ_i:

$$G = \frac{\rho_i \dot{V}}{A_o} = \frac{1.1626 \times 1}{0.15} = 7.7507\,\text{kg/m}^2\,\text{s}$$

The values of coefficients k_c and k_e can be taken as 0.3 each (in the absence of any other information), which is their typical value for usual operating conditions. Using Eq. (4.127) we can now get

$$\Delta p = \frac{G^2}{2\rho_i} \cdot \left[(1 - \sigma^2 + k_c) + \frac{4fL}{D_h}\frac{\rho_i}{\rho_m} + 2\left(\frac{\rho_i}{\rho_o} - 1\right) - (1 - \sigma^2 - k_e)\frac{\rho_i}{\rho_o}\right]$$

$$= \frac{7.7507^2}{2 \times 1.1626}\left[(1 - 0.6^2 + 0.3) + \frac{4 \times 0.05 \times 0.8}{0.004} \cdot \frac{1.1626}{0.998}\right.$$

$$\left. + 2\left(\frac{1.1626}{0.8742} - 1\right) - (1 - 0.6^2 - 0.3)\frac{1.1626}{0.8742}\right]$$

$$= 25.835778[0.94 + 46.5972 + 0.6598 - 0.4522]$$

$$= 1233.5\,\text{Pa} = 1.2335\,\text{kPa}$$

Since the pressure drop in the heat exchanger is only about 1.2% of the inlet pressure, the assumption used in the calculation of ρ_o, that is, $P_i \approx P_o$ is justified. Had this pressure drop been very significant, we would calculate ρ_o at the "correct" exit pressure and iterate to get more accurate results.

Example 4.9

A plate heat exchanger is being used to heat cold water with the help of hot oil. Determine the pressure drop on the water side from the following design and operating data:

1. Number of flow passages: 20   Plate dimensions: 0.6 m (width) ×1.2 m (height)
2. Port diameter: 0.1 m   Channel spacing: 4 mm
3. Cold water flow rate: 15 kg/s   Mean dynamic viscosity: 750×10^{-6} Pa s
4. Mean density: 998 kg/m^3

The water is flowing vertically upward in the heat exchanger. Estimate the friction factor for flow of water through the plates from the equation $f = 0.5\ Re^{-0.3}$, where Re is the Reynolds number.

Solution

To determine the pressure drop we first find the mass velocity through the plates, G as:

$$G = \frac{\dot{m}}{A_o} = \frac{\dot{m}}{\text{No. of passages} \times \text{Width} \times \text{Plate spacing}}$$

$$= \frac{15}{20 \times 0.6 \times 0.004} = 312.5\ \text{kg/m}^2\ \text{s}$$

The hydraulic diameter of the channel passages ≈ 2 Channel spacing $= 8$ mm

$$\therefore \quad Re = \frac{GD_h}{\mu} = \frac{312.5 \times 8 \times 10^{-3}}{750 \times 10^{-6}} = 3333.3$$

$$\text{and} \quad f = \frac{0.5}{Re^{0.3}} = \frac{0.5}{(3333.3)^{0.3}} = 0.04386$$

We can now determine the total pressure drop using Eq. (4.131). Since change in the density of water due to heating is very small, we can neglect the acceleration term in Eq. (4.131). To calculate the first term, that is, the pressure drop in the port, we first find the mass velocity in the part, G_p using Eq. (4.132):

$$G_p = \frac{15}{\frac{\Pi}{4} \times 0.1^2} = 1909.86\ \text{kg/m}^2\ \text{s}$$

Thus giving

$$\Delta p = \frac{1.5 \times 1909.86^2 \times 1}{2 \times 998} + \frac{4 \times 0.04386 \times 1.2 \times (312.5^2)}{(2 \times 0.004) \times 998 \times 2} + 998 \times 9.8 \times 1.2$$

$$= 2741.16 + 1287.54 + 11{,}736.48$$

$$= 15{,}765.18\ \text{Pa}$$

The total pressure drop in the heat exchanger is 15.77 kPa.

4.1.6 Microchannel Heat Exchangers

The use of microchannels to improve the rate of heat dissipation from very large-scale integrated circuits was suggested by Tuckerman and Pease in 1981. They predicted that heat fluxes of the order of $1000\,\text{W/cm}^2$ would occur in future high–power laser and electronic radar systems. The successful development and use of such microchannel heat sinks encouraged their introduction in automotive and aerospace applications where minimization of space and weight of heat exchangers has always been a challenge. Now microchannel heat exchangers where liquid flows in small hydraulic diameter passages are widely used in automotive and refrigeration industry. While the basic concepts like that of heat exchanger effectiveness and the equations relating it to the overall heat-transfer coefficient discussed previously remain valid even in these microchannel heat exchangers, the equations governing the heat-transfer coefficient of the fluid flowing through them differ from those of conventional channels (i.e., with hydraulic diameter $> 3\,\text{mm}$). A number of books and research papers bring out these differences. We shall present the equations recommended by Kandilkar et al. [11].

Flow Through Microchannels

Kandilkar et al. [11] define microchannels as flow passages with hydraulic diameter (D_h) less than $200\,\mu$, and introduce the category of minichannels with $d_h > 200\,\mu\text{m}$, but less than $3\,\text{mm}$. *For gas flows* through microchannels, the continuum assumption may become inapplicable. This is governed by the Knudsen number Kn defined as

$$Kn = \frac{\lambda}{L} \tag{4.134}$$

where λ is the mean free path and L the characteristics length of the control volume. For $Kn < 10^{-3}$, the continuum assumption, and the Navier-Stokes equations based on it, is valid, but as Kn rises with reduction in L, it is no longer true. For $Kn > 10^{-3}$ but less than 10^{-1}, the most commonly encountered range in microchannels with gas flow, slip occurs at the boundaries. The Navier-Stokes equations can still be used, provided a velocity slip and temperature jump at the walls are accounted for. For $Kn > 10^{-1}$, the continuum approach is no longer valid and for $Kn > 10$, the flow is essentially a free molecular flow.

Considering the case of microchannels with slip flow, the first step in the analysis is to assume a slip model. Over the years numerous intuitive models have been proposed ranging from a simple first-order model first proposed by Kundt and Warburg based on experiments performed in 1875 to complex second-order models proposed by numerous researchers [11]. The general form of many second-order models for the flow between parallel planes is

$$\text{Slip velocity} = u_g - u_w = A_1\lambda \left(\frac{\partial u}{\partial \gamma}\right)_w + A_2\lambda^2 \left(\frac{\partial^2 u}{\partial \gamma^2}\right)_w \tag{4.135}$$

The velocity gradients are taken at wall normal to the flow direction. The values of coefficients A_1 and A_2 have been empirically determined by various researchers and

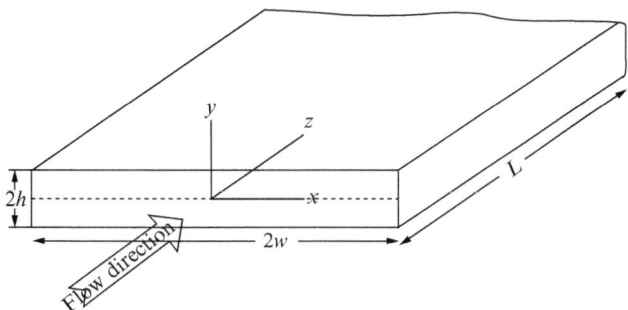

Fig. 4.18 Plane microchannel section.

typical values are given in Kandilkar et al. [11]. Clearly if in the above equation $A_2 = 0$, then it becomes a first-order model.

Next step is to solve the Navier-Stokes equations keeping Eq. (4.135) as a boundary condition. For the simple case of flow through a channel limited by parallel planes (see Fig. 4.18) the Navier-Stokes equation simplifies to:

$$\frac{d^2 U_z}{dy^2} = \frac{1}{\mu}\frac{dP}{dz} \tag{4.136}$$

This can be easily solved using the boundary condition of Eq. (4.135) along with the condition of symmetry, that is

$$\left.\frac{dU_z}{dy}\right|_{y=0} = 0 \tag{4.137}$$

to get the velocity distribution along with the flow direction (z) as:

$$U_z^* = 1 - y^{*2} + 8A_1\ Kn - 32A_2\ Kn^2 \tag{4.138}$$

where $U_z^* = U_z/U_{zo}$, $y^* = y/h$, and $U_{zo} = U_z(y = 0, Kn = 0)$ is the velocity at the center of the microchannel when there is no slip. Its value can be obtained using Eqs. (3.128), (3.129) as

$$U_{z0} = \frac{h^2}{2\mu}\left(-\frac{dP}{dz}\right) \tag{4.139}$$

The Knudsen number Kn is defined on the basis of the hydraulic diameter which, for $w \gg h$, is $\approx 4\,h$.

Integrating Eq. (4.138) across the channel we can get the average velocity $\overline{U}_z$ and from there an equation for the mass flow rate as:

$$\dot{m} = \rho\overline{U}_z A = \frac{2wh^3 P_o^2}{\mu RTL}\left[\frac{\Pi^2 - 1}{3} + 8A_1 K_{no}(\Pi - 1) - 32A_2 K_{no}^2 \ln \Pi\right] \tag{4.140}$$

where $\Pi = \frac{P_i}{P_o}$; P_o is the pressure at the channel outlet, P_i is the pressure at the channel inlet, and Kn_0 is the Knudsen number at channel outlet.

We shall illustrate the method with the help of an example.

Example 4.10

Arkilic et al. [12] report experimental data on flow of helium through a plain microchannel of length 7500 μm, width 52.25 μm, and height 1.33 μm. A typical reading is

$$\text{Outlet pressure } P_o = 100.8\,\text{kPa}$$

$$\text{Pressure ratio } \Pi = 1.85$$

$$\text{Temperature } T = 314\,\text{K}$$

$$\text{Outlet Knudsen number (based on channel height) } Kn_O = 0.165\,\text{K}$$

$$\text{Viscosity } \mu = 20.66 \times 10^{-6}\,\text{Ns/m}^2$$

$$\text{Specific gas constant } R = 2077\,\text{J/kg K}$$

Estimate the mass flow rate using Eq. (4.140) and compare it with the experimental value of 2.2×10^{-12} kg/s.

Solution

Rewriting the given data in terms of the nomenclature of Fig. 4.18, we note that

$$h = \frac{1.33}{2} = 0.665\,\mu\text{m}$$

$$w = \frac{52.25}{2} = 26.125\,\mu\text{m}$$

$$L = 7500\,\mu\text{m}$$

We first do the calculations with the value of A_1 and A_2, recommended by Maxwell [13], that is, $A_1 = 1$ and $A_2 = 0$. This implies use of a first-order model. The multiplier term of the expression for mass flow rate is

$$\frac{2wh^3 P_o^2}{\mu RTL} = \frac{2 \times (26.125 \times 10^{-6})(0.665 \times 10^{-6})^3 (100.8 \times 10^3)^2}{20.66 \times 10^{-6} \times 2077 \times 314 \times (7500 \times 10^{-6})}$$

$$= 1.545 \times 10^{-12}$$

This gives for $A_1 = 0$ and $A_2 = 0$

$$\dot{m} = 1.545 \times 10^{-12} \left[\frac{1.85^2 - 1}{3} + 8 \times 0.0825 \times (1.85 - 1) \right]$$

$$= 2.11 \times 10^{-12}\,\text{kg/s}$$

This is only about 4% lesser than the experimental value.

Clearly the predictions based on first-order slip models also agree very closely with experimental values since the Knudsen number is quite low. If we had chosen a second-order model with A_2 a small negative value, we may get even better agreement between the experimental and predicted values.

   Comment: If we neglect slip, the predicted flow rate would be

$$\dot{m}_{no\ slip} = 1.545 \times 10^{-12} \left[\frac{1.85^2 - 1}{3} \right] = 1.25 \times 10^{-12}\,\text{kg/s}$$

This is clearly grossly in error, being about 43% lower than the experimental value.

   Liquid flow in a microchannel is not much different from that in a macrochannel. The concept of mean free path is not valid since the liquid molecules are closely bound. If we use lattice spacing to be an equivalent concept, the Knudsen numbers are very small. Thus for water typically the lattice spacing is 0.3 nm, so in a microchannel with 1 μm gap, Knudsen number is 3×10^{-4}, a value small enough to ensure applicability of assumptions of continuum flow and incompressibility. Therefore, in principle all the equations discussed in Section 3.2 for flow through pipes could be used for flow through microchannels also. Since rectangular channels are popular in microchannel heat exchangers, the equation given by Shah and London is recommended by Kandilkar et al. [11] to determine that friction factor f in laminar flow:

$$f \cdot Re = 6(1 - 3.553\alpha_c + 1.9467\alpha_c^2 - 1.7012\alpha_c^3 + 0.9564\alpha_c^4 - 0.2537\alpha_c^5) \quad (4.141)$$

Here α_c is the aspect ratio of the channel and equals the ratio of length of the short side to the long side. While calculating the Re, the hydraulic diameter should be multiplied by a factor ϕ, given by Jones [14] to get better prediction of friction factor:

$$\phi = \frac{2}{3} + \frac{11}{24}\alpha_c(2 - \alpha_c) \quad (4.142)$$

For turbulent flow Eq. (3.131) can be used with the Reynolds number based on hydraulic diameter multiplied by Jones' factor ϕ (Eq. 4.142) as the characteristic dimension.

Heat Transfer in Microchannels

For gas flow through microchannels, we come across a phenomenon similar to that of slip, that is, temperature jump at the wall. This is usually represented by a first-order model equation:

$$T - T_{wall} = (B_1\lambda) \left| \frac{\partial T}{\partial y} \right|_w \quad (4.143)$$

where B_1 is a coefficient akin to the coefficient A_1 for slip velocity.

Considering the simple case of a plain microchannel, the energy equation can be simplified to:

$$k\frac{\partial^2 T}{\partial y^2} = \rho U_z C_p \frac{\partial T}{\partial \tau} \tag{4.144}$$

$$\text{with } \frac{\partial T}{\partial z} = \frac{d\bar{T}}{dz} = \frac{dT_w}{dz} \tag{4.145}$$

where $\bar{T}$ is the bulk temperature of the gas.

Considering the case when the lower wall is adiabatic and the upper wall has a constant heat flux, an additional boundary condition can be written as

$$\left.\frac{dT}{dy}\right|_{y=-h} = 0 \tag{4.146}$$

Further the energy balance over an element length dz gives

$$\frac{d\bar{T}}{dz} = \frac{q_h}{2h\rho C_p \bar{U}_z} \tag{4.147}$$

where $\bar{U}_z$ is the average bulk velocity of the fluid, and q_h is the heat flux input at the upper wall. Solving Eq. (4.144) along with the two boundary conditions, (4.143), (4.146), we get an equation for the Nusselt number as

$$Nu = \left(\frac{B_1\,Kn}{4} + \frac{26 + 147\,A_1\,Kn + 210\,A_1^2\,Kn^2}{140(1 + 3\,A_1\,Kn)^2}\right)^{-1} \tag{4.148}$$

Similarly, the solution for symmetrically heated microchannel can be obtained as

$$Nu = \left(\frac{B_1\,Kn}{4} + \frac{17 + 84\,A_1\,Kn + 105\,A_1^2\,Kn^2}{140(1 + 3\,A_1\,Kn)^2}\right)^{-1} \tag{4.149}$$

In case the compressibility effects are significant, simple solution to the energy equation is not possible and numerical methodologies have to be used. For liquid flow through microchannels, in principle, equations given in Section 3.3.3 can be used to predict the heat-transfer coefficient. For rectangular channels (the most common geometry of microchannels), Kandilkar et al. recommend following correlations for commonly encountered conditions in fully developed laminar flow:

(i) Constant wall temperature

$$Nu_T = 7.54(1 - 2.610\alpha_c + 4.970\alpha_c^2 - 5.119\alpha_c^3 + 2.7020\alpha_c^4 - 0.548\alpha_c^5) \tag{4.150}$$

(ii) Constant wall heat flux, both circumferentially and axially:

$$Nu_H = 8.235(1 - 10.6044\alpha_c + 61.1755\alpha_c^2 - 155.1803\alpha_c^3 + 176.9203\alpha_c^4 - 72.923\alpha_c^5) \tag{4.151}$$

For other cases, reference may be made to the book by Kandilkar et al. [11]

In turbulent region, the following modifications to the Gnielinski's correlation (Eq. 3.251) are suggested:

$$Nu = Nu_{GN}(1 + F)$$

$$F = C\,Re(1 - (D/Do)^2)$$

where Nu_{GN} is the Nusselt number value given by Gnielinski correlation and C and D_o are empirical constants, whose typical values are

$$C = 7.6 \times 10^{-5} \quad \text{and} \quad D_o = 1.164\,\text{mm}$$

4.2 HEAT AND MASS EXCHANGERS

Many thermal systems employ equipment where simultaneous transfer of heat and mass (usually water vapor) occurs. These include equipments like cooling and dehumidifying coils, spray washers, cooling towers, absorbers and regenerators of desiccant cooling and absorption refrigeration systems, etc. The simulation of these equipments is done using the principles of simultaneous heat and mass transfer presented in Section 3.4.1. In this section we shall study how these fundamental principles, especially the concept of enthalpy potential, can be used to predict the performance of some of these equipment.

4.2.1 Cooling and Dehumidifying Coils

Cooling and dehumidifying coils are generally plate-fin tube heat (and mass) exchangers consisting of corrugated or plain fin plates assembled over a bank of tubes. The air is made to pass between the fin plates while the coolant (cold water or the refrigerant) passes through the tubes. These coils generally remove both moisture and sensible heat from air, and the total heat transfer needed for this purpose is picked up by the coolant. Usually the coolant is circulated through the coils arranged in complex circuits to make the design compact. An accurate analysis of the performance of such coils is thus quite complex [15] and involves subdivision of the entire coil into a three-dimensional array of "minuscule" heat exchangers which are properly analyzed keeping in view the directions of flow of the coolant and the air. We shall first describe a simplified one-dimensional approach and then discuss its extension to more realistic three-dimensional simulation.

Fig. 4.19 shows a simplified schematic diagram of such a coil, as also the temperature and enthalpy changes along its length. Moist air passing over the outside surface of the coil is sensibly cooled, and sheds moisture when the surface temperature is lower than the dew point temperature of the entering air. We thus have simultaneous heat and mass transfer at this surface, and the total heat transfer needed for this purpose can be estimated using the concept of enthalpy potential (Section 3.4.1). This total heat transfer is conducted through the thin water layer on the outer surface of the coil and its metallic (usually copper or aluminum) walls to the coolant circulating through the tubes, usually counter-current to the flow of air, as indicated in Fig. 4.19. In the temperature profiles shown in Fig. 4.19, it is presumed that this coolant does not change phase, and therefore gets warm as it picks up heat from the air which gets cooled. Considering a small elemental surface area of the wetted surface, dA_w, the successive transport processes can be mathematically expressed as follows:

(a) Total heat-transfer rate due to simultaneous heat and mass transfer from moist air (Eq. 3.317):

$$dq_t = \frac{\alpha_{aw}dA_w(h_a - h_s)}{c_{pm}} \qquad (4.152)$$

where α_{aw} is the convective heat-transfer coefficient at air-wet surface interface; c_{pm}: moist air-specific heat; h_a: enthalpy of air; and h_s: enthalpy of saturated air at wet surface temperature, T_s.

(b) Heat conduction through the metal wall and convection through the coolant (Eq. 3.219):

$$dq_t = \frac{(T_s - T_r)dA_w}{R_{mw} + R_r} \qquad (4.153)$$

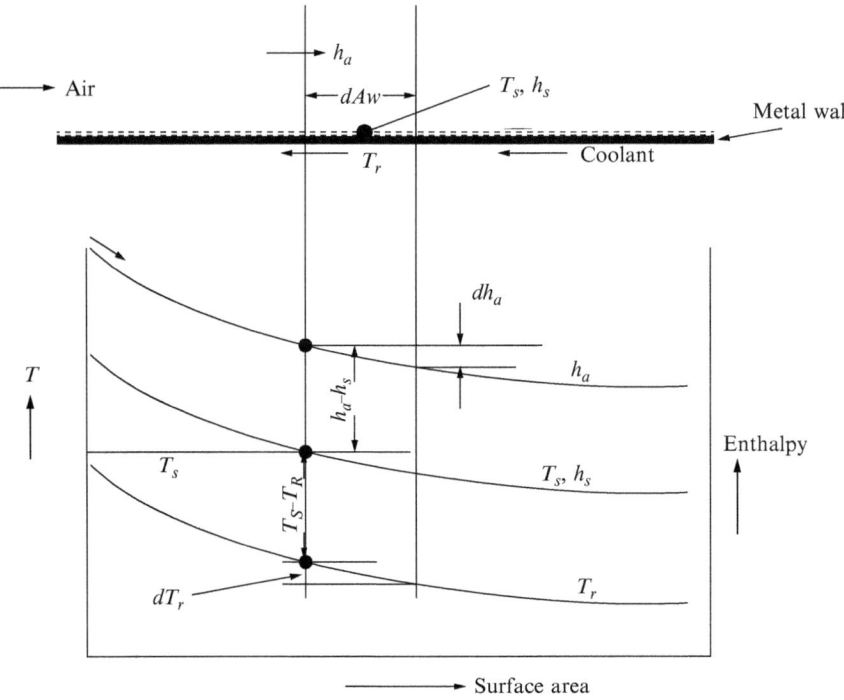

Fig. 4.19 One-dimensional analysis of a cooling and dehumidifying coil.

where R stands for the thermal resistance in $(m^2 \ K/W)$ referred to the external area, and subscripts mw and r refer to the metal wall and the coolant (or the refrigerant), respectively.

In this equation we have neglected the thermal resistance of the water film on the outer surface since usually this film is very thin due to constant drainage of water from the coils.

(c) Warming up of the coolant due to this heat transfer:

$$dq_t = -\dot{m}_r c_r dT_r \tag{4.154}$$

where $\dot{m}_r$ is the mass flow rate and c_r the specific heat of the coolant. Negative sign in this equation arises from the flow direction of the coolant being opposite to that of the air, as shown in Fig. 4.19.

Combining Eqs. (4.152), (4.153), and rearranging terms we get

$$\frac{T_s - T_r}{h_a - h_s} = \frac{R_{mw} + R_r}{c_{pm} \cdot (1/\alpha_{aw})} = \frac{R_{mw} + R_r}{c_{pm} R_{aw}} = C \tag{4.155}$$

where $R_{aw}(= 1/\alpha_{aw})$ is the convective thermal resistance at the air–wet surface interface. The term involving the ratio of various thermal resistances and moist air-specific heat, usually called the coil characteristic, C, depends primarily on the coil design and the operating conditions, mainly the flow rates of the entering air and the coolant. Thus the value of C can be determined using the basic laws of heat transfer and the equations for convective heat-transfer coefficient (Section 3.3.3). Using this value of C in Eq. (4.155), we can determine the wet surface temperature T_s and the corresponding saturated air enthalpy h_s at any location along the coil surface, where the condition of air (and hence enthalpy h_a) and the coolant temperature T_r are known.

Enthalpy of saturated air h_s (in kJ/kg) can be expressed as a function of the temperature T_s (in °C) by a cubic equation [16]

$$h_s = 9.3625 + 1.7861 T_s + 0.01135 T_s^2 + 0.0098855 T_s^3 \tag{4.156}$$

which gives a good estimate for T_s between 2 and 30°C.

Substituting Eq. (4.156) in Eq. (4.155), we get a nonlinear equation

$$\frac{T_s}{C} - \frac{T_r}{C} - h_a + 9.3625 + 1.7861 T_s + 0.01135 T_s^2 + 0.00098855 T_s^3 = 0 \tag{4.157}$$

which can be solved for T_s, for known values of T_r, h_a, and C.

In the counter-flow configuration of Fig. 4.19, the values of T_r and h_a are not known simultaneously at any position along the coil. Therefore, an iterative procedure becomes necessary. We could assume, for example, the outlet temperature of the coolant leaving the coil. With this assumption, we can carry out the calculations in following steps:

(i) Assume coolant outlet temperature, $T_{r,out}$

(ii) Subdivide the coil into N smaller areas such that in each area the coolant temperature rise is $\Delta T_r = \frac{T_{r,out} - T_{r,in}}{N}$

(iii) The total heat transfer in each subarea is

$$\Delta Q_t = \dot{m}_r c_r \Delta T_r$$

This can be equated to the total enthalpy change of air in each section.

$$\Delta Q_t = \dot{m}_a \cdot \Delta h_a$$

(iv) Thus the values of air enthalpies and the coolant temperatures at the entry and the exit of each of these subdivisions can be computed.

(v) The values of wet surface temperature T_s, at the midpoint of each of these sections can be found by solving Eq. (4.157) using h_a and T_r as the mean of the values at the exit and the entrance of each section.

(vi) The coil surface area needed for each of these sections can then be found using Eq. (4.153).

(vii) The total surface area needed for all these sections can now be compared with the actual surface area of the coil, and if the two values do not agree within permissible limits, second iteration is done by making an appropriate assumption for the coolant outlet temperature and repeating steps (ii) to (vii) till convergence is obtained.

The procedure is illustrated in Example 4.11.

Example 4.11

In a counter-flow chilled water coil, 2 kg/s air enters at 30°C DBT and 60% RH and the cooling water enters at 10°C. Estimate the outlet condition of air and the cooling water temperature. Given the design data for the coil:

Total outside surface area of the coil: 100 m²

Ratio of coil outside to inside surface area: 15

Convective heat-transfer coefficient on air side: 50 W/m² K

Convective heat-transfer coefficient for the coolant: 2000 W/m² K

Tube wall thermal resistance: 5×10^{-6} m² K/W

Cooling water flow rate: 2.5 kg/s

Take mean moist air-specific heat as: 1020 J/kg K

Solution

As a first approximation, we assume that the cooling water temperature rises by 5°C as it cools and dehumidifies the air. We subdivide the whole coil into five sections, in each of which, we assume, the water temperature rises by 1°C. The total heat transfer in each section is

$$\Delta Q_t = 2.5 \times 4.186 \times 1 = 10.465 \,\text{kW}$$

The air enthalpy change in each section is therefore

$$\Delta h_a = \frac{\Delta Q_t}{\dot{m}_a} = \frac{10.465}{2} = 5.232 \,\text{kJ/kg}$$

The enthalpy of air at the inlet conditions of 30°C DBT and 60% RH can be seen from the psychrometric chart as 71.8 kJ/kg. Thus the air enthalpy values and the coolant temperatures in the five sections of the coil can be calculated as shown in Fig. 4.20. The mean surface temperature in each of these sections can be found using Eq. (4.157). To do so, we first determine the value of the coil characteristic C using the given design data:

$$R_{aw} = \frac{1}{\alpha_{aw}} = \frac{1}{50}$$
$$= 0.02 \,\text{m}^2 \,\text{K/W}$$
$$R_{mw} = (5 \times 10^{-6}) \times 15$$
$$= 75 \times 10^{-6} \,\text{m}^2 \,\text{K/W}$$

Since the tube wall resistance is usually given on the basis of tube inner diameter, and all thermal resistances in Eq. (4.155) are based on the external surface area, we have multiplied the given value of thermal resistance by the area ratio. The same factor needs to be taken into account while calculating R_r, the convective resistance on the coolant side.

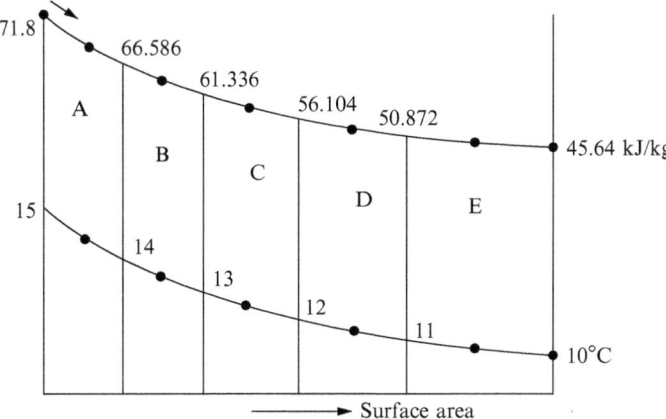

Fig. 4.20 Coolant temperature and air enthalpy variation along the coil.

$$R_r = \frac{1}{h_r}\left(\frac{A_o}{A_i}\right) = \frac{1}{2000} \times 15 = 7.5 \times 10^{-3} \text{ m}^2 \text{ K/W}$$

$$\Rightarrow C = \frac{R_{mw} + R_r}{c_{pm} \cdot R_{aw}} = \frac{75 \times 10^{-6} + 7.5 \times 10^{-3}}{1020 \times 0.02} = 3.713 \times 10^{-4} \text{ kg K/J}$$

$$= 0.3713 \text{ kg K/kJ}$$

Using this value of C in Eq. (4.157), we can find the values of mean surface temperature, T_s, at the midpoint of various sections using h_a and T_r values from Fig. 4.20. The results of these iterative calculations are summarized in following table.

Section	Mean h_a	Mean T_r	T_s	h_s
A	69.184	14.5	19.5	55.8
B	63.952	13.5	18.2	51.6
C	58.72	12.5	16.8	47.3
D	53.488	11.5	15.4	43.2
E	48.256	10.5	13.9	39.0

The total coil surface area actually needed to effect these changes can now be found by calculating the surface area of each section by using Eq. (4.153), and then summing these up. This gives

$$\Sigma \, dA_w = \sum \frac{\Delta Q_t (R_{mw} + R_r)}{T_s - T_r}$$

$$= \sum \frac{(10.465 \times 10^3)(75 \times 10^{-6} + 7.5 \times 10^{-3})}{T_s - T_r}$$

$$= 79.272375 \left\{ \frac{1}{19.5 - 14.5} + \frac{1}{18.2 - 13.5} + \frac{1}{16.8 - 12.5} \right.$$

$$\left. + \frac{1}{15.4 - 11.5} + \frac{1}{13.9 - 10.5} \right\}$$

$$= 15.854 + 16.866 + 18.435 + 20.326 + 23.315$$

$$= 94.8 \text{ m}^2$$

The actual area of the coil as given in the data is 100 m². Thus our assumption about the coolant exit temperature is quite near the mark. Since the actual coil surface area is slightly (~5%) higher, we could do another iteration with a slightly higher coolant exit temperature to get a more accurate result. It is left as an exercise for the students.

To find the outlet condition of air, we need one more property, besides the enthalpy value. We could, for example, find the sensible heat transfer in each section and thus get the outlet DBT of air. If $T_{a,A}$ represents the DBT of air at the end of section A, we can write the following energy balance equation for this section.

$$2 \times 1020 \times (30 - T_{a,A}) = \Delta A_A \cdot \alpha_{aw} \cdot \left(\frac{30 + T_{a,A}}{2} - T_s\right)$$

$$= 15.854 \times 50 \left(\frac{30 + T_{a,A}}{2} - 19.5\right)$$

which gives $T_{a,A} = 26.58°\text{C}$

Proceeding sequentially, and using the values of T_s and ΔA for each section calculated above we can determine the air temperature at the end of each section as follows:

1. *Section B*

$$2 \times 1020 \times (26.58 - T_{a,B}) = 16.866 \times 50 \left(\frac{26.58 + T_{a,B}}{2} - 18.2 \right)$$

which gives $T_{a,B} = 23.71°C$

2. *Section C*

$$2 \times 1020 \times (23.71 - T_{a,C}) = 18.435 \times 50 \left(\frac{23.71 + T_{a,C}}{2} - 16.8 \right)$$

which gives $T_{a,C} = 21.16°C$

3. *Section D*

$$2 \times 1020 \times (21.16 - T_{a,D}) = 20.326 \times 50 \left(\frac{21.16 + T_{a,D}}{2} - 15.4 \right)$$

which gives $T_{a,D} = 18.86°C$

4. *Section E*

$$2 \times 1020 \times (18.86 - T_{a,E}) = 23.315 \times 50 \left(\frac{18.86 + T_{a,E}}{2} - 13.9 \right)$$

which gives $T_{a,E} = 16.66°C$

Thus the exit DBT of air is $= 16.66°C$, Enthalpy $= 45.64 \, kJ/kg$ which corresponds to a humidity ratio, w of 11.4 g/kg da, as seen from the psychrometric chart.

Comment

We can cross—check the value of exit moisture content estimated above from DBT and air enthalpy, by independently calculating the moisture removal from air from first principles. For this purpose we use the mass transfer rate (Eq. 3.318), to estimate the moisture removal in each section, starting from section A. The moisture content of inlet air at A can be found from the psychrometric chart for the given conditions of $30°C$ DBT and 60% RH as $w = 16.2 \, g/kg$ da. Using this, we can write an equation for the moisture balance involving w_A, the humidity ratio at the end of this section as:

$$\dot{m}_a \cdot (16.2 - w_A) \times 10^{-3} \, (kg/s) = \frac{\alpha_{aw} \Delta A_A}{c_{pm}} \left(\frac{16.2 + w_A}{2} - w_S \right) \times 10^{-3}$$

where w_s is the humidity ratio of saturated air at the mean surface temperature, T_s. The value of w_s at $T_s = 19.5°C$, the mean wet surface temperature of section A calculated above, can be obtained, by using psychrometric chart, as $w_s = 14.2 \, g/kg$. Substituting this in the above equation we get

$$2 \times (16.2 - w_A) = \frac{50 \times 15.854}{1020} \left(\frac{16.2 + w_A}{2} - 14.2 \right)$$

which gives $w_A = 15.54 \, g/kg$ da

Proceeding similarly, we can sequentially determine the humidity ratio of air at the end of each section as follows, using T_s and ΔA values for each section calculated above:

1. *Section B*

$$2 \times (15.54 - w_B) = \frac{50 \times 16.866}{1020}\left(\frac{15.54 + w_B}{2} - 13.2\right)$$

which gives $w_B = 14.74\,\text{g/kg da}$

2. *Section C*

$$2 \times (14.74 - w_C) = \frac{50 \times 18.435}{1020}\left(\frac{14.74 + w_C}{2} - 12\right)$$

which gives $w_C = 13.73\,\text{g/kg da}$

3. *Section D*

$$2 \times (13.73 - w_D) = \frac{50 \times 20.326}{1020}\left(\frac{13.73 + w_D}{2} - 11\right)$$

which gives $w_D = 12.64\,\text{g/kg da}$

4. *Section E*

$$2 \times (12.64 - w_E) = \frac{50 \times 23.315}{1020}\left(\frac{12.64 + w_E}{2} - 9.95\right)$$

which gives $w_E = 11.44\,\text{g/kg da}$

This compares very well with the value of exit humidity ratio ($w = 11.49\,\text{kg da}$) estimated above on the basis of the leaving air DBT and air enthalpy.

In the above analysis, and in the example too, we have made an implicit assumption that whole of the coil surface is wet. Under certain operating conditions, for example, with warmer coolant entry temperature, the coil surface temperature at the exit may be higher than the dew point temperature T_{dp} of the entering air. The coil may thus be partially dry, and we need to find the surface area of this dry–portion of the coil where only sensible heat transfer is taking place. This can be done by locating the point along the air flow direction where the surface temperature reaches the dew point (Fig. 4.21).

Considering an infinitesimal area around this dry-wet boundary, we can write following equations for the heat-transfer dq (cf. Eqs. 4.152, 4.153):

$$\begin{aligned} dq &= \frac{\alpha_{aw}dA_w}{c_{pm}}(h_{ab} - h_{dp}) \\ &= \alpha_{aw} \cdot dA_w(T_b - T_{dp}) \\ &= \frac{(T_{dp} - T_{rb})dA_w}{R_{mw} + R_r} \end{aligned} \qquad (4.158)$$

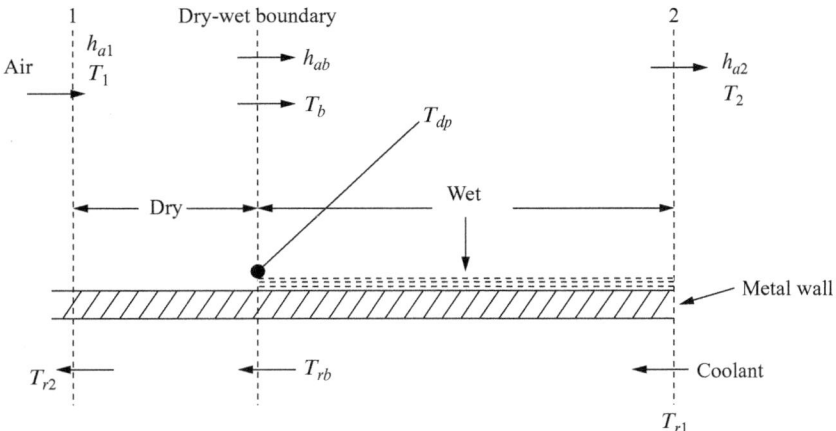

Fig. 4.21 Partially dry coil.

where we have assumed that the heat-transfer coefficients over the dry portion and the wet portion near the boundary are equal.

This gives (cf. Eq. 4.154):

$$\frac{T_{dp} - T_{rb}}{h_{ab} - h_{dp}} = \frac{R_{mw} + R_r}{R_{aw}c_{pm}} = C \tag{4.159}$$

However, since both T_{rb} and h_{ab} are not known, we need to identify one more relationship between these. That relationship comes from the consideration of energy balance of section 1-boundary, which gives

$$\dot{m}_r c_r (T_{r2} - T_{rb}) = \dot{m}_a (h_{a1} - h_{ab})$$

Rearranging this, we get

$$\frac{T_{r2} - T_{rb}}{h_{a1} - h_{ab}} = \frac{\dot{m}_a}{\dot{m}_r c_r} = Y, \quad \text{say} \tag{4.160}$$

Eliminating T_{rb} from Eqs. (4.159), (4.160) we get

$$h_{ab} = \frac{T_{dp} + Ch_{dp} + Yh_{a1} - T_{r2}}{C + Y} \tag{4.161}$$

In the above equation for h_{ab}, all the terms on the RHS except T_{r2} are known. Since during simulation of a coil the very first step involves assuming a suitable value for the coolant exit temperature, we can introduce in the procedure outlined above an additional step of calculating h_{ab}, using Eq. (4.161). If this value of h_{ab} is greater than h_{a1}, the enthalpy of air at entrance, it implies that there is no place in the coil where wet and dry surfaces co-exist. Thus the coil is fully wet, and the rest of the calculations can be done as before. However, if $h_{ab} < h_{a1}$, but greater than h_{a2}, the enthalpy of air at the coil exit, it implies that the coil is partially dry. In case h_{ab} turns out to be even smaller than h_{a2}, the coil surface is completely dry.

The analysis of the dry portion of the coil can be carried out like a simple counterflow heat exchanger, and then merged with the analysis of the wet–portion of coil (if any) carried out as explained above. Example 4.12 illustrates the method of analysis of a partially wet coil.

Example 4.12

In the chilled water coil of Example 4.10, if the coolant temperature at entry is raised to 20°C estimate the rate of heat transfer and moisture transfer from air, and the exit condition of air.

Solution

Since the coolant temperature now is much higher than in Example 4.10, the total heat transfer is expected to be much lesser. As a first approximation, let us presume that the coolant temperature now rises by 3°C (as against 5°C in Example 4.10); and carry out the coil analysis as before by subdividing the coil. However, we first check whether the coil is fully wet or not by determining the value of h_{ab} using Eq. (4.161). For this we need to first find the values of various terms on the right-hand side of this equation. These are

$$Y = \frac{\dot{m}_a}{\dot{m}_r c_r} = \frac{2}{2.5 \times 4.186} = 0.1911 \text{ kg K/kJ}$$

$T_{dp} = 21.3°C$ (from psychrometric chart at 30°C DBT and 60% RH)

$h_{dp} = 62.4 \text{ kJ/kg}$ (from psychrometric chart)

$T_{r2} = 20 + 3 = 23°C$

Substituting in Eq. (4.161), we get

$$h_{ab} = \frac{21.3 + 0.3713 \times 62.4 + 0.1911 \times 71.8 - 23}{0.3713 + 0.1911}$$

$$= 62.57 \text{ kJ/kg}$$

Since $h_{ab} < h_{a1}$, it follows that a portion of coil is dry. Further,

$$h_{a2} = h_{a1} - \frac{\dot{m}_r c_r (T_{r2} - T_{r1})}{\dot{m}_a} = 71.8 - \frac{2.5 \times 3 \times 4.186}{2} = 56.10$$

Since h_{ab} is greater than h_{a2}, it follows that the coil is not fully dry.

To analyze the dry position of the coil, we determine T_{rb} using Eq. (4.159)

$$\frac{21.3 - T_{rb}}{62.57 - 62.4} = C = 0.3713$$

which gives $T_{rb} = 21.3 - 0.3713 \times 0.17 = 21.24°C$

Thus, in the dry portion of the coil, the air enthalpy drops from 71.8 to 62.57 kJ/kg and the coolant temperature is rising from 21.24 to 23°C. The surface area of this portion of the coil can be found using standard counter-flow heat exchanger analysis procedure. The temperature of air at the inlet is given as 30°C, and that at the exit of this dry-coil, T_b, can be found using Eq. (4.158), which gives

$$T_b - T_{dp} = \frac{h_{ab} - h_{dp}}{c_{pm}}$$

$$\Rightarrow \quad T_b = T_{dp} + \frac{h_{ab} - h_{dp}}{c_{pm}} = 21.3 + \frac{62.57 - 62.4}{1.02} = 21.47°C$$

The LMTD in this "heat exchanger" is

$$LMTD = \frac{(30 - 23) - (21.47 - 21.24)}{\ln\left(\frac{30-23}{21.47-21.24}\right)} = 1.982°C$$

and therefore the surface area of the dry portion of the coil is

$$A_{dry} = \frac{\text{Heat transfer} \times 1.0 / (\text{Overall heat-transfer coefficient})}{(LMTD)}$$

$$= 2.5 \times 4186 \times (23 - 21.24)\left\{\frac{75 \times 10^{-6} + 7.5 \times 10^{-3} + 0.02}{1.982}\right\}$$

$$= 256.2 \text{ m}^2$$

Since the total surface area of the coil is only 100 m², it follows that our original assumption about cooling water outlet temperature is incorrect. As second guess, we $T_{r2} = 22°C$ and repeat above calculations.

$$T_{r2} = 22°C$$

$$h_{ab} = \frac{21.3 + 0.3713 \times 62.4 + 0.1911 \times 71.8 - 22}{0.3173 + 0.1911} = 64.35 \text{ kJ/kg}$$

To find T_{rb}, we again use Eq. (4.159)

$$\frac{21.3 - T_{rb}}{64.35 - 62.4} = C = 0.3713$$

$$\Rightarrow \quad T_{rb} = 20.58°C$$

The DBT of air at dry-wet boundary is (from Eq. 4.158)

$$T_b = 21.3 + \frac{64.35 - 62.4}{1.02} = 23.2°C$$

LMTD in the dry portion of the coil is

$$LMTD = \frac{(30 - 22) - (23.2 - 20.58)}{\ln\left(\frac{30-22}{23.2-20.58}\right)} = 4.82°C$$

$$A_{dry\ surface} = \frac{2.5 \times 4186 \times (22 - 20.58)[75 \times 10^{-6} + 7.5 \times 10^{-3} + 0.02]}{4.82}$$

$$= 84.8\ m^2$$

In the wet portion of the coil the air enthalpy changes from 64.35 kJ/kg to h_{a2}. The value of h_{a2} can be found from the energy balance:

$$h_{a2} = h_{a1} - \frac{\dot{m}_r c_r (T_{r2} - T_{r1})}{\dot{m}_a} = 71.8 - \frac{2.5 \times 4.186 \times 2}{2}$$

$$= 61.335\ kJ/kg$$

The coolant temperature changes from 20 to 20.58°C. To estimate the area required in this wet portion of the coil, we first determine the wet surface temperature T_s at the air exit position using Eq. (4.157), that is

$$\frac{T_s}{C} - \frac{T_r}{C} - h_a + 9.3625 + 1.7861\,T_s + 0.01135\,T_s^2 + 0.00098855\,T_s^3 = 0$$

$$\frac{T_s}{0.3713} - \frac{20}{0.3713} - 61.335 + 9.3625 + 1.7861\,T_s + 0.01135\,T_s^2$$

$$+ 0.00098855\,T_s^3 = 0$$

which gives (by trial and error):

$$T_s \approx 20.6°C \quad \text{and} \quad h_s = 59.6\ kJ/kg$$

The wet surface area can now be estimated using Eq. (4.153) as:

$$\Delta A_w = \frac{dq_t(R_{mw} + R_r)}{(T_s - T_r)_{mean}} = \frac{2.5 \times 4186 \times 0.58 \times (7.5 \times 10^{-3} + 75 \times 10^{-6})}{\left[\frac{(20.6-20)+(21.3-20.58)}{2}\right]}$$

$$= 69.7\ m^2$$

The total surface area requirement is thus 154.5 m²(= 69.7 + 84.8) which is still much larger than the actual area available. This implies that actual heat transfer is much smaller. As the next iteration we, therefore, assume the water outlet temperature as 21.5°C and repeat all the calculations, as follows:

$$T_{r2} = 21.5°C$$

$$h_{ab} = \frac{21.3 + 0.3713 \times 62.4 + 0.1911 \times 71.8 - 21.5}{0.3713 + 0.1911}$$

$$= 65.24\ kJ/kg$$

$$h_{a2} = h_{a1} - \frac{\dot{m}_r c_r (T_{r2} - T_{r1})}{\dot{m}_a} = 71.8 - \frac{2.5 \times 4.186 \times 1.5}{2}$$

$$= 63.95\ kJ/kg$$

$$\left(\frac{21.3 - T_{rb}}{65.24 - 62.4}\right) = C = 0.3713$$

$$\Rightarrow T_{rb} = 20.25°C$$

$$T_b = 21.3 + \frac{65.24 - 62.4}{1.02} = 24.08°C$$

$$(LMTD)_{dry\ surface} = \frac{(30 - 21.5) - (24.08 - 20.25)}{\ln\left(\frac{30-21.5}{24.08-20.25}\right)} = 5.86°C$$

$$A_{dry\ surface} = \frac{2.5 \times 4186 \times (21.5 - 20.25)(75 \times 10^{-6} + 7.5 \times 10^{-3} + 0.02)}{5.86}$$

$$= 61.6\ m^2$$

In the wet portion of the coil the air enthalpy changes from 65.24 to 63.95 kJ/kg; and the coolant temperature is changing from 20 to 20.25°C. The wet surface temperature at the dry/wet boundary is the dew point temperature, that is, 21.3°C, and that at the coolant entry position can be determined by solving Eq. (4.157), that is

$$\frac{T_s}{0.3713} - \frac{20}{0.3713} - 63.95 + 9.3625 + 1.7861\,T_s + 0.01135\,T_s^2 + 0.00098855\,T_s^3 = 0$$

which gives $T_s \approx 21°C$.

The wet surface area required to achieve the requisite heat transfer is

$$\Delta A_w = \frac{2.5 \times 4186 \times 0.25 \times (7.5 \times 10^{-3} + 75 \times 10^{-6})}{[(21 - 20) + (21.3 - 20.25)]/2} = 19.33\ m^2$$

The total area requirement thus works out at $61.6 + 19.33 = 80.93\ m^2$ which is about 20% lesser than the actual area of the coil. It therefore follows that the actual cooling will be more than that presumed in the last iteration. Another iteration with assumed coolant exit temperature of 21.7°C gives the dry surface area requirement of 69.7 m^2 and the wet surface area of 33.4 m^2, amounting to a total of 103.1 m^2 which is quite near the given coil surface area of 100 m^2. The important results from this iteration are

$$h_{ab} = 64.88\ kJ/kg \quad h_{a2} = 62.9\ kJ/kg \quad T_{rb} = 20.38°C$$

$$T_b = 23.73°C \quad A_{dry\ surface} = 69.7\ m^2 \quad A_{wet\ surface} = 33.4\ m^2$$

$$T_s(at\ air\ exit) = 20.88°C$$

The total heat transfer $= 2.5 \times 4.186 \times 1.7 = 17.79\ kW$

$$\text{The rate of moisture transfer} = \frac{\alpha_{aw}}{c_{pm}}\ \Delta A_{wet}\left(\frac{w_{ab} + w_2}{2} - \frac{w_s + w_{dp}}{2}\right)$$

$$= \dot{m}_a(w_{ab} - w_2)$$

From the psychrometric chart we get

$$w_{ab} = w_1 = 16.2\ g/kg\ da$$

$$w_{dp} = 16.2\ g/kg\ da \quad w_s = 15.5\ g/kg\ da$$

Substituting values we get

$$\frac{50 \times 33.4}{1020} \left(\frac{16.2 + w_2}{2} - \frac{16.2 + 15.5}{2} \right) = 2(16.2 - w_2)$$

$$\text{which gives} \quad w_2 \simeq 16\,\text{g/kg da}$$

So the rate of moisture transfer is $= 2(16.2 - 16) = 0.4\,\text{g/s}$

The exit condition of air is: Enthalpy $= 69.6\,\text{kJ/kg}$

Humidity ratio $= 16\,\text{g/kg da}$

In the above discussion we have presumed that the coolant is a single-phase fluid (such as water or some brine) which warms up as it picks up heat from the air to be cooled and dehumidified. In some designs of the coil, especially those in the window air conditioners, the refrigerant itself is circulated through the tubes. The analysis of these DX-coils (dry or direct-expansion coils) can also be carried out in a similar manner, the only difference being that here the refrigerant temperature remains constant (in a major portion of the coil), as it vaporizes during passage through the coil, provided the pressure drop is not very significant. However, as the refrigerant vaporizes, the inside heat-transfer coefficient changes appreciably, decreasing drastically in the high vapor-fraction regions (dryness-fraction > 0.9). Usually these coils are designed so that the refrigerant vapors leaving the coil are superheated. In this region the inside heat-transfer coefficient values are very low. Thus a rigorous analysis of such coils demands that all these changes be taken into account. Further complication in analysis of these, and also many water-cooled coils, arises from the complex circuiting arrangements used for compactness. In such coils it becomes necessary to determine the distribution of the coolant in various circuits, which in turn depends on the rate of heat transfer, thus necessitating multinested iterative calculations. Vardhan and Dhar [15] give simple numerical model for detailed computer simulation of such coils which can analyze complex circuiting arrangements and take into account variation in heat-transfer coefficient along the coil. A brief account of the method is presented below.

The coil is discretized into a three-dimensional array of mini heat exchangers, referred to as the nodes, each of which is formed around a portion of the coolant tube as shown in Fig. 4.22. These nodes are then numbered from one corner of the coil with the help of three indices showing the relative position of a node in the three directions (i.e., along the length, height, and the depth of the coil). Each node is then analyzed independently, assuming that the total surface area of the fin plates is divided equally among all the nodes, while the total air flow rate is assumed to be equally divided among the nodes in the face area of the coil.

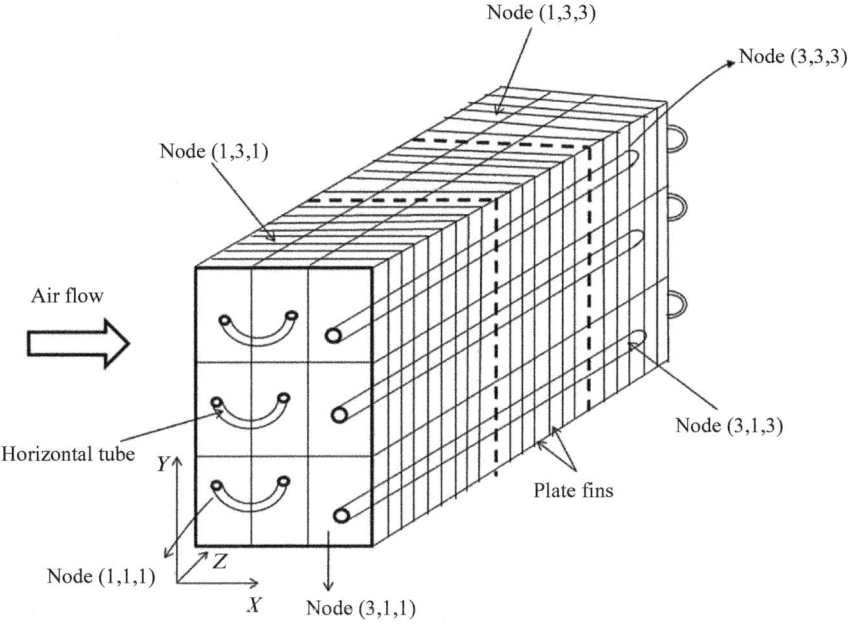

Fig. 4.22 Discretizing the coil—showing only $3 \times 3 \times 3$ nodes.

The nodes are taken as either completely dry or completely wet and analyzed using the basic equations discussed above, making the usual assumption of negligible water film thickness and neglecting the heat carried away by the condensate. However, in order to account for variation in the temperature along the finned surface, the concept of fin efficiency is introduced. This is used to relate the mean surface temperature of the fins and the outside surface temperature of the tubes carrying the coolant, for both dry surface conditions and, after appropriate modifications, for wet surface conditions, as explained below.

Under dry surface conditions, the overall fin efficiency for a node, including both the base surface and the fins, is calculated as (cf. Eq. 3.229)

$$\eta = \frac{(A_p + A_f \phi)}{A_{node}} \tag{4.162}$$

$$\text{where} \quad A_{node} = A_p + A_f \tag{4.163}$$

Here A_p denotes the prime (tube) surface area, A_f the fin area, and ϕ the fin efficiency as calculated by standard heat-transfer principles based on fin geometry (Section 3.3.1). Each node can then be visualized as a simple cross-flow heat exchanger with an air-side surface area A_{node}, air side heat-transfer coefficient α_a (found using appropriate correlation from Section 3.3.3), coolant side surface area A_i (based on tube inside diameter) and heat-transfer coefficient α_r (found using appropriate correlations for chilled water coils/dry-expansion coils, as the case may be). The overall heat-transfer coefficient U is then calculated as:

$$\frac{1}{UA_{node}} = \frac{1}{\eta \alpha_a A_{node}} + \frac{\ln(d/d_i)}{2\pi k_t L_t} + \frac{1}{\alpha_r A_i} \qquad (4.164)$$

where d and d_i are the outer/inner diameters of the tube; L_t the tube length in the node, and k_t the thermal conductivity of tube material. The NTU of each heat exchanger node can then be determined. Since the air flow is subdivided into a large number of parts, while the coolant flow rate through each node is the same as that in the actual heat exchanger, the heat capacity of the coolant, which is thoroughly mixed, is very high in comparison to that of the air, which is not mixed due to the presence of fin plates. The effectiveness of each node can therefore be calculated using Eq. (4.67). This is then used to calculate the heat transfer in the node and to find the outlet air conditions.

Under wet surface conditions, both the heat and mass transfer occur simultaneously and we can not use the concept of overall efficiency η, defined above on the basis of heat transfer alone. The total heat-transfer rate is now governed by Eq. (4.152), which can be further approximated, following Threlkeld [17], as follows:

$$dq_t = \frac{\alpha_{aw}\, dA_w}{c_{pm}}(h_a - h_s) \approx \frac{\alpha_{aw}\, dA_w}{c_{pm}} \cdot m''(T_{a,wb} - T_s) \qquad (4.165)$$

where m'' is the slope of the enthalpy-saturation temperature curve for moist air at the mean fin surface temperature T_s, and $T_{a,wb}$ is the free stream wet bulb temperature, that is, the saturation air temperature corresponding to the actual air enthalpy h_a.

$$\text{Defining} \quad \alpha_{eff} = \frac{\alpha_{aw} m''}{c_{pm}} \qquad (4.166)$$

Eq. (4.165) can be written as

$$dq_t = \alpha_{eff}\, dA_w(T_{a,wb} - T_s) = \frac{(T_{a,wb} - T_s)}{1/(\alpha_{eff}\, dA_w)} \qquad (4.167)$$

This equation suggests that the wet fin can be considered equivalent to a dry fin with an effective thermal resistance R_{wet}, defined below, between an effective free stream temperatures of $T_{a,wb}$ and the mean fin surface temperature T_s. The magnitude of R_{wet}, for a finite node area A_{node}, can be derived following Eq. (4.167), as

$$R_{wet} = \frac{1}{\alpha_{eff} A_{node}} \qquad (4.168)$$

This mean fin surface temperature T_s can now be related to the tube outer surface temperature T_{tos} using the concept of overall finned surface efficiency η, Eq. (4.162), since the total heat transfer in each step can also be expressed as:

$$Q_{node} = \frac{\alpha_{aw} A_{node}}{c_{pm}}(h_a - h_s) \approx \alpha_{eff} A_{node}(T_{a,wb} - T_s)$$

$$= \eta \alpha_{eff} A_{node}(T_{a,wb} - T_{tos}) \qquad (4.169)$$

The total heat transfer under wet surface conditions can thus be visualized to take place in the following manner (see Fig. 4.23). Sensible and latent heat transfer from air to the wet surface at a mean surface temperature T_s, which is equivalent to "pure heat transfer" between an effective air temperature $T_{a,wb}$ and mean surface temperature T_s through a thermal resistance R_{wet}; conduction from the fin surface at mean temperature T_s to the tube outer surface at temperature T_{tos} (thermal resistance R_f); conduction from the tube outer surface to the tube inner surface through the tube wall (thermal resistance R_t); and finally convection from the tube inner wall to the coolant flowing in the tube (thermal resistance R_i). The values the resistances R_t and R_i can be found using following standard equations (see Section 3.3.1):

$$R_t = \frac{\ln(d/d_i)}{2\pi k_t L_t}; \quad R_i = \frac{1}{\alpha_r A_i} \qquad (4.170)$$

The value of R_f can be calculated using Eq. (4.169) and the definition of R_f:

$$Q_{node} = \frac{T_{a,wb} - T_s}{1/(\alpha_{eff} A_{node})} = \frac{T_{a,wb} - T_{tos}}{1/(\eta \alpha_{eff} A_{node})} = \frac{T_s - T_{tos}}{R_f} \qquad (4.171)$$

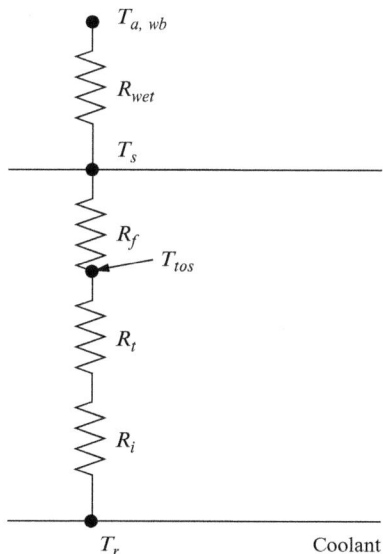

Fig. 4.23 Resistance network for wet finned surface.

which gives

$$R_f = \left(\frac{1-\eta}{\eta}\right) \frac{1}{\alpha_{eff} A_{node}} \tag{4.172}$$

The resistance network for the wet-finned surface can thus be formed between the "effective air temperature" and the coolant temperature, as shown in Fig. 4.23. This gives

$$Q_{node} = \frac{T_{a,wb} - T_s}{R_{wet}} = \frac{\alpha_{aw} A_{node}}{c_{pm}} (h_a - h_s) = \frac{T_s - T_r}{R_f + R_t + R_i} \tag{4.173}$$

Rearranging terms, we can convert Eq. (4.173) into a familiar form as

$$\frac{T_s - T_r}{h_a - h_s} = \frac{R_f + R_t + R_i}{c_{pm}/(\alpha_{aw} A_{node})} = C, \quad \text{the coil characteristic} \tag{4.174}$$

which is similar to Eq. (4.155), the only difference being in the manner of defining various thermal resistances. In Eq. (4.155), all thermal resistances are referred to the external surface area, while in Eq. (4.174), these are defined based on the pertinent area, by Eqs. (4.170), (4.172). The above equation can be solved, as before, in conjunction with the psychrometric relationship between T_s and h_s. If the calculated T_s turns out to be less than the dew point temperature of air, wet surface conditions are justified, otherwise the dry surface equations are used for the analysis of the node.

The analysis of various nodes into which the coil is discretized is carried out along the coolant path. However, in view of the complexity of tube circuiting arrangement (see, e.g., Fig. 4.24), the condition of air at entrance to various nodes may not be known at the beginning of the calculations. Therefore, to start the iterative calculations, Vardhan and Dhar [15] assume the air conditions at the inlet to each node to be the same as at the inlet to the cooling coil. The nodes which form the starting point of the coolant circuit (where the coolant enters the cooling coil) are first analyzed, using the aforementioned procedure, and the total heat transfer calculated. For each of the starting nodes, the coolant outlet conditions for the nodes, which are the coolant inlet conditions for the next node on the coolant path can be found. The air conditions at the exit of these nodes are also found. Once the starting nodes are analyzed, the next set of nodes on the coolant path are taken up for analysis. In this manner, a march is made along each circuit of the coolant simultaneously until the end of the circuit is encountered. Once one march is completed, the air conditions at various nodes are changed, in accordance with the coil circuitry, based on the air exit conditions calculated in this march. These changed conditions are then used for the second march, as shown schematically in Fig. 4.25.

These marches are repeated until the air and the coolant conditions do not change significantly between two successive marches. It has been found that even for a close tolerance of 0.01°C the iterative scheme converges fairly rapidly (see Vardhan and Dhar [15] for details). Since the air flow direction is perpendicular to that of the coolant,

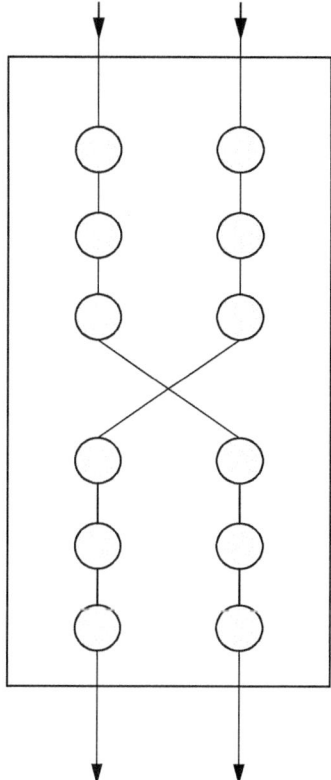

Fig. 4.24 A tube circuiting arrangement.

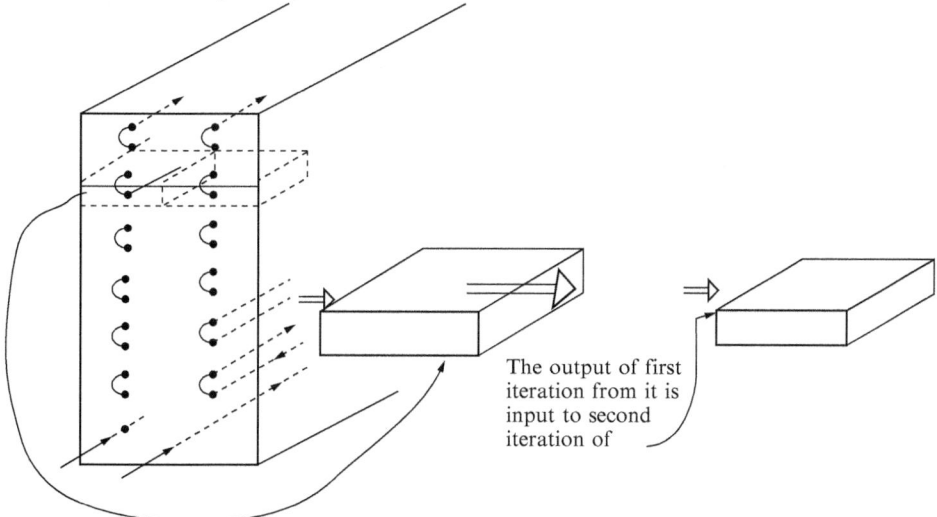

Fig. 4.25 Schematic representation of the iterative scheme.

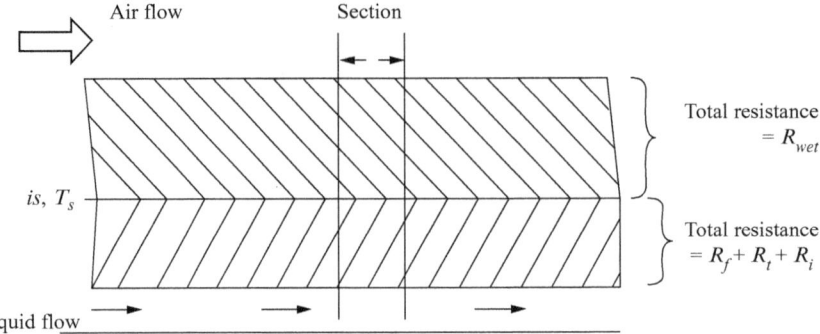

Fig. 4.26 Discretization of a node along the path of air flow.

its enthalpy drop in each node is quite significant. Thus in a two-row coil, the enthalpy drop in each node would be approximately half of the total enthalpy drop of air, which is usually quite large. Clearly in such cases, the calculation of heat transfer using the difference between the inlet air enthalpy and the enthalpy of saturated air at inlet surface temperature in Eq. (4.173), would lead to large errors. To remove this shortcoming, the nodes are further subdivided into a number of sections in the air flow direction. Since the effect of fin efficiency has already been incorporated in the resistance R_f, each node can be visualized us a heat exchanger with uniform surface temperature, the heat transfer also being assumed to take place uniformly across the width of the fin as shown in Fig. 4.26.

Parallel flow can be assumed instead of cross flow without much error since the coolant heat capacity is extremely large as compared to that of air flowing through one node. This conceptual heat exchanger is broken into a number of sections along the air flow direction as shown in Fig. 4.26. At each section the heat transfer is calculated first on the basis of the difference between free stream and surface enthalpy at the inlet of the section. The exit enthalpy for this section is then found from this heat transfer and the known mass flow rate of the air. The heat transfer is now corrected by recalculating it at the average of the inlet and outlet enthalpies. The exit air enthalpy is again calculated and becomes the inlet enthalpy for the next section. The outlet dry bulb temperature (DBT) of air for each section can be found from the energy balance for sensible heat transfer, that is

$$\dot{m}_{da}c_{pm}dT_a = \alpha_{aw} \cdot dA(T_s - T_a) \tag{4.175}$$

which gives on integration over section area

$$T_{a,out} = T_s + (T_{ain} - T_s)e^{-C^*} \tag{4.176}$$

$$\text{where} \quad C^* = \frac{A_{sec} \cdot \alpha_{aw}}{\dot{m}_{da}c_{pm}} \tag{4.177}$$

is a dimensionless number, akin to the NTU, for the wet surface and A_{sec} is the surface area of each of the sections into which the node is subdivided.

This $T_{a,out}$ is used as the inlet air temperature for the next section and for the estimation of outlet air humidity $w_{a,out}$ using the psychrometric relationship between enthalpy $h_{a,out}$, temperature $T_{a,out}$, and humidity ratio $w_{a,out}$. Proceeding sequentially in this manner, analysis of all the subsections can be carried out to get a more accurate estimation of the air enthalpy change in a node. Vardhan and Dhar [15] indicate that this procedure predicts the performance of the coils with good accuracy, a claim also borne out by Sharma [18], who used the same approach to predict the performance of DX-coils and condenser coils manufactured by a reputed Indian company.

4.2.2 Cooling Towers and Spray Washers

Most central air-conditioning plants, thermal power stations and industrial processes where water is used to carry away heat to be rejected from the system, need a device to cool the water back to its initial temperature so that it can be continuously used in the process. This device, called the cooling tower (Fig. 4.27), uses the principle of forced evaporation of a small part of the water in order to cool the rest. This is effected by bringing water and atmospheric air in the intimate contact inside a tower filled with suitable packings. In a typical design of cooling tower, water is sprayed from the top

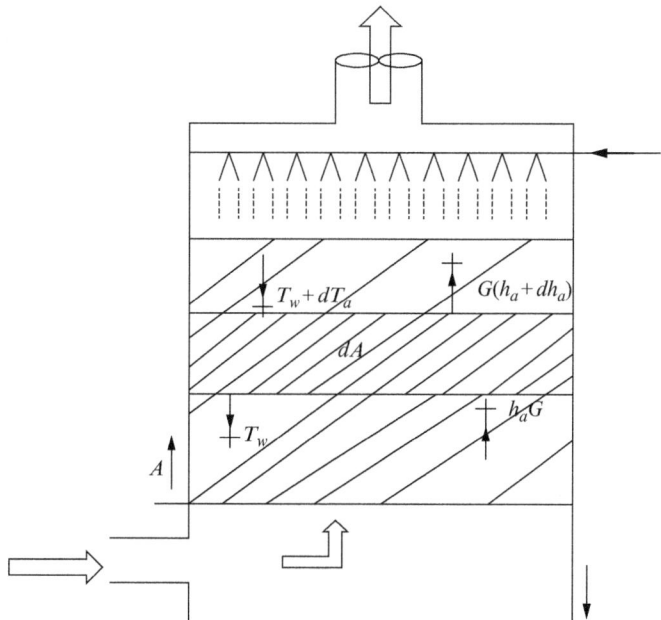

Fig. 4.27 Counter-flow cooling tower.

of the tower through an array of nozzles while air is made to flow upward, often with the help of a fan. Thus here too we have simultaneous heat and mass transfer occurring between air and water. Since the water entering the cooling tower is warm, the enthalpy of air saturated at this temperature is higher than that of the incoming air. The total heat transfer is, therefore, from water to air, resulting in cooling of water and humidification of the air.

In spray washers, used in ventilation and air-conditioning systems, the temperature of water may even be low enough to cause total heat transfer from air to water, resulting in cooling, and even dehumidification, of air. Irrespective of actual change in the thermodynamic states of air and water both air washers and cooling towers can be analyzed by using the principles of simultaneous heat and mass transfer (Section 3.4.1).

To illustrate the approach, we consider a differential volume of a typical counter-flow cooling tower, as shown in Fig. 4.27. Let dA be the total contact area between air and water, which includes the surface area of the droplets as also of the wetted fill material in the tower. Let L kg/s be the mass flow rate of water at any cross section, and G the mass flow rate of air (on dry air basis). The total heat transfer in this elemental volume can be calculated using the concept of enthalpy potential (Section 3.4.1) and coupled with energy balance equations for air and water to get

$$dq = \frac{\alpha_c dA}{c_{pm}}(h_w - h_a) \tag{4.178}$$

$$dq = Lc_L \, dT_w = G \, dh_a \tag{4.179}$$

where h_w is the enthalpy of saturated air at water temperature T_w. Combining these two equations, we get, on integration over the entire cooling tower height:

$$\int_0^A \frac{\alpha_c \, dA}{c_{pm}} = \int_{T_{w,out}}^{T_{w,in}} \frac{Lc_L}{h_w - h_a} \, dT_w$$

The convective heat-transfer coefficient α_c is usually assumed to remain constant over the height of the cooling tower. This gives

$$\frac{\alpha_c A}{c_{pm}} = \int_{T_{w,out}}^{T_{w,in}} \frac{Lc_L}{h_w - h_a} \, dT_w \tag{4.180}$$

$$\approx Lc_L \int_{T_{w,out}}^{T_{w,in}} \frac{dT_w}{h_w - h_a} \tag{4.181}$$

where we have written the approximate equation, (4.181), by neglecting the change in the water mass flow rate L due to evaporation, which indeed, is quite small in most practical situations.

Since both h_w and h_a vary with the temperature, explicit integration of the RHS of Eq. (4.181) is not possible. We therefore take recourse to numerical integration by subdividing the cooling tower into a number of elemental volumes. Usually the subdivisions are done such that the water temperature drops an equal amount ΔT, in each elemental volume, and thus Eq. (4.181) becomes

$$\frac{\alpha_c A}{L c_{pm}} = c_L \Delta T \sum \frac{1}{(h_w - h_a)_m} \tag{4.182}$$

where $(h_w - h_a)_m$ is the mean enthalpy difference in the elemental volume. The dimensionless quantity of Eq. (4.182) is called the NTU of a cooling tower and gives the number of times the average enthalpy potential goes into the overall "cooling duty" of the cooling tower. It is thus a measure of the degree of difficulty of the task.

The complete procedure for determining the NTU of a cooling tower on the basis of its performance ratings is explained below:

(i) Divide the whole cooling tower into a number of elemental volumes in each of which the water temperature changes by an equal amount.

Thus, for example, in Fig. 4.28, the tower is divided into 10 elemental volumes and the water temperatures at various positions are

$$T_{11} = T_{w,in} \quad \text{and} \quad T_1 = T_{w,out}$$

$$\Delta T = \frac{T_{w,in} - T_{w,out}}{10}$$

$$T_2 = T_1 + \Delta T$$

$$T_3 = T_2 + \Delta T \quad \text{and so on}$$

(ii) The total heat transfer in each control volume is

$$\Delta q = L c_L \Delta T$$

(iii) Considering the energy balance of each control volume we can write

$$\Delta q = G \cdot \Delta h_a$$

and therefore air enthalpies at various sections of the cooling tower can be determined as follows:

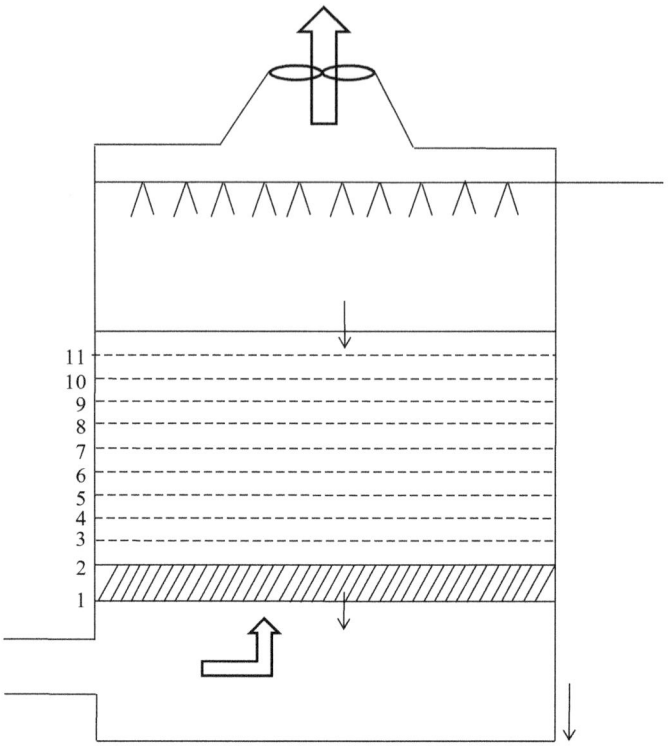

Fig. 4.28 Analyzing a counter-flow cooling tower.

$$h_{a1} = h_{a_{in}} \quad \text{(from known inlet air conditions)}$$
$$h_{a2} = h_{a1} + \Delta h_a$$
$$h_{a3} = h_{a2} + \Delta h_a \quad \text{and so on}$$

(iv) The enthalpy of saturated air at the water temperature at various sections can be found using saturated air property tables or psychrometric chart. Let these be indicated as $hw_1, hw_2, \ldots, hw_{11}$.

(v) Considering the first control volume encountered by the air, that is, 1–2, we can now write the mean enthalpy difference in it as

$$(h_w - h_a)_{m,12} = \frac{hw_1 + hw_2}{2} - \frac{ha_1 + ha_2}{2}$$

(vi) Substituting it in Eq. (4.182), applied to one control volume, we can write

$$\frac{\alpha_c \Delta A_{12}}{Lc_{pm}} = \frac{c_L \cdot \Delta T}{(h_w - h_a)_{m,12}}$$

where ΔA_{12} is the area of contact between air and water in this control volume.

(vii) Proceeding similarly we can calculate mean enthalpy difference in various elemental control volumes as

$$(h_w - h_a)_{m,23} = \frac{hw_2 + hw_3}{2} - \frac{ha_2 + ha_3}{2}$$

$$(h_w - h_a)_{m,34} = \frac{hw_3 + hw_4}{2} - \frac{ha_3 + ha_4}{2}$$

and so on.......

and then find the total NTU for the whole cooling tower, considering all elemental control volumes from 1–2 to 10–11, as:

$$NTU = \frac{\alpha_c A}{L c_{pm}} = c_L \Delta T \sum_{1-2}^{10-11} \frac{1}{(h_w - h_a)_{m,ij}}$$

(viii) If α_c value is known, we can find the area of contact between the air and water as

$$A = \frac{NTU \cdot L \cdot c_{pm}}{\alpha_c}$$

In the above calculations, the enthalpy of air along the cooling tower is computed. However, this by itself is not sufficient to specify the thermodynamic state of air; one more property needs to be determined at each section. The property which can be determined most conveniently is its DBT. For this purpose, we consider the sensible heat balance in each elemental control volume. Thus for the first section encountered by air, we can write

$$G c_{pm}(T_{a2} - T_{a1}) = \alpha_c \Delta A_{12} \left(\frac{T_{w1} + T_{w2}}{2} - \frac{T_{a1} + T_{a2}}{2} \right) \tag{4.183}$$

Having determined ΔA_{1-2} above, and knowing T_{w1}, T_{w2}, and T_{a1} (and other quantities, that is, G, c_{pm}, α_c) we can determine T_{a2}, the DBT at the exit of this first elemental control volume. Proceeding similarly, we can find the DBT of air at the exit of each succeeding control volumes. Using the values of air enthalpy and the DBT, the thermodynamic state of air as it flows through the cooling tower can be indicated on a psychometric chart.

Thus the design of a cooling tower can be done on the basis of its performance ratings and the thermodynamic state of air within the tower can also be determined. The conjugate problem of determining the performance of a cooling tower of given design, that is, a specified NTU value, for any given water and air entry conditions, however, needs iterative solution.

We assume a suitable value of the water outlet temperature and then, following the above procedure, determine the NTU required to achieve that. This is then compared with the given value, and based on the difference between the two, a better guess is made for the water outlet temperature. The above procedure is repeated till convergence is obtained. The following example illustrates this procedure through the analysis of a typical cooling tower.

Example 4.13

The catalog of a counter-flow cooling towers gives the following ratings at design conditions:

$$\text{Water flow rate} = 6500\,\text{kg/h}$$
$$\text{Air flow rate} = 5500\,\text{kg/h}$$
$$\text{Water inlet temperature} = 34°C$$
$$\text{Water outlet temperature} = 28°C$$
$$\text{Air entry conditions} = 35°C\ DBT \quad \text{and} \quad 25°C\ WBT$$

If the water flow rate falls to 6000 kg/h, and the inlet temperature increases to 35°C, what will be the water outlet temperature, other conditions remaining the same? Assume that the NTU of the cooling tower remains unaltered.

Solution

We first need to find the NTU of the cooling tower from its given performance at the design conditions. Since we are doing calculations manually, we subdivide the tower into elemental volumes in each of which, the water temperature changes by 1°C. Thus the temperature at various sections are (see Fig. 4.28):

$$T_1 = 28°C \quad T_2 = 29°C \quad T_3 = 30°C \quad T_4 = 31°C \quad T_5 = 32°C$$
$$T_6 = 33°C \quad T_7 = 34°C$$

The total heat transfer in each control volume is

$$\Delta q = L c_L \Delta T = \frac{6500}{3600} \times 4.186 \times 1\,\text{kW} = 7.558\,\text{kW}$$

The change in air enthalpy in each control volume Δh_a is therefore

$$\Delta h_a = \Delta q / G = \frac{7.558}{5500/3600} = 4.947\,\text{kJ/kg}$$

The enthalpy of air at inlet conditions of 35°C DBT/25°C WBT, as seen from the psychometric chart, is 76.5 kJ/kg da. The air enthalpy at various sections can now be found by adding Δh_a to the air enthalpy at the entrance to each elemental volume of the cooling tower. Subsequent calculations are indicated in following table (h_w values taken from ASHRAE, Handbook of Fundamentals, 2005).

Section no.	Water temp. (°C)	h_w (kJ/ kg da)	h_a (kJ/ kg da)	$(h_w - h_a)$	$(h_w - h_a)_m$ (mean diff.)	$NTU = \frac{4.186 \times 1}{(h_w - h_a)_m}$
1	28	89.976	76.5	13.476		
					13.4535	0.3111
2	29	94.878	81.447	13.431		
					13.5215	0.3096
3	30	100.006	86.394	13.612		
					13.820	0.3029
4	31	105.369	91.341	14.028		
					14.3595	0.2915
5	32	110.979	96.288	14.691		
					15.1565	0.2762
6	33	116.857	101.235	15.622		
					16.2255	0.2580
7	34	123.011	106.182	16.829		
						$\Sigma = 1.7493$

Thus NTU of the cooling tower is 1.7493.

To find the outlet water temperature under altered conditions, we note that water inlet temperature has increased, and its flow rate is reduced. Both these changes would tend to increase the water temperature drop (called cooling tower range) in the tower. We assume that this drop increases by 1°C, so that the cooling tower range is now 7°C, and therefore the water outlet temperature is 28°C. We repeat above calculations to find the NTU value needed to achieve this temperature change, by subdividing tower into seven segments where ΔT is 1°C each.

Section no.	Water temp. (°C)	h_w (kJ/ kg da)	h_a (kJ/ kg da)	$(h_w - h_a)$	$(h_w - h_a)_m$ (mean diff.)	$NTU = \frac{4.186 \times 1}{(h_w - h_a)_m}$
1	28	89.976	76.500	13.476		
					13.6437	0.3068
2	29	94.878	81.0666	13.8114		
					14.0921	0.2970
3	30	100.006	85.6332	14.3728		
					14.771	0.2834
4	31	105.369	90.1998	15.1692		
					15.6909	0.2668
5	32	110.979	94.7664	16.2126		
					16.8683	0.2482
6	33	116.857	99.333	17.524		
					18.3177	0.2285
7	34	123.011	103.8996	19.1114		
					20.0501	0.2088
8	35	129.455	108.4662	20.9888		
						$\Sigma = 1.8395$

In view of change in the water flow rate, the new Δh_a value can be found as:

$$\Delta h_a = \frac{Lc_l\Delta T}{G} = \frac{6000 \times 4.186 \times 1}{5500} = 4.5666\,\text{kJ/kg}$$

As can be seen from the detailed calculations shown in the previous table, the NTU value required to achieve this temperature drop is 1.8395.

Since the actual NTU of the cooling tower is less than this, it implies that water can not be cooled by 7°C in this cooling tower. We therefore redo the calculations after changing this assumption to 6.5°C, which corresponds to a water outlet temperature of 28.5°C.

To find the NTU required to achieve this water outlet temperature, we again subdivide the tower, this time, into seven elemental volumes. In the first element the temperature drops from 29.0 to 28.5°C, with $\Delta T = 0.5°C$; in all other elemental volumes the temperature drops by 1.0°C, that is, $\Delta T = 1°C$. The change in air enthalpy for the first subdivision is therefore half of that in the rest, that is, 2.2833 kJ/kg.

The detailed calculations are shown in following table from where we can see that the NTU needed to achieve this temperature drop is 1.47392.

This is lesser than the available NTU of the cooling tower. It implies that the actual temperature drop in the cooling tower will be larger than 6.5°C. Earlier iteration showed it to be lesser than 7°C, and thus it is likely to be somewhere in between, say 6.8°C. The students are encouraged to carry out another iteration with this assumption and check whether the required NTU matches with the NTU of the given cooling tower.

Section no.	Water temp. (°C)	h_w (kJ/ kg da)	h_a (kJ/ kg da)	$(h_w - h_a)$	$(h_w - h_a)_m$ (mean diff.)	$NTU = \frac{4.186 \times 1}{(h_w - h_a)_m}$
1	28.5	92.427	76.5	15.927		
					16.0108	0.1307
2	29.0	94.878	78.7833	16.0947		
					16.3754	0.2556
3	30.0	100.006	83.3499	16.6561		
					17.0543	0.2455
4	31.0	105.369	87.9165	17.4525		
					17.9742	0.2329
5	32.0	110.979	92.4831	18.4959		
					19.1516	0.2186
6	33.0	116.857	97.0497	19.8073		
					20.601	0.2032
7	34.0	123.011	101.6163	21.3947		
					22.3334	0.1874
8	35.0	129.455	106.1829	23.2721		
						$\Sigma = 1.4739$

Example 4.14

Determine the thermodynamic state of the air leaving the cooling tower of Example 4.13, when it is operating at design conditions.

Solution

In Example 4.13, we determined the enthalpy of air as it traverses the cooling tower. However, to get a complete description of its state, we need one more property. The change in the dry-bulb temperature of air ΔT_a can be easily calculated by determining the sensible heat transfer in each elemental volume, using Eq. (4.183), which can be rewritten in a generalized manner applicable to each elemental volume, as

$$\Delta T_a = \frac{\alpha_c \Delta A}{GCp_m} \left(T_{w,mean} - T_{a,mean} \right)$$

$$T_{a,exit} - T_{a,entry} = \frac{L}{G} \Delta NTU \left(T_{w,mean} - \frac{T_{a,entry} + T_{a,exit}}{2} \right) \qquad (4.184)$$

where ΔNTU is the NTU of the elemental volume calculated above. Rearranging this equation, we can get an explicit expression for air temperature at the exit of each control volume as

$$T_{a,exit} = \frac{2 \cdot z \cdot T_{w,mean} + (2 - z) T_{a,entry}}{2 + z} \qquad (4.185)$$

where $z = \frac{L}{G} \cdot \Delta NTU$. These calculations can now be conveniently tabulated as follows:

Section no.	Water temp. (°C)	Tw_{mean}	ΔNTU	z	Air temp. (°C)
1	28				35
		28.5	0.3111	0.36766	
2	29				32.98
		29.5	0.3096	0.36589	
3	30				31.90
		30.5	0.3029	0.35797	
4	31				31.475
		31.5	0.2915	0.3445	
5	32				31.482
		32.5	0.2762	0.32642	
6	33				31.768
		33.5	0.2580	0.30491	
7	34				32.226

The enthalpy of air leaving various elemental volumes as calculated in Example 4.12, along with the temperature calculated above give a complete description of the thermodynamic state of air as it traverses through the cooling tower. The state of air at cooling tower exit is thus:

1. DBT: 32.226°C   Enthalpy: 106.182 kJ/kg da

Locating the state on psychrometric chart we can get the value of humidity ratio as 28.8 g/kg da. The humidity ratio of the air at entry is 15.8 g/kg da. This gives the total amount of moisture picked up air, $\Delta \dot{m}_w$ to be

$$\Delta \dot{m}_w = 5500 \times \frac{(28.8 - 15.8)}{1000} = 71.5 \, \text{kg/h}$$

This must be equal to the decrease in the mass flow rate of water as it cooled in the cooling tower. It can be seen that it is approximately about 1% of the water flow rate.

The same approach can be employed to predict the performance of air washers used in heating, ventilating, and air-conditioning systems. In Example 4.15 we shall illustrate the procedure by considering the typical case of a parallel flow air washer being used to clean, cool, and dehumidify air in a hot and humid environment.

Example 4.15

In a parallel flow spray air washer, cold water at 15°C is being sprayed at the flow rate of 10 kg/s into hot and humid air entering the washer at 35°C DBT and 27°C WBT at a rate of 12 kg/s. Estimate the condition of air at the exit of the washer and the temperature and flow rate of water leaving the washer (Fig. 4.29).

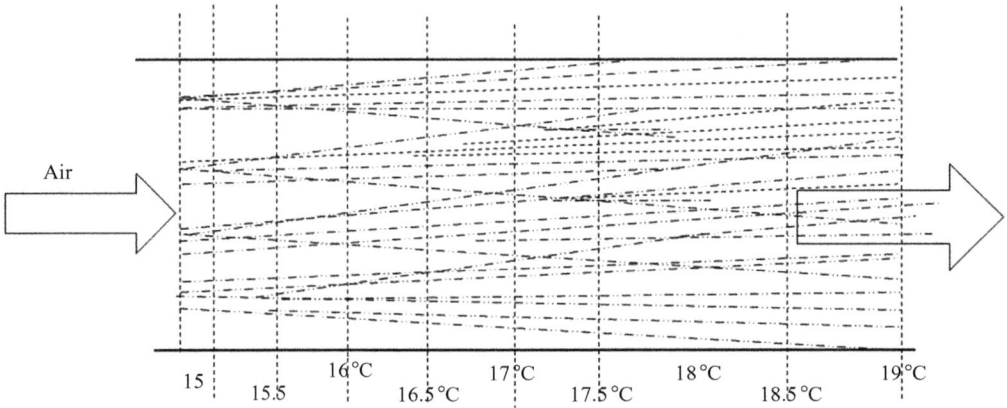

Fig. 4.29 Spray air washer simulation, Example 4.14.

Given:
1. Heat-transfer coefficient between air and water: 50 W/m² K
2. Effective area of contact: 100 m²

Solution

This, being a simulation problem, would need iterative solution. We find the area of contact needed to bring about a small change of 0.5°C, in the water temperature in successive elemental volumes. This needs the value of change in the enthalpy of air in each of these elemental volumes which is calculated by using Eqs. (4.179), (4.182) as:

$$\Delta h_a = \frac{L c_l \Delta T_w}{G} = 1.744 \, \text{kJ/kg}$$

The following table shows the detailed area calculations as also the cumulative area requirement.[a]

Section no.	Water temp. (°C)	h_w	h_a	$h_w - h_a$	$(h_w - h_a)$ mean	$\Delta A (m^2)$	Cum. area m^2
1	15.0	42.113	85.2	43.087			
					41.403	10.353	10.353
2	15.5	43.538	83.256	39.718			
					38.134	11.241	21.594
3	16.0	44.963	81.512	36.549			
					34.936	12.269	33.863
4	16.5	46.444	79.768	33.324			
					31.711	13.517	47.380
5	17.0	47.926	78.024	30.098			
					28.456	15.064	62.444
6	17.5	49.467	76.280	26.813			
					25.170	17.030	79.474
7	18.0	51.008	74.536	23.528			
					21.854	19.614	99.088
8	18.5	52.612	72.792	20.18			
					18.506	23.163	122.251
9	19.0	54.216	71.048	16.832			

As can be seen from this table, if the area of contact is 99.08 m², water would get warmed up to 18.5°C, while if the area of contact is 122.25 m², it would reach 19.0°C. Since the given area of contact is 100 m², it follows that the water temperature at washer exit would be slightly higher than 18.5°C, ≈18.52°C. The air enthalpy at exit would be slightly lower than 72.792 kJ/kg da, and using linear interpolation, it can be estimated as 72.716 kJ/kg da.

To find the temperature of air at washer exit, we calculate the sensible heat transfer in each of the elemental volumes and determine the successive drops in air temperature, as follows, using Eq. (4.183). Since, we have calculated the area of contact, ΔA, in each elemental volume above, we can rearrange this equation as before to get (cf. Eq. 4.185):

$$T_{a2} = \frac{T_{a1}(2-z) + T_{w,mean}2z}{2+z}$$

$$\text{where} \quad z = \frac{\alpha_c \Delta A}{G c_{pm}}$$

The calculations are summarized in following table.

Section no.	Water temp. (°C)	Tw_{mean}	ΔAm^2	z	Air temperature
1	15.0				35
		15.25	10.353	0.04213	
2	15.5				34.185
		15.75	11.2407	0.045738	
3	16.0				33.361
		16.25	12.2695	0.049925	
4	16.5				32.528
		16.75	13.5173	0.05500	
5	17.0				31.683
		17.25	15.0635	0.06129	
6	17.5				30.825
		17.75	17.030	0.069295	
7	18.0				29.949
		18.25	19.614	0.079809	
8	18.5				29.051
		18.75	23.163	0.09425	
9	19.0				28.124

The exit air temperature can be found by interpolation between the last two values keeping in view the fact that actual area of contact is 100 m² while the area of contact for these values are 99.08 and 122.25 m², respectively. This gives the exit air temperatures as 29.014°C. The specific humidity of air at this temperature and having an enthalpy of 72.716 kJ/kg da can be found from the psychrometric chart as 17 g/kg da. The specific humidity of entering air (35°C DBT, 27°C WBT) is 19.4 g/kg da. Thus the moisture shed off by air is = $12 \times (19.4 - 17) = 28.8$ g/s, and therefore the mass flow rate of water at exit of the air washer is 10.0288 kg/s.

[a]We have taken the average value of c_{pm} as 1.024 kJ/kg da.

4.2.3 Heat and Mass Exchangers Using Desiccants

In many industrial applications where low moisture content has to be maintained in the working atmosphere (e.g., in drying of powders and films, seasoning of wood, etc.) liquid desiccants like aqueous solution of lithium chloride, calcium chloride, various glycols, etc. or solid desiccants like silica gel, molecular sieves, etc. are often used to dehumidify ambient air before sending it to the processing unit/plant. The designs of these dehumidifiers employed in practice vary depending upon the nature of desiccant, process requirements, and other design constraints. In many designs of the liquid desiccant dehumidifiers, cold water is also circulated to remove the heat of sorption and thus increase the efficiency of dehumidification. Thus both heat and mass transfer occur in these dehumidifiers. The liquid desiccant, which gets diluted due to absorption of water, needs to be regenerated back to the original concentration so that it can be reused for dehumidification. This is achieved by heating the diluted solution to sufficiently high temperature so that the moisture can now migrate from the solution to the ambient air.

The most common design of solid desiccant-based systems is that using rotary desiccant wheel, with the desiccant impregnated in the matrix material. As the wheel rotates, a part of it is exposed to the process air, which gets dehumidified and sheds its moisture on to the matrix. The remaining portion of the wheel is exposed to hot air, which derives off the moisture condensed in it earlier. Thus these equipments also have simultaneous heat and mass transfer.

Of late, desiccant-based air-conditioning systems have also been explored as an ecofriendly alternative to traditional air-conditioning systems based on vapor-compression refrigeration. Here too both solid and liquid desiccant-based systems are being suggested. We shall, in this section, study the methodology for simulation of typical dehumidifiers/regenerators used in these systems.

Liquid Desiccant Dehumidifier/Regenerator

A variety of designs of liquid desiccant-based dehumidifiers and regenerators are available. All the designs aim at providing a large area of contact between the desiccant and air, without increasing the air-side pressure drop inordinately. Some of the designs also enable removal of heat of sorption, which considerably improves the performance of the dehumidifier.

One such design is that of falling film dehumidifier (Fig. 4.30) wherein the desiccant flows down a vertical surface (usually a plate or tube surface) in a thin film and comes in direct contact with air (usually) flowing upward.

In a tubular falling film configuration, cooling water is also circulated through the shell surrounding the tubes. It removes the heat of sorption, thus preventing rise in temperature of the desiccant, which can therefore retain its moisture removing capability in spite of drop in its concentration as it flows down. Fig. 4.30A shows the sectional view of such a dehumidifier and Fig. 4.30B shows the various streams entering/leaving a typical control volume. Considering the z-direction as the direction of flow of desiccant, with $z = 0$ corresponding to the top of the dehumidifier, we can write the conservation equations as follows:

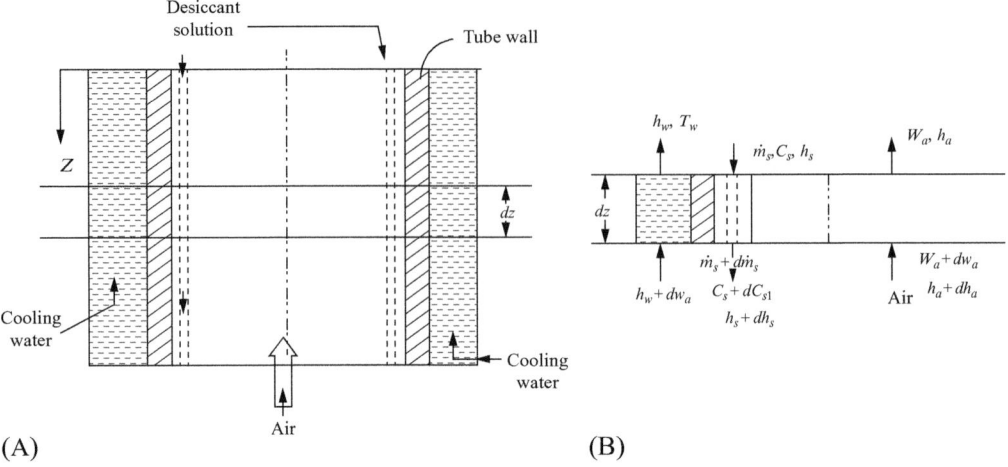

Fig. 4.30 A typical liquid desiccant-based dehumidifier. (A) Sectional view of dehumidifier. (B) Control volume of dehumidifier.

Mass Balance of Moisture Content

Air

$$\dot{m}_a(w_a + dw_a) - h_m \cdot P_a(w_a - w_e) \cdot dz = \dot{m}_a w_a$$

$$\text{which gives} \quad \dot{m}_a \frac{dw_a}{dz} = h_m P_a(w_a - w_e) \qquad (4.186)$$

where w_e is humidity ratio of air in equilibrium with the desiccant; w_a the local humidity ratio of air; h_m the mass-transfer coefficient between air and the desiccant; $\dot{m}_a$ the mass flow rate of air in kilogram dry air/second; and P_a the perimeter of contact between air and the desiccant.

Desiccant Solution

$$\dot{m}_s(1 - c_s) + h_m P_a(w_a - w_e)dz = (\dot{m}_s + d\dot{m}_s)(1 - c_s - dc_s)$$

which gives on simplification:

$$\frac{d\{\dot{m}_s(1 - c_s)\}}{dz} = h_m P_a(w_a - w_e) \qquad (4.187)$$

where $\dot{m}_s$ is the desiccant flow rate and c_s its concentration.

Energy Balance

Sensible Heat Transfer From Air Considering only the sensible heat transfer from air, we can write the energy balance as:

$$\dot{m}_a c_{pm}(T_a + dT_a) - \alpha_a P_a(T_a - T_s)dz = \dot{m}_a c_{pm} T_a$$

where α_a is the convective heat-transfer coefficient between the air (temperature T_a) and the desiccant (temperature T_s). Simplifying we get

$$\dot{m}_a c_{pm} \frac{dT_a}{dz} = \alpha_a P_a (T_a - T_s) \tag{4.188}$$

Desiccant

$$\dot{m}_s h_s - U_{sw} P_w dz(T_s - T_w) - \alpha_a P_a dz(T_s - T_a)$$
$$+ h_m P_a dz(w_a - w_e)h_{uv} = \dot{m}_s h_s + \frac{d(\dot{m}_s h_s)}{dz} dz$$

where the second term on LHS accounts for heat transfer to water through the desiccant film and metal wall, the third term represents sensible heat transfer to air, and the fourth term represents the energy content of water vapors that diffuse from air into the desiccant. Often the water vapor enthalpy h_{uv} is taken as the latent heat of evaporation of water h_{fg}. Simplifying we get the energy equation as:

$$\frac{d}{dz}(\dot{m}_s h_s) = h_m P_a (w_a - w_e)h_{fg} - U_{sw} P_w (T_s - T_w) - \alpha_a P_a (T_s - T_a) \tag{4.189}$$

where U_{sw} is the overall heat-transfer coefficient between the desiccant solution (temperature T_s) and water (temperature T_w), and P_w is the wetted perimeter of metal surface.

Water The energy balance for water which receives heat from the desiccant film through the walls of the tube (or plate) separating the two streams can be directly written as

$$m_w c_{pw} \frac{dT_w}{dz} = -U_{sw} P_w (T_s - T_w) \tag{4.190}$$

To predict the performance of the dehumidifier, the simultaneous differential equations (4.186)–(4.190) have to be solved. Since these equations involve complicated nonlinear functions (e.g., relationship of h_s and w_e to desiccant concentration c_s and temperature T_s), analytical solution is not possible. Numerical solution can be obtained by dividing the dehumidifier into a large number of elemental volumes and using any of the standard methods (like fourth-order Runge-Kutta method, Section 2.4.1) to solve the differential equations. To start the calculations, the values of air DBT and humidity, and cooling water temperature at $z = 0$, Fig. 4.30 (i.e., the exit conditions of air and water), are assumed and then step-by-step calculations done along z-direction till we reach the bottom of the dehumidifier where the desiccant solution is leaving. The calculated values of the air temperature and humidity, and water temperature are compared with the prescribed air-entry and water-entry conditions, and based on the deviations new assumptions for their properties at dehumidifier exit are made; and the whole process is repeated till convergence is obtained.

This approach has been successfully used by Jain [19], Jain et al. [20], Asati [21], and Ritunesh [22] to predict the performance of both dehumidifiers and regenerators. While in the dehumidifier mode the system is cooled by circulating cold water over the tubes, in the regenerator mode hot oil is circulated outside the tubes to help moisture transfer from the desiccant to air. Since the above formulation is general it takes care of change in the direction of heat and mass transfer caused by change in the desiccant temperature and the circulating fluid temperature in the regenerator. Thus in the regenerator the equilibrium humidity ratio w_e would be greater than w_a, and therefore $\frac{dW_a}{dz}$ would be negative (Eq. 4.186). Since air flows counter to $+z$-direction, it implies its moisture content would increase as it flows upward, due to contact with hot desiccant solution. Similarly, we can see from Eq. (4.190) that $\frac{dT_w}{dz}$ will become +ve for regenerator, thus indicating that the heating fluid temperature would fall as it flows upward. Thus these equations would also be applicable to the regenerator too.

Solid Desiccant Dehumidifier

The most common design of solid desiccant dehumidifier cum regenerator is that of rotary desiccant wheel (Fig. 4.31). The desiccant is impregnated on a suitable material (i.e., special type of thick paper), which is usually shaped into a long-corrugated sheet. This sheet is then wound into a spiral along with two flat sheets, as shown in Fig. 4.31, so as to form a honeycomb-type passage for flow of air (Figs. 4.31 and 4.32).

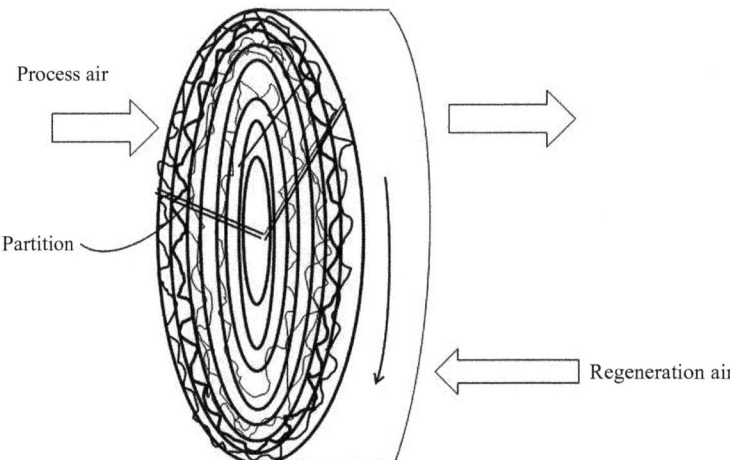

Fig. 4.31 The rotary desiccant wheel.

Process air

Partition

Regeneration air

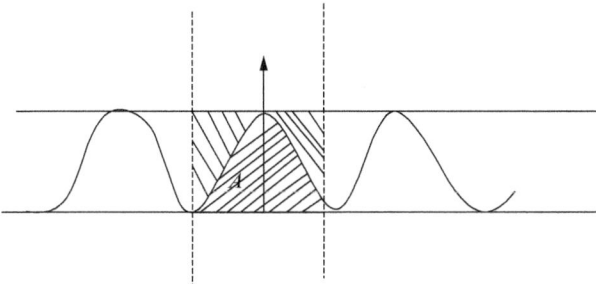

Fig. 4.32 Details of air passage in the matrix.

A partition with suitable seals divides the wheel face area into two parts, one through which the process air flows and the other through which hot air for regeneration of the desiccant flows (Fig. 4.31). Since the air passage is quite small and so is the desiccant layer thickness, simplified one-dimensional model for the simultaneous heat and mass transfer processes occurring in the wheel, gives reasonable results. We shall present one such model based on the paper by Zhang et al. [23] neglecting second-order phenomena like heat transfer across air passages, axial conduction, and mass diffusion in both the air stream and the desiccant, and the effect of absorption/desorption on the thermal and concentration boundary layers.

Considering a differential element of the air passage, of length dz (Fig. 4.33), we can write various conservation and transfer equations keeping in view the fact that the desiccant impregnated on the walls of the passage, comes in contact with air on both its sides (Fig. 4.32). The equation for conservation of moisture can be stated in words as:

(Rate of ingress of moisture with air at $x = z$)

   $-$ (Rate of outflow of moisture with air at $x = z + dz$)

 $=$ (Rate of increase of moisture content of desiccant)

   $+$ (Rate of increase of moisture content of air within the control volume)

In mathematical terms this can be written as:

$$[2A\rho_a Vw]_z - [2A\rho_a Vw]_{z+dz} = \frac{\partial}{\partial t}[M_d \, dz \cdot (1 - C_s)] + \frac{\partial}{\partial t}[2 \cdot A \cdot \rho_a dz \cdot w]$$

where we have considered the total air flowing in the hatched area shown in Fig. 4.32 for conservation equation since this much of air comes in contact with the desiccant material impregnated on the passage walls.

In the above equation A is the cross-sectional area of a single air passage; V the velocity of air flowing through it, ρ_a its density, w its humidity ratio; M_d is the mass of

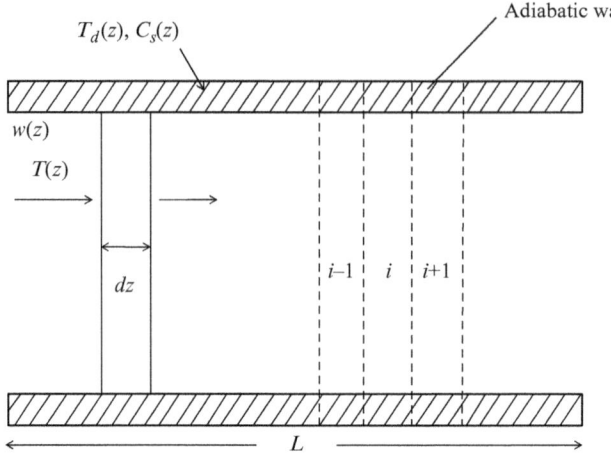

Fig. 4.33 Computational grid inside the wheel.

desiccant material on the passage walls per unit length and C_s the concentration of solid desiccant in it (which implies $1 - C_s$ is the mass of water per kilogram of total desiccant material). Simplifying we get

$$V\frac{\partial w}{\partial z} + \frac{\partial w}{\partial t} = a_1 \frac{\partial C_s}{\partial t} \tag{4.191}$$

where

$$a_1 = \frac{M_d}{2A\rho_a} \tag{4.192}$$

The rate at which moisture is transferred to the desiccant material would be governed primarily by the mass-transfer coefficient h_m (in kg/s per m^2). This transfer equation can be written as:

$$\frac{\partial}{\partial t}[M_d(1 - C_s)] \cdot dz = h_m(2P\,dz)(w - w_d)$$

where P is the perimeter of a single air passage and w_d the humidity ratio of air in equilibrium with the desiccant. Simplifying we get

$$\frac{\partial C_s}{\partial t} = a_2(w - w_d) \tag{4.193}$$

$$\text{where}\quad a_2 = \frac{2Ph_m}{M_d} \tag{4.194}$$

Proceeding similarly we can write *sensible heat balance* for the air stream, considering its interaction with the desiccant impregnated walls, as follows:

$$\left[|(2Adz\rho_a V)c_{pm}T|_z - |(2Adz\rho_a V)c_{pm}T|_{z+dz} \right]$$

$$= \begin{bmatrix} \text{Rate of increase} \\ \text{of the energy} \\ \text{content of walls} \end{bmatrix} + \begin{bmatrix} \text{Rate of increase} \\ \text{of energy of air} \\ \text{due to temperature} \\ \text{change} \end{bmatrix} - \begin{bmatrix} \text{Rate of evolution} \\ \text{of heat of} \\ \text{adsorption} \end{bmatrix}$$

$$= dz[M_d c_{pwd} + M_s c_{ps}]\frac{\partial T_d}{\partial t} + (2Adz\rho_a)c_{vm}\frac{\partial T}{\partial t} - 2Pdz h_m(w - w_d)Q$$

where c_{pm} is the specific heat of the moist air at constant pressure ($= c_{pa} + wc_{pv}$), c_{pv} is the specific heat of water vapor, c_{vm} is the moist air-specific heat at constant volume; c_{pwd} is the specific heat of wet desiccant material which can be approximated as $c_{pwd} = c_{pd}C_s + (1-C_s)c_{pl}$, where c_{pd} is the specific heat of dry solid desiccant and c_{pl} is the specific heat of liquid water; M_s is the mass of substrate (in which the desiccant is impregnated) per unit length, c_{ps} its specific heat; and Q is the heat of absorption per unit mass of water absorbed by the desiccant.

Simplifying it we get

$$\frac{c_{vm}}{c_{pm}}\frac{\partial T}{\partial t} + V\frac{\partial T}{\partial z} + a_3\frac{\partial T_d}{\partial t} = a_4(w - w_d) \tag{4.195}$$

$$\text{where} \quad a_3 = \frac{M_d c_{pwd} + M_s c_{ps}}{2A\rho_a c_{pm}} \tag{4.196}$$

$$\text{and} \quad a_4 = \frac{Ph_m Q}{A\rho_a c_{pm}} \tag{4.197}$$

The energy balance for the matrix walls can be stated in words as:

$$\begin{pmatrix} \text{Rate of increase of} \\ \text{thermal energy} \\ \text{content of walls} \end{pmatrix} = \begin{pmatrix} \text{Rate of convective} \\ \text{heat transfer from} \\ \text{air} \end{pmatrix} + \begin{pmatrix} \text{Rate of energy} \\ \text{transfer due to} \\ \text{migration of} \\ \text{moisture} \end{pmatrix}$$

$$+ \begin{pmatrix} \text{Rate of evolution} \\ \text{of heat of} \\ \text{adsorption} \end{pmatrix}$$

Mathematically this can be written as

$$dz(M_d c_{pwd} + M_s c_{ps})\frac{\partial T_d}{\partial T} = (2Pdz)\alpha(T - T_d) + (2Pdz)h_m(w - w_d)c_{pv}(T - T_d)$$

$$+ (2Pdz)h_m(w - w_d)Q$$

Simplifying we get

$$\frac{\partial T_d}{\partial T} = a_5(T - T_d) + a_6(w - w_d)(T - T_d) + a_7(w - w_d) \tag{4.198}$$

$$\text{where} \quad a_5 = \frac{2P\alpha}{M_d c_{pwd} + M_s c_{ps}} \tag{4.199}$$

$$a_6 = \frac{2Ph_m c_{pv}}{M_d c_{pwd} + M_s c_{ps}} \tag{4.200}$$

$$a_7 = \frac{2Ph_m Q}{M_d c_{pwd} + M_s c_{ps}} \tag{4.201}$$

Eqs. (4.191), (4.193), (4.195), (4.198) form the set of four equations whose solution determines the performance of a rotary desiccant wheel. The initial conditions (for all z, and $t = 0$) can be taken as the entry conditions of the process air and the initial state of the desiccant wheel, before air flows through it. The boundary conditions are suitably taken for the process air and the regenerative air. The partial differential equations can be solved by using standard numerical techniques. Zhang et al. [23] have represented the time and space derivatives using backward difference scheme and solved the above equations through an implicit up-wind formulation. Their numerical predictions are in reasonable agreement with the experimental data.

4.3 RECIPROCATING DEVICES

Reciprocating pumps and compressors are two devices commonly employed in many thermal systems. Reciprocating pumps are used for pumping liquids, especially when small quantities of fluid have to be raised to high pressures, for example, in the fuel injection pumps of diesel engines. Similarly reciprocating compressors are used for compressing air, gases like CNG, and refrigerants where high pressures are needed at modest mass flow rates.

The simulation of the reciprocating devices has four different facets, that is, the kinematics of the reciprocating piston to determine how the cylinder volume changes as the crank rotates; thermodynamics of the working fluid inside the cylinder to determine the change in its temperature and pressure with time, and the heat and work transfer; the dynamics of the motion of the valves to determine when the valves would open and close; and finally the mechanics of flow through the valves. While general equations for the kinematics and thermodynamics of the reciprocating devices can be formulated, the last two aspects depend strongly on the valve design. We shall develop these equations in Section 4.3.1 for a typical refrigeration compressor with ring-type valves.

Another very important reciprocating device is the ubiquitous internal combustion engine, the heart of automobiles. The IC engine simulation can be viewed as an extension of the reciprocating compressor simulation, with two main differences, that is the occurrence of combustion of the fuel, and fixed valve opening and closing times (decided by the cam design) independent of whatever happens in the cylinder. The later removes consideration of the valve dynamics from IC engine simulation; while the former introduces additional thermodynamic equations. Usually in SI engines the combustion is assumed to be at constant volume, while in CI engines it is assumed to be at constant pressure and the basic energy equation is used to determine the temperature of the products of combustion. This is illustrated in Section 4.3.2 for a typical SI engine.

4.3.1 Reciprocating Compressor

Kinematic Equation

The movement of the piston due to the rotation of the crank can be determined from purely geometric considerations by referring to Fig. 4.34. This gives piston displacement from the top dead center, S_2 as:

$$S_2 = R(1 - \cos\theta) + L_c\left(1 - \sqrt{1 - \frac{R^2}{L_c^2}\sin^2\theta}\right) \qquad (4.202)$$

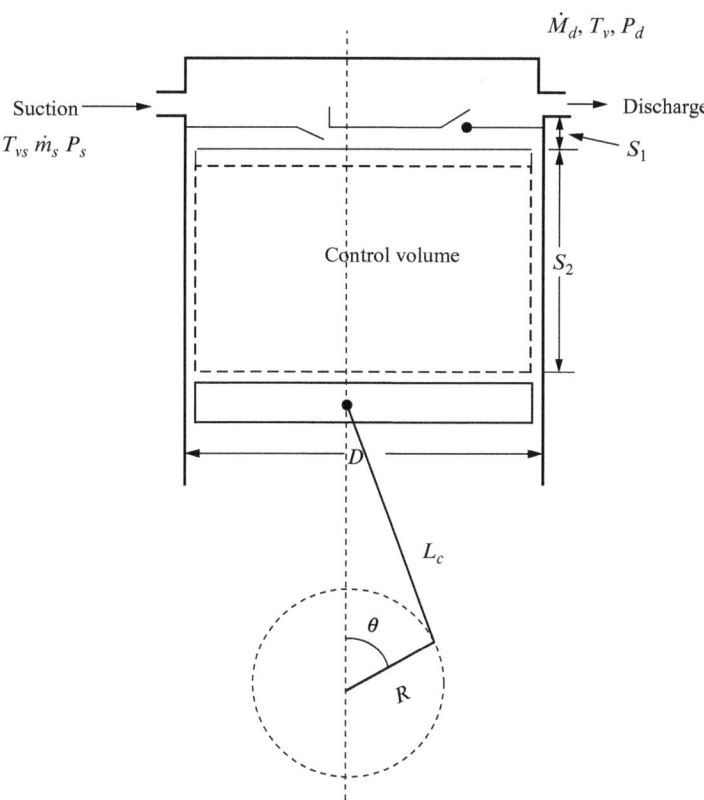

Fig. 4.34 Reciprocating compressor.

where R is the crank radius, L_c the length of the connecting rod, and θ the angle of rotation of the crank after passing through the TDC position.

Referring to Fig. 4.34 again, the instantaneous volume of gases in the cylinder equals the sum of the clearance volume and the volume swept by the piston as the crank rotates. This gives

$$V = V_c + \frac{\pi}{4}D^2 S_2 \tag{4.203}$$

where D is the cylinder diameter and V_c the clearance volume. The distance S_1 shown in the figure is equal to $V_c/(\frac{\pi}{4}D^2)$.

Thermodynamic Equation

Making the usual assumptions of uniform state, uniform flow, and neglecting the contribution of kinetic and potential energies in the flowing streams, the first law statement for the control volume shown in Fig. 4.34 can be written as (Eq. 3.11):

$$\dot{Q} = \dot{W}_{sh} + \frac{dU}{dt} + \dot{m}_d h_d - \dot{m}_s h_s \tag{4.204}$$

where $\dot{Q}$ is the rate of heat input to the control volume, $\dot{W}_{sh}$ is the rate of work output through moving boundaries, $\dot{m}_d$ and $\dot{m}_s$ are the mass flow rates of the streams at discharge and suction, respectively; and h_d and h_s are their respective enthalpies.

Each of these terms can be further expressed in terms of the basic design parameters of the compressors. Thus the total heat input to the compressor can be subdivided into two components, as

$$\dot{Q} = \dot{Q}_{cw} + \dot{Q}_{ch} \tag{4.205}$$

where $\dot{Q}_{cw}$ is the rate of heat transfer through cylinder walls, and $\dot{Q}_{ch}$ the rate of heat transfer through the cylinder head. This distinction is often necessary not only in view of their different orientations but also because in many compressors the cylinder walls are surrounded by water jackets. Their magnitudes (heat input) can be written, assuming quasisteady state, as:

$$\dot{Q}_{cw} = U_i A_{cw}(T_{surr} - T) \tag{4.206}$$

$$\dot{Q}_{ch} = U_h A_h(T_a - T) \tag{4.207}$$

where U_i is the overall heat-transfer coefficient for heat transfer through cylinder walls based on inner surface area A_{cw}; T_{surr} is the temperature of the fluid surrounding the outer walls; and T is the instantaneous temperature of the working fluid inside the cylinder. Similarly U_h represents the overall heat-transfer coefficient for heat transfer through cylinder head of surface area A_h; T_a being the ambient temperature, that is, the temperature outside the cylinder head. The value of these overall heat-transfer coefficients can be found out by combining the conductive resistance of the wall with the convective resistance on its both sides (Eq. 3.226) [24].

The second term in energy equation (4.204), viz. $\dot{W}_{sh}$ can be easily computed using Eqs. (4.202), (4.203) in the basic thermodynamic Eq. (3.4) (with $W_{sh} = -W_{cv}$), that is

$$\dot{W}_{sh} = P\frac{dV}{dt} \tag{4.208}$$

The rate of change of total internal energy of the control volume can be expressed as:

$$\frac{dU}{dt} = \frac{d}{dt}(Mu) = M\frac{du}{dt} + u\frac{dM}{dt} \tag{4.209}$$

where M is the instantaneous mass of working fluid in the control volume. The rate of change of this mass is obviously related to the rates of inflow and outflow from the control volume, and can be written using mass conservation law (Eq. 3.10) as:

$$\frac{dM}{dt} = \dot{m}_s - \dot{m}_d \tag{4.210}$$

where $\dot{m}_s$ is the mass flow rate of fluid admitted through the suction valve and $\dot{m}_d$ is the mass flow rate through the discharge valve.

The changes in the specific internal energy can be related to the changes in the temperature and volume by using the basic thermodynamic relation, Eq. (3.34):

$$du = c_v\, dT + \left(T\left(\frac{\partial P}{\partial T}\right)_v - P\right)dv \tag{4.211}$$

Substituting Eq. (4.211) in Eq. (4.204) we get

$$\frac{dU}{dt} = Mc_v\frac{dT}{dt} + M\left(T\left(\frac{\partial P}{\partial T}\right)_v - P\right)\frac{dv}{dt} + u\frac{dM}{dt} \tag{4.212}$$

Substituting expressions for various terms in the energy equation from Eqs. (4.208), (4.209), (4.211), and simplifying we can get following equation for the rate of change of control volume temperature:

$$\frac{dT}{dt} = \frac{\dot{m}_s h_s - \dot{m}_d h_d + \dot{Q} - \frac{PdV}{dt} - \frac{udM}{dt}}{Mc_v} - \frac{1}{c_v}\left[\left(T\left(\frac{\partial P}{\partial T}\right)_v - P\right)\right]\frac{dv}{dt} \tag{4.213}$$

Note that V is the instantaneous volume of the gases in the CV and v, their specific volume.

Equation of Motion for the Valves

A variety of valves are used in reciprocating compressors, and clearly, their equations of motion will vary with design. Considering the high speed with which modern compressors operate, the valves experience a highly dynamic loading, and therefore the inertia and damping should play an important role in their motion. However, for steady-state simulation, simplifying assumptions are often made and the motion of valves is analyzed on the basic of static equilibrium. This assumption is usually justified only for large compressors using ring-type plate valve with heavy-duty external springs to hold the plate tight against the opening in the cylinder head. Figs. 4.35 and 4.36 show the valve assembly and forces acting on the valve plate of a typical large refrigeration compressor with ring valves. To estimate the forces acting on the valve plate in the valve passage due to the flow of refrigerant a linear pressure distribution is assumed, as shown in Fig. 4.36B. During the flow of refrigerant through the valve, the pressure first drops from the suction port pressure P_u to P_i and then across the valve passage from P_i to P. The pressure P_i is determined to ensure that the flow rate through the port and the valve passage are equal. The equation of motion for the valve plate can now be obtained by considering the balance of various forces on it, as:

$$M_v \frac{d^2 h}{dt^2} + \xi \frac{dh}{dt} + N_{sp} S_k h = P_i A_p + \left(\frac{P_i + P}{2} \right) (A_{vp} - A_p) - P A_{vp} - N_{sp} S_{pl} \quad (4.214)$$

where M_v is the mass of valve plate, ξ the damping coefficient, N_{sp} the total number of springs, S_{pl} the spring preload, S_k the spring constant, A_{vp} the surface area of valve plate, and A_p the cross-sectional area of the port; and h is the valve lift. If we neglect the inertia and the damping effects, the above equation simplifies to an explicit expression for the valve lift, h as:

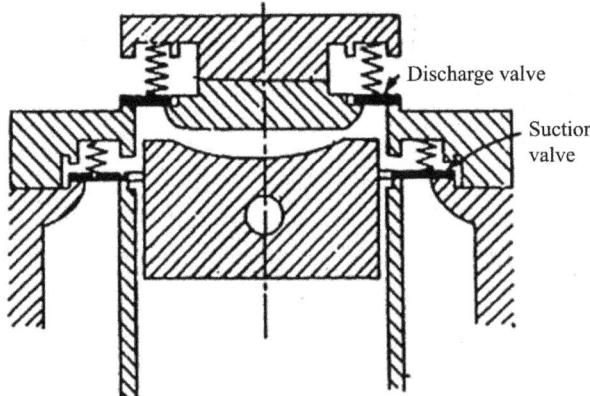

Fig. 4.35 Compressor cylinder with ring valves. *(Taken from G.R. Saraf, P.L. Dhar, Computer simulation of reciprocating refrigerant compressors, in: Proceedings of Sixth National Symposium on Refrigeration and Air Conditioning, IIT Bombay, 1978, pp. 115–124).*

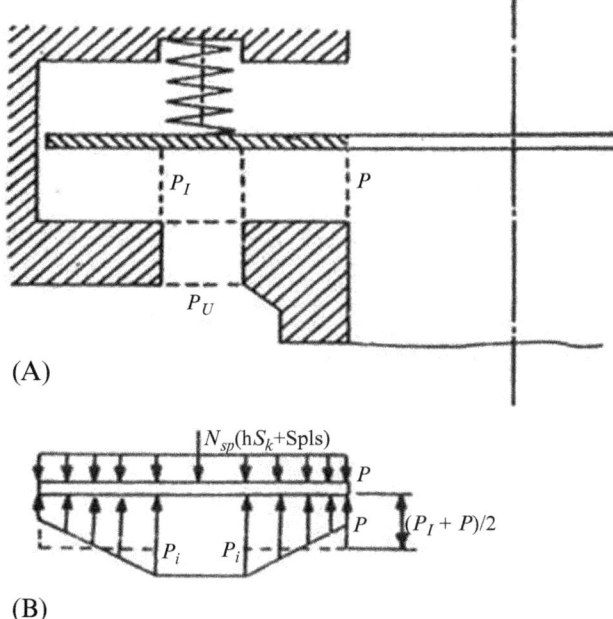

Fig. 4.36 Valve assembly and forces acting on valve plate. (A) Ring valve assembly. (B) Flow forces acting on the valve plate. *(Taken from G.R. Saraf, P.L. Dhar, Computer simulation of reciprocating refrigerant compressors, in: Proceedings of Sixth National Symposium on Refrigeration and Air Conditioning, IIT Bombay, 1978, pp. 115–124).*

$$h = \frac{\left[\frac{P_i - P}{2}(A_p + A_{vp}) - N_{sp}S_{pl}\right]}{S_k N_{sp}} \qquad (4.215)$$

$$\text{for} \quad h \le h_{max}$$
$$= h_{max} \quad \text{otherwise (when valve plate is at the stop)}$$

Equation for Flow Through the Valves

The flow of working fluid through the valves can be modeled as compressible flow through an orifice. Using the basic equations for such compressible flow (cf. Eq. 3.155), we can write

$$\dot{m} = C_d A P_u \left(\frac{2\gamma}{\gamma - 1}\frac{1}{R_g T_u}\right)^{1/2}\left(r^{2/\gamma} - r^{\frac{\gamma+1}{\gamma}}\right)^{1/2} \quad \text{for } r_c \le r \le 1 \qquad (4.216)$$

and

$$\dot{m} = C_d A P_u \left(\frac{2\gamma}{\gamma - 1} \frac{1}{R_g T_u} \right)^{1/2} \left(r_c^{2/\gamma} - r_c^{\frac{\gamma+1}{\gamma}} \right)^{1/2} \quad \text{for } 0 \leq r \leq r_c \tag{4.217}$$

where r_c is the critical pressure ratio, given by:

$$r_c = \left(\frac{2}{\gamma + 1} \right)^{\frac{\gamma}{\gamma+1}} \tag{4.218}$$

where r is the prevailing pressure ratio

$$r = \frac{\text{Pressure downstream of the valve}}{\text{Pressure upstream of the valve}} \tag{4.219}$$

In the above equations, R_g is the gas constant, T_u the upstream fluid temperature, A the flow area, and C_d the flow coefficient which depends on the valve design, and may even vary with valve lift. Typical values of C_d are 0.6 for flexing strip valve, 0.73 for ring valve, and 0.9 for cantilever reed valve [24].

Computational Method

To simulate the performance of a reciprocating compressor all the above equations need to be solved simultaneously. The computational scheme followed by Saraf and Dhar [24], is described below.

The computations are initiated from the top dead center (point A in Fig. 4.37) by assuming both the pressure and the temperature of the gases inside the cylinder. The calculations are simplified greatly if we assume a pressure value which keeps both the suction and the discharge valves closed at this point. The instantaneous values of the temperature and the pressure of the control volume, after each small rotation (typically

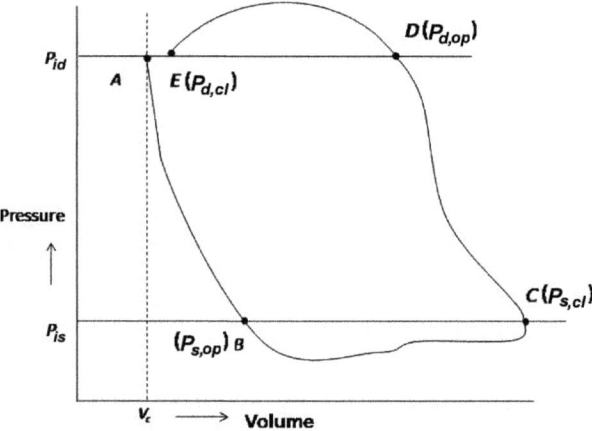

Fig. 4.37 Pressure-volume diagram of compressor.

1 degree or 2 degree) of the crank are then determined, using Eq. (4.213) expressed in finite-difference form as:

$$\Delta T = T_2 - T_1 = \frac{\delta m_s h_s - \delta m_d h_d + \Delta Q_w - P_1 \Delta V - v_1 \Delta M}{M_1 c_{v1}}$$

$$- \frac{1}{c_{v1}} \left[T_1 \left(\frac{\partial P}{\partial T} \right)_{v1} - P_1 \right] \left(\frac{V_2}{M_2} - \frac{V_1}{M_1} \right) \tag{4.220}$$

where

$$\Delta Q_w = \dot{Q} \delta t; \quad \Delta M = (\dot{m}_s - \dot{m}_d) \delta t = \delta m_s - \delta m_d$$
$$M_2 = M_1 + \Delta M; \quad \Delta V = V_2 - V_1$$

and the suffixes 1 and 2 refer, respectively, to the positions of the piston before and after this small rotation.

Eq. (4.220) enables us to find the temperature T_2, provided values of all the terms on the RHS are known. During the period when both the suction and the discharge valves are closed, $\delta m_s = 0$, $\delta m_d = 0$, and $\Delta M = \delta m_s - \delta m_d = 0$ and all other terms can be easily computed using the equations discussed above. The temperature T_2 can thus be determined. The volume V_2 can be determined using Eqs. (4.202), (4.203) and then the specific volume v_2 can be computed. The pressure P_2 in the cylinder, corresponding to these values of T_2 and v_2, can then be found using the equation of state of the working fluid.

However, during the period when either of the valves is open, the mass flow rate through the valve has to be computed using Eq. (4.216) or (4.217). This needs the value of flow area A, which is directly related to the value lift h, given by Eq. (4.215). All these equations have therefore to be solved simultaneously, along with Eq. (4.220), and the equation of state of the working fluid, to determine T_2 and P_2.

The conditions for valve opening and closing can be obtained from the valve lift equation (4.215) by putting $h = 0$. This gives

$$P_{is} = P_s - \frac{2 S_{pls} N_{sps}}{(A_{ps} + A_{vps})} \tag{4.221}$$

$$P_{id} = P_d + \frac{2 S_{pld} N_{spd}}{(A_{pd} + A_{vpd})} \tag{4.222}$$

where P_{is} is the cylinder pressure below which suction valve will open, and P_{id} the pressure above which the discharge valve will open. The valve opening (subscript op) and closing pressures (subscript cl) can therefore be written as

$$P_{s,op} = P_{is} - \delta; \quad P_{s,cl} = P_{is} + \delta \tag{4.223}$$

$$\text{and} \quad P_{d,op} = P_{id} + \delta; \quad P_{d,cl} = P_{id} - \delta \tag{4.224}$$

where δ is an arbitrary small value (taken as 0.000035 bar $\backsim$ 0.005 psi by Saraf and Dhar [24]). In order to locate valve opening/closure accurately step size is reduced and successive bisection method employed (Section 2.2) to converge to the solution.

The above procedure is continued till we complete calculation for one complete rotation (i.e., 360 degree) of the crank. If the pressure in the cylinder equals $P_{d,cl}$ before the piston reaches the top dead center (point E in Fig. 4.37), the discharge of vapor from E to the clearance volume is assumed to take place at constant pressure and constant temperature. The cylinder pressure and temperature values thus obtained after computations for 360 degree of crank rotation are then compared with the values assumed in the beginning of the computation. If these do not tally within the desired accuracy, another cycle of computations is begun using these computed values as the new starting assumption, and the procedure repeated till convergence is obtained.

The above-simplified simulation procedure has been validated by Saraf and Dhar [24] by comparing its predictions with the ratings of an industrial refrigerant compressor over a wide range of operating conditions. Table 4.2 shows some of the results. It is seen that the deviation between the predicted and the rated values of power consumption is $\pm 6\%$ and that of refrigerant mass flow rate is $\pm 7\%$.

Table 4.2 Rated and predicted performance of a refrigerant compressor

Section no.	Operating conditions			Power input		Ref. flow rate		Discharge T	
	Suc. P (bar)	Suc. T (°C)	Dis. P (bar)	Rated BHP	Pred. BHP	Rated (kg/min)	Pred. (kg/min)	Rated (°C)	Pred. (°C)
1	3.29	−6.68	16.78	95.9	93.09	79.71	83.91	90.0	83.5
2	4.00	−1.98	16.78	106.3	105.6	102.4	107.3	83.5	78.8
3	4.81	2.79	16.78	115.1	116.85	128.3	133.8	77.4	74.7
4	3.29	−8.07	12.81	87.7	83.97	87.46	89.79	75.2	67.2
5	6.85	11.39	12.81	102.3	107.5	200.1	207.1	50.5	50.9
6	3.29	−7.38	14.66	92.0	88.8	83.53	87.14	82.3	75.2
7	3.29	−7.03	15.69	94.0	91.0	81.67	85.71	85.9	79.4
8	4.00	−1.30	19.12	111.7	110.8	97.77	104.0	91.8	87.0

4.3.2 IC Engine

The simplified procedure for modeling of reciprocating compressors presented above can be extended to simulation of an internal combustion engine by making certain simplifying assumptions. Two of these are neglecting combustion kinetics and assuming the combustion process to proceed at constant volume/constant pressure in an SI/CI engine. The other processes that occur in a typical four stroke engines can be directly modeled by extending the general procedure outlined above for reciprocating devices, but taking care of the fact that the residual gases trapped in the clearance volume of an IC engine are different from the working fluid which is taken in during the suction stroke.

Further, in SI engine the working fluid itself is a mixture of air and fuel, while it is pure air in a CI engine. The formation of air-fuel mixture in the carburettor of an SI engine (Fig. 4.38) can be analyzed by applying the SFEE (Eq. 3.109). This gives

$$N_a h_a + h_f = N_a h_{am} + h_{vm}$$

where N_a is the *molar* air-fuel ratio; h_a the specific *molar* enthalpy of air at inlet, h_{am} at the outlet; h_f the specific *molar* enthalpy of (liquid) fuel entering the carburettor, and h_{vm} the specific *molar* enthalpy of fuel vapors in the mixture. Rearranging we get

$$N_a(h_a - h_{am}) = h_{vm} - h_f = h_{vm} - h_v + h_v - h_f$$
$$= h_{vm} - h_v + h_{fg} \tag{4.225}$$

where h_v is the enthalpy of saturated fuel vapors at temperature T_f and h_{fg} is the latent heat of vaporization of the fuel. If we assume ideal gas behavior for air and fuel vapors, the above equation can be written in the terms of the temperatures as:

$$N_a c_{pa}(T_a - T_m) = c_{pf}(T_m - T_f) + h_{fg}$$
$$T_m = \frac{N_a c_{pa} T_a + c_{pm} T_f - h_{fg}}{N_a c_{pa} + c_{pf}} \tag{4.226}$$

where c_{pf} is the molar-specific heat of fuel vapors and c_{pa} that of the air. Thus T_m the temperature of the air-fuel mixture inducted into the cylinder can be determined.

The pressure at which the mixture is admitted depends on the setting of the throttle valve regulating the flow. If the valve is fully open, the pressure is nearly atmospheric; but if it is partly open the pressure in the intake manifold is below atmospheric. The magnitude of this pressure drop can be determined using appropriate value of C_d in Eqs. (4.216), (4.217).

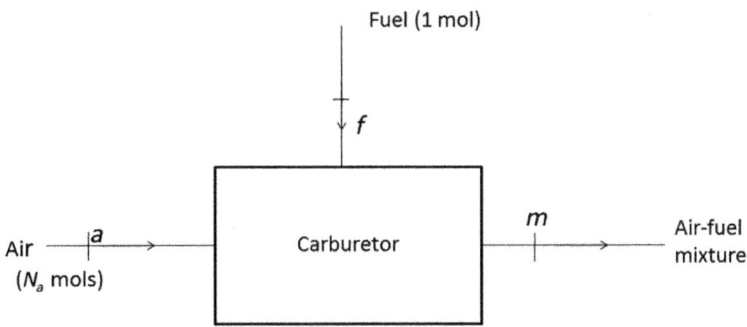

Fig. 4.38 Formation of air-fuel mixture.

In an IC engine the suction and the discharge valves are operated by a set of cams mounted on the cam shaft, and therefore their opening and closing is externally controlled, and we do not have to concern ourselves with the equation governing their motion. Thus, in a typical SI engine, the suction valve opens a few degrees before the TDC (i.e., during the exhaust stroke, when flue gases are being thrown out into the discharge manifold) (see Fig. 4.39), even before the exhaust valve closes. The flow field during this overlap period is quite complex. To simplify, we usually assume for simulation purpose that till the TDC the flue gases (from previous cycle) are being thrown out, and thereafter the fresh air-fuel mixture is being inducted. This fresh air will mix with the residual flue gases in the clearance volume and this needs to be accounted for in the thermodynamic equation while analyzing the suction stroke of the engine.

The compression process begins when the suction valve is closed (state 1, Fig. 4.39) and can be analyzed, as in the reciprocating compressor. Often, the calculations are begun by assuming the temperature of the working fluid at state 1, since the state of the flue gases trapped in the clearance volume at the end of exhaust stroke is not known a priori. This analysis can be done till state point 2 (Fig. 4.39), where the spark ignition of the fuel takes place.

The combustion process is then analyzed using the basic thermodynamics of reactive mixtures (Section 3.1.2). As mentioned above, in an SI engine, the combustion is assumed to be at constant volume, and since it occurs extremely rapidly, it can also be assumed to be adiabatic. The energy equation (cf. Eq. 3.118) can therefore be written as:

$$\Sigma U(T_3)_p = \Sigma U(T_2)_r \tag{4.227}$$

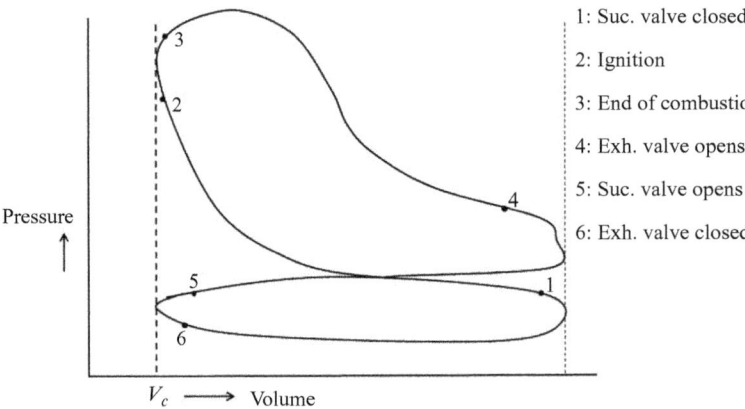

1: Suc. valve closed

2: Ignition

3: End of combustion

4: Exh. valve opens

5: Suc. valve opens

6: Exh. valve closed

Fig. 4.39 Indicator diagram of an SI engine.

where subscript p refers to the products of combustion, and subscript r to the reactants. The temperature T_3, at the end of combustion, can thus be found by using above equation along with the chemical equation for the combustion of the working fluid, as explained through Example 3.3 in Section 3.1.2. The pressure P_3 can then be found using ideal gas assumption as:

$$P_3 = P_2 \left(\frac{T_3}{T_2}\right)\left(\frac{N_p}{N_1}\right) \tag{4.228}$$

where N_1 is the total number of moles of the working fluid at the beginning of the compression, is found using ideal gas assumption:

$$N_1 = \frac{P_1 V_1}{R T_1} \tag{4.229}$$

V_1 being the cylinder volume and T_1 the temperature at state 1. The expansion process 3–4 can again be analyzed, as in the case of compressor by using the thermodynamic equation along with the kinematic equation. Often, the exhaust valve opens a few degrees before the BDC and this needs to be properly accounted for in the analysis by using equation for flow through valves to find instantaneous value of $\dot{m}_d$, the mass flow rate through the exhaust valve.

Similarly, the analysis of the exhaust stroke can be carried out to determine the actual state of working fluid trapped in the clearance volume at the TDC, state point 5. The molar mass of these gases can be found, making ideal gas assumption as:

$$N_5 = \frac{P_5 V_c}{R T_5} \tag{4.230}$$

A proper analysis of the process of mixing of these gases with the fresh air-fuel mixture can now be carried out by substituting values in the energy equation (4.213), and the state of working fluid at the beginning of compression, (state 1) can be found. This is compared with the earlier assumption and if the two differ significantly, another iteration is carried out, till successive cycles give almost the same values. Thus the state of working fluid in the cylinder as the crank rotates through two revolutions (during which the four strokes get completed) can be determined. The net work output per cycle, and thereby the thermal efficiency of the engine, can thus be predicted.

The analysis of CI engines can also be carried out in a similar manner, the only difference being that the combustion is now assumed to proceed at constant pressure. Eq. (4.227) is therefore replaced by

$$\Sigma H(T_3)_p = \Sigma H(T_2)_r \tag{4.231}$$

and Eq. (4.229) is no longer valid since the pressure remains constant during the combustion. The temperature can be assumed to be constant for the entire duration of the fuel injection, and thereafter the analysis of the expansion and exhaust process can be carried out as before. For more details, and solved examples based on this procedure see books by Ganesan [25, 26].

4.4 ROTATING DEVICES

A variety of rotating devices are used in thermal systems. Most common are turbines and the compressors. The compressors can be further subdivided into two categories, that is, positive displacement and dynamic compressors. Among the positive displacement rotary compressors the roller, screw- and scroll-type compressors are popular. In the dynamic compressors we have centrifugal and axial flow compressors. Among the various types of positive displacement-type units, we shall discuss here the simulation procedures of the scroll-type rotary compressors, which are now widely used in small and medium capacity refrigeration systems. Similarly between the two dynamic compressors we shall discuss in detail the simulation of the centrifugal compressors.

   The basic approach of simulation of the dynamic compressors is also applicable to turbines, and therefore we will not discuss turbines separately. In addition we also find rotary wheels used for energy recovery. We have already discussed the method of simulating the performance of rotary desiccant wheels in Section 4.2.3.

4.4.1 Centrifugal Compressors

The heart of a centrifugal compressor is the rotor assembly consisting of a series of impeller wheels mounted on a steel shaft and enclosed in a casing. These impellers are driven by an external device (usually an electric motor) and impart kinetic energy to the working substance as also increase its static pressure. This kinetic energy is converted into further pressure rise in the diffuser. In turbines, the working fluid at high pressure and temperature is expanded while passing through the impeller producing power output at the shaft.

   An accurate simulation of these rotating machines needs determining the flow field within the rotating blades by solving the basic three-dimensional Navier-Stokes equations. This can only be done numerically using CFD softwares. In the view of the enormous computational effort needed for such simulation, the performance of systems involving such turbo machines, whether obtained experimentally or through such comprehensive CFD simulations, is usually represented in terms of graphs or equations between dimensionless parameter defined below:

$$\text{Specific speed} \quad N_s = \frac{N\sqrt{Q}}{H^{3/4}}$$

$$\text{Specific diameter} \quad D_s = \frac{DH^{1/4}}{\sqrt{Q}}$$

$$\text{Flow coefficient} \quad \phi = \frac{Q}{ND^3}$$

$$\text{Head coefficient} \quad \psi_p = \frac{H}{N^2D^2}$$

$$\text{Power coefficient} \quad \Pi = \frac{P}{\rho N^3 D^5}$$

where N, Q, D, and H are the impeller speed, the volume flow rate, the impeller diameter, and the head, respectively.[1]

However, if we wish to predict the design point performance of a turbo machine, simple one-dimensional analysis can give reasonable results. This is the procedure used in the preliminary design, which can then be further refined in the light of CFD analysis. We shall discuss here one such simple one-dimensional analysis of a single impeller stage.

Assuming that the working fluid moves with a uniform velocity in between the impeller blades, its entering and leaving being guided by the blade angles, we can draw the velocity diagram of Fig. 4.40. Here U indicates the velocity of fluid relative to that of the blade, and V its total velocity. Subscripts r and t indicate the radial and the tangential components of the fluid velocity, respectively.

The power required to bring about the change in the velocities shown in Fig. 4.40 can be estimated as

$$P = \text{Torque } (T) \times \text{Angular velocity } (w) \tag{4.232}$$

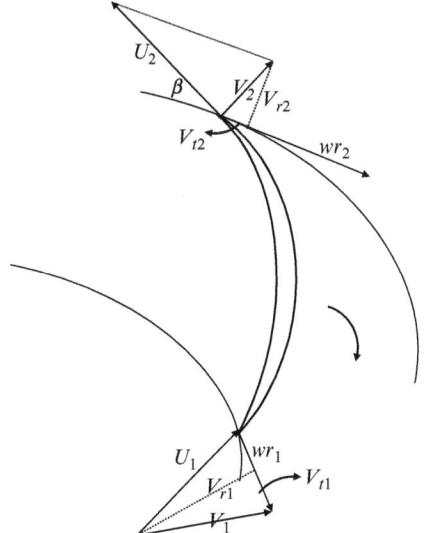

Fig. 4.40 Velocity diagram for a centrifugal compressor.

---

[1] Head = Energy transfer per unit mass

where the torque T is given by the rate of momentum change as

$$T = \dot{m}(V_{t2}r_2 - V_{t1}r_1) \tag{4.233}$$

Combining Eqs. (4.232), (4.233) we get

$$P = \dot{m}(V_{t2}wr_2 - V_{t1}wr_1) \tag{4.234}$$

and the energy input per unit mass flow rate of the working substance is

$$W = w(V_{t2}r_2 - V_{t1}r_1) \tag{4.235}$$

We can also write the following relationships between the various components of the velocity triangles at the inlet and the exit:

$$(V_{t2})(wr_2) = \frac{1}{2}(w^2 r_2^2 + V_2^2 - U_2^2) \tag{4.236}$$

$$(V_{t1})(wr_1) = \frac{1}{2}(w^2 r_1^2 + V_1^2 - U_1^2) \tag{4.237}$$

Using Eqs. (4.236), (4.237) in Eq. (4.235) we get following expression for work transfer per kilogram of fluid, usually called the head H.

$$H \equiv W = \frac{1}{2}\left[(V_2^2 - V_1^2) + (w^2 r_2^2 - w^2 r_1^2) + (U_1^2 - U_2^2)\right] \tag{4.238}$$

In Eq. (4.238) the first term on RHS represents the increase in kinetic energy of the fluid which has to be recovered in the diffuser. The rest of the two terms thus indicate the work required to cause a static pressure rise in the impeller itself.

We can also get an expression for the specific work from thermodynamic considerations. Thus, if the process in assumed to the isentropic, we get from Eq. (3.38)

$$\Delta h_{is} = W_{is} = C_p T_1 \left[\left(\frac{P_2}{P_1}\right)^{\frac{\gamma-1}{\gamma}} - 1\right] \tag{4.239}$$

If it is not isentropic, but with an isentropic efficiency η, we get

$$\Delta h = W = \frac{C_p T_1}{\eta}\left[\left(\frac{P_2}{P_1}\right)^{\frac{\gamma-1}{\gamma}} - 1\right] \tag{4.240}$$

If we assume that the velocity of the fluid at diffuser exit is negligible, we can equate the expression for specific work obtained above, Eqs. (4.235) or (4.238) and (4.240). This gives an expression for the pressure rise in the compressor:

$$\frac{P_2}{P_1} = \left(\frac{W \cdot \eta}{C_p T_1} + 1\right)^{\frac{\gamma}{\gamma-1}} \tag{4.241}$$

where W is given by Eq. (4.235) or (4.238).

The flow rate through the compressor can be calculated on the basis of the radial component of the fluid exit velocity:

$$Q = (2\pi r_2)b_2 \cdot V_{r2} \qquad (4.242)$$

where b_2 is the width of the blades at the outlet.

In many centrifugal compressor designs the flow at the inlet is guided so that it enters radially. This implies $V_{t1} = 0$, and Eq. (4.235) simplifies to

$$W = wr_2 V_{t2} \qquad (4.243)$$

Further V_{t2} can be written as the "vector" sum of wr_2 and the tangential component of the exit velocity relative to the blade. Thus, referring to Fig. 4.40, we get

$$V_{t2} = wr_2 - V_{r2} \cot \beta \qquad (4.244)$$

Substituting it in Eq. (4.243) we get

$$Head = W = w^2 r_2^2 - wr_2 V_{r2} \cot \beta \qquad (4.245)$$

Eqs. (4.242), (4.245) can be used to draw the performance characteristic of the compressor in terms of head versus flow rate. Following Eq. (4.245), the theoretical curve is a straight line, with head falling with increase in flow (V_{r2}). However, the actual head–flow characteristic is much different (Fig. 4.41). This is due to various losses that occur within the compressor. These include losses due to friction on the impeller and on the surface of the rotor between the hub and the casing, the losses arising from negative velocity gradients in the boundary layer, leakage across the clearance between the rotating impeller and the stationary casing, recirculation loss due to backflow into the impeller (due to adverse pressure gradient in flow direction), the incidence loss due to the flow not entering blade smoothly at off-design conditions, and various losses in the diffuser section.

Over the last few decades many empirical loss models have been developed and their combined predictions validated against experimental data. One of the most recent of such studies by Gang and Chen [27] has investigated the effect of nine loss components on the compressor performance using available loss models. They have also given models to estimate losses in the diffuser and the volute casing. The details of the models recommended by them are presented below.

Fig. 4.42, adopted from Gang and Chen, shows the meridional view of the centrifugal compressor and indicates various parts as also the cardinal dimensions needed in analysis. Thus the impeller inlet area is

$$A_1 = \frac{\pi}{4} \left(D_{t1}^2 - D_{h1}^2 \right) \qquad (4.246)$$

with subscripts t and h referring to the "tip" and the "hub," respectively.

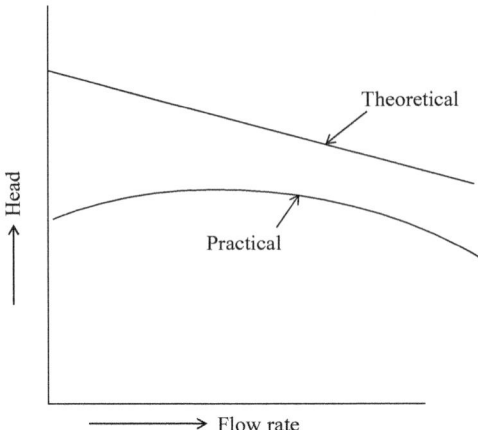

Fig. 4.41 Centrifugal compressor characteristics.

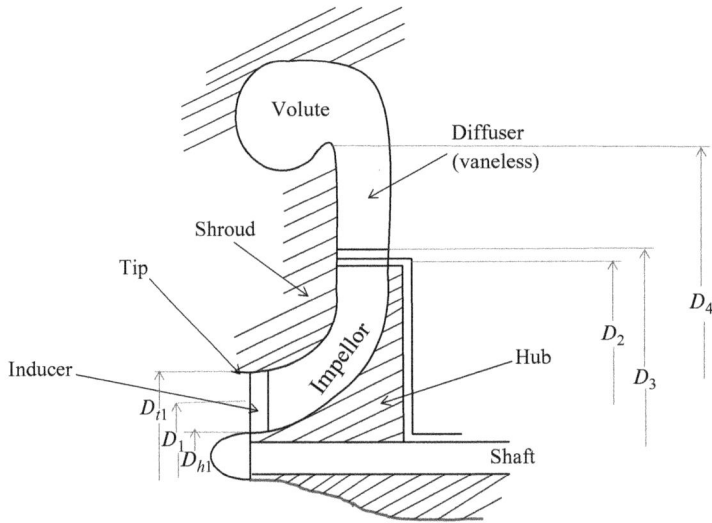

Fig. 4.42 Meridional view of centrifugal compressor.

All the pressure losses are nondimensionalized with respect to the total relative pressure loss. The nondimensional losses Δp_i are defined so that the total relative pressure loss can be calculated as:

$$\Delta P_{tr} = fc \, (P_{tr1} - P_1) \, \Sigma_i \overline{(\Delta p_i)} \tag{4.247}$$

where P_{tr1} and P_1 are the total relative pressure and the static pressure at the impeller inlet, respectively; f_c is a correction factor defined as

$$fc = \frac{\rho_{tr2} T_{tr2}}{\rho_{tr1} T_{tr1}} \qquad (4.248)$$

where ρ_r is the total relative density and T_r is the total relative temperature.

The total relative pressure P_{tr} and total relative temperature T_{tr} can be calculated using Eqs. (3.147), (3.149) with Mach number M replaced by the relative Mach number M_r defined as

$$M_r = \frac{U}{c} \qquad (4.249)$$

where U is the relative velocity and c is the velocity of sound given by Eq. (3.146). Writing these equations here again, we can say

$$T_{tr} = T\left[1 + \left(\frac{\gamma - 1}{2}\right) M_r^2\right] \qquad (4.250)$$

$$P_{tr} = P\left[1 + \frac{\gamma - 1}{2} M_r^2\right]^{\frac{\gamma}{\gamma-1}} \qquad (4.251)$$

The actual total relative pressure at the exit is determined as:

$$P_{tr2,actual} = P_{tr2,ideal} - \Delta P_{tr} \qquad (4.252)$$

$$P_{tr2,ideal} = P_{2,ideal}\left[1 + \left(\frac{\gamma - 1}{2}\right) M_{r2}^2\right]^{\frac{\gamma}{\gamma-1}} \qquad (4.253)$$

where $P_{2,ideal}$ is the pressure at the impeller exit had the compression been isentropic. Gang and Chen identify nine impeller losses that influence the performance significantly. These are discussed below in brief.

Shock Losses

This loss is common is compressors used in aircrafts and arises from compression shock at the inlet. The equation suggested for estimating its magnitude is

$$\overline{\Delta p_{sh}} = 1 - \left(\frac{U_{th}}{U_1}^2\right) - \left(\frac{2}{(\gamma - 1)M_{r1}^2}\right)\left[\left(\frac{P_{th}}{P_1}\right)^{\frac{\gamma-1}{\gamma}} - 1\right] \qquad (4.254)$$

where U_{th} is the impeller throat relative velocity, given by

$$U_{th} = \frac{V_{rth}}{\sin(\beta_{th})} \qquad (4.255)$$

where V_{rth} is throat redial velocity and β_{th} is the relative blade angle at throat.

Incidence Loss

This occurs due to difference in the angle of incidence of flow at impeller inlet and the optimal incidence angle. As a result the fluid does not enter the blades smoothly resulting in pressure loss. Following equation is recommended for estimating it.

$$\overline{\Delta p_{inc}} = 0.8 \left[1 - \frac{V_{r1}}{U_1 \sin(\theta_{m1})} \right]^2 + \left[\frac{t_{b1} Z}{2\pi r_1 \sin(\theta_{m1})} \right]^2 \tag{4.256}$$

where t_{b1} is the compeller inlet blade thickness and θ_{m1} is the mean blade angle at the inlet.

Clearance Loss

This is the loss due to fluid leakage through the clearance gaps and the labyrinth seals and has been modeled as

$$\overline{\Delta p_{cl}} = \frac{2\dot{m}_{cl} \Delta P_{gap}}{\dot{m} \rho_1 U_1^2} \tag{4.257}$$

where ΔP_{gap} is the pressure difference across the gap and $\dot{m}_{cl}$ is the flow rate through it.

Skin Friction Loss

This is the loss arising from the adhesive forces between the solid surfaces and the fluid. The equation recommended for its calculation is similar to the Darcy equation (3.128) for pipe flow:

$$\overline{\Delta p_{sf}} = 4C_f \left(\frac{\bar{U}}{U_1} \right)^2 \frac{L_B}{D_H} \tag{4.258}$$

$$\bar{U}^2 = \frac{U_1^2 + U_2^2}{2} \tag{4.259}$$

where L_B is the blade mean camberline length, D_H is the hydraulic diameter which is taken as the average of the throat and discharge hydraulic diameters. The skin friction factor C_f is correlated using standard equations for rough pipes.

Blade Loading Loss

This is the loss due to secondary flow caused by pressure difference between adjacent blades and can be estimated by using a simple correlation:

$$\overline{\Delta p_{bl}} = \frac{(\Delta U / U_1)^2}{24} \tag{4.260}$$

Hub-to-Shroud Loading Loss

This arises from the pressure gradient in the hub-to-shroud direction, which produces secondary flows leading to dissipative losses. The recommended equation is

$$\overline{\Delta p_{hs}} = \frac{(\bar{K}m\bar{b}\bar{U}/U_1)^2}{6} \tag{4.261}$$

where $\bar{U}$ is the average relative velocity, $\bar{K}m$ is the average passage curvature, $\bar{b}$ its average width.

Blockage Loss

This represents the losses arising from mixing of flow due to distorted meridional velocities, and can be estimated from equation

$$\overline{\Delta p_{\exp}} = \left[\frac{(\lambda - 1) V_{r2}}{U_1} \right]^2 \tag{4.262}$$

where λ is the tip distortion factor.

Mixing Loss

Mixing loss term refers to the losses due to mixing of blade wake flows with the free stream flow, and can be calculated as

$$\overline{\Delta p_{mix}} = \left[\frac{V_{r,wake} - V_{r,mix}}{U_1} \right]^2 \tag{4.263}$$

where $V_{r,wake}$ and $V_{r,mix}$ are the radial components of the wake velocity and the mixed flow, respectively.

Supercritical Mach Number Loss

Supercritical Mach number loss occurs when blade surface velocities reach supersonic values, and is estimated by using following equation

$$\overline{\Delta p_{scmn}} = 0.4 \left[\frac{(M_{r1} - M_{cr})}{U_1} U_{max} \right]^2 \tag{4.264}$$

where M_{cr} is the inlet critical Mach number. Putting all these dimensionless pressure loss terms in Eq. (4.247), we can calculate total relative pressure loss ΔP_{tr}; substituting this in Eq. (4.252) the actual total relative pressure at impeller exit can be determined.

Similar approach is suggested for analysis of the losses in the diffuser/volute, and details can be seen in the research paper by Gang and Chen [27]. This model has been validated by them against experimental data from the literature, and a good agreement of prediction with the data is reported. Their results show that the most significant contributors to the losses in the impeller are the incidence loss and the skin friction loss; while in the diffusers the skin friction loss contributes the most.

4.4.2 Scroll Compressors

A scroll compressor consists of two involute spiral scroll, Fig. 4.43 assembled at a relative angle of 180 degree so that they touch at several points forming a series of crescent shaped pockets (Fig. 4.44). While one scroll is fixed, the other orbits pushing these pockets toward the center and progressively reducing their volume. Fig. 4.44[2] shows a complete compression cycle of the scroll compressor, completed over three orbits of the moving scroll. The first orbit (position A) begins with the ends of both scrolls fully open, creating the interstitial space to fill the working fluid (typically refrigerant vapor or air). The lower scroll's orbit eventually closes the first pockets of the gas, and the first pair of crescent-shaped pockets have migrated inwards to a middle position. The scroll's outer ends now begin to open again, position D, and admit more fresh change. The second orbit pushes the first gas pockets toward the center of the scroll set, continually decreasing the gas volume and increasing its pressure.

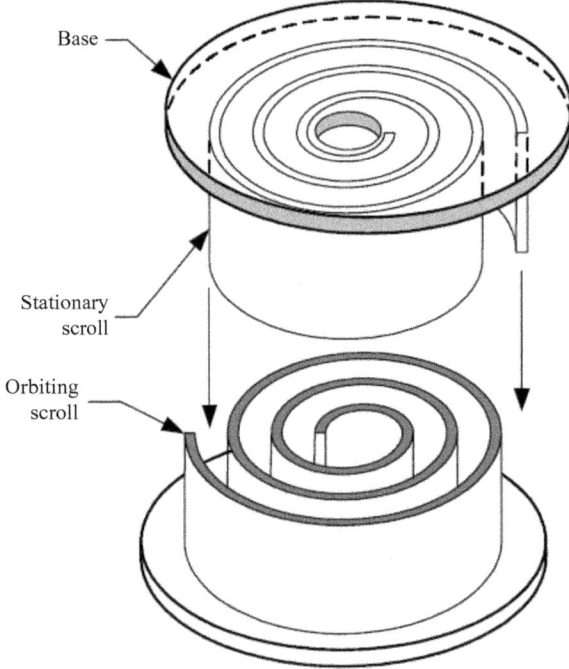

Fig. 4.43 Scroll compressor geometry.

---

[2] This figure and following description from Carrier Corporation's brochure.

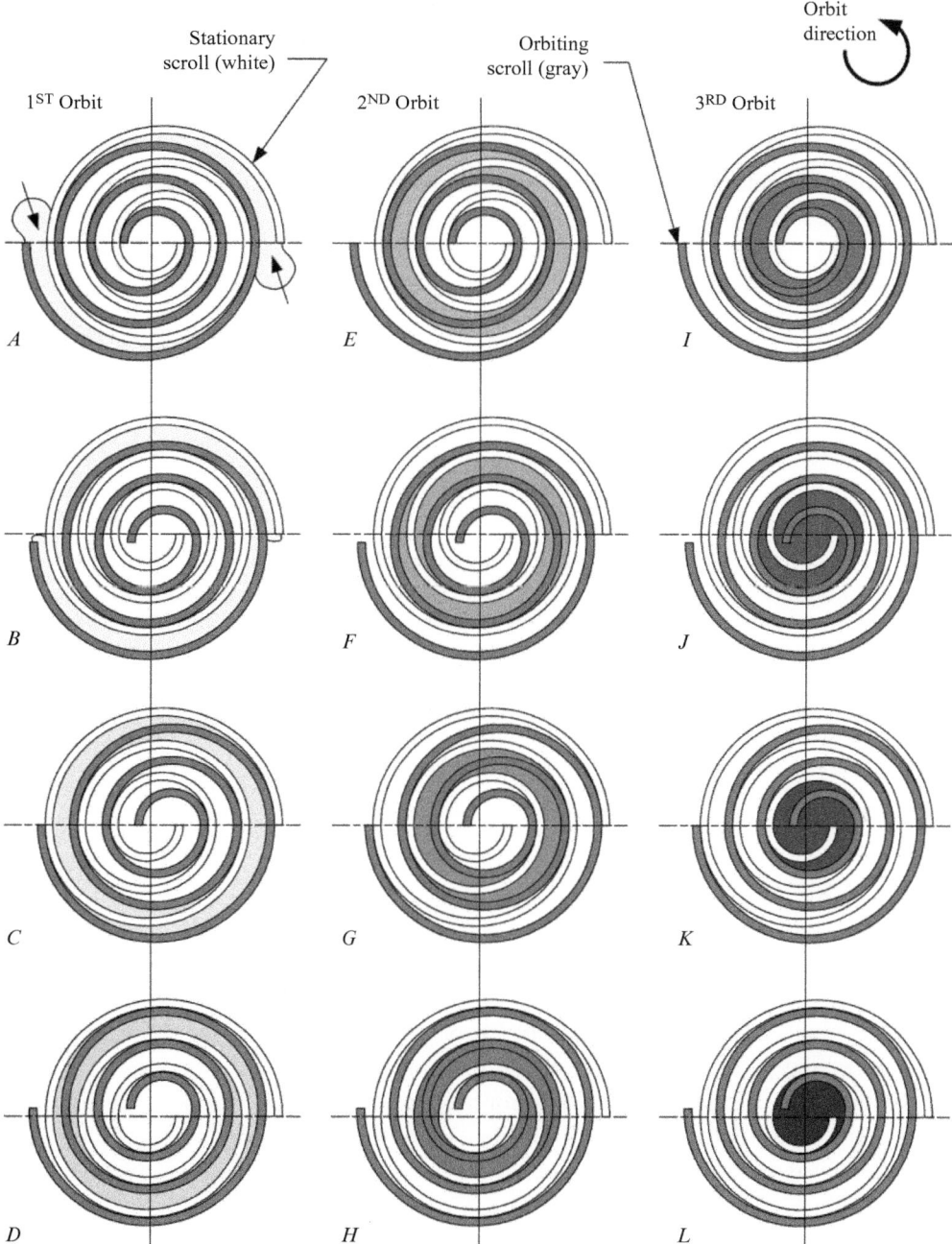

Fig. 4.44 Scroll compressor complete cycle.

The third orbit begins with crescent-shaped pockets just outside of the scroll set center. As the third orbit continues, the inner ends of the scrolls break contact (position J), admitting the compressed gas to the center discharge port. The third orbit continues the compression cycle, discharging the working fluid now at high pressure (position L).

As should be clear from the above description of the compression process, it is a smooth and continuous process without vibration, excessive noise or strong pulsations as in the reciprocating compressors. This is the main advantage of such compressors, which are now widely used in small- and medium-sized refrigeration and air-conditioning units.

Performance Prediction

Just as we carried out a detailed analysis of the reciprocating compressor, considering the instantaneous position of the piston with crank rotation, a similar analysis can also be carried out for the scroll compressor. However, in view of the complex geometry of the control volumes, and the presence of six different chambers at any instant,[3] such a comprehensive analysis becomes extremely complex (see, e.g., [28]). We shall present here a simplified model of simulating the performance of a scroll compressor developed by Winandy et al. [29].

In order to avoid complex geometrical analysis of various crescents, the model assumes that the design volume ratio of the compressor ϵ, defined as follows, is provided by the manufacturer (or estimated on the basis of experimental results along with other parameters):

$$\epsilon = \frac{V_s}{V_f} \tag{4.265}$$

where V_s is the initial volume of the crescent at suction and V_f is the final volume before discharge.

Winandy et al. [29] adopt a simplified model for simulating various heat transfers in the compressor (Fig. 4.45), since the actual processes are very complex with continually changing areas across which heat transfer occurs between the incoming gases, the hot compressed gases, as also the surrounding atmosphere. Since in steady state the temperature everywhere in the compressor would reach a constant value, it is presumed that a fictitious wall of uniform temperature T_w is able to represent all the heat-transfer modes involved, that is, the heating of the suction gas ($\dot{Q}_{su}$), the cooling of the exhaust gas ($\dot{Q}_{ex}$), the heat loss to the ambient ($\dot{Q}_{amb}$), and the heat gain from electro-mechanical losses in the compressor ($\dot{W}_{loss}$). These heat transfers change the state of the working

---

[3] That is, two suction chambers, two compression chambers, and two discharge chambers.

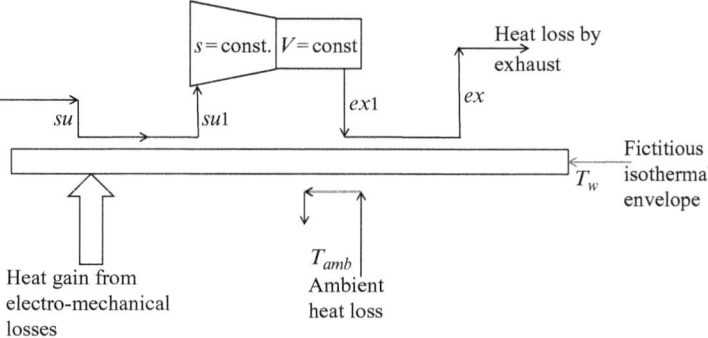

Fig. 4.45 Conceptual scheme of the compressor model. *(From E. Winandy, O.C. Saavedra, J. Lebrun, Experimental analysis and simplified modelling of a hermetic scroll refrigeration compressor, Appl. Therm. Eng. 22 (2) (2002) 107–120).*

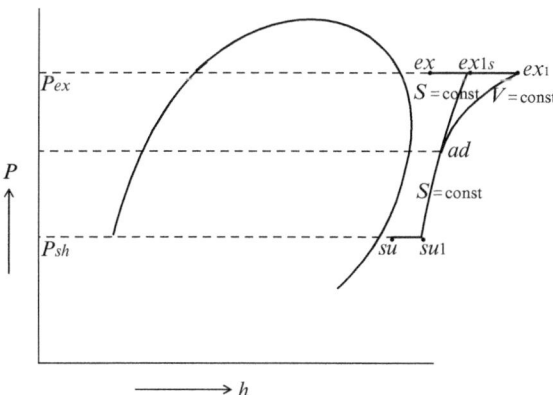

Fig. 4.46 Compression process on $p - h$ chart.

fluid during the entire journey through the scroll as shown in Figs. 4.45 and 4.46. Thus suction gas heating is indicated in the $p - h$ diagram of Fig. 4.46 as a constant pressure process $su \rightarrow su1$. This is followed by pressure rise inside the compressor and the exit state at the end of compression is indicated as $ex1$ (detailed explanation of the processes given below). The cooling of the exhaust gas then changes its state from $ex1 \rightarrow ex$, again at constant pressure.

Considering all the energy transfers, we can write the following heat balance equation for the compressor:

$$\dot{W}_{loss} + \dot{Q}_{ex} - \dot{Q}_{su} - \dot{Q}_{amb} = 0 \qquad (4.266)$$

Using the basic heat exchanger effectiveness concept (Section 4.1.1), the heat transfer to suction gases can be computed as:

$$\dot{Q}_{su} = \epsilon_{su}\dot{M}C_p(T_w - T_{su}) \tag{4.267}$$

$$\text{where} \quad \epsilon_{su} = 1 - e^{-AU_{su}/\dot{M}C_p} \tag{4.268}$$

and the symbols have their usual meanings. The temperature of the heated suction gas can be found from the equation:

$$\dot{Q}_{su} = \dot{M}C_p(T_{su1} - T_{su}) \Rightarrow T_{su1} = T_{su} + \frac{\dot{Q}_{su}}{\dot{M}C_p} \tag{4.269}$$

The mass flow rate $\dot{M}$ is given by the equation:

$$\dot{M} = \frac{NV_s}{v_{su1}} \tag{4.270}$$

where v_{su1} is the specific volume at the compressor inlet, and the N the compressor revolutions per second.

Eqs. (4.267)–(4.269) can also be used to compute the heat loss from exhaust gases, Q_{ex}, after substituting appropriate temperatures and overall heat-transfer coefficient values.

The ambient losses are (Eq. 3.227)

$$\dot{Q}_{amb} = AU_{amb}(T_w - T_{amb}) \tag{4.271}$$

The discharge temperature at the end of compression process ($su1 \rightarrow ex_1$), Fig. 4.46, is found by considering it as adiabatic. This gives

$$\dot{W}_{in} = \dot{M}(h_{ex1} - h_{su1}) \tag{4.272}$$

$\dot{W}_{in}$ needed here is calculated using Eq. (4.273).

A unique aspect of scroll compressor modeling arises from it being a fixed geometry machine, which compresses the fluid over a certain volume ratio, ϵ, Eq. (4.265). As a result, for any fluid admitted at certain conditions, the final discharge pressure gets fixed. Now if it is a refrigeration compressor, the discharge pressure has to vary in accordance with the condenser design and the operating conditions. Thus at off-design conditions the compressor exit pressure may not match with the prevailing pressure in the condenser. The compressor is then not adapted to the task and as a result power losses arise due to the adaptation process (Fig. 4.47). These are indicated by the hatched area in this figure. To determine the work input needed in the adapted compression process, we divide the compression in two parts, the first an adiabatic-reversible process up to the adapted pressure, the second one an adiabatic constant volume process arising from flow of the working fluid after the exposure to discharge plenum. This gives

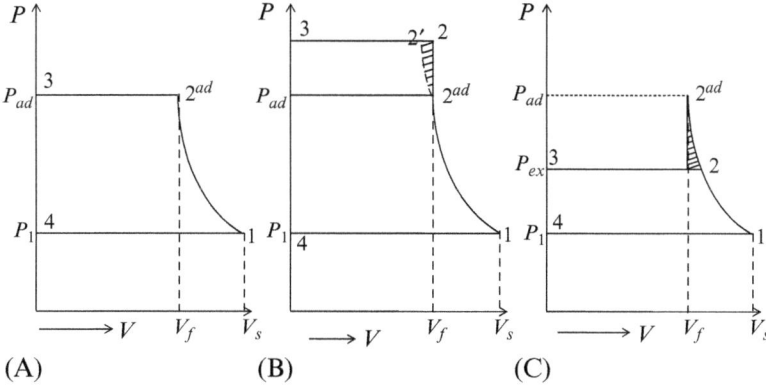

Fig. 4.47 Adaptation of compressor to external pressure: (A) adapted; (B) over pressure $P_2 > P_{ad}$; (C) under pressure $P_2 < P_{ad}$.

$$\dot{W}_{in} = \dot{M}(h_{ad} - h_{su1}) + N(P_{ex1} - P_{ad})V_f \qquad (4.273)$$

where P_{ad} is determined on the basis of the compression process $su_1 \rightarrow ad$ being isentropic.

Winandy et al. [29] recommended using ASHRAE Toolkit [30] approach for calculating the compressor shaft power. This toolkit suggests that the total shaft power required is the sum of internal compression power requirement $\dot{W}_{in}$, the constant electromechanical loss term $\dot{W}_{const\ loss}$, and the electromechanical losses proportional to the internal compression power $\alpha \dot{W}_{in}$, where $\dot{W}_{const\ loss}$ and α are parameters to be identified from experimental data. Thus

$$\dot{W} = \dot{W}_{in} + \dot{W}_{const\ loss} + \alpha \dot{W}_{in} \qquad (4.274)$$

The values of various parameters, that is, AU_{su}, AU_{ex}, AU_{amb}, $\dot{W}_{const\ loss}$, α and also the values of V_s and ϵ have been estimated by Winandy et al. [29] by using nonlinear optimization software, the criteria for selection of their values being minimization of the error on the experimentally measured values of mass flow rate, shaft power, and discharge temperature for a few sets of operating conditions. The main strength of this approach is its ease of use in simulating refrigeration systems using scroll compressor as one of the components.

4.5 THERMOELECTRIC MODULES

As the search for environment-friendly energy conversion devices intensifies, thermo-electric devices are finding increasing application some 200 years after the discovery of Seebeck effect in 1821. The Seebeck effect refers to the creation of a small voltage difference between the junctions of two dissimilar metals (or semiconductors) due to temperature between them. It is quantified as:

$$V = \alpha_{ab}(T_a - T_e) \tag{4.275}$$

where V is the voltage; T_a and T_e are the temperatures of the heat absorbing and heat emitting junctions, respectively; and α_{ab} is the Seebeck coefficient (in V/K) corresponding to that pair of materials. By combining a number of such thermoelectric couples as shown in Figs. 4.48 and 4.49 [31], a thermoelectric module (TEM) can be built for direct conversion of heat into electricity.

A TEM can also be used as a cooling device by passing electric current through it (see Fig. 4.49), one of the junctions would then absorb heat and the other would emit heat. The magnitudes of these heat transfers are

$$Q_a = \alpha_{ab} T_a I \tag{4.276}$$

$$Q_e = \alpha_{ab} T_e I \tag{4.277}$$

Thus in a typical TEM the energy transfers are the Joulean heating inside the module due to the passage of electric current, the Peltier effect at the two ends, and the heat conduction through the materials because of the temperature difference. Since the Peltier effect is confined to the boundaries, we could analyze the conductive heat transfer, along with the heat generation due to electric current, using Eq. (3.212). Usually, a simplified analysis is done assuming the complete module to be one-dimensional thermal system. The governing equation for steady-state temperature profile in each element is

$$\frac{d^2 T}{dX^2} + \frac{I^2 R}{ALk} = 0 \tag{4.278}$$

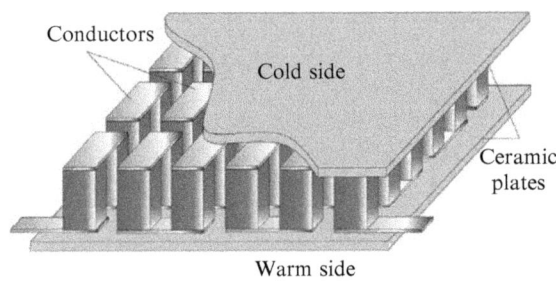

Fig. 4.48 Thermoelectric module. *(Adopted from S.L. Lineykin, S. Ben-Yaakov, Modelling and analysis of thermoelectric modules, IEEE Trans. Ind. Appl. 43 (2) (2007) 505–512).*

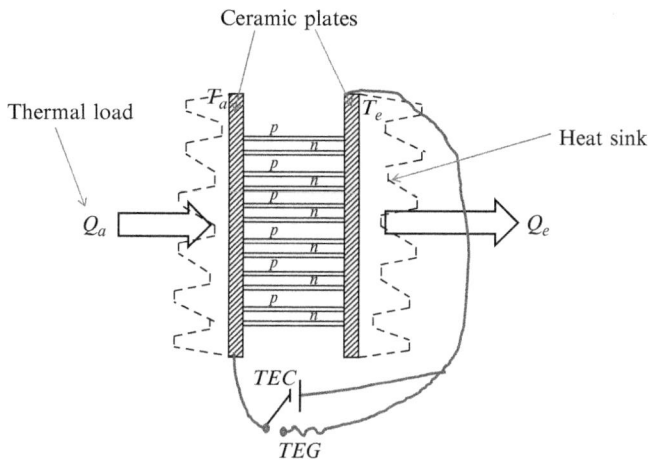

Fig. 4.49 Thermoelectric module schematic.

where R is the electric resistance, A the cross-sectional area, L the length, and k the thermal conductivity of the material. The boundary conditions are

$$T = T_a \quad \text{at} \quad x = 0 \tag{4.279}$$

$$T = T_e \quad \text{at} \quad x = L \tag{4.280}$$

Eq. (4.278) can be solved directly, and using the specified boundary conditions, the temperature profile can be obtained as:

$$T = T_a + (T_e - T_a)\frac{x}{L} + \frac{I^2 R}{2LAk}(Lx - x^2) \tag{4.281}$$

The heat flows, arising from conduction at the two ends Q_{ac} and Q_{ec}, are

$$Q_{ac} = -kA\left(\frac{\partial T}{\partial x}\right)_{x=0} = -\frac{T_e - T_a}{L/kA} - \frac{I^2 R}{2} \tag{4.282}$$

$$Q_{ce} = -kA\left(\frac{\partial T}{\partial x}\right)_{x=L} = -\frac{T_e - T_a}{L/kA} + \frac{I^2 R}{2} \tag{4.283}$$

Each element of the module consists of a couple of dissimilar materials, typically p- and n-type semiconductors (Fig. 4.49). For each material the above equation would be applicable. Indicating their properties by subscripts p and n, we can write for each couple constituting the module:

$$Q_{ac,p} = -\frac{T_e - T_a}{(L/kA)_p} - \frac{I^2 R_p}{2} \tag{4.284}$$

$$Q_{ec,p} = -\frac{T_e - T_a}{(L/kA)_p} + \frac{I^2 R_p}{2} \tag{4.285}$$

$$Q_{ac,n} = -\frac{T_e - T_a}{(L/kA)_n} - \frac{I^2 R_n}{2} \tag{4.286}$$

$$Q_{ec,n} = -\frac{T_e - T_a}{(L/kA)_n} + \frac{I^2 R_n}{2} \tag{4.287}$$

Combining we can write for a couple:

$$Q_{ac} = Q_{ac,n} + Q_{ac,p} = -\frac{(T_e - T_a)}{R_{th}} - \frac{I^2 R_e}{2} \tag{4.288}$$

$$Q_{ce} = Q_{ec,n} + Q_{ec,p} = -\frac{T_e - T_a}{R_{th}} + \frac{I^2 R_e}{2} \tag{4.289}$$

where

$R_e = R_n + R_p$ is the total electrical resistance of a couple

$R_{th} =$ is the total thermal resistance of a couple

$$= \frac{1}{(kA/L)_p + (kA/L)_n}$$

Combining this with the Peltier effect at the two ends we get

$$Q_a = \alpha_{ab} \cdot T_a I - \frac{T_e - T_a}{R_{th}} - \frac{I^2 R_e}{2} \tag{4.290}$$

$$Q_e = \alpha_{ab} \cdot T_e I - \frac{T_e - T_a}{R_{th}} + \frac{I^2 R_e}{2} \tag{4.291}$$

and the electromotive force to be applied, to overcome the potential drop due to electrical resistance and the Seebeck effect, is

$$V = IR_e + \alpha_{ab}(T_e - T_a) \tag{4.292}$$

For the complete TEM consisting of N such couples which are electrically in series but thermally in parallel the governing equations would be similar to Eqs. (4.290)–(4.292), with R_e replaced by the total electrical resistance of the module, that is, $R_e^* = NR_e$ and R_{th} is replaced by $R_{th}^* = R_{th}/N$. These equations form the basic equations for simulating the performance of a TEM.

Often, the temperature of the two ends of the TEM are not known, but the temperature of the two fluids in contact with these ends are known. Usually there are finned surfaces at the two ends to enhance the rate of heat transfer. The relationship between the fluid temperatures T_{af} and T_{ef} and the corresponding surface temperatures T_a and T_e can be written on the basis of the efficiency of these finned surfaces (Eq. 3.229) as

$$Q_a = \eta_a \cdot A_a \, h_a (T_{af} - T_a) \tag{4.293}$$

$$Q_e = \eta_e \cdot A_e h_e (T_e - T_{ef}) \tag{4.294}$$

Using Eqs. (4.290)–(4.295) it should be possible to predict the performance of any TEM.

Thus considering first the case of a thermoelectric cooler (TEC), the operating parameters fixed exogenously are the voltage applied V, the temperature T_{ef} (usually the ambient temperature) and T_{af}, the temperature of the cold space. The design parameters to be specified are the finned areas A_a and A_e the fin efficiencies η_a and η_e, the heat-transfer coefficients at these surface (which can be determined using appropriate equations from Section 3.3.3), the Seebeck coefficient α_{ab}, the thermal and the electrical resistance of the TEM, that is, T_{th}^* and R_e^*.

To solve these equations, we use Eqs. (4.293), (4.294) to express T_a and T_e in terms of other variables:

$$T_a = T_{af} - \frac{Q_a}{\eta_a A_a h_a} \tag{4.295}$$

$$T_e = T_{ef} + \frac{Q_e}{\eta_e A_e h_e} \tag{4.296}$$

Substituting these in Eqs. (4.290), (4.291) we transform these into equations involving only three variables, that is, Q_a, Q_e, and I, all remaining variables being known from operating conditions or the TEM design. Coupling these two equations with Eq. (4.292), we have three equations in three variables and thus the performance of the module working as a TEC can be obtained.

In a similar manner the analysis of a thermoelectric generator (TEG) can also be done. Here, both V (which is now generated by the TEM, and not externally applied) and I are unknowns, but we have an additional equation which relates V and I though the resistance of the external electrical load.

In the above-simplified model the influence of Thomson effect arising because of the dependence of the Seebeck coefficient on the temperature has been neglected. As mentioned by Zhao and Tan [32] incorporation of this effect gives remarkable improvement in predicting the TEM performance. The new equations for the two heat transfers are

$$Q_a = \alpha_{ab,a} I T_a - 0.5 I^2 R_e - \frac{T_e - T_a}{R_{th}} + 0.5 \overline{\tau} I (T_e - T_a)$$

$$Q_e = \alpha_{ab,e} I T_e - 0.5 I^2 R_e - \frac{T_e - T_a}{R_{th}} - 0.5 \overline{\tau} I (T_e - T_a)$$

where $\overline{\tau}$ is the average value of Thomson coefficient, and given by equation:

$$\overline{\tau} = T_{mean} \frac{\alpha_{ab,e} - \alpha_{ab,a}}{T_e - T_a}$$

$\alpha_{ab,e}$ and $\alpha_{ab,a}$ being the values of the Seebeck coefficient at temperatures T_e and T_a, respectively, $T_{mean}(= \frac{T_e+T_a}{2})$ being the mean temperature in the TEM. The value of the thermal resistance R_{th} is also evaluated at this mean temperature.

Example 4.16 illustrates the application of this method for a typical thermoelectric cooling module.

Example 4.16

Lineykin and Ben–Yaakov [31] give the following data of a typical thermoelectric cooling module.

$$\alpha_{ab} = 0.053 \, \text{V/K} \quad R_{th} = 1.5 \, \text{K/W} \quad R_e = 1.6 \, \text{ohm}$$

What will be current needed to extract 20 W of heat from a space at temperatures of (a) 270 K, (b) 260 K, and (c) 250 K, when the ambient temperature is 300 K.

Solution

Since Q_a value is given, we can directly substitute the given data in Eq. (4.290) to get an equation for I:

$$20 = 0.053 \times 270I - \frac{300 - 270}{1.5} - \frac{I^2 \times 1.6}{2}$$

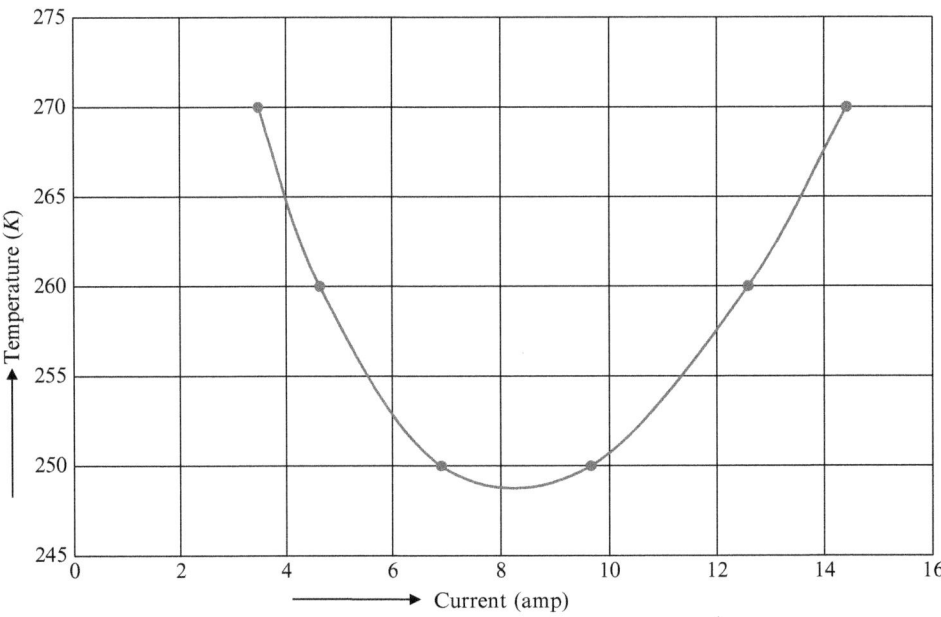

Fig. 4.50 Figure for Example 4.16.

which simplifies to

$$0.8I^2 - 14.31I + 40 = 0$$

The solutions of this quadratic equation are

$$I = 14.42 \text{ and } 3.47 \text{ amps}$$

Proceeding similarly we get, for $T_a = 260$ K the equation for I

$$0.8I^2 - 13.78I + 46.67 = 0$$

which gives

$$I = 12.59 \text{ and } 4.633 \text{ amp}$$

For $T_a = 250$ K, the equation becomes

$$0.8I^2 - 13.25I + 53.33 = 0$$
$$\text{which gives} \quad I = 9.66 \text{ and } 6.9 \text{ amp}$$

On plotting these results, Fig. 4.50, we notice that there must be a minimum temperature that can be attained while maintaining a cooling effect of 20 W. The students are advised to check by calculation that this minimal temperature is $\sim$249 K.

If the TEM module is very thick and the one-dimensional heat-transfer assumption is not valid, we need to determine the temperature using three-dimensional heat conduction equation (3.212). This complicates the analysis and simple analytical solution can not be obtained. Usually finite element modeling is done, which is reported in many research publications, see for example Dziurdzia [33].

The same set of equations can be used to analyze a thermoelectric generator taking care that the voltage is now being generated (i.e., V is negative, with respect to that in TEC) and the electric power is output (and would again have negative value). Example 4.17, based on data given by Lineykin and Ben-Yaakov [31], illustrates the calculation procedure.

Example 4.17

A thermodynamic generator has following design characteristics:

$$\alpha_{ab} = 0.0238 \text{ V/K} \quad R_{th} = 0.589 \text{ K/V} \quad R_e = 0.298 \, \Omega$$

What is the maximum electric power that it can deliver when the heat source is at 230°C and the sink at 30°C? What is the efficiency of conversion of heat into electricity?

Solution

The power input in TEM is given by equation

$$P = V \cdot I = I^2 R_e + \alpha_{ab}(T_e - T_a)I$$

Substituting the data we get

$$P = I^2 \cdot 0.298 + 0.0238(30 - 230)I$$

$$= 0.298I^2 - 4.76I$$

$$P_{max}, \frac{dP}{dI} = 0 \Rightarrow 0.298 \times 2I = 4.76 \quad \text{or} \quad I = 7.986 \, \text{amp}$$

$$\text{and} \quad P = -19 \, \text{W} \quad \text{or} \quad \text{Output} = 19 \, \text{W}$$

To find the thermal efficiency, we need to find the heat absorbed Q_a. Using Eq. (4.290) we get

$$Q_a = 0.0238 \times 503 \times 7.986 + \frac{200}{0.589} - \frac{7.986^2 \times 0.298}{2} = 425.6 \, \text{W}$$

This gives the efficiency η as

$$\eta = \frac{19}{425.6} = 0.0446 \text{ or } 4.46\%$$

Note: This low conversion efficiency has been the inhibiting factor in popularization of the TEGs.

4.6 OTHER APPLICATIONS

The basic principles of heat, mass, and momentum transfer have also been applied in the analysis and design of equipment used in other "nonthermal" applications like VLSI systems, manufacturing processes, heat treatment, food dehydration, etc. In this section, we shall illustrate the methods used for simulating the performance of some of these equipment.

4.6.1 Cooling of Electronic Equipment

The rate at which heat has to be dissipated from an electronic equipment depends upon the switching rates and the physical size of the integrated circuit. As the switching rates increase, with increasing demand for faster signal processing, and the physical sizes of the equipment decrease due to increasing demand for miniaturization, the heat flux dissipation requirements have increased manifold over the last two decades. For a long-time satisfactory cooling of the electronic equipment could be achieved by passive devices like arrays of fins where heat dissipation is primarily by free convection of heat conducted through the fins. However, electronic equipment used in modern high-speed computers need forced convection cooling over finned surfaces and even heat pipes.

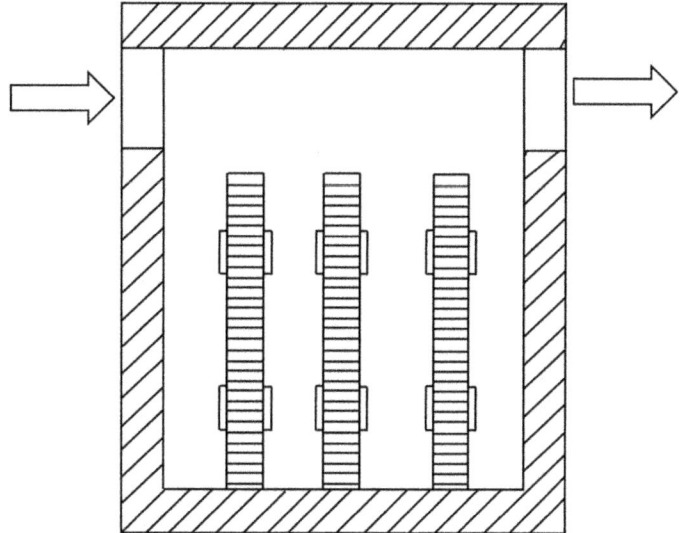

Fig. 4.51 An array of PCBs being cooled.

Since the process model of such an equipment would depend upon the geometrical shape, we shall illustrate the approach by considering the simple case of an array of vertical printed circuit boards containing electronic components, following Jaluria [34].

As the air is blown through the enclosure (Fig. 4.51), it will flow around the PCBs in a manner governed by the flow passages available and the rate at which it picks up heat from them. The convective transport in the air is thus coupled with heat conduction through the PCBs. A rigorous analysis would therefore need simultaneous solution of the partial differential equations governing flow and temperature fields in air and the conduction equations governing the temperature distribution in the PCB.

However, a simplified analysis can be done by assuming that the value of convective heat-transfer coefficient along a PCB, α_c, is known; the thickness of the PCB is small enough to neglect temperature change across it, and the temperature of the bottom of the fin (being in contact with the base of the equipment) is known, say T_b. Further as a simplifying assumption the heat generated in the electronic components is taken to be uniformly distributed over the whole PCB volume. It can thus be considered as a volumetric energy source, q''' (W/m^3) distributed over the board. The governing equation for temperature distribution can now be obtained by writing the energy balance for a small element, as shown in Fig. 4.52, as:

$$
\begin{bmatrix} \text{Energy entering the} \\ \text{element by} \\ \text{conduction per unit} \\ \text{time} \end{bmatrix} + \begin{bmatrix} \text{Energy generated per} \\ \text{unit time} \end{bmatrix} = \begin{bmatrix} \text{Energy leaving per} \\ \text{unit time due to} \\ \text{conduction and} \\ \text{convection} \end{bmatrix}
$$

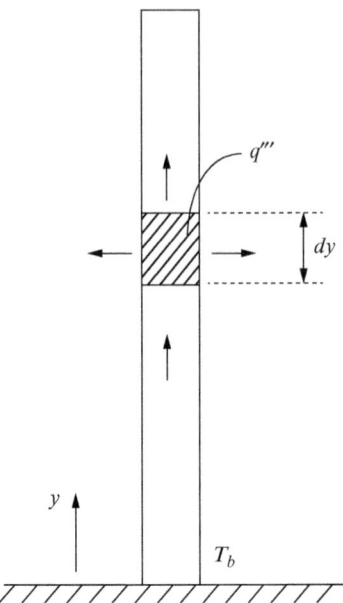

Fig. 4.52 Simplified model of a PCB.

This gives

$$\left[-k\frac{dT}{dy}\right]_y \cdot A + q''' \cdot (A \cdot dy) = \left[-k\frac{dT}{dy}\right]_{y+dy} (A) + \alpha_c(P \cdot dy)(T - T_\infty)$$

where A is the cross-sectional area of the PCB, P its perimeter, k the thermal conductivity of its material, and q''' the average volumetric heat generation rate which can be computed as:

$$q''' = \frac{Q}{A \cdot L} \tag{4.297}$$

where L is the total height of the PCB and Q its total heat generation rate.

Rearranging the energy balance equation, we get

$$q'''(A \, dy) = -k\frac{d^2 T}{dy^2} \cdot (A \, dy) + \alpha_c \cdot P \cdot dy(T - T_\infty)$$

which, on simplification, gives

$$\frac{d^2 T}{dy^2} - \frac{\alpha_c P}{kA}(T - T_\infty) + \frac{q'''}{k} = 0 \tag{4.298}$$

The boundary conditions are

$$\text{at} \quad y = 0, \quad T = T_b \quad \text{and at} \quad y = L, \quad -k\frac{dT}{dY} = \alpha_c(T - T_\infty)$$

Defining $\theta = T - T_\infty$ and $x = y/L$, the above equation transforms to:

$$\frac{d^2\theta}{dx^2} - m^2\theta + n = 0 \tag{4.299}$$

and the boundary conditions become:

$$x = 0, \quad \theta = \theta_b = T_b - T_\infty; \quad \text{at} \quad x = 1, \quad \frac{d\theta}{dx} = -\frac{\alpha_c L}{k} \cdot \theta = -B\theta \tag{4.300}$$

$$\text{where} \quad m^2 = \frac{\alpha_c P L^2}{kA}; \quad n = \frac{q''' L^2}{k}, \quad B = \frac{\alpha_c L}{k} \tag{4.301}$$

Eq. (4.299) is a standard second-order differential equation, whose general solution is

$$\theta = C_1 e^{mx} + C_2 e^{-mx} + \frac{n}{m^2} \tag{4.302}$$

Using the boundary conditions of Eq. (4.300), we get complicated expressions for the two constants:

$$C_1 = \frac{\theta_b m e^{-m} + \frac{n}{m} e^m - \theta_b B e^{-m} + \frac{Bn}{m^2} e^m - \frac{Bn}{m^2}}{m e^m + m e^{-m} + B e^m - B e^{-m}} - \frac{n}{m^2} \tag{4.303}$$

and

$$C_2 = \frac{\theta_b m e^m - \frac{n}{m} e^m + \theta_b B e^m - \frac{Bn}{m^2} e^m + \frac{Bn}{m^2}}{m e^m + m e^{-m} + B e^m - B e^{-m}} \tag{4.304}$$

Thus the temperature profile in the PCB can be determined.

Example 4.18

Determine the temperature profile, and the maximum temperature in a PCB of following specifications:

    Heat dissipation: 300 W
    PCB dimensions: 150 × 100 × 10 mm
    PCB thermal conductivity: 200 W/m K
    Convective heat-transfer coefficient: 30 W/m^2 K

    Assume the temperature of fin base is approximately equal to the room temperature of 35°C.

Solution

The temperature profile is given by Eq. (4.302). We first find the various constants used in that equation.

$$m^2 = \frac{\alpha_c P L^2}{kA} = \frac{30 \times (220/1000) \times 0.15^2}{200 \times (0.1 \times 0.01)} = 0.7425$$

$$\Rightarrow m = 0.86168 \quad e^m = 2.36713 \quad e^{-m} = 0.42245$$

$$q''' = \frac{300}{0.15 \times 0.1 \times 0.01} = 2 \times 10^6 \text{ W/m}^3$$

$$n = \frac{q''' L^2}{k} = \frac{2 \times 10^6 \times 0.15^2}{200} = 225.0$$

$$B = \frac{\alpha_c L}{k} = \frac{30 \times 0.15}{200} = 0.0225$$

$$\frac{n}{m^2} = \frac{225}{0.86168^2} = 303.03 \quad \frac{n}{m} = 261.12 \quad \frac{Bn}{m} = 5.8752 \quad \frac{Bn}{m^2} = 6.81818$$

Since $T_b = T_\infty$, this implies $\theta_b = 0$. Putting in Eqs. (4.303), (4.304) we get

$$C_1 = \frac{\frac{n}{m} e^m + \frac{Bn}{m^2} e^m - \frac{Bn}{m^2}}{m e^m + m e^{-m} + B e^m - B e^{-m}} - \frac{n}{m^2}$$

and

$$C_2 = \frac{\frac{-n}{m} e^m - \frac{Bn}{m^2} e^m + \frac{Bn}{m^2}}{m e^m + m e^{-m} + B e^m - B e^{-m}}$$

Substituting values, gives

$$C_1 = \frac{261.12 \times 2.36713 + 6.81818 \times 2.36713 - 6.81818}{0.86168 \times (2.36713 + 0.42245) + 0.0225 \times (2.36713 - 0.42245)} - 303.03$$

$$= 256.356 - 303.03 = -46.674$$

$$C_2 = \frac{-261.12 \times 2.36713 - 6.81818 \times 2.36713 + 6.81818}{0.86168 \times (2.36713 + 0.42245) + 0.0225 \times (2.36713 - 0.42245)} = -256.356$$

The temperature profile is thus given by equation:

$$\theta = -46.674 \, e^{0.86168x} - 256.356 \, e^{-0.86168x} + 303.03$$

and the maximum temperature at $x = 1$ is

$$\theta = -46.674 \times 2.36713 - 256.356 \times 0.44245 + 303.03$$

$$= 79.12°C$$

which gives $\quad T = 79.12 + 35 = 114.12°C$

4.6.2 Thermal Processing of Moving Materials

In many manufacturing processes like rolling, extrusion, wire drawing, continuous casting and in heat treatment, the manufactured part moves through a thermal environment. The temperature distribution in the moving workpiece depends on its temperature difference with the surroundings as also on the convective heat-transfer coefficient. The later in turn is dependent both on the workpiece movement and the circulation of air surrounding it. If the net air movement over the workpiece is not significant, free convection governs the heat transfer, in which case the convective heat-transfer coefficient depends on the workpiece temperature also; and the solution to the problem of heat conduction through the workpiece becomes coupled with the problem of convection over the surface. We shall consider here only the simpler situation where the convective heat-transfer coefficient (α_c) over the moving workpieces is specified. We assume that the material is moving at a uniform velocity (V), its cross-sectional area (A) is constant and there is negligible temperature difference across it. Now, considering a small control volume, of length dx (Fig. 4.53), we can write the energy balance as:

$$
\begin{bmatrix} \text{Energy entering} \\ CV \text{ per unit} \\ \text{time at face } x \end{bmatrix} - \begin{bmatrix} \text{Energy leaving} \\ CV \text{ per unit time} \\ \text{at face } x+dx \end{bmatrix} - \begin{bmatrix} \text{Heat lost per} \\ \text{unit time due} \\ \text{to convection} \end{bmatrix} = \begin{bmatrix} \text{Energy going} \\ \text{into storage per} \\ \text{unit time} \end{bmatrix}
$$

$$
\Rightarrow \left[(VA\rho cT)_x + \left(-kA\frac{\partial T}{\partial x} \right)_x \right] - \left[(VA\rho cT)_{x+dx} + \left(-kA\frac{\partial T}{\partial x} \right)_{x+dx} \right]
$$

$$
-\alpha_c P \, dx(T - T_\infty) = (\rho A \, dx)c\frac{\partial T}{\partial t} \tag{4.305}
$$

Combining terms and simplifying, we get

$$
-VA\rho c\frac{\partial T}{\partial x} + kA\frac{\partial^2 T}{\partial x^2} - \alpha_c P(T - T_\infty) = \rho Ac\frac{\partial T}{\partial t} \tag{4.306}
$$

where P is the perimeter, c the specific heat of the material, and T_∞ the environment temperature.

Dividing both sides by kA gives the governing differential equation for temperature as:

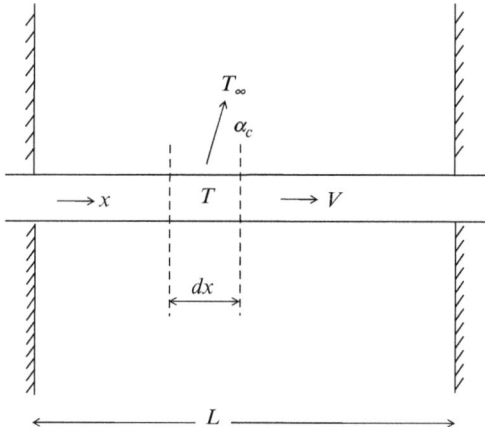

Fig. 4.53 Heat transfer from a continuously moving material.

$$\frac{\partial^2 T}{\partial x^2} - \frac{V}{\alpha}\frac{\partial T}{\partial x} - \frac{\alpha_c P}{kA}(T - T_\infty) = \frac{1}{\alpha}\frac{\partial T}{\partial t} \tag{4.307}$$

where α is the thermal diffusivity of the material of the work piece. This equation needs two boundary conditions for solution.

A set of simplified BCs is

$$T = T_o \quad \text{at} \quad x = 0 \tag{4.308}$$

$$\frac{\partial T}{\partial x} = 0 \quad \text{at} \quad x = L \tag{4.309}$$

Defining a characteristic length λ as:

$$\lambda = \frac{A}{P} \tag{4.310}$$

We transform the basic differential equation and the boundary conditions by introducing dimensionless variables defined as follows:

$$x^* = \frac{x}{\lambda}, \quad t^* = \frac{t\alpha}{\lambda^2}, \quad Pe = \frac{V\lambda}{\alpha}, \quad Bi = \frac{\alpha_c \lambda}{k} \tag{4.311}$$

Further we define $\theta = T - T_\infty$.

Introducing these new variables the partial differential equation (4.307) and the boundary conditions (4.308), (4.309) get transformed to:

$$\frac{\partial^2 \theta}{\partial x^{*2}} - Pe\frac{\partial \theta}{\partial x^*} - Bi\theta = \frac{\partial \theta}{\partial t^*} \tag{4.312}$$

$$\text{subject to} \quad \theta = \theta_o \text{ at } x^* = 0 \quad (\theta_o = T_0 - T_\infty) \tag{4.313}$$

$$\frac{\partial \theta}{\partial x^*} = 0 \text{ at } x^* = L^*(= L/\lambda) \tag{4.314}$$

While the solution of the PDE is best done numerically, the ordinary differential equation to which it reduces for steady-state condition, that is

$$\frac{d^2 \theta}{dx^{*2}} - Pe\frac{d\theta}{dx^*} - Bi\theta = 0 \tag{4.315}$$

has a simple analytic solution, that is

$$\frac{\theta(x)}{\theta_o} = \frac{m_1 e^{(m_1 L^* + m_2 x^*)} - m_2 e^{(m_2 L^* + m_1 x^*)}}{m_1 e^{m_1 L^\star} - m_2 e^{m_2 L^\star}} \tag{4.316}$$

where

$$m_1, m_2 = \frac{Pe \pm \sqrt{Pe^2 + 4Bi}}{2} \tag{4.317}$$

Eq. (4.316) can be used to predict the steady-state temperature distribution in the moving workpiece.

4.6.3 Temperature Distribution During Welding

Determining the temperature distribution in a workpiece during welding process is of great practical importance. As the welding electrode (or gas torch) is moved over the workpiece the temperature distribution changes both spatially and temporally. From thermal perspective, this process is modeled as a heat conduction problem with a moving heat source. The modeling is greatly facilitated by viewing the process from the perspective of an observer sitting on the welding torch as it is moving over the workpiece. Considering this observer to be at the center of a moving coordinate system (ξ, η, ζ), its relationship to the fixed coordinate system (x, y, z) will be as shown in Fig. 4.54. Thus considering the movement to be in x-direction, with a velocity V, the relation between the two coordinate systems can be written as:

$$\xi = x - Vt; \quad t = t' \quad \eta = y \quad \zeta = Z \tag{4.318}$$

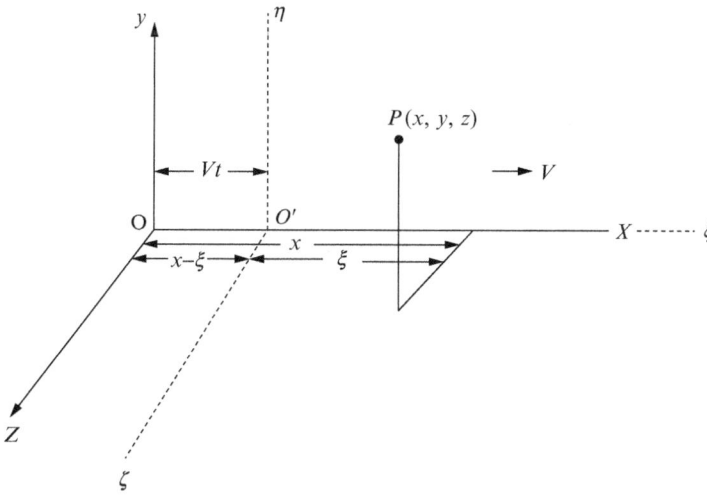

Fig. 4.54 Coordinate system for analyzing welding process.

Transformation of quantities from stationary coordinate system to moving coordinate system can then be done as follows:

$$\frac{\partial \xi}{\partial x} = 1 \quad \frac{\partial \xi}{\partial t} = -V \quad \frac{\partial t'}{\partial t} = 1 \quad t, t' \text{ being time} \tag{4.319}$$

Further

$$\frac{\partial T}{\partial x} = \frac{\partial T}{\partial \xi}; \quad \frac{\partial^2 T}{\partial x^2} = \frac{\partial^2 T}{\partial \xi^2}; \quad \frac{\partial^2 T}{\partial y^2} = \frac{\partial^2 T}{\partial \eta^2}; \quad \frac{\partial^2 T}{\partial z^2} = \frac{\partial^2 T}{\partial \zeta^2} \tag{4.320}$$

$$\frac{\partial T}{\partial t} = \frac{\partial T}{\partial \xi} \cdot \frac{\partial \xi}{\partial t} + \frac{\partial T}{\partial t'} \cdot \frac{\partial t'}{\partial t} = -V \frac{\partial T}{\partial \xi} + \frac{\partial T}{\partial t'} \tag{4.321}$$

The general three-dimensional heat conduction equation (cf. Eq. 3.210) for the moving coordinate system can now be written as:

$$\alpha \left(\frac{\partial^2 T}{\partial \xi^2} + \frac{\partial^2 T}{\partial \eta^2} + \frac{\partial^2 T}{\partial \zeta^2} \right) + \frac{Q'}{\rho c} = \frac{\partial T}{\partial t'} - V \frac{\partial T}{\partial \xi} \tag{4.322}$$

To greatly simplify the model, it is often assumed that a body of sufficiently large dimensions would appear to be in steady state from the point of view of the observer riding along the welding torch. This implies, $\frac{\partial T}{\partial t'} = 0$, and for this quasisteady situation Eq. (4.322) becomes

$$\frac{\partial^2 T}{\partial \xi^2} + \frac{\partial^2 T}{\partial \eta^2} + \frac{\partial^2 T}{\partial \zeta^2} + \frac{Q'}{k} = -\frac{V}{\alpha} \frac{\partial T}{\partial \xi} \tag{4.323}$$

where Q' is the volumetric heat input rate into the workpiece.

Eq. (4.323) governs the temperature distribution during the welding of a workpiece.

Considering the simplest possible case of a long thin rod of constant cross-section area A, the temperature can be assumed to be constant across its cross-section. Further the convective heat loss at any section can be modeled as a "volumetric heat input" Q' of magnitude:

$$Q' = -\frac{P \cdot dx \cdot \alpha_c (T - T_\infty)}{A dx} = -\frac{P \alpha_c}{A} (T - T_\infty) \tag{4.324}$$

Eq. (4.323) thus reduces to

$$\frac{\partial^2 T}{\partial \xi^2} - \frac{P \alpha_c}{kA} (T - T_\infty) = -\frac{V}{\alpha} \frac{\partial T}{\partial \xi} \tag{4.325}$$

Defining a characteristic length λ, as in Eq. (4.310), and other dimensionless variables as in Eq. (4.311), the above equation can be written as:

$$\frac{\partial^2 \theta}{\partial \xi^{*2}} + Pe \frac{\partial \theta}{\partial \xi^*} - Bi\theta = 0 \tag{4.326}$$

where $\xi^* = \xi / \lambda$.

Since the quasisteady approximation is applicable only when the workpiece is very long, the boundary conditions in the moving coordinate system are

$$\theta = 0 \quad \text{as} \quad \xi^* \to \pm\infty \tag{4.327}$$

The general solution to the (now) ordinary differential equation (4.326) can directly written as

$$\theta = C_1 e^{m_1 \xi^*} + C_2 e^{m_2 \xi^*} \tag{4.328}$$

$$\text{where} \quad m_1, m_2 = \frac{-Pe \pm \sqrt{Pe^2 + 4Bi}}{2} \tag{4.329}$$

Eqs. (4.327)–(4.329) clearly demand that values of C_1 and C_2 should be different in the regions $\xi^* < 0$ and $\xi^* > 0$. Hence we get two different equations for temperature distribution in these regions as follows:

$$\text{for} \quad \xi^* > 0; \quad \theta = C_1 \quad \exp\left\{ \left[-\frac{Pe}{2} - \sqrt{\left(\frac{Pe}{2}\right)^2 + Bi} \right] \xi^* \right\} \tag{4.330}$$

$$\text{and for} \quad \xi^* < 0; \quad \theta = C_2 \quad \exp\left\{ \left[-\frac{Pe}{2} + \sqrt{\left(\frac{Pe}{2}\right)^2 + Bi} \right] \xi^* \right\} \tag{4.331}$$

At $\xi^* = 0$, the temperatures of both the sides of the welding torch are equal, and represent the maximum temperature θ_{max}. Thus both C_1 and C_2 are equal to θ_{max}, and above equations can be written as

$$\theta = \theta_{max} \quad \exp\left\{ \left[-\frac{P_e}{2} - \sqrt{\left(\frac{P_e}{2}\right)^2 + B_i} \right] \xi^* \right\} \text{for} \quad \xi^* > 0 \tag{4.332}$$

$$\theta = \theta_{max} \quad \exp\left\{ \left[-\frac{P_e}{2} + \sqrt{\left(\frac{P_e}{2}\right)^2 + B_i} \right] \xi^* \right\} \text{for} \quad \xi^* < 0 \tag{4.333}$$

Eqs. (4.332), (4.333) can be plotted to show the temperature distribution along the rod at any instant (Fig. 4.55).

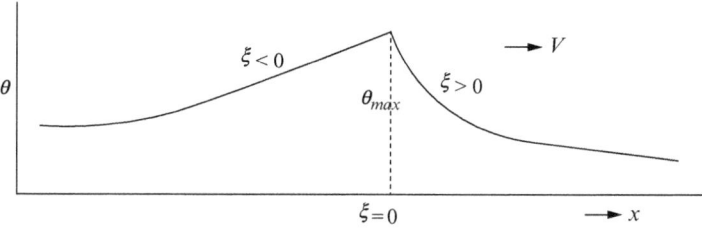

Fig. 4.55 Temperature distribution in a rod during welding.

The heat-transfer rate at the position $\xi = 0$ (the welding torch location) in the two directions can be found using these equations. The total heat input from the torch being the source of these heat transfers, we can get an expression for this Q, in terms of θ_{max} as

$$Q = -\frac{kA}{\lambda}\left\{\frac{\partial\theta}{\partial\xi^*}\right\}_{\xi^*=0^+} + \frac{kA}{\lambda}\left(\frac{\partial\theta}{\partial\xi^*}\right)_{\xi^*=0^-} \tag{4.334}$$

$$= \frac{\theta_{max}\,kA}{\lambda}\left[\frac{Pe}{2} + \sqrt{\left(\frac{Pe}{2}\right)^2 + Bi} - \frac{Pe}{2} + \sqrt{\left(\frac{Pe}{2}\right)^2 + Bi}\right]$$

$$= \frac{2kA\theta_{max}}{\lambda}\left(\sqrt{\left(\frac{Pe}{2}\right)^2 + Bi}\right) \tag{4.335}$$

from which we get an expression for θ_{max} in terms of the strength of the moving heat source Q (in watts), as

$$\theta_{max} = \frac{Q\lambda}{2kA\sqrt{\left(\frac{Pe}{2}\right)^2 + Bi}} \tag{4.336}$$

Using this value of θ_{max} in Eqs. (4.332), (4.333), we can get the instantaneous temperature profile in the rod at any time.

For a more realistic simulation of the welding process, the heat flow in all the three dimensions needs to be considered, which can be done by numerically solving Eq. (4.323) with appropriate boundary conditions.

Example 4.19

Two steel flats 20 mm × 3 mm cross section and 4 m long are being welded along the length to form a thicker flat of 6 mm thickness. The heat input from the welding torch is 6 kW. Estimate the maximum temperature likely to be reached during welding. Given

   Velocity of welding torch: 1 cm/s
   Properties of steel: ρ: 7272 kg/m³ c: 0.420 kJ/kg K k: 52 W/m K
Take atmospheric temperature as 35°C and convective heat-transfer coefficient as 20 W/m².

Solution

The overall cross section being welded is 6 mm × 20 mm. This gives

$$A = 20 \times 6 = 120\,\text{mm}^2 = 120 \times 10^{-6}\,\text{m}^2 \quad P = (20+6) \times 2 = 52\,\text{mm}$$

$$\lambda = \frac{A}{P} = \frac{120}{52} = 2.308\,\text{mm} = 2.308 \times 10^{-3}\,\text{m}$$

Further $\alpha = \dfrac{k}{\rho c} = \dfrac{52}{7272 \times (0.420 \times 10^3)} = 1.702 \times 10^{-5} \text{ m}^2/\text{s}$

$$Pe = \dfrac{V\lambda}{\alpha} = \dfrac{(1 \times 10^{-2})(2.308 \times 10^{-3})}{1.702 \times 10^{-5}} = 1.356$$

$$Bi = \dfrac{\alpha_c \lambda}{k} = \dfrac{20 \times 2.308 \times 10^{-3}}{52} = 0.8877 \times 10^{-3}$$

Substituting in Eq. (4.336), we get

$$\theta_{max} = \dfrac{(6000) \times (2.308 \times 10^{-3})}{2 \times 52 \times (120 \times 10^{-6}) \times \sqrt{\left(\frac{1.356}{2}\right)^2 + 0.8877 \times 10^{-3}}}$$

$$= 1635°\text{C}$$

which gives the maximum temperature as $1635 + 35 = 1670°\text{C}$.

4.6.4 Heat and Mass Transfer During Drying of Solids

Dehydration of the wet solids is an important process encountered in a wide range of industries: pharmaceuticals and chemicals, food processing, paper drying, wood seasoning, etc. A variety of methods are employed to dry different products and their modeling would thus be dependent on the methodology of drying. As an illustration of the modeling approach we shall focus here on drying of food products, specifically tropical fruits, with the help of hot air, which is one of the most commonly employed methods. Most fruits contain a large amount of moisture, typically 70% to 95% of their gross weight. Therefore, in the initial phase of drying, as the moisture evaporates from the surface, the diffusion of moisture from inside the fruit to the surface quickly replenishes it. Thus a continuous film of water is maintained over the surface of the fruit in contact with hot air. During this period the drying rate remains constant, and can be calculated using the basic equations for simultaneous heat and mass transfer (Section 2.4.1) as applied to evaporation from a wet surface. As the moisture content inside the solid falls down, it becomes difficult to maintain the surface "wet" and the drying rate falls down. Now the diffusion of moisture and heat within the solid governs the rate of drying. Further, most agricultural products, especially fruits, shrink appreciably on drying, and this influences both the diffusion coefficients as also the mechanism of heat and mass transport.

As an illustration of the modeling of drying of such products, we shall develop, following Karim and Hawlader [35], a simple mathematical model for drying of a fruit, like banana, which shrinks appreciably on drying. The product being dried is taken as a thin slab of thickness $2L$ (Fig. 4.56) at a uniform initial temperature T_o and moisture content w_p, defined as mass of water per unit mass of dry solid. It is being dried from

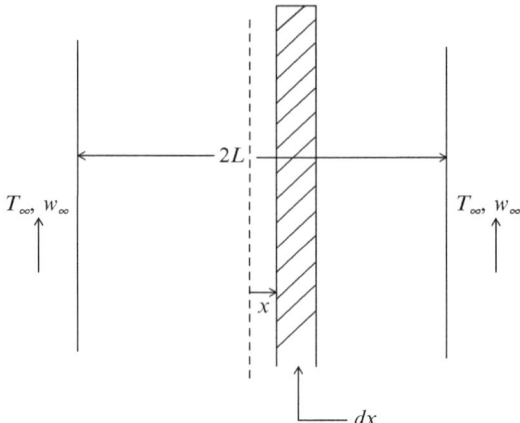

Fig. 4.56 Drying of a product.

both its sides by a current of warm dry air at a temperature T_∞ and humidity ratio w_∞ (Fig. 4.56). We assume that as a result of drying the product surface shrinks at a velocity $u(x)$. With these simplifying assumptions we can write the various conservation equations as follows.

Considering a small control volume of thickness dx (Fig. 4.56), the equation for conservation of moisture can be written as:

$$
\begin{bmatrix} \text{Rate of moisture} \\ \text{entry at } x \\ \text{due to diffusion} \end{bmatrix} - \begin{bmatrix} \text{Rate of moisture} \\ \text{leaving at } x + dx \\ \text{due to diffusion} \end{bmatrix} + \begin{bmatrix} \text{Rate of moisture} \\ \text{entry at } x \\ \text{due to velocity } u(x) \end{bmatrix}
$$

$$
- \begin{bmatrix} \text{Rate of moisture} \\ \text{leaving at } x + dx \\ \text{due to velocity } u(x) \end{bmatrix}
$$

$$
= [\text{Rate of increase of moisture in the control volume}]
$$

Writing mathematical expressions for each of the above terms we get

$$
\left| -(\rho \cdot A \cdot dx)D\frac{\partial w_p}{\partial x}\right|_x - \left| -(\rho \cdot A \cdot dx)D\frac{\partial w_p}{\partial x}\right|_{x+dx}
$$

$$
+ |\rho \cdot u \cdot A \cdot w_p|_x - |\rho \cdot u \cdot A \cdot w_p|_{x+dx} = (\rho \cdot A \cdot dx)\frac{\partial w_p}{\partial t} \tag{4.337}
$$

where ρ is the density of the product, and A the surface area of the control volume. Simplifying we get

$$D\frac{\partial^2 w_p}{\partial x^2} - u\frac{\partial w_p}{\partial x} = \frac{\partial w_p}{\partial t} \tag{4.338}$$

Proceeding, in exactly analogous manner, the energy equation can be obtained as

$$\alpha\frac{\partial^2 T}{\partial x^2} - u\frac{\partial T}{\partial x} = \frac{\partial T}{\partial t} \tag{4.339}$$

The initial and boundary conditions to solve Eqs. (4.338), (4.339) are

$$(w_p)_{t=0} = w_0 \quad (T)_{t=0} = T_0 \tag{4.340}$$

$$\text{at } x = 0 \left(\frac{\partial w_p}{\partial x}\right) = 0 \quad \text{and} \quad \left(\frac{\partial T}{\partial x}\right) = 0 \tag{4.341}$$

along with the energy and moisture balance at the surface $x = L$, which can be written as:

$$-D\left(\frac{\partial w_p}{\partial x}\right)_{x=L} + (uw_p)_{x=L} = h_m\rho\{(w_p - w_{p,e})\}_{x=L} \tag{4.342}$$

and

$$\left(k\frac{\partial T}{\partial x} - \rho c_p uT\right)_{x=L} = h(T_\infty - T)_{x=L} - h_m\rho\{(w_p - w_{p,e})\}h_{fg} \tag{4.343}$$

where h_m is the mass-transfer coefficient and h the heat-transfer coefficient at the surface of the product; $w_{p,e}$ the equilibrium moisture content of the product at the environment conditions; h_{fg} the latent heat of vaporization of moisture and c_p the specific heat of the product. The solution of these equations also needs specification of the shrinkage velocity u, which is either input from experimental data or predicted on the basis of a simplified model, details of which can be seen in the paper by Karim and Hawlader [35].

PROBLEMS

Note

In the problems where the thermophysical properties of the working substances are not given, take these from internet/handbooks.

4.1 A parallel flow heat exchanger having a surface area of 0.2 m^2 is operating under following conditions:

$$\text{Cold fluid heat capacity rate} = 200 \text{ W/K}$$
$$\text{Hot fluid heat capacity rate} = 100 \text{ W/K}$$

If the overall heat-transfer coefficient is 200 W/m^2 K, determine the rate of heat transfer for cold fluid inlet temperature of 20°C and hot fluid inlet temperature of 120°C.

4.2 If the above heat exchanger is operated in counter-flow mode, what would be the rate of heat transfer. Compare the effectiveness of the heat exchanger in this mode with that in parallel flow.

4.3 Determine the rate of heat transfer if a cross-flow heat exchanger with the same design and operating conditions as in Problem 4.1 were operated with
   (a) both fluids unmixed
   (b) hot fluid mixed, cold unmixed
   (c) cold fluid mixed, hot unmixed

4.4 If the heat exchanger of Problem 4.1 were a two-pass shell and tube heat exchanger with surface area equally divided between the two passes, what would the heat-transfer rate be?

4.5 Derive an expression for the effectiveness of a two-pass shall and tube exchanger in which the surface area and the overall heat-transfer coefficients in the two passes are U_1, A_1 and U_2, A_2, respectively.

4.6 Consider a two-pass heat exchanger having a surface area of 0.08 m^2 in the first pass and 0.12 m^2 in the second pass. The average overall heat-transfer coefficients in the two passes are 100 and 300 W/m^2 K. If the rest of the operating conditions are the same as in Problem 4.1, determine the rate of heat transfer and the heat exchanger effectiveness.

4.7 Fig. 4.57 shows a three-fluid heat exchanger where a hot fluid is being used to heat two cold fluids. Assuming the various heat-transfer coefficients to be invariant with x, write the basic equations governing its performance and indicate how will you determine the heat gained by the two cold fluids.

4.8 Consider a concentric tube single-pass heat exchanger in which water is flowing through the inner tube and hot gases, being used to boil water, are flowing in the annulus. Water enters the heat exchanger at 90°C and starts boiling at 110°C. Hot gases enter at 200°C, with a flow rate of 1 kg/s. If the flow rate of water is 0.1 kg/s, determine the total heat transfer in the heat exchanger and the fraction of steam in outgoing steam-water mixture. Given the following design data for the heat exchanger.

   Area: 5 m^2 Gas side heat-transfer coefficient: 200 W/m^2 K
   Water side heat-transfer coefficient in W/m^2 K $= 800 - 300x + 8000x^2 - 8000x^3$

where x is the stream dryness fraction. Neglect the thermal resistance offered by the tube walls.

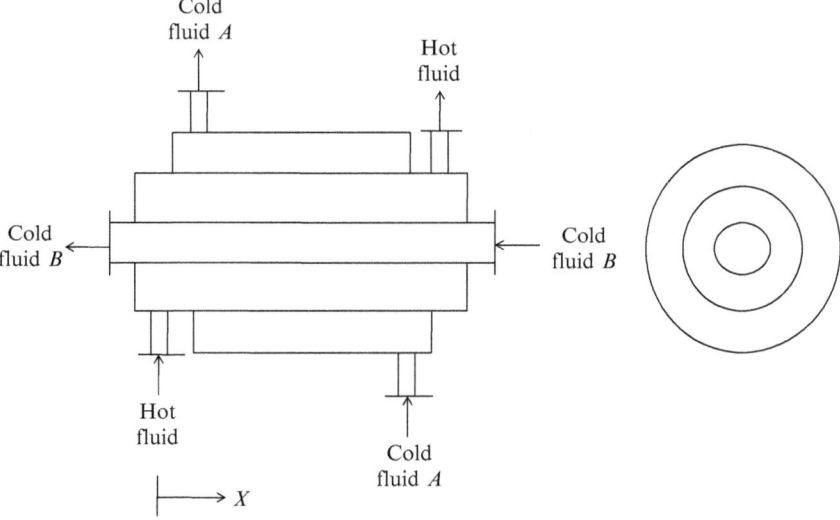

Fig. 4.57 Three-fluid heat exchanger.

4.9 In a counter-flow heat exchanger, cold water enters at 20°C and the hot fluid at 80°C. The two flow rates are m_1 and m_2 (kg/s) and the specific heats are 4.2 and 4.5 kJ/kg K, respectively. The value of the overall heat-transfer coefficient in kW/m² K is given by the equation:

$$\frac{1}{UA} = 0.2 + \frac{0.1}{m_1^{0.8}} + \frac{0.15}{m_2^{0.8}}$$

where A is the heat exchanger surface area in m². The total pressure drop in kPa is given by the equation:

$$\Delta p = 5m_1^{1.25} + 7.5m_2^{1.25}$$

Determine the values of m_1 and m_2 if during a trial it is found that the total heat transfer is 180 kW and the pressure drop is 20 kPa

4.10 In a counter-flow heat exchanger hot oil ($c_p = 3$ kJ/K) at 110°C is being used to heat water entering at 30°C ($c_p = 4.2$ kJ/K). The overall heat-transfer coefficient between the two is 300 W/m² and the surface area of the exchanger is 2.5 m². The outer surface of the outer pipe is not insulated and therefore heat is lost to the atmosphere through that. The overall heat-transfer coefficient between the hot oil and the atmosphere can be taken as 20 W/m². Using the formulation of Problem 4.7, determine the heat transfer from the hot fluid to the cold fluid and the atmosphere. *Given*: flow rate of oil: 1 kg/s; flow rate of water: 0.5 kg/s; outer surface area of heat exchanger: 5 m².

4.11 Determine the rate of heat transfer in a two-pass fire-tube boiler of following specifications:

First pass tube inner diameter = 25 cm;   length = 3 m

Second pass has 20 tubes of 5 cm ID

Water inlet temperature: 40°C

Steam generation pressure: 3 bar

Temperature of gases leaving the burner: 800°C

Flow rate of hot gases: 0.5 kg/s.

4.12 Determine the heat transfer and pressure drop in a single-pass flooded chiller of following specifications:

Tube diameter: 2.5 cm   Tube length: 1.5 m

No. of tubes: 150   Velocity of water in tubes: 1.5 m/s

Water inlet temperature: 15°C

Refrigerant (R-134a) temperature: 3°C

4.13 The "radiator" of a car designed to keep the engine "cool" can be treated as a single-pass cross-flow plate fin heat exchanger. A typical radiator has the following specifications:

40 Tubes (flattened) of cross section 2 mm × 18 mm

Gross face area: 300 mm × 300 mm

Water flow rate: 5 kg/s   Air flow rate: 0.5 kg/s

Number of tube rows: 4   Tube pitch: 25 mm × 6 mm

Fins spacing: 1 mm   Fin thickness: 0.5 mm

Estimate the rate of heat transfer from it when air inlet temperature is 40°C and water inlet temperature is 80°C. What will be the power input in the pump (efficiency: 60%) circulating water through the "radiator." Estimate also the air side pressure drop taking both expansion and contraction coefficients (K_c and K_e) as 0.2.

4.14 Estimate the pressure drop on the air side of a plate-fin type compact heat exchanger from following operating conditions and design data.

Plate spacing: 6.35 mm   Hydraulic diameter: 3 mm

Ratio of core-free area to frontal area, σ: 0.5

Velocity of air inside the heat exchanger plates: 5 m/s

Heat exchanger length: 80 cm   Total frontal area: 60 cm × 40 cm

Friction factor for flow inside plates: 0.01

Air temperature at entry: 30°C   Air exit temperature: 80°C

Take both contraction coefficient K_c and the expansion coefficient K_e to be 0.3.

4.15 In a parallel flow chilled water coil, air enters with a velocity of 4 m/s at 25°C DBT and 50% RH, and the cooling water enters at 8°C. Estimate the outlet condition of air and the cooling water temperature from the following design data.

 Coil frontal area: 1 m (height) × 1.5 m
 No. of tubes in the face: 50 Number of such tube banks: 6
 Tube outer diameter: 14 mm
 Tube pitch along air flow direction: 20 mm
 Total no. of fins: 1000 Fin thickness: 0.3 mm
 Take heat-transfer coefficient on air side as 40 W/m² K and on water side as 1000 W/m². Neglect tube wall thermal resistance. The cooling water flow rate is 8 kg/s

4.16 If the above coil had counter-flow of air and water, how would its performance change? Would the coil be completely wet?

4.17 A counter-flow cooling tower has following ratings at design conditions:

 Water flow rate: 600 kg/h
 Air flow rate: 500 kg/h
 Water inlet temperature: 40°C
 Water outlet temperature: 30°C
 Air entry conditions: 35°C DBT and 26°C WBT

Estimate its performance when air entry conditions are 30°C DBT and 23°C WBT, other conditions remaining the same.

4.18 What is the thermodynamic state of air leaving the above cooling tower under design conditions? How much is the water loss due to evaporation?

4.19 In a parallel flow spray air washer cold water at 12°C is being sprayed at a flow rate of 5 kg/s into a stream of hot air at 40°C and 30% RH flowing at the rate of 10 kg/s. Estimate the condition of air at the exit of the washer, given:

 Heat-transfer coefficient between air and water: 40 W/m² K
 Effective area of contact: 200 m²

4.20 A carpet is washed and then spread out on a lawn to dry in the Sun. It loses moisture due to combined effect of solar radiation and atmospheric air blowing gently over it. During the period that its surface remains wet, the carpet temperature at any location reaches a steady value. Write down the general equations governing heat and mass transfer during this period, to enable determination of carpet temperature at any position along the length. Neglect heat loss from carpet to the ground.

4.21 Determine the temperature profile and the maximum temperature in a PCB of following specifications:

 Heat dissipation: 500 W
 PCB dimensions: 200 × 100 × 10 mm
 PCB thermal conductivity: 250 W/m K
 Convective heat-transfer coefficient: 20 W/m² K

Assume the temperature of the fin base to be approximately equal to the room temperature of 30°C.

4.22 In order to restrict the maximum temperature in the PCB of the above problem to 80°C, it is proposed to increase the air velocity over it by using a large fan. Estimate the value of convective heat-transfer coefficient needed to achieve this.

4.23 Two steel flats 20 mm × 2 mm cross section, each 2 m long, are being welded along the length to form a wider flat of 40 mm × 2 mm cross section. Estimate the torch power input necessary to achieve a maximum temperature of 1500°C. Given

Welding rate: 3 mm/s   Atmospheric temperature: 35°C
Convective heat-transfer coefficient: 15 W/m² K
For steel: density: 7500 kg/m³   Specific heat: 0.4 kJ/kg K
Thermal conductivity: 50 W/m K

4.24 Fig. 4.58 shows the schematic diagram of a regenerative type of air cooler. The primary air enters the lower passage and is cooled by the partition plate which, in turn, is kept cool by water evaporating on its top side. A part of this cooled air is diverted back into the upper passage where it flows over the water film maintained over the metallic partition plate, and keeps it cool by evaporation. Starting from the basic principles of heat and mass transfer, write down the energy balance of the two streams as they flow across a small control volume of thickness dz (Fig. 4.58). Therefrom develop the equations to predict the exit conditions of both the primary and the secondary air.

4.25 A stretch of a long steam pipe has been left uninsulated and loses heat to surroundings by convection and radiation. The high-pressure steam is at a temperature of 200°C and a dryness fraction of 0.99. Estimate the dryness

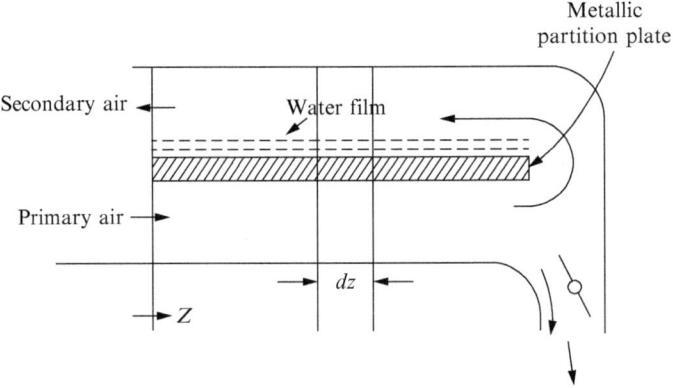

Fig. 4.58 Regenerative air cooler.

fraction [x] of the steam at the end of this uninsulated stretch of steam pipe, from the following data:

Pipe diameter: 15 cm   Pipe length: 200 m   Pipe thickness: 3 mm
Steam flow rate: 10 kg/min   Latent heat: 2000 kJ/kg

Convective heat-transfer coefficient at pipe outer surface is $h_o = 20(T - T_\infty)^{0.25}$ W/m^2 K, where T is the pipe outer surface temperature (in °C) and T_∞, the ambient temperature is 25°C.

The convective heat-transfer coefficient at the pipe inner surface is given by equation

$$h_i = -10{,}258x^2 + 13{,}487x - 3032.5 \text{ W/m}^2\text{ K}$$

Given: pipe outer surface emissivity: 0.8   Pipe thermal conductivity: 50 W/m K

4.26 In order to rapidly dissipate heat from a hot surface, it is provided with a trapezoidal fin kept wet with the help of thin wicks dipped in water (Fig. 4.59). Starting from the fundamentals write down the energy balance of a small control volume, and thereby develop the differential equation for predicting the temperature at any position along the fin. Neglect temperature variation across fin cross–section.

4.27 Develop a software to simulate an SI engine making following assumptions:
Working medium is prepared in the carburettor by mixing of air and fuel.

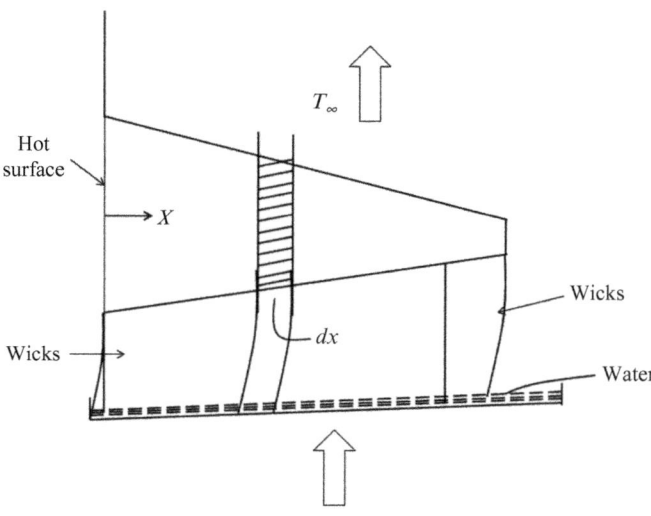

Fig. 4.59 A wet-finned surface.

The intake valve opens at TDC and the pressure remains constant throughout the intake stroke. This pressure equals atmospheric pressure at full throttle and falls down when throttling is done.

The combustion process takes place adiabatically and instantaneously with piston at TDC.

The exhaust valve opens at BDC and then the pressure drops instantaneously to atmospheric pressure. As the piston moves to TDC, most of the exhaust gases are pushed out except for what remains in the clearance volume.

The exhaust valve closes at TDC and the intake valve opens and fuel-air mixture is drawn into the engine, till the piston moves back to BDC.

The compression process begins as the piston starts moving up from the BDC.

The program should be able to predict the indicated power output and draw the indicator diagram and the pressure-crank angle diagram of the engine. Use this program to carry out parametric studies to study the effect of fuel-air ratio, throttle position, engine speed, etc. and compare these results with typical experimental data from literature.

4.28 A two-pass shell and tube type D-X chiller is being used to cool water entering its shell at 10°C. The refrigerant enters the first pass tubes as a liquid–vapor mixture (dryness fraction $x = 0.2$) at 2°C, and evaporates as it picks up heat from the water surrounding the tubes. Write down a computer program to solve the equations governing the change in water temperature and the refrigerant dryness fraction in the heat exchanger and use this program to predict the performance of the chiller under following rating conditions:

Refrigerant flow rate: 10 kg/min   Latent heat: 200 kJ/K

Water flow rate: 50 kg/min

Given: Heat exchanger tube length: 1.5 m

No. of tubes in the first pass: 20,   in the second pass: 30

Tube diameter: ID 15 mm   OD: 17 mm

Tube material thermal conductivity: 350 W/m K

Water side heat-transfer coefficient: 500 W/m² K

Refrigerant side heat-transfer coefficient is given by the equation

$$h_r = (1 - 5x + 15x^2 - 10.9x^3)\, \text{kW/m}^2\, \text{K}$$

where x is the dryness fraction.

4.29 The evaporator coil of a typical window air conditioner using R-22 has following specifications:

No. of tube rows: 3   No. of tubes in each rows: 16

Tube spacing: 12.7 mm   Diameter: 7 mm   Length: 66 cm

Air side fin density: 6 fin/cm   Air mass flow rate: 0.18 kg/s

Refrigerant mass flow rate: 2 kg/min;   Temperature: 2°C

Refrigerant condensation temperature: 50°C

Write a computer program to simulate the performance of this coil for varying air inlet conditions and flow rates, using appropriate equations (from Chapter 3) for predicting air-side and refrigerant-side heat-transfer coefficients.

4.30 In a two-pass shell and tube condenser steam entering at 50°C is being condensed by water entering at 30°C. It has following specifications:

Tube length: 2 m   Tube diameter: 22 mm/25 mm

No. of tubes in each pass: 200   Water velocity inside tube: 2 m/s

Write a computer program to predict the performance of this condenser under varying conditions of steam temperature, water temperature, velocity, etc. Use appropriate correlations to predict the heat-transfer coefficients of water and steam.

4.31 In a liquid desiccant dehumidifier, aqueous solution of calcium chloride flows downward while the process air flows upward. Write a computer program to simulate its performance to determine the dehumidification rate as a function of air flow rate, inlet humidity, desiccant flow rate, and desiccant concentration at the inlet. Given design data:

Air flow rate: 2 kg/min   Desiccant flow rate: 3 kg/min

Total surface area of contact between the air and the desiccant: 4 m^2

Concentration of calcium chloride at inlet: 40%   Temperature: 30°C

Air inlet conditions: 25°C DBT, 20°C WBT

4.32 How would the performance of above dehumidifier be influenced if arrangement is also made to cool the desiccant solution as it flows downward with the help of cold water (at 25°C) flowing upward. Take the heat-transfer coefficient between water and the tube surface as 500 W/m^2, and neglect the thermal resistance of the tubes separating water from the desiccant.

4.33 If in Problem 4.31 the desiccant solution entering the equipment has a concentration of 30%, and a temperature of 70°C, what will be its concentration at the exit? Study the effect of inlet temperature of desiccant on the increase in its concentration.

REFERENCES

[1] P.L. Dhar, On derivation of equations for heat exchanger analysis, Int. J. Appl. Eng. Educ. 6 (1) (1990) 55–60.

[2] F.P. Incropera, D.D. Dewitt, Fundamentals of Heat and Mass Transfer, fifth ed., Wiley, New Delhi, India, 2006, pp. 917.

[3] R.K. Shah, D.P. Sekulic, Fundamentals of Heat Exchanger Design, John Wiley, New Jersey, 2003.

[4] W.M. Kays, A.L. London, Compact Heat Exchangers, McGraw Hill, New York, 1964.

[5] P.L. Dhar, C.P. Arora, Computer simulation of DX-type liquid chillers, in: Proceedings of Third National Symposium on Refrigeration and Air Conditioning, CFTRI, Mysore, 1974, pp. 191–199.

[6] P.L. Dhar, A.K. Jain, Computer simulation of flooded liquid chillers, Proceedings, Fifth National Symposium of Refrigeration and Airconditioning. IIT Madras, 1976, pp. 143–151.

[7] P.L. Dhar, G.R. Saraf, Computer Simulation and Design of Refrigeration Systems, Khanna Publishers, Delhi, 1987.

[8] D. Gorenflo, Pool boiling, in: VDI Heat Atlas, VDI-Verlag, Düseldorf, Germany, 1993.

[9] A. Bejan, A.D. Kraus (Eds.), Heat Transfer Hand Book, John Wiley, New Jersey, 2003.

[10] P.L. Dhar, Engineering Thermodynamics: A Generalized Approach, Elsevier, New Delhi, 2008.

[11] S.K. Kandilkar, S. Garimella, D. Li, S. Colin, M.R. King, Heat Transfer and Fluid Flow in Minichannels and Microchannels, Butterwart Heinemann, Oxford, UK, 2006.

[12] E.B. Arkilic, M.A. Schmidt, K.S. Breuer, Gaseous slip flow in long microchannels, IEEE J. Microelectromech. Syst. 6 (1997) 167–178.

[13] J.C. Maxwell, On stresses in rarefied gases arising from inequalities of temperature. Philos. Trans. R. Soc. 170 (1879) 231–256.

[14] O.C. Jones, An improvement in the calculation of turbulent friction in rectangular ducts, J. Fluids Eng. 98 (2) (1976) 173–181.

[15] A. Vardhan, P.L. Dhar, A new procedure for performance prediction of air conditioning coils, Int. J. Refrig. 21 (1) (1998) 77–83.

[16] W.F. Stoecker, J.W. Jones, Refrigeration and Air Conditioning, McGraw Hill, New York, 1983.

[17] J.L. Threlkeld, Thermal Environment Engineering, Practice Hall, New Jersey, 1970, pp. 253–265.

[18] A.K. Sharma, Optimization of split type air conditioners, MTech thesis, Mechanical Engineering Department, IIT Delhi, 2007.

[19] S. Jain, Studies on desiccant augmented evaporative cooling systems, PhD thesis, IIT Delhi, 1994.

[20] S. Jain, P.L. Dhar, S.C. Kaushik, Optimal design of liquid desiccant cooling systems, ASHRAE Trans. 106 (2000) 79–86.

[21] A.K. Asati, Performance studies of falling film absorber and regenerator using triethylene glycol as desiccant, PhD thesis, IIT Delhi, 2007.

[22] R. Kumar, Studies on stand-alone liquid desiccant based air conditioning systems, PhD thesis, IIT Delhi, 2008.

[23] Z.J. Zhang, Y.J. Dai, R.Z. Wang, A simulation study of heat and mass transfer in a honeycombed rotary desiccant dehumidifier, Appl. Therm. Eng. 23 (2003) 989–1003.

[24] G.R. Saraf, P.L. Dhar, Computer simulation of reciprocating refrigerant compressors, in: Proceedings of Sixth National Symposium on Refrigeration and Air Conditioning, IIT Bombay, 1978, pp. 115–124.

[25] V. Ganesan, Computer Simulation of Spark Ignition Engine Processes, Universities Press, Hyderabad, 1996.

[26] V. Ganesan, Computer Simulation of Compression Ignition Engine Processes, Universities Press, Hyderabad, 2000.

[27] Z. Gang, R. Chen, Total press loss mechanism of centrifugal compressors, Mech. Eng. Res. 4 (2) (2014) 45–59.

[28] B. Blunier, et al., A new analytical and dynamical model of a scroll compressor with experimental validation, Int. J. Refrig. 32 (2000) 874–891.

[29] E. Winandy, O.C. Saavedra, J. Lebrun, Experimental analysis and simplified modelling of a hermetic scroll refrigeration compressor, Appl. Therm. Eng. 22 (2) (2002) 107–120.

[30] J.P. Bourdhouxhe, et al., HVAC1 TOOLKIT: A Toolkit for Primary HVAC System Energy Calculation, ASHRAE, 1999.

[31] S.L. Lineykin, S. Ben-Yaakov, Modelling and analysis of thermoelectric modules, IEEE Trans. Ind. Appl. 43 (2) (2007) 505–512.

[32] D. Zhao, G. Tan, A review of the thermoelectric cooling: materials, modeling and applications, Appl. Therm. Eng. 66 (2014) 15–24.

[33] P. Dziurdzia, Modelling and Simulation of Thermoelectric Energy Harvesting Processes, Intechopen, 2011, www.intechopen.com/download/pdf/25370.

[34] Y. Jaluria, Design and Optimization of Thermal Systems, McGraw Hill, Singapore, 1998.

[35] M.A. Karim, M.N.A. Hawlader, Mathematical modelling and experimental investigation of tropical fruit drying, Int. J. Heat Mass Transfer 48 (2005) 4914–4925.

CHAPTER 5

System Simulation

A system, by definition, is a collection of components which are connected in such a manner that they influence each others' performance. System simulation thus implies predicting the performance of such a system for any set of operating conditions using the information about the design of its constituent components. Thus system simulation essentially consists of combining the models for predicting the performance of its various components into a comprehensive procedure which takes care of the interconnections between the components and also ensures that various conservation laws (like those of mass, momentum, energy, and species conservation) are satisfied by the whole system. We have already studied in Chapter 4 the methods for simulation of various typical components found in thermal systems. System simulation would thus involve combining these procedures suitably so that the performance of the overall system can be predicted for any set of values of the operating parameters.

Thus, for example, consider a simple vapor compression refrigeration system, Fig. 5.1. It has four components, namely, the compressor, the condenser, the expansion device, and the evaporator connected together to form a closed loop. Here each component influences the performance, not only of the component that follows it in the cycle, as one would normally expect, but even of that one that precedes it. Thus simulating this system involves combining the component simulation procedures so that all the interrelationships, and the conservation laws, are satisfied. For example, under steady-state conditions, the mass flow rate through all the components should be the same, and that becomes an important condition to be satisfied.

There is also a thermodynamic requirement, namely the net heat transferred in the cycle should be equal to the net work input in the cycle. Besides these, there also are other conditions arising from the obvious interconnections, viz. the state of the working fluid leaving a component is the same as that of the fluid entering the succeeding component. From a mathematical perspective, we could consider each component simulation model as nonlinear equations which enable prediction of the conditions of the outgoing stream, and the pertinent energy transfers (like work or heat transfer) occurring in the component, from the known conditions of the incoming stream. Thus the system simulation problem could be viewed as solution of all these nonlinear equations representing various components. Since often these nonlinear equations are

Thermal System Design and Simulation
http://dx.doi.org/10.1016/B978-0-12-809449-5.00005-X

297

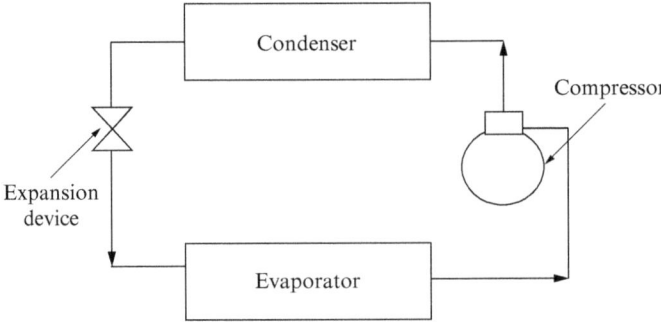

Fig. 5.1 A simple vapor compression refrigeration system.

in reality detailed process models involving solution of differential equations also (as we saw in Chapter 4), it is essential to evolve a suitable strategy for system simulation. Information flow diagrams (IFDs) are very helpful in this regard. This diagram, which is based on an understanding of the detailed process model of each component, is in the form of a black box, which indicates the minimum number of input variables necessary to obtain the desired output information. By suitably choosing the input and the output variables, and combining the individual component IFDs judiciously, it is possible to evolve strategies which significantly reduce the computational effort for system simulation. We shall be discussing these concepts in detail in this chapter.

5.1 INFORMATION FLOW DIAGRAM

In engineering numerous types of diagrams are used to represent various aspects of real systems. Some of these are schematic diagram, block diagram, flow diagram, process flow diagram, flow chart, circuit diagram, data flow diagram, indicator diagram, and various kinds of diagrams ($p-h$ diagram, $T-s$ diagram, $h-s$ diagram, exergy diagram, $p-h$ diagrams, etc.) encountered in thermodynamics. Fig. 5.1 could be called a schematic diagram since it represents various components by abstract geometrical symbols which do not resemble real parts of the system. The process flow diagrams, very popular in chemical engineering systems, use a host of standardized symbols for commonly encountered equipment. Fig. 5.1 could also be termed as a block diagram, since most components are represented by interconnected blocks showing the relationships of various blocks. IFDs are, however, quite different from these since these do not represent the physical configuration of the system, but represent its mathematical simulation procedure. Thus the IFD of a component is a block showing the input information needed in its simulation procedure and the output information that can be obtained there-from. Fig. 5.2 shows, for example, the schematic diagram, the exergy flow

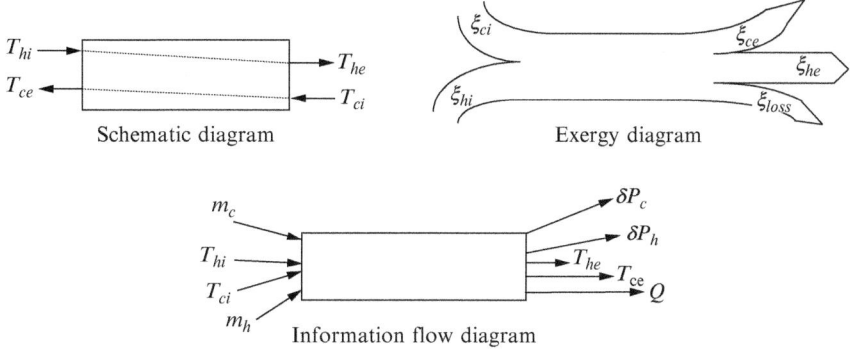

Fig. 5.2 Schematic, exergy, and information flow diagrams of a heat exchanger.

diagram, and the IFD of a typical counter flow heat exchanger. The schematic diagram indicates that a counterflow heat exchanger is an equipment in which two streams (one hot, properties indicated by subscripts h; another cold, properties indicated by subscript c) enter from opposite sides, and interact thermally (without mixing) before exiting from it. The exergy diagram shows the exergy (ξ) of the two streams at inlet (i) to the heat exchanger, and their exergies at exit (e) from the heat exchanger, and the exergy loss (ξ_{loss}) inside the heat exchanger due to irreversible nature of the heat-transfer process. This diagram thus gives us a quantitative picture of the thermodynamics of the process. The IFD is quite different from both of them. It indicates that if we know the mass flow rates and the temperatures of the two streams, we can determine their temperatures at exit from the heat exchanger and the drop in the pressure of the two streams (δP), from the known heat exchanger design (viz. the details of its configuration, the area of contact between the two fluids, etc.) by using the pertinent simulation procedure from Chapter 4. The diagram also indicates the other output information, viz. the total heat transfer Q, which can be obtained from the heat balance. The IFD of Fig. 5.2 can be said to represent the complex mathematical relationships between the outputs and the inputs. We could write these, in a general form as:

$$f_1(m_c, m_h, T_{ci}, T_{hi}, \delta P_h) = 0 \tag{5.1}$$

$$f_2(m_c, m_h, T_{ci}, T_{hi}, \delta P_c) = 0 \tag{5.2}$$

$$f_3(m_c, m_h, T_{ci}, T_{hi}, T_{he}) = 0 \tag{5.3}$$

$$f_4(m_c, m_h, T_{ci}, T_{hi}, T_{ce}) = 0 \tag{5.4}$$

$$f_5(m_c, m_h, T_{ci}, T_{hi}, Q) = 0 \tag{5.5}$$

Unlike the other two diagrams, the IFD of any equipment is obviously not unique, for it depends on the problem specification—what is the input information given and what are the output values sought. For example, Fig. 5.3 shows two other possible

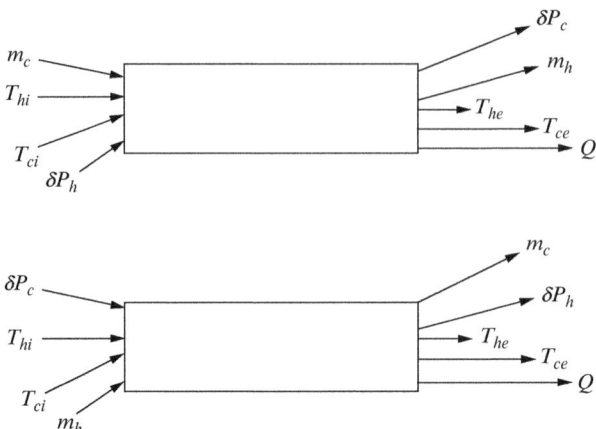

Fig. 5.3 Alternative information flow diagrams for a heat exchanger.

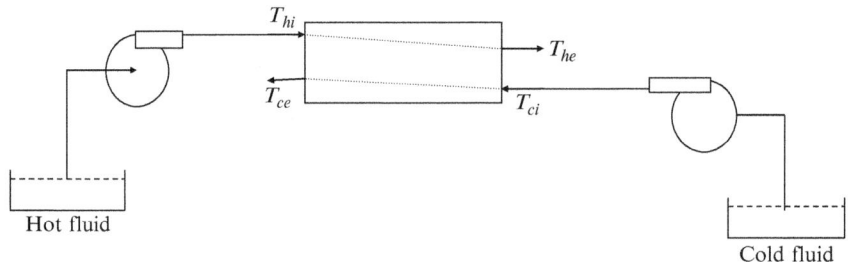

Fig. 5.4 A heat exchanger and its two pumps.

IFDs for a heat exchanger. In these two diagrams the information about the total pressure drop of one of the two fluids is provided and its mass flow rate is now an "output" sought from the simulation. Similarly, we could visualize other situations where some other variables are known and the rest are to be determined, and draw the IFDs for these. These blocks can also be conceived of as mathematical relationships like Eqs. (5.1)–(5.5).

The real utility of the IFD lies in evolving procedures for system simulation by identifying the input and output variables for each component based on their IFDs and combining them judiciously. To illustrate this, let us consider a simple example of a heat exchanger coupled to two pumps which circulate the cold and hot fluid through it (Fig. 5.4). To develop a method for simulation of this simple system, we need to identify a suitable IFD for the pump. The simulation of a pump considered in isolation can be perceived in at least two different ways, illustrated by the IFDs of Fig. 5.5. In

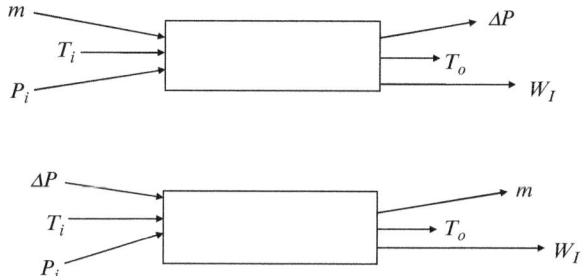

Fig. 5.5 Alternative information flow diagrams for a pump.

one perspective the mass flow rate of the fluid is known as also its inlet pressure. The quantities to be calculated are the pressure rise in the pump (usually called its pressure head) and the power input required. In another perspective shown in Fig. 5.5, the pressure head is specified as input information and the mass flow rate through the pump is the desired output information. These diagrams also can be visualized as representing mathematical relationships between various variables (cf. Eqs. 5.1–5.5):

$$f_a(m, P_i, T_i, W_i) = 0 \tag{5.6}$$

$$f_b(m, P_i, T_i, T_o) = 0 \tag{5.7}$$

$$f_c(m, P_i, T_i, \Delta P) = 0 \tag{5.8}$$

$$f_d(\Delta P, P_i, T_i, W_i) = 0 \tag{5.9}$$

$$f_e(\Delta P, P_i, T_i, T_o) = 0 \tag{5.10}$$

Eqs. (5.6)–(5.8) represent the functional relationships corresponding to the first of the two diagrams and Eqs. (5.8)–(5.10) represent the later diagram.

Now to evolve a method for simulation of the pump-heat exchanger system of Fig. 5.4, we now have to decide which of the IFDs of the heat exchanger and the pump should be picked up. To do this we first write the functional relationships for the system under consideration, taking into account the input information. Thus in the system of Fig. 5.4, the inlet pressure to both the pumps, and the pressure of both the fluids while leaving the heat exchanger is atmospheric, which would be known. Similarly, the temperatures of the two fluids at inlet to the pumps must also be known from the respective sump temperatures. The functional relationships for various components can now be written as:

Pump (hot fluid):

$$f_{ah}(m_{hp}, \Delta P_h) = 0 \tag{5.11}$$

$$f_{bh}(m_{hp}, W_{ih}) = 0 \tag{5.12}$$

$$f_{ch}(m_{hp}, T_{oh}) = 0 \tag{5.13}$$

Pump (cold fluid):

$$f_{ac}(m_{cp}, \Delta P_c) = 0 \qquad (5.14)$$

$$f_{bc}(m_{cp}, W_{ic}) = 0 \qquad (5.15)$$

$$f_{cc}(m_{cp}, T_{oc}) = 0 \qquad (5.16)$$

Heat exchanger:

$$f_1(m_c, m_h, T_{ci}, T_{hi}, \delta P_h) = 0 \qquad (5.17)$$

$$f_2(m_c, m_h, T_{ci}, T_{hi}, \delta P_c) = 0 \qquad (5.18)$$

$$f_3(m_c, m_h, T_{ci}, T_{hi}, T_{he}) = 0 \qquad (5.19)$$

$$f_4(m_c, m_h, T_{ci}, T_{hi}, T_{ce}) = 0 \qquad (5.20)$$

$$f_5(m_c, m_h, T_{ci}, T_{hi}, Q) = 0 \qquad (5.21)$$

Further since the fluids enter the heat exchanger directly after leaving the respective pumps, the mass flow rate of liquid through a pump must equal the mass flow rate of that stream in the heat exchanger, and if we neglect the small pressure and temperature drops in the connecting pipe lines, we can write following additional equations:

$$\delta P_h = \Delta P_h \qquad (5.22)$$

$$\delta P_c = \Delta P_c \qquad (5.23)$$

$$T_{oh} = T_{hi} \qquad (5.24)$$

$$T_{oc} = T_{ci} \qquad (5.25)$$

$$m_{hp} = m_h \qquad (5.26)$$

$$m_{cp} = m_c \qquad (5.27)$$

Thus the system simulation problem can be seen as a problem-involving simultaneous solution of the above 17 equations involving 17 variables, namely m_c, m_h, T_{ci}, T_{hi}, δP_h, δP_c, T_{he}, T_{ce}, Q, ΔP_h, W_{ih}, T_{oh}, ΔP_c, W_{ic}, T_{oc}, m_{hp}, and m_{cp}. If we write down explicitly the functional relationships of Eqs. (5.11)–(5.21), we will get a set of nonlinear (and a few linear) equations which can be solved using appropriate method Section 2.2.

However, we can simplify the calculations greatly, by using the IFDs of various components drawn above, representing these functional relationships, and combine them to get a IFD for the whole system. One such IFD for the system, using the IFD of Fig. 5.2 for the heat exchanger, is shown in Fig. 5.6. This diagram is a closed loop, indicating the need for simultaneous solution of the functional relationships. However, the diagram also suggests a method by which the number of simultaneous equations to be solved can be reduced, viz. by assuming the values of some of the variables, using the functional relations to find other variables, and then identifying the conditions necessary to close the loop. For example, it is evident from Fig. 5.2 that if we assume the values of four "input information" variables needed for simulation of the heat exchanger, namely

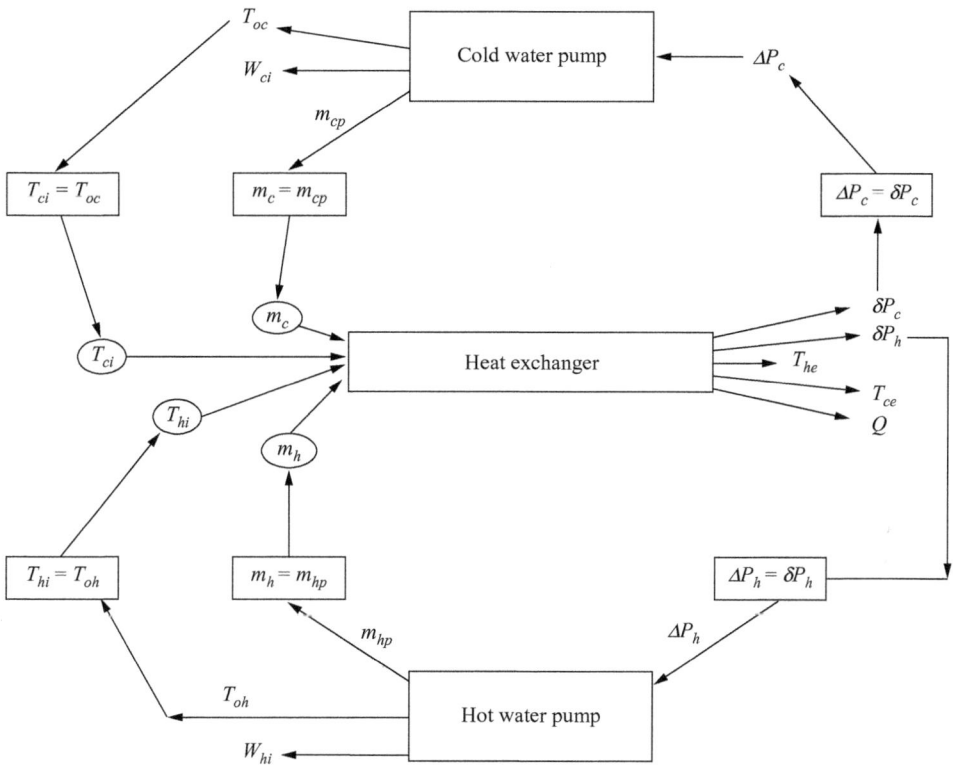

Fig. 5.6 Information flow diagram for the heat exchanger pump system.

m_c, m_h, T_{ci}, and T_{hi},[1] we can predict the values of the variables indicated as its output information, namely δP_h, δP_c, T_{he}, T_{ce}, and Q. Using the values of the two pressure drops, δP_h and δP_c, and using Eqs. (5.22), (5.23), we get the values of pressure heads developed by the two pumps. The pump simulation relationships then provide us with the values of the mass flow rates of the streams passing through them and their outlet temperatures. These calculated mass flow rates through the pumps must be equal to those assumed at the beginning of the calculations as the mass flow rates through the heat exchanger, Eqs. (5.26), (5.27). Similarly the calculated temperatures at the exit of the pumps must be equal to those assumed as temperatures of the two fluids entering the heat exchanger, Eqs. (5.24), (5.25). The whole system simulation procedure can thus be seen as: finding the values of four variables m_c, m_h, T_{ci}, and T_{hi} such that the four equations, (5.24), (5.27), termed compatibility conditions, are satisfied. In effect we

---

[1] Each assumed value has been shown in Fig. 5.6 inside an ellipse.

have thus reduced the problem of simultaneous solution of 17 equations for 17 variables to that of solution of only 4 equations for 4 variables.

It is evident that there could be alternate ways of combining/using the component IFDs, some of which may further reduce the computational effort. One such alternative scheme is shown in Fig. 5.7. Here, only two assumptions are made, namely those of ΔP_c and ΔP_h, the pressure heads developed by the two pumps. Using these as input information for the pumps, the flow rates through these (i.e., m_{hp} and m_{cp}), and the exit temperatures of the fluids pumped by them (i.e., T_{oh} and T_{oc}) are determined. Using these values in Eqs. (5.24)–(5.27), we get all the input information necessary for simulation of the heat exchanger. The heat transfer in the heat exchanger and the pressure drops of the two fluids (δP_h and δP_c) can thus be found out. According to Eqs. (5.22), (5.23) these should be equal to the pressure heads developed by the respective pump (ΔP_h and ΔP_c) assumed at the beginning of the calculations. The whole system simulation procedure is thus effectively reduced to finding the values of two variables ΔP_c and ΔP_h such that at the end of all calculations enumerated above, Eqs. (5.22), (5.23) are satisfied. In effect we have thus reduced the problem of simultaneous solution of 17 equations for 17 variables to that of solution of only 2 equations for 2 variables. The computational effort in system simulation is thus drastically reduced by judicious use of the IFDs. This is especially important in real industrial systems with a large number of components where direct mathematical enunciation of the problem may demand simultaneous solution of hundreds of complex nonlinear equations. The number of equations could be reduced by an order of magnitude by

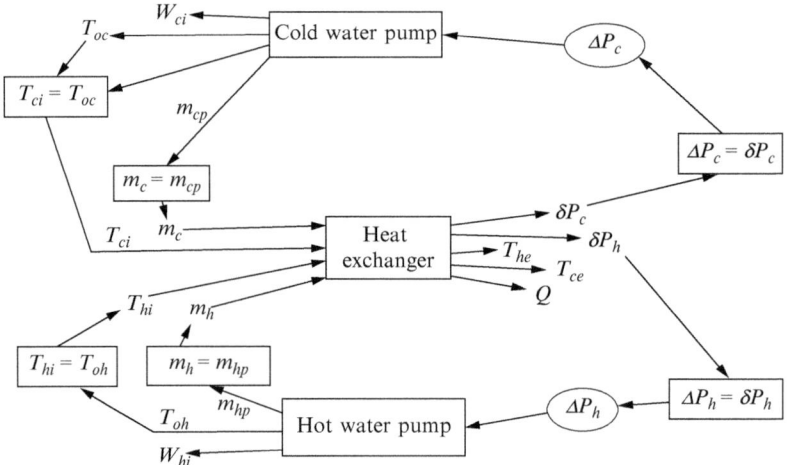

Fig. 5.7 Alternate IFD for the heat exchanger pump system.

evolving suitable strategy with the help of information flow diagrams. We shall see such case studies in Chapter 6.

In this chapter we shall study the application of IFDs with the help of a few more simple examples.

Example 5.1

Fig. 5.8 shows the condenser cooling water circuit of a steam power plant. Indicate, with the help of an IFD, how can we find the steady-state condenser pressure, and other operating conditions, from the given design of various components for certain environmental conditions and prescribed steam mass flow rate.

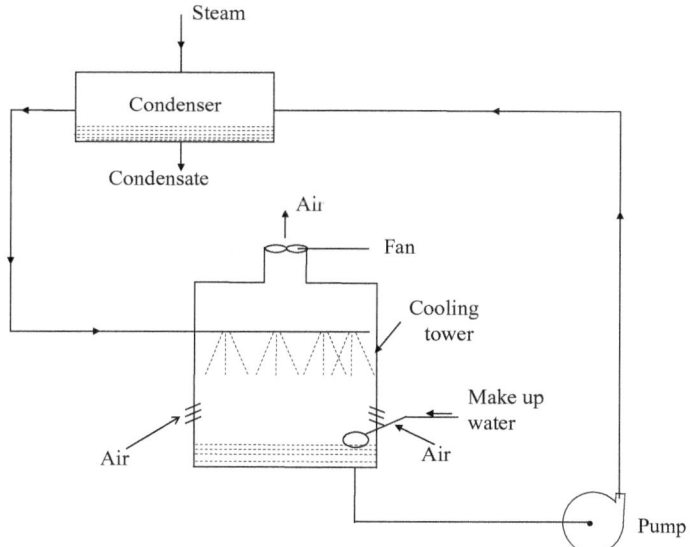

Fig. 5.8 Steam condenser cooling water circuit.

Solution

This system consists of three components, namely the condenser (basically a heat exchanger), the cooling tower (basically a heat-mass exchanger), and the cooling water pump. We have already seen above various possible IFDs for each of these equipment (for the cooling tower, the diagram would be similar to that of the heat exchanger, with additional input variable of air-specific humidity, and an output variable of the make-up water requirement). However, it would be helpful to introduce a minor change in these. Rather than defining the water pressure drop/rise as a distinct variable, we use the actual pressures of various streams as variables. This is helpful because the pressure inside the cooling tower is atmospheric, and hence known a priori; it thus reduces the number of variables needed in the simulation. Neglecting the pressure drop in the connecting piping we can combine the IFDs of the three components as shown in Fig. 5.9. In this diagram we have indicated the variables, whose values are externally input by a *rectangular box* (I), and the three variables whose values need to be assumed to start the solution procedure are enclosed in an *elliptical boxes*.

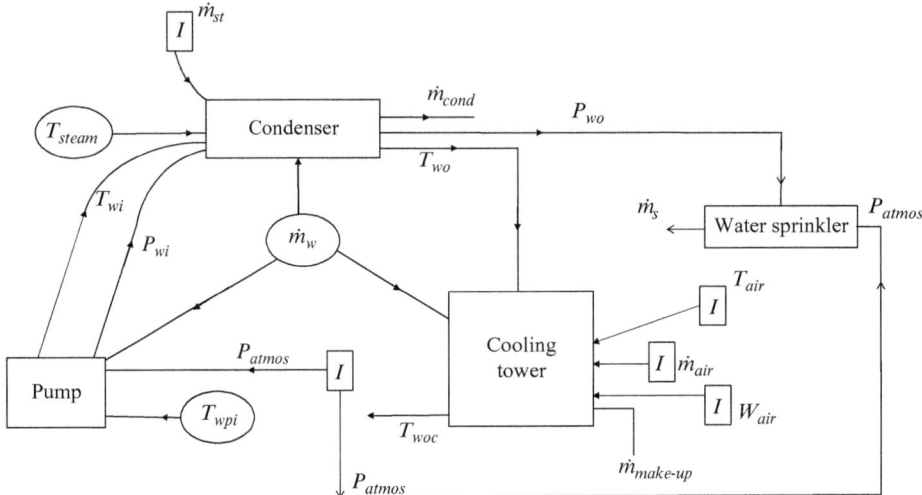

Fig. 5.9 Information flow diagram for steam condenser cooling water circuit.

The solution procedure could start from the pump. Using the assumed values of the water mass flow rate ($\dot{m}_w$) and its inlet temperature to the pump (T_{wpi}), the pressure and the temperature of the water leaving the pump can be determined. Using these values along with the assumed steam temperature (or pressure) and the prescribed steam mass flow rate ($\dot{m}_{st}$), the condenser simulation can be done and the mass flow rate of the condensate, and the temperature and the pressure of water leaving the condenser can be determined. These form the input to the cooling tower simulation, along with the assumed water mass flow rate, the specific environmental conditions and the air flow rate. The cooling tower simulation then gives us the water outlet temperature, and the make-up water requirement. The flow rate of water sprayed into the cooling tower m_s is governed by the design of the sprinkler and the pressure drop across it. The three compatibility conditions checking the correctness of the assumed values are

$$\dot{m}_{cond} = \dot{m}_{st}$$

$$T_{woc} = T_{wpi}$$

$$\dot{m}_w = m_s$$

Thus the simulation of this system is effectively reduced to a mathematical problem of solving three nonlinear equations for three unknown variables assumed in the beginning, namely $\dot{m}_w$ (the cooling water mass flow rate), T_{steam} (the condenser), and T_{wpi}, the pump water inlet temperature.

The students are advised to write all the equations for simulation of the three components and verify that the direct solution of these equations would have involved at least seven variables and seven nonlinear equations.

Example 5.2
Fig. 5.10 shows the schematic diagram of a biogas cleaning and compression system. The gas generated in the biogas plant at 1.1 bar is first pressurized to 10 bar in the LP compressor. It is then passed through a gas cleaner where CO_2 is scrubbed by a spray of water. The enriched biogas leaving the cleaner is then compressed to 100 bar in HP compressor and sent to cylinder for storage. Both the compressors are operated by engines working on a part of the raw biogas tapped from the main header.

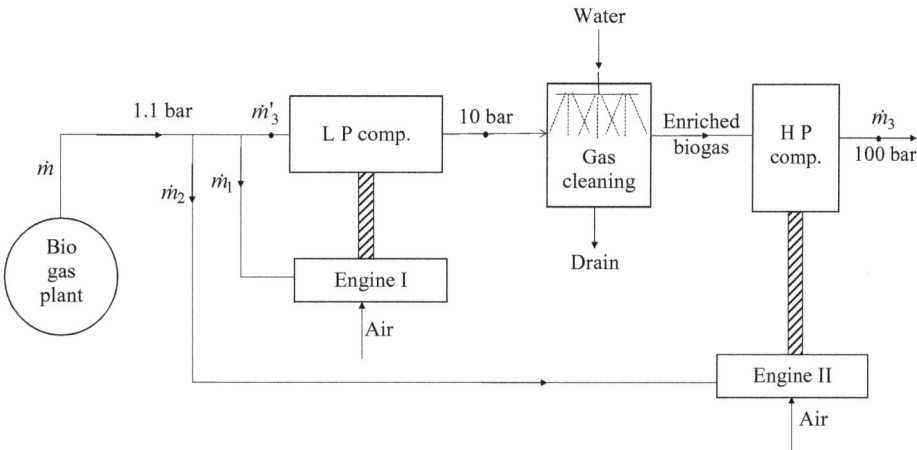

Fig. 5.10 Biogas cleaning and compression system.

Show with the help of an IFD, how can the mass flow rates of various streams be determined for a specified value of the mass flow rate drawn from the biogas plant.

Solution
Since the task is to estimate the various mass flow rates, it is evident that by assuming values of $\dot{m}_1$ and $\dot{m}_2$, it should be possible to evolve a solution strategy whereby the correctness of these value is checked and an iterative procedure set up to arrive at the converged solution. However, it is also possible to solve the problem by making only one assumption, as is brought out in the IFD shown in Fig. 5.11.

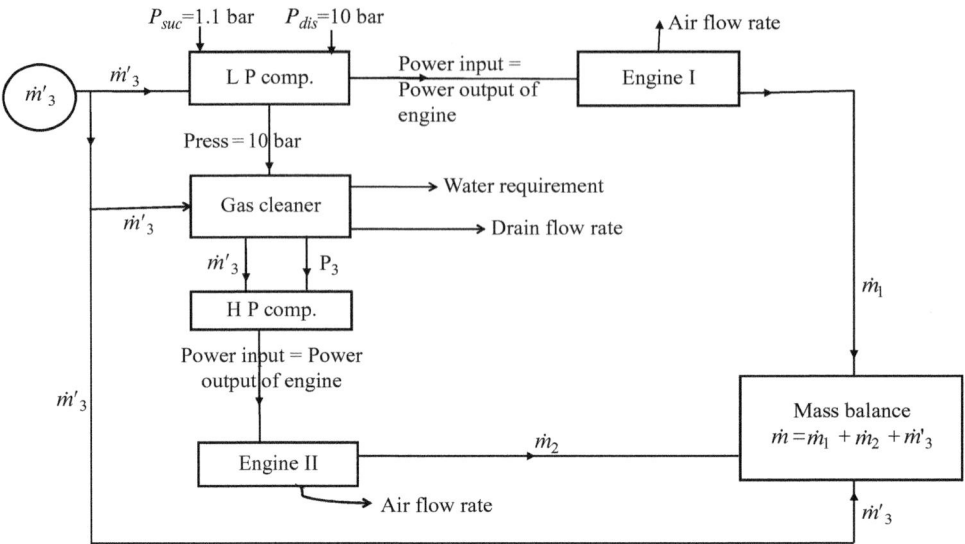

Fig. 5.11 Information flow diagram for bio-gas cleaning and compression system.

Once the mass flow rate of gas entering the LP compressor, $\dot{m}'_3$ is assumed, its analysis can be done since both the inlet and the exit pressure are known. The power required to compress the gas can be determined from the compressor characteristics. This should be the power output of the engine *I* driving this compressor. Knowing the engine output, its fuel requirement (i.e., $\dot{m}_i$) can be estimated from the engine characteristics. Proceeding again along the gas stream leaving the LP compressor its pressure at the exit of the gas cleaner, and the mass flow rate leaving the cleaner can be determined on the basis of cleaner characteristics. The HP compressor characteristics will then enable us to determine the power requirement for compressing the gas to 100 bar. Now, since this power has to come from the engine II, the gas mass flow rate $\dot{m}_2$ can be found from the engine characteristics. The mass balance equation gives us a check on the correctness of the initial assumption of $\dot{m}'_3$. Thus the simulation problem is reduced to iterating for the value of only one variable.

Example 5.3

The bagasse from a sugar mill (Fig. 5.12) is being used as a fuel in its captive steam power plant. Explain, with the help of an IFD how can one determine the power output and the operating conditions of the steam power plant using the following input information:

- capacity of sugar mill (in tonnes/day of sugar);
- bagasse production per kilogram of sugar;

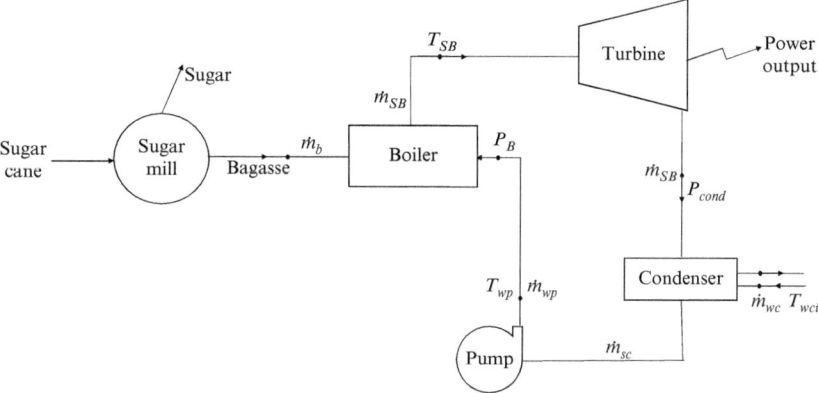

Fig. 5.12 Captive power plant of a sugar mill.

- water flow rate and the temperature at the entry to the condenser; and
- calorific value of bagasse.

The performance characteristics of the turbine and the pump are given, as also the detailed design of the boiler and the condenser.

Solution

The analysis of the sugar mill directly gives us the bagasse production rate as

$\dot{m}_b =$ (Capacity of sugar mill in kilogram per hour) $\times$ (Bagasse production per kilogram of sugar)

The steam power plant being a closed loop, we need to analyze each component carefully, recognizing its interdependence with the component that precedes it, as also the one that follows it. To illustrate the utility of IFD, we shall develop the mathematical formulation first.

Starting from the boiler we note that the state of steam leaving it—whether it is dry saturated or superheated—needs to be specified. In the later case the "boiler" has to be analyzed in two parts, namely the boiler proper, which produces saturated steam, and the superheater. For the sake of simplicity, we take the boiler as producing saturated steam at pressure P_B. Its temperature T_{SB} is therefore known from the thermodynamic properties of stream. Thus we can write

$$T_{SB} = f_i(P_B)$$

Further, the boiler is basically a heat exchanger where water pumped into it at temperature T_{wp} and pressure P_B is heated and converted into steam, the energy for this coming from burning bagasse. The heat-transfer rate in the boiler Q_B can thus be mathematically related to various operated parameters as:

$$Q_B = f_2(\dot{m}_b, T_{flame}, \dot{m}_{wp}, P_B, T_{wp})$$

where T_{flame}, the temperature of burning bagasse is dependent on its moisture content and chemical composition, and can be taken as an input information, and $\dot{m}_{wp}$ is the given mass flow rate of water sent into boiler by the pump at temperature T_{wp}. This heat-transfer rate can be directly related to the mass flow rate of steam leaving the boiler:

$$Q_B = \dot{m}_{SB} \times h_{fg}$$

where h_{fg}, the latent heat of evaporation at the boiler pressure, can be determined from the thermodynamic property table.

$$h_{fg} = f_3(P_B)$$

The power output from the turbine and the condition of steam at the exit can be obtained from its characteristics, as a function of steam mass flow rate, its enthalpy, and the exit pressure, which is the condenser pressure, that is

$$W_{turb} = f_4(\dot{m}_{SB}, T_{SB}, P_B, P_{cond})$$
$$h_{exit} = f_5(T_{SB}, P_B, P_{cond})$$

where h_{exit} is the enthalpy of steam at turbine exit. The steam condenser being another heat exchanger we can relate the heat transfer in it as:

$$Q_{cond} = f_6(\dot{m}_{SB}, P_{cond}, h_{exit}, \dot{m}_{wc}, T_{wci})$$

Here the condenser cooling water flow rate $\dot{m}_{wc}$, and its temperature at entry to the condenser, T_{wci}, are input information. The condensate flow rate can be determined from this heat transfer as:

$$\dot{m}_{sc} = \frac{Q_{cond}}{h_{exit} - h_{cond}}$$

where h_{cond}, the condensate enthalpy is a function of the condenser pressure only:

$$h_{cond} = f_7(P_{cond})$$

The pump characteristics determine the mass flow rate of water delivered by it to the boiler as also it exit temperature as a function of the suction and discharge pressures:

$$\dot{m}_{wp} = f_8(p_{cond}, P_B) \quad T_{wp} = f_9(P_{cond}, P_B)$$

Since steady-state operation demands that the mass flow rate circulating through various components of the steam power plant be the same, we have

$$\dot{m}_{wp} = \dot{m}_{SB} = \dot{m}_{sc}$$

We can thus see that the simulation problem reduces to a mathematical problem of solving 14 equations given above for variables, namely $\dot{m}_b$, T_{SB}, P_B, $\dot{m}_{sp}$, T_{wp}, Q_B, W_{tube}, h_{fg}, $\dot{m}_{SB}$, P_{cond}, h_{exit}, Q_{cond}, $\dot{m}_{sc}$, and h_{cond}; the rest of the variables involved in these equations (such as T_{flame}, $\dot{m}_{wc}$, T_{wci}) being input information specified by the problem. Rather than solving these 14 equations simultaneously considerable simplification can be done by preparing a suitable IFD for the system, like the one shown in Fig. 5.13.

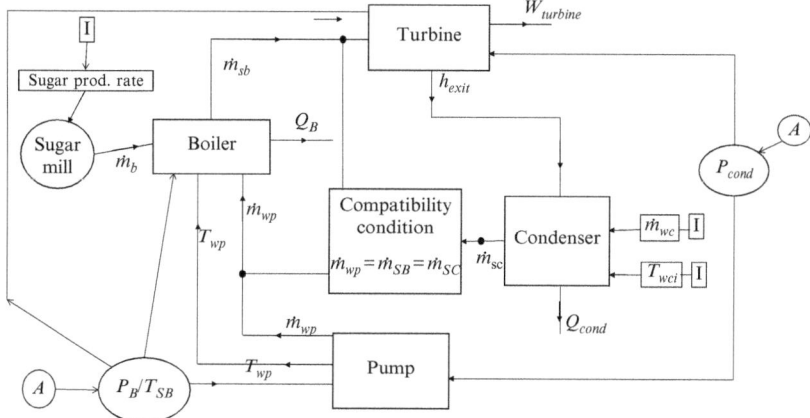

Fig. 5.13 Information flow diagram for captive power plant of a sugar mill.

As can be seen from Fig. 5.13, assuming the boiler and the condenser pressures, it should be possible to go around the whole power plant and simulate the performance of each component. We start calculations from the pump, and for the assumed values of P_{cond} and P_B, determine the water mass flow rate $\dot{m}_{wp}$ and its temperature T_{wp}.

The bagasse flow rate is directly known from the sugar mill production data, and thus all the information needed for the boiler simulation is now available. The steam flow rate leaving the boiler $\dot{m}_{SB}$ can thus be found. The turbine simulation is done next to get the net turbine output and the steam exit condition, which in turn is fed as input to the condenser simulation procedure. The condensate flow rate $\dot{m}_{sc}$ can thus be found. The compatibility condition, namely equality of the three mass flow rates calculated independently (in the boiler, the condenser, and the pump) provides the necessary check about the correctness of the assumed values of the two pressures. Thus the system simulation problem is effectively reduced to the solution of the two compatibility equations, which can be seen as complicated function of the condenser and the boiler pressures. Thus, instead of 14 equations, we need to solve only 2 equations simultaneously, to complete the system simulation.

5.2 SOLUTION METHODOLOGY

As indicated above through different examples all system simulation problems can be finally transformed into solution of complex "nonlinear" equations. IFDs can help in identifying strategies by which the number of these "nonlinear" equations can be reduced. The methods of solving such equations discussed in Section 2.2 can then be used to get the solution. In this section we shall illustrate the use of these methods through typical simple examples.

Example 5.4

Fig. 5.14 shows the schematic diagram of an indirect heating system where hot flue gases are being used to heat process air indirectly through a heat-transfer oil. Determine the steady-state operating conditions of the system using information given below:

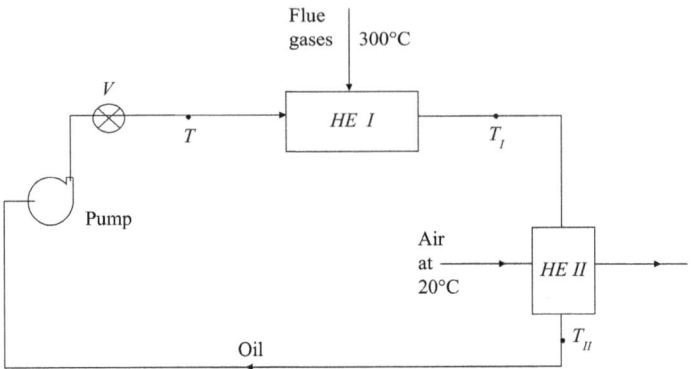

Fig. 5.14 Indirect heating system.

1. Flue gas flow rate: 20 kg/min
2. Air flow rate: 15 kg/min
3. Specific heat of air/flue gas: 1.0 kJ/kg°C
4. Average specific heat of heat-transfer oil: 2.5 kJ/kg in the likely temperature range
5. Oil pressure drops:

$$HE\ I = 1.2m^{1.8}\ kPa$$

$$HE\ II = 1.3m^{1.7}\ kPa$$

where m is the mass flow rate of oil in kilogram per minute
Pump head-flow characteristic:

$$\Delta P(\text{in kPa}) = 100 - 5m - 0.5m^2$$

6. Oil temperature rise in pump (in °C) $= 5 + (m/5)^{1.3}$
   HE effectiveness:

$$\text{Effectiveness of } HE\ I = 1 - e^{-(m/10)^{0.8}}$$

$$\text{Effectiveness of } HE\ II = 1 - e^{-(m/10)^{0.7}}$$

Solution

The main task in the simulation of the system is to determine the oil mass flow rate. This will be governed by the ability of the pump to circulate oil through the heat exchangers and the interconnecting pipes. Neglecting the later pressure drop and that in the control valve, we can write the pressure-drop balance equation as:

$$\Delta P = 100 - 5m - 0.5m^2 = 1.2m^{1.8} + 1.3m^{1.7}$$

which can be rearranged as

$$f(m) = 100 - 5m - 0.5m^2 - 1.2m^{1.8} - 1.3m^{1.7} = 0$$

To solve this nonlinear equation we could use any of the methods discussed in Section 2.2. Let us use the Newton-Raphson method, with a starting assumption of $m = 5$ kg/min. The various steps of solution are given in detail below:

$$f(m) = 100 - 5m - 0.5m^2 - 1.2m^{1.8} - 1.3m^{1.7}$$

$$f'(m) = -5 - 2 \times 0.5m - 1.2 \times 1.8m^{0.8} - 1.3 \times 1.7m^{0.7}$$

$$= -5 - m - 2.16m^{0.8} - 2.21m^{0.7}$$

First assumption $m = 5$ $f(m) = 20.7$ $f'(m) = -24.65$

Next assumption $m = 5 - \dfrac{f(m)}{f'(m)} = 5 + \dfrac{20.7}{24.65} = 5.84$

$$f(m) = -1.12 \quad f'(m) = -27.3$$

Next assumption $m = 5.84 - \dfrac{1.12}{27.3} = 5.799$ kg/min

$$f(m) = -0.0039 \approx 0$$

Hence the solution to the nonlinear equation is $m = 5.799$ kg/min.

To find the system operating conditions, we first determine the values of effectiveness of the two heat exchangers.

Heat exchanger I $(HE\ I):\ \epsilon_I = 1 - e^{-(m/10)^{0.8}} = 1 - e^{-(0.5799)^{0.8}} = 0.4762$

Heat exchanger II $(HE\ I):\ \epsilon_{II} = 1 - e^{-(m/10)^{0.7}} = 1 - e^{-(0.5799)^{0.7}} = 0.4948$

To apply the definition of heat exchanger effectiveness, Eq. (4.22), we first need to find the heat-capacity values of all the fluids. Thus

$$C_{oil} = (m \times Cp)_{oil} = 5.799 \times 2.5 = 14.498 \text{ kJ/min}\,°C$$

$$C_{air} = (m \times Cp)_{air} = 15 \times 1 = 15 \text{ kJ/min}\,°C$$

$$C_{fluegas} = (m \times Cp)_{fluegas} = 20 \times 1 = 20 \text{ kJ/min}\,°C$$

Thus in Eq. (4.22), $C_{min} = C_{oil} = 14.498$ kJ/min $°C$.

Indicating by symbol T the temperature of the oil entering HE I, and by T_I and T_{II}, the temperature of the oil leaving the HE I and HE II, respectively, we get using Eq. (4.23);

$$\epsilon_I = \frac{T_I - T}{300 - T} = 0.4762$$

$$\epsilon_{II} = \frac{T_I - T_{II}}{T_I - 20} = 0.4948$$

Further, using the given equation for temperature rise in the pump, we get

$$T - T_{II} = 5 + (5.799/5)^{1.3} = 6.2°C$$

The above three linear equations in three variables T, T_I, and T_{II} can be easily solved by substitution of variables in terms of the other.

Thus substituting the value of T from the above equation, in terms of T_{II} in the earlier two equations we get two linear equations for variables T_I and T_{II}, which can be easily solved by usual procedure. The final result which we get is

$$T = 120°C \quad T_I = 205.7°C \quad T_{II} = 113.8°C$$

Example 5.5

Fig. 5.15 shows the schematic diagram of a simple refrigeration plant where the saturated condensate leaving the condenser is subcooled with the help of dry-saturated vapor leaving the evaporator. Determine its steady-state operating conditions using the following design data:

Evaporator: Coolant flow rate: 2.0 kg/s; inlet temperature: 10°C; Effectiveness: 0.7; specific heat: 4.0 kJ/kg K

Condenser: Coolant flow rate: 1.0 kg/s; inlet temperature: 30°C; Effectiveness: 0.8; specific heat: 4.2 kJ/kg K

Subcooler: Effectiveness: 0.3

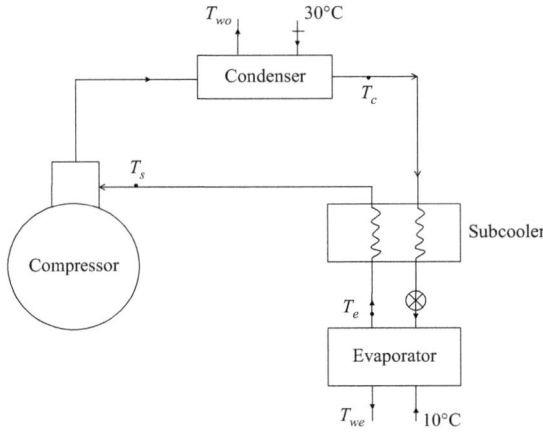

Fig. 5.15 Refrigeration system.

The *compressor* characteristic equations are

Refrigeration capacity: $Q_{comp} = 20.0 + 5.0T_e - 0.02T_e \cdot T_s - 0.3T_c - 0.02T_eT_c$
Compressor power: $P = 1.2T_c - 3.0T_e - 0.004T_c^2 - 0.01T_cT_s$
Both Q_e and P in the above equations are in kilowatt.

Solution

We shall first develop the IFD for the system in order to evolve the basic strategy for system simulation. The refrigeration system, Fig. 5.15, consists of three heat exchangers, a compressor, and an expansion device. Since no information about the expansion device is given, it follows that the device is of the type which automatically adjusts to varying mass flow rates. Thus for system simulation we can ignore the expansion device. The possible IFDs for the heat exchangers have already been discussed previously and we could use these directly. However, since the effectiveness of the heat exchangers is given, we can simplify the IFDs for the three heat exchangers as shown in Fig. 5.16.

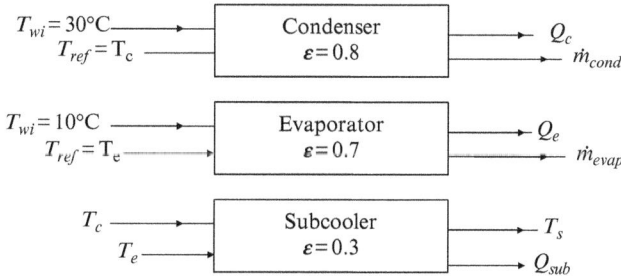

Fig. 5.16 Information flow diagrams for the heat exchangers.

The refrigerant compressor raises the pressure of vapors. From a purely energetic perspective its function is similar to that of a pump, the only difference being in the state of the working fluid. Hence, its IFD is similar to that of a liquid pump, Fig. 5.5. Fig. 5.17 shows the IFD of compressor most suited for our task. Here the pressures of the incoming and the outgoing stream are directly related to the saturation temperatures and so the two inputs variables could either be the evaporator and the condenser pressure (P_{evap} and P_{cond}) or the corresponding saturation temperatures T_e and T_c. Since the IFDs of other components involve the two temperatures we shall prefer the use of temperatures in the IFD of the compressor.

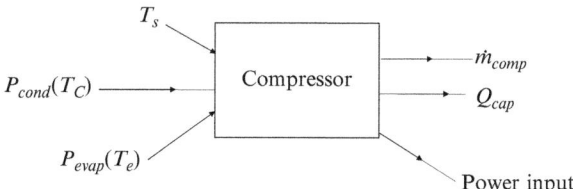

Fig. 5.17 Information flow diagram for a refrigeration compressor.

The four-component IFDs can be combined to evolve a strategy for system simulation. Fig. 5.18 shows one such IFD for the complete system. As is evident from the diagram, by making two assumptions about variables T_c and T_e, the simulation of all the components can be done in the following sequence: Subcooler → compressor → condenser → evaporator. The compatibility conditions are

$$\dot{m}_{cond} = \dot{m}_{evap} = \dot{m}_{compressor}$$

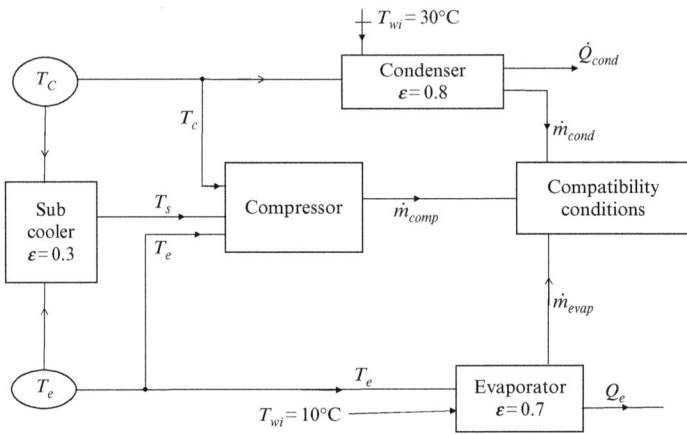

Fig. 5.18 Information flow diagrams for refrigeration system simulation.

Thus the task of system simulation is effectively reduced to solution of these two equations (in which the terms involved are complex functions of T_c and T_e) for the two variables T_c and T_e.

However, there is a difficulty in the use of the above compatibility conditions. The data provided in the problem enable us to calculate the heat transfers in the evaporator and the condenser, and the "cooling" capacity of the compressor. To calculate the three mass flow rates from these heat transfers, we need to know the corresponding enthalpy differences. But these enthalpy differences depend upon the unknown temperature T_c and T_e and so are not known a priori. Of course, theoretically speaking, it should be possible to express the enthalpy differences in terms of T_c and T_e by using the thermodynamic property tables for the refrigerant and fitting equation into the computed results. To avoid this complication, we change the compatibility conditions in terms of these heat transfers as:

$$Q_{evap} = Q_{comp} \text{ (from compressor)}$$
$$Q_{evap} + P_{(compressor)} = Q_{cond}$$

The first equation is equivalent to equating the refrigerant mass flow rate in the evaporator and the compressor. The second equation results from the application of the first law of thermodynamics to the whole system, and can be seen as equivalent to equating the mass flow rates of the evaporator (and the compressor) to that in the condenser. Now, since all the terms in these equations can be directly expressed in terms of T_c and T_e, system simulation would entail solving two nonlinear equations by using the given information about the various components of the refrigeration system.

Now, from the given value of effectiveness of the evaporator we can write the following equation.

$$\epsilon = 0.7 = \frac{T_{wi} - T_{we}}{T_{wi} - T_e} = \frac{10 - T_{we}}{10 - T_e}$$

where T_{we} is the temperature of the "coolant" at evaporator exit.

This gives

$$T_{we} = 3 + 0.7T_e$$
$$\text{and} \quad Q_e = 2.0 \times 4.0 \times (10 - T_{we}) = 8(7 - 0.7T_e)$$
$$= 56 - 5.6T_e \text{ (kW)}$$

Equating it to the equation for cooling capacity of the compressor, we get

$$56 - 5.6T_e = 20.0 + 5.0T_e - 0.02T_eT_s - 0.3T_e - 0.02T_eT_c$$

Besides T_e and T_c, this equation has the variable T_s, that is, the temperature of vapors entering the compressor after getting "heated" in the subcooler. We need to express this also in term of T_c and T_e. This can be done using the IFD for the subcooler, which gives (since vapor stream has lower heat capacity)

$$\epsilon = 0.3 = \frac{T_s - T_e}{T_c - T_e}$$
$$\Rightarrow T_s = 0.3T_c + 0.7T_e$$

Substituting this in the above equation we get

$$f_1(T_c, T_e) = 10.6T_e - 36 - 0.3T_c - 0.026T_eT_c - 0.014T_e^2 = 0$$

This is the first equation involving T_c and T_e resulting from the "modified" compatibility condition.

To develop the second equation, we first use the IFD of the condenser to find the heat transfer Q_{cond}. Since its effectiveness is given as 0.8, we get the equation:

$$\epsilon = 0.8 = \frac{T_{wc} - 30}{T_c - 30}$$

where T_{wc} is the temperature of the coolant leaving the condenser. Rearranging terms, we get

$$T_{wc} = 0.8T_c + 6$$

The heat transfer in the condenser is

$$Q_{cond} = 1 \times 4.2 \times (T_{wc} - 30)$$
$$= 4.2(0.8T_c - 24) = 3.36T_c - 100.8 \text{ kW}$$

The second "modified" compatibility condition can now be written as:

$$Q_{cond} - Q_{evap} = P_{compressor}$$

which gives

$$(3.36T_c - 100.8) - (56 - 5.6T_e) = 1.2T_c - 3.0T_e - 0.004T_c^2 - 0.01T_cT_s$$

Again expressing T_s in terms of T_c and T_e, as done above, we get

$$f_2(T_c, T_e) = 2.16T_c - 156.8 + 8.6T_e + 0.004T_c^2 + 0.01T_c(0.3T_c + 0.7T_e) = 0$$

Combining various terms, we can write the two equation f_1 and f_2 as:

$$f_1(T_c, T_e) = 10.6T_e - 36 - 0.3T_c - 0.026T_eT_c - 0.014T_e^2 = 0$$
$$f_2(T_c, T_e) = -156.8 + 2.16T_c + 8.6T_e + 0.007T_c^2 + 0.007T_cT_e = 0$$

These equations can be solved by using any of the methods discussed in Section 2.2. We shall illustrate the use of Warner's method here. Since there are two variables, we need $2 + 1 = 3$ sets of assumptions for T_c and T_e to start the procedure. We know from thermodynamic considerations that $T_c > 30°C$ and $T_e < 10°C$. Thus we can easily select three sets of values of T_c and T_e, and determine the values of functions f_1 and f_2 for these values (using Excel software) as given in following table.

T_c	T_e	f_1	f_2
45	2	−30.696	−27.595
40	4	−9.984	−23.68
40	6	8.856	−5.92

Thus following the notations of Warner's method (Section 2.2.1) we have

$$[\rho] = \begin{bmatrix} -30.696 & -27.595 & 1 \\ -9.984 & -23.68 & 1 \\ 8.856 & -5.92 & 1 \end{bmatrix} \quad \text{and} \quad [\chi] = \begin{bmatrix} 45 & 2 \\ 40 & 4 \\ 40 & 6 \end{bmatrix}$$

To get the new assumption for variables, we have to use Eq. (2.29). For this we need to get inverse of the function matrix $[\rho]$. This can be obtained easily using online matrix calculator[a] as:

$$[\rho]^{-1} = \begin{bmatrix} -0.06 & 0.074 & -0.013 \\ 0.064 & -0.134 & 0.07 \\ 0.914 & -1.449 & 1.535 \end{bmatrix}$$

The matrix containing the new estimate for the values of the variables can now be obtained by using Eq. (2.29):

$$[\psi] = [\rho]^{-1}[\chi] = \begin{bmatrix} -0.06 & 0.074 & -0.013 \\ -0.064 & 0.134 & 0.07 \\ 0.914 & -1.449 & 1.535 \end{bmatrix} \begin{bmatrix} 45 & 2 \\ 40 & 4 \\ 40 & 6 \end{bmatrix}$$

The values (c_1 and c_2) in the last row of $[\psi]$ matrix containing the new estimates of variables, can be calculated from above using matrix multiplication rules as (cf. Eq. 2.28):

$$c_1 = 0.914 \times 45 - 1.449 \times 40 + 1.535 \times 40 = 44.57$$
$$c_2 = 0.914 \times 2 - 1.449 \times 4 + 1.535 \times 6 = 5.242$$

For these values of variables x_1 and x_2, the function values are $f_1 = -0.265$ and $f_2 = 0.0932$.

Low values of the functions indicate that the correct solution is quite near the c_1 and c_2 values obtained in the first iteration itself. To get a better estimate following Warner's method, we replace the "worst" among the earlier assumed values, by this new set of values of the variables. The "worst" set is identified on the basis of the criterion of Eq. (2.30). Here the values of the "overall error" for the three sets of values of variables assumed initially are

$$T_1 = 30.696 + 27.595 = 58.291$$
$$T_2 = 9.984 + 23.68 = 33.664$$
$$T_3 = 8.856 + 5.92 = 14.776$$

clearly set 1 has the largest error, and so in the next iteration we replace it by the new values of the variables obtained above in the first iteration. This gives

$$[\rho] = \begin{bmatrix} -0.265 & 0.0932 & 1 \\ -9.984 & -23.68 & 1 \\ 8.856 & -5.92 & 1 \end{bmatrix}$$

and

$$[\chi] = \begin{bmatrix} 44.57 & 5.242 \\ 40.0 & 4 \\ 40 & 6 \end{bmatrix}$$

We now carry out the second iteration by first finding the inverse of matrix $[\rho]$ and multiplying that with variable matrix $[\chi]$. This gives

$$[\rho]^{-1} = \begin{bmatrix} -0.065 & 0.022 & 0.086 \\ 0.068 & -0.033 & -0.035 \\ 0.977 & -0.003 & 0.026 \end{bmatrix}$$

and $[\psi] = [\rho]^{-1}[\chi]$ gives

$$c_1 = 0.977 \times 44.57 - 0.003 \times 40 + 0.026 \times 40 = 44.465$$
$$c_2 = 0.977 \times 5.242 - 0.003 \times 4 + 0.026 \times 6 = 5.265$$

The values of functions f_1 and f_2 for this set of values of variables x_1 and x_2 are

$$f_1 = -0.0054 \quad \text{and} \quad f_2 = 0.00211$$

Since these values are ≈ 0, we accept the latest c_1 and c_2 values as the values as the final solution, that is

$$T_c \equiv c_1 = 44.465°C$$
$$T_e \equiv c_2 = 5.265°C$$

The values of various heat transfers can be found by substituting these values of T_c and T_e in the equations developed above. This gives

$$Q_e = 8(7 - 0.7 \times 5.265) = 26.5 \text{ kW} \quad Q_c = 3.36 \times 44.465 - 100.8 = 48.6 \text{ kW}$$
$$T_s = 0.3 \times 44.465 + 0.7 \times 5.265 = 17.025°C$$

We can verify the correctness of the solution by calculating refrigeration capacity from the compressor characteristics. This give

$$Q_{comp} = 20 + 5 \times 5.265 - 0.02 \times 5.265 \times 17.025 - 0.3$$
$$\times 44.465 - 0.02 \times 5.265 \times 44.465$$
$$= 26.51 \text{ kW}$$

Similarly, $\quad P_{comp} = 1.2 \times 44.465 - 3.0 \times 5.265 - 0.004 \times 44.465^2 - 0.01$
$$\times 17.025 \times 44.465 = 22.08 \text{ kW}$$

and $Q_{comp} + P_{comp} = 48.59$ kW, which matches with the Q_{cond} calculated above from the condenser performance characteristics.

[a] For example, at www.bluebit.gr/matrix-calculator/default.aspx.

Example 5.6

Fig. 5.19 shows the system for heating of water from a pond with the help of hot flue gases available at some distance from the pond. Determine the steady-state operating conditions of the system from the following design information.

Pipe Dimensions

Pipe	Length (m)	Diameter (mm)
1–2	100	50
2–3	50	30
4–5	50	30
5–6	100	50

Heat Exchanger Design Details

Both the heat exchangers have water flowing through 15 tubes of 10 mm diameter each and a total effective length of 50 m, in a direction opposite to that of the flue gases.

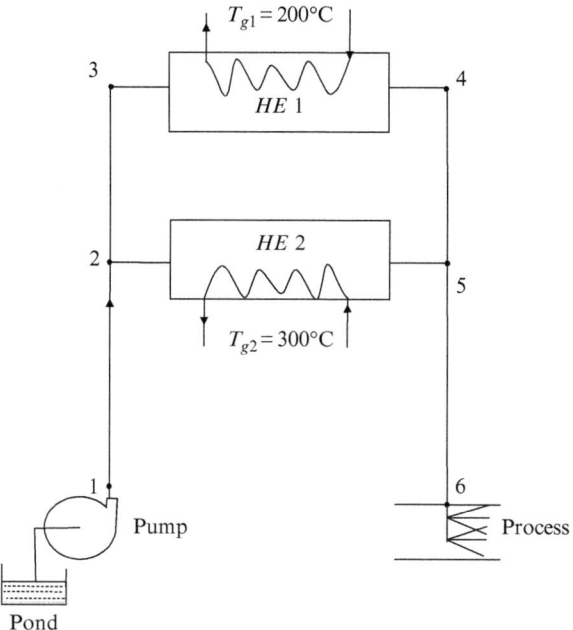

Fig. 5.19 Water heating system.

The heat-transfer coefficient on the flue gas side is 50 W/m^2 K and the external surface area is 12 times of internal surface area. Take the flue gas mass flow rate as 2 kg/s, and its specific heat as 1000 J/kg K.

Pump Characteristics

$$\Delta P = 400 - 30\dot{m}^{1.1} - 2.5\dot{m}^2$$

where the water flow rate $\dot{m}$ is in kilogram per second and the pressure difference ΔP is in kilopascal.

Assume pressure at inlet to the pump and the exit of the heating system (i.e., point 6) is atmospheric. Neglect temperature rise in the pumps.

Take water properties in the pipe lengths 1–2 and 2–3 as:

$\mu = 797.4 \times 10^{-6}$ Pa s $\rho = 995.6$ kg/m^3 $k = 0.6155$ W/m K $c_p = 4.18$ kJ/kg K

and the average properties inside the heat exchangers, and the pipe lengths 4–5 and 5–6 as

$\mu = 354.3 \times 10^{-6}$ Pa s $\rho = 971.8$ kg/m^3 $k = 0.67$ W/m K $c_p = 4.197$ kJ/kg K

Take $T_1 = T_2 = T_3 = 30°C$.

Solution

We shall first write the equations relating the flow rate and the pressure drop in various components of the system, namely pipe sections 1–2, 2–3, 4–5, and 5–6; and the heat exchangers 1 and 2. By balancing the pressure drop in these components with the pressure head developed by the pump, we can determine the flow rates in the various pipes. The heat-transfer calculations can only be done thereafter. This segregation of heat-transfer and flow-rate calculations is possible since the thermophysical properties of water in various subsections of the system have already been given. Otherwise an iterative solution would be necessary since the properties of water inside the heat exchanger, and in the pipe sections downstream would depend upon the magnitude of the heat transfer in the two heat exchangers.

The pressure drop in various pipe sections can be calculated using the equations given in Section 3.2:

$$\Delta P = \frac{f\,L}{D}\,\frac{\rho V^2}{2}$$

$$f = (0.79\ \ln Re - 1.64)^{-2}$$

$$Re = \frac{\rho\,Vd}{\mu}$$

Substituting the data available, we can write the following equations for the four pipe sections

Pipe 1–2

$$P_1 - P_2 = \frac{f_{12} \times 100}{0.05} \cdot \frac{995.6 \cdot V_{12}^2}{2} = 995.6 \times 10^5 f_{12} V_{12}^2$$

$$f_{12} = (0.79 \ln Re_{12} - 1.64)^{-2}$$

$$Re_{12} = \frac{995.6 \times V_{12} \times 0.05}{797.4 \times 10^{-6}} = 6.2428 \times 10^4 V_{12}$$

$$\dot{m}_{12} = 995.6 \times \left(\frac{\pi}{4} \times 0.05^2\right) \times V_{12} = 1.954856\,V_{12}$$

Pipe 2–3

$$P_2 - P_3 = \frac{f_{23} \times 50}{0.03} \cdot \frac{995.6 \cdot V_{23}^2}{2} = 8.29667 \times 10^5 f_{23} V_{23}^2$$

$$f_{23} = (0.79 \ln Re_{23} - 1.64)^{-2}$$

$$Re_{23} = \frac{995.6 \times V_{23} \times 0.03}{797.4 \times 10^{-6}} = 3.7447 \times 10^4 V_{23}$$

$$\dot{m}_{23} = 995.6 \times \left(\frac{\pi}{4} \times 0.03^2\right) \times V_{23} = 0.703748\,V_{23}$$

Pipe 4–5

$$P_4 - P_5 = \frac{f_{45} \times 50}{0.03} \times \frac{971.8}{2} V_{45}^2 = 8.09833 \times 10^5 f_{45} V_{45}^2$$

$$f_{45} = (0.79 \ln Re_{45} - 1.64)^{-2}$$

$$Re_{45} = \frac{971.8 \times V_{45} \times 0.03}{354.3 \times 10^{-6}} = 8.22862 \times 10^4 V_{45}$$

Pipe 5–6

$$P_5 - P_6 = \frac{f_{56} \times 100}{0.05} \times \frac{971.8}{2} V_{56}^2 = 9.718 \times 10^5 f_{56} V_{56}^2$$

$$f_{56} = (0.79 \ln Re_{56} - 1.64)^{-2}$$

$$Re_{56} = \frac{971.8 \times V_{56} \times 0.05}{354.3 \times 10^{-6}} = 13.714366 \times 10^4 V_{56}$$

To write the equations for the pressure drop in the heat exchangers, we need to first find the mass flow rate through each of these, subdivide it into the number of tubes through which it flows, and then find the velocity of water flowing through the heat exchanger tubes.

Heat Exchanger 1

$$\dot{m}_{HE1} = \dot{m}_{23}$$

$$V_{HE1} = \frac{\dot{m}_{HE1}}{15 \times \left(\frac{\pi}{4} \times 0.01^2\right) \times 971.8} = 0.873458 \, \dot{m}_{HE1} \text{ m/s}$$

$$Re_{HE1} = \frac{971.8 \times V_{HE1} \times 0.01}{354.3 \times 10^{-6}} = 2.74287 \times 10^4 V_{HE1}$$

$$f_{HE1} = (0.79 \ln Re_{HE1} - 1.64)^{-2}$$

$$P_3 - P_4 = \frac{f_{HE1} \times 20}{0.01} \frac{(971.8) V_{HE1}^2}{2} = 9.718 \times 10^5 f_{HE1} V_{HE1}^2$$

Heat Exchanger 2

$$\dot{m}_{HE2} = \dot{m}_{12} - \dot{m}_{23}$$

$$V_{HE2} = \frac{\dot{m}_{HE2}}{15 \times \left(\frac{\pi}{4} \times 0.01^2\right) \times 971.8} = 0.873458 \, \dot{m}_{HE2} \text{ m/s}$$

$$Re_{HE2} = \frac{971.8 \times V_{HE2} \times 0.01}{354.3 \times 10^{-6}} = 2.74287 \times 10^4 V_{HE2}$$

$$f_{HE2} = (0.79 \ln Re_{HE2} - 1.64)^{-2}$$

$$P_2 - P_5 = \frac{f_{HE2} \times 20}{0.01} \frac{(971.8) V_{HE2}^2}{2} = 9.718 \times 10^5 f_{HE2} V_{HE2}^2$$

Pump
Taking atmospheric pressure as 10^5 Pa, we can write pump characteristic equation as

$$(P_1 - 10^5) = (400 - 30\dot{m}_{12}^{1.1} - 2.5\dot{m}_{12}^2) \times 1000$$

Mass Flow Rate Balance
Further, since mass flow rates through sections 2–3, 3–4 and 4–5 are equal, and that through sections 1–2 and 5–6 are equal, we get additional equations relating the variables, that is

$$\dot{m}_{56} = \dot{m}_{12} = 971.8 \times \left(\frac{\pi}{4} \times 0.05^2\right) \times V_{56} = 1.908125 V_{56}$$

$$\dot{m}_{45} = \dot{m}_{23} = 971.8 \times \left(\frac{\pi}{4} \times 0.03^2\right) \times V_{45} = 0.686925 V_{45}$$

Since all mass flow rates can be expressed in terms of $\dot{m}_{12}$ and $\dot{m}_{23}$, we consider only these two-mass flow rate as variables. Further since pressure P_6 is given to be atmospheric, we have in all 25 variables namely, 6 friction factors, 6 Reynolds numbers, 6 velocities, 5 pressures, and 2 mass flow rates. We can also see that in all we have 25 independent equations in these variables, namely 3 equations each for the four pipe sections; 4 equations each for heat exchangers 1, and 2, 1 equation for the pump, 2 equations for the two mass flow rates and 2 additional equations for mass flow rate balance. Out of these equations 12 are linear algebraic equations, while the remaining 13 are nonlinear equations. If we were to take a pure mathematical perspective, the system simulation would involve solution of these 25 equations (some linear, some nonlinear) for 25 variables, which would surely be an arduous task needing extensive computational effort. However, if we draw the IFD for each component and combine these judiciously, the task can be greatly simplified.

One such possibility is shown in Fig. 5.20. Here starting from the assumption of the velocity of hot water in the pipe sections 5–6 (i.e., V_{56}), the simulation of various components is carried out in the sequence shown in the diagram and the correctness of the assumption is checked by comparing the values of pressure P_5 calculated in two independent ways. Of course, iteration would also be needed while simulating heat exchanger 2. Since the nature of equations does not permit explicit calculation of the mass flow rate through it (i.e., $\dot{m}_{25}$) on the basis of the input information of the pressure drop across it. Thus effectively the task is reduced to solving two separate nonlinear equations for single variable each. These are the equation for mass flow rate $\dot{m}_{25}$ that could be obtained by combining various equations for heat exchanger simulation for known value of $(P_2 - P_5)$; and the "overall equation" that we could visualize as have been obtained by combining all the equations into a function of V_{56}, by eliminating all other variables. Thus the task of system simulation is considerably simplified by the insights

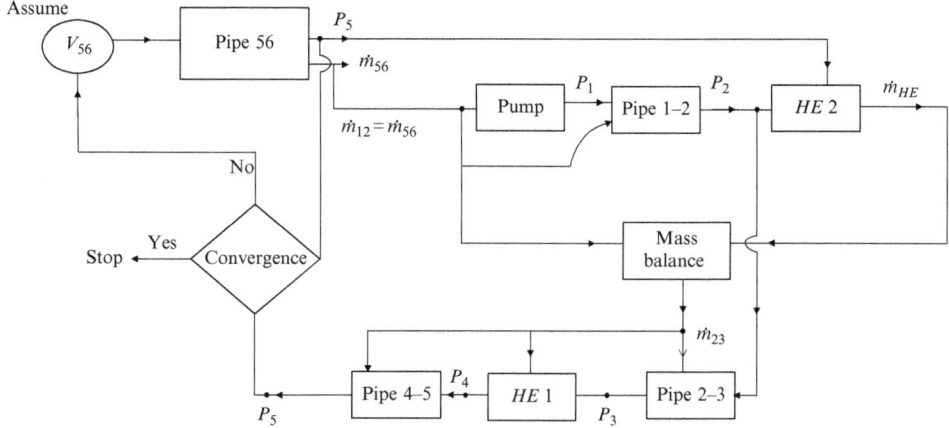

Fig. 5.20 IFD for simulating water heating system of Fig. 5.19.

provided by the IFD. Now, to solve the problem, we begin by assuming velocity V_{56}, as 2 m/s, and carry out the various steps in accordance with the IFD, using the equations obtained above:

Pipe 5–6

$$Re_{56} = 13.714366 \times 10^4 \times 2 = 274{,}287.3$$
$$f_{56} = (0.79\ln[274{,}287.3] - 1.64)^{-2} = 0.014684$$
$$P_5 - P_6 = 9.718 \times 10^5 \times (0.014684)(2)^2 = 5.708 \times 10^4 \text{ Pa}$$

Since P_6 is given as 10^5 Pa, we get

$$P_5 = (10^5 + 5.708 \times 10^4) = 1.5708 \times 10^5 \text{ Pa}$$
$$\text{also} \quad \dot{m}_{56} = 1.908125 \times 2 = 3.816 \text{ kg/s} = \dot{m}_{12}$$

Pump

$$(P_1 - 10^5) = [400 - 30(3.816)^{1.1} - 2.5(3.816)^2] \times 1000$$
$$= 232.7 \times 10^3 \text{ Pa}$$
$$\Rightarrow P_1 = 3.327 \times 10^5 \text{ Pa}$$

Pipe 1–2

$$\dot{m}_{12} = \dot{m}_{56} = 3.816 \text{ kg/s} \quad V_{12} = \frac{\dot{m}_{12}}{1.954856} = 1.952 \text{ m/s}$$
$$Re_{12} = 6.2428 \times 10^4 \times 1.952 = 12.186 \times 10^4$$

$$f_{12} = (0.79 \ln[12.186 \times 10^4] - 1.64)^{-2} = 0.01726$$

$$P_1 - P_2 = 9.956 \times 10^5 \times (0.01726) \times (1.952^2) = 6.5481 \times 10^4 \text{ Pa}$$

$$\Rightarrow P_2 = P_1 - 0.65481 \times 10^5 = 2.67219 \times 10^5 \text{ Pa}$$

HE2

$$P_2 - P_5 = 2.67219 \times 10^5 - 1.5708 \times 10^5 = 1.10139 \times 10^5 \text{ Pa}$$

To get $\dot{m}_{HE2}$ or V_{HE2} we need to do iterations. We combine all the equations to get a single nonlinear equation for the pressure drop $(P_2 - P_5)$ in terms of V_{HE2}, as indicated below:

$$P_2 - P_5 = 1.10139 \times 10^5 = 9.718 \times 10^5 f_{HE2} V_{HE2}^2$$

$$= 9.718 \times 10^5 \times (0.79 \ln Re_{HE2} - 1.64)^{-2} V_{HE2}^2$$

$$= 9.718 \times 10^5 \times (0.79 \ln[2.74287 \times 10^4 V_{HE2}] - 1.64)^{-2} V_{HE2}^2$$

Thus final equation can be written as

$$f(V_{HE2}) = 9.718(0.79 \ln[27{,}428.7 V_{HE2}] - 1.64)^{-2} V_{HE2}^2 - 1.10139 = 0$$

which is a nonlinear equation in a single variable V_{HE2}. We can solve it by any of the methods discussed in Section 2.2. For the sake of illustration we choose the bisection method. The first step is to bracket the solution, which can be easily done. Taking $V_{HE2} = 1$ m/s we get $f(V_{HE2}) = -0.86658$; and $V_{HE2} = 4$ m/s gives $f(V_{HE2}) = 1.641985$.

Having bracketed the solution, we follow the bisection rule. Various steps are summarized below:

V_{HE2}	$f(V_{HE2})$
2.5	+0.0843
1.75	−0.4718
2.125	−0.2131
2.3125	−0.06914
2.4062	0.00634

We stop the calculations since a very low value of the function f has been obtained, and round off the solution to $V_{HE2} = 2.4$ m/s.

This gives $\dot{m}_{HE2} = \frac{V_{HE2}}{0.873458} = 2.7477$ kg/s.

From the mass balance we get

$$\dot{m}_{23} = 3.816 - 2.7477 = 1.0683 \text{ kg/s}$$

Pipe 2–3

$$V_{23} = \frac{\dot{m}_{23}}{0.703748} = \frac{1.0683}{0.703748} = 1.518 \, \text{m/s}$$
$$Re_{23} = 3.7447 \times 10^4 \times 1.518 = 5.68445 \times 10^4$$
$$f_{23} = (0.79 \ln[56,844.5] - 1.64)^{-2} = 0.020355$$
$$P_2 - P_3 = 8.29667 \times 10^5 \times 0.020355 \times 1.518^2 = 3.8917 \times 10^4 \, \text{Pa}$$
$$\Rightarrow P_3 = P_2 - 3.8917 \times 10^4 = 2.67219 \times 10^5 - 3.8917 \times 10^4 = 2.28302 \times 10^5 \, \text{Pa}$$

HE1

$$\dot{m}_{HE1} = \dot{m}_{23} = 1.0683 \, \text{kg/s}$$
$$V_{HE1} = 0.873458 \times 1.0683 = 0.933115 \, \text{m/s}$$
$$Re_{HE1} = 2.74287 \times 10^4 \times 0.933115 = 2.559413 \times 10^4$$
$$f_{HE1} = (0.79 \ln[25,594.1] - 1.64)^{-2} = 0.024578$$
$$P_3 - P_4 = 9.718 \times 10^5 \times (0.024578) \times 0.933115^2 = 0.207967 \times 10^5 \, \text{Pa}$$
$$\Rightarrow P_4 = P_3 - 0.207967 \, 10^5 = 2.075 \times 10^5 \, \text{Pa}$$

Pipe 4–5

$$V_{45} = \frac{\dot{m}_{23}}{0.686925} = \frac{1.0683}{0.686925} = 1.55519 \, \text{m/s}$$
$$Re_{45} = 8.22862 \times 10^4 \times 1.55519 = 12.79708 \times 10^4$$
$$f_{45} = (0.79 \ln(127,970.8) - 1.64)^{-2} = 0.017087$$
$$P_4 - P_5 = 8.09833 \times 10^5 \times (0.017087) \times 1.55519^2 = 0.334679 \times 10^5 \, \text{Pa}$$
$$P_5 = P_4 - 0.334679 \times 10^5 = (2.075 - 0.334679) \times 10^5$$
$$= 1.74032 \times 10^5 \, \text{Pa}$$

Comparing this value, with that obtained earlier from calculation of pipe length 5–6, which was 1.5708×10^5 Pa, we find that there is considerable difference between the two values. An iteration is therefore necessary. We need to decide on the next assumption for V_{56}. Since the P_5 value obtained above from the analysis of pressure drop in sections 5–6 was lower, we should increase the value of V_{56} so that the pressure drop in sections 5–6 increases and so does the value of P_5 (since P_6 is given as constant $= 10^5$ Pa). Let us repeat the above calculation with $V_{56} = 2.5 \, \text{m/s}$.

Calculations With $V_{56} = 2.5\,m/s$
Pipe 5–6

$$Re_{56} = 13.714366 \times 10^4 \times 2.5 = 342{,}859.15$$
$$f_{56} = (0.79 \times 12.745075 - 1.64)^{-2} = 0.014076$$
$$P_5 - P_6 = 9.718 \times 10^5 \times (0.014076) \times 2.5^2 = 0.854959 \times 10^5\,Pa$$
$$P_5 = 1.854959 \times 10^5\,Pa$$
$$\dot{m}_{56} = 1.908125 \times 2.5 = 4.7703\,kg/s$$

Pump

$$(P_1 - 10^5) = [400 - 30(4.7703)^{1.1} - 2.5(4.7703)^2] \times 1000\,Pa$$
$$= 1.758 \times 10^5$$
$$P_1 = 2.758 \times 10^5\,Pa$$

Pipe 1–2

$$\dot{m}_{12} = \dot{m}_{56} = 4.7703\,kg/s \quad V_{12} = \frac{4.7703}{1.954856} = 2.44023\,m/s$$
$$Re_{12} = 6.2428 \times 10^4 \times 2.44023 = 1.523387 \times 10^5$$
$$f_{12} = 0.016488$$
$$P_1 - P_2 = 9.956 \times 10^5 \times 0.016488 \times 2.44023^2 = 0.977495 \times 10^5\,Pa$$
$$P_2 = (2.758 - 0.977495) \times 10^5 = 1.7805 \times 10^5\,Pa$$

HE2

We find that $P_2 - P_5$ is $-$ve. This implies that our initial assumption of V_{56} is too high.[a] So we repeat the calculations with a lower value, $V_{56} = 2.2\,m/s$:

Calculations With $V_{56} = 2.2\,m/s$
Pipe 5–6

$$Re_{56} = 13.714366 \times 10^4 \times 2.2 = 30.1716 \times 10^4$$
$$f_{56} = 0.0144198$$
$$P_5 - P_6 = 9.718 \times 10^5 \times 0.0144198 \times 2.2^2 = 0.678235 \times 10^5\,Pa$$
$$P_5 = 1.678235 \times 10^5\,Pa$$
$$\dot{m}_{56} = 1.908125 \times 2.2 = 4.197875\,kg/s = \dot{m}_{12}$$

Pump

$$(P_1 - 10^5) = [400 - 30(4.197875)^{1.1} - 2.5(4.197875)^2] \times 1000$$
$$= 210.5817 \times 1000$$
$$P_1 = 3.105817 \times 10^5 \, \text{Pa}$$

Pipe 1–2

$$\dot{m}_{12} = 4.197875 \quad V_{12} = \frac{\dot{m}_{12}}{1.954856} = 2.14741 \, \text{m/s}$$
$$Re_{12} = 6.2428 \times 10^4 \times 2.14741 = 13.40584 \times 10^4$$
$$f_{12} = 0.01692438$$
$$P_1 - P_2 = 9.956 \times 10^5 \times (0.01692438) \times (2.14741)^2 = 0.777012 \times 10^5 \, \text{Pa}$$
$$P_2 = P_1 - 0.777012 \times 10^5 = 2.328805 \times 10^5 \, \text{Pa}$$

HE2

$$P_2 - P_5 = (2.328805 - 1.678235) \times 10^5 \, \text{Pa}$$
$$= 0.65057 \times 10^5 \, \text{Pa}$$

Equation to determine V_{HE2}:

$$9.718(0.79 \ln[27{,}428.7 \; V_{HE2}] - 1.64)^{-2} V_{HE2}^2 - 0.777012 = 0$$

Its solution, found as before, is $V_{HE2} = 1.9705 \, \text{m/s}$

$$\dot{m}_{HE2} = \frac{1.9705}{0.873458} = 2.25597 \, \text{kg/s}$$

From mass balance

$$\dot{m}_{23} = \dot{m}_{12} - \dot{m}_{HE2}$$
$$= 4.197875 - 2.25597$$
$$= 1.9419 \, \text{kg/s}$$

Pipe 2–3

$$V_{23} = \frac{\dot{m}_{23}}{0.703748} = 2.759 \, \text{m/s}$$
$$Re_{23} = 3.7447 \times 10^4 \times 2.759 = 10.333 \times 10^4$$

$$f_{23} = [0.79 \ln(10.333 \times 10^4) - 1.64]^{-2} = 0.017868$$

$$P_2 - P_3 = 8.29667 \times 10^5 \times (0.017868) \times 1.9419^2 = 0.559 \times 10^5 \, \text{Pa}$$

$$\Rightarrow P_3 = P_2 - 0.559 \times 10^5 = (2.328805 - 0.559) \times 10^5 = 1.76978 \times 10^5 \, \text{Pa}$$

HE1

$$\dot{m}_{HE1} = \dot{m}_{23} = 1.9419 \, \text{kg/s}$$

$$V_{HE1} = 0.873458 \times 1.9419 = 1.696168 \, \text{m/s}$$

$$Re_{HE1} = 2.74287 \times 10^4 \times 1.696168 = 5.32638 \times 10^4$$

$$f_{HE1} = 0.020658$$

$$P_3 - P_4 = 9.718 \times 10^5 \times (0.020658) \times 1.696168^2 = 0.577575 \times 10^5 \, \text{Pa}$$

$$P_4 = P_3 - 0.577575 \times 10^5 = (1.76978 - 0.577575) \times 10^5$$

$$= 1.192205 \times 10^5 \, \text{Pa}$$

Pipe 4–5

$$V_{45} = \frac{\dot{m}_{23}}{0.686925} = \frac{1.9419}{0.686925} = 2.82695 \, \text{m/s}$$

$$Re_{45} = 8.22862 \times 10^4 \times 2.82695 = 23.26186 \times 10^4$$

$$f = 0.0151586$$

$$P_4 - P_5 = 8.09833 \times 10^5 \times (0.0151586) \times 2.82695^2$$

$$= 0.9810 \times 10^5 \, \text{Pa}$$

$$P_5 = P_4 - 0.9810 \times 10^5 = 0.21116 \times 10^5 \, \text{Pa}$$

Comparing this with the value P_5 obtained directly from the analysis of pipe 5–6 (which is 1.67×10^5 Pa), we find that it is much smaller.

Recalling that with $V_{56} = 2 \, \text{m/s}$, we had an opposite situation, it follows that the correct solution lies somewhere between 2.0 and 2.2 m/s. To evolve a proper procedure for estimating the next assumption for V_{56}, we define an "error function" as follows:

$$\text{Error function:} \quad f(V_{56}) = (P_5)_{56} - (P_5)_{12345}$$

Here the term $(P_5)_{56}$ indicates that this value of P_5 is calculated on the basis of analysis of pipe sections 5–6, and the term $(P_5)_{12345}$ refers to the value obtained after analysis of all the elements in the path 1–5. The results obtained with two assumptions can now be written as:

$$V_{56} = 2.2 \, \text{m/s} \quad f(V_{56}) = 1.678 \times 10^5 - 0.21116 \times 10^5 = 1.46684 \times 10^5$$

$$V_{56} = 2.0 \, \text{m/s} \quad f(V_{56}) = 1.5708 \times 10^5 - 1.74032 \times 10^5 = -0.16952 \times 10^5$$

The next assumption for V_{56} can be found from any of the methods of solution of nonlinear equations indicated in Section 2.2. Thus, using regula falsi method, we get

$$V_{56} = \frac{1.46684 \times 2 + 0.16952 \times 2.2}{1.46684 + 0.16952} = 2.02 \, \text{m/s}$$

Repeating the calculations with this value of V_{56} we get the following results:

$$(P_5)_{56} = 1.5812 \quad \text{and} \quad (P_5)_{12345} = 1.539 \text{ giving:}$$

$$\text{for } V_{56} = 2.02 \, \text{m/s} \quad f(V_{56}) = (1.5812 - 1.5397) \times 10^5$$

$$= (0.0415) \times 10^5$$

Clearly the exact solution lies between 2.0 and 2.02 m/s. Let us take it as 2.01 m/s. The mass flow rate of water in various sections can be found as above. The values are

$$\dot{m}_{12} = 3.835 \, \text{kg/s} \quad \dot{m}_{23} = 1.12 \, \text{kg/s} \quad \dot{m}_{HE1} = 1.12 \, \text{kg/s}$$

$$\dot{m}_{HE2} = 2.715 \, \text{kg/s}$$

Using these values of mass flow rates we can do the heat-transfer calculations.

HE1

$$\dot{m}_{HE1} = 1.12 \, \text{kg/s} \quad Re_{HE1} = 2.683 \times 10^4$$

$$f = 0.024293 \quad Pr = \frac{354.3 \times 10^{-6} \times 4197}{0.67} = 2.219$$

We use Gnielinski correlation (Eq. 3.251) to find the heat-transfer coefficient. This gives

$$Nu_D = \frac{(0.024293/8)[26,832.7 - 10^3]2.219}{1 + 12.7(0.024293/8)^{1/2}(2.219^{2/3} - 1)}$$

$$= 116.76$$

$$h = \frac{116.76 \times 0.67}{0.010} = 7822.9 \, \text{W/m}^2 \, \text{K}$$

Internal surface area $= \pi \times \frac{10}{1000} \times 50 = 1.5708 \, \text{m}^2$.

Overall heat-transfer coefficient U can now be calculated as:

$$\frac{1}{UA} = \frac{1}{7822.9 \times 1.5708} + \frac{1}{50 \times 12 \times 1.5708}$$

which gives $UA = 875.34 \, \text{W/K}$

Heat capacities of two streams are

$$C_{gas} = 2 \times 1000 = 2000 \, \text{W/K}$$

$$C_{water} = 1.12 \times 4197 = 4700.6 \, \text{W/K}$$

which gives $C_{min} = 2000 \, \text{W/K}$ and $C_R = 0.42547$.

Now we can determine the effectiveness of the heat exchanger using Eq. (4.28), which gives

$$\epsilon = \frac{1 - \exp\left(-\frac{875.34}{2000}(1 - 0.42547)\right)}{1 - 0.42547 \cdot \exp\left(-\frac{875.34}{2000}(1 - 0.42547)\right)}$$
$$= 0.33227$$

Therefore, heat transfer is

$$Q_{HE1} = 0.33227 \times 2000 \times (200 - 30)\,\text{W}$$
$$= 112.97\,\text{kW}$$

The temperature of water at 4 can now be found using energy balance (given, $T_3 = 30°\text{C}$)

$$(T_4 - T_3) \times 1.12 \times 4.197 = 112.97$$
$$\Rightarrow T_4 = 54.03°\text{C}$$

Similarly we carry out the analysis of heat exchanger 2. We first find the water-side heat-transfer coefficient:

$$\dot{m}_{HE2} = 2.715\,\text{kg/s} \quad Re_{HE2} = 6.50454 \times 10^4$$
$$f = 0.019751 \quad Pr = 2.219$$
$$Nu_D = \frac{(0.019751/8)[65{,}045.4 - 10^3]2.219}{1 + 12.7(0.019751/8)^{1/2}(2.219^{2/3} - 1)} = 243.23$$
$$h = \frac{243.23 \times 0.67}{0.01} = 16{,}296.4\,\text{W/m}^2\,\text{K}$$
$$\frac{1}{UA} = \frac{1}{16{,}296.4 \times 1.5708} + \frac{1}{50 \times 12 \times 1.5708}$$
$$\Rightarrow UA = 909.01\,\text{W/K}$$
$$C_{gas} = 2000\,\text{W/K}$$
$$C_{water} = 2.715 \times 4197 = 11{,}394.85\,\text{W/K}$$
$$C_{min} = 2000\,\text{W/K} \quad C_R = 0.1755$$
$$\epsilon = \frac{1 - \exp\left(-\frac{909.01}{2000}(1 - 0.1755)\right)}{1 - 0.1755 \cdot \exp\left(-\frac{909.01}{2000}(1 - 0.1755)\right)}$$
$$= 0.35541$$
$$Q_{HE2} = 0.35541 \times 2000 \times (300 - 30)\,\text{W}$$
$$= 191.92\,\text{kW}$$

Energy balance gives

$$(T_5 - T_2) \times 2.715 \times 4.197 = 191.92$$
$$T_5 = 46.84°\text{C}$$

The temperature of the stream at 6, that is, after mixing of the streams leaving the two heat exchangers, can be found using the energy conservation principle[b]:

$$T_6 = \frac{\dot{m}_4 T_4 + \dot{m}_5 T_5}{\dot{m}_4 + \dot{m}_5} = \frac{1.12 \times 54.03 + 2.715 \times 46.84}{1.12 + 2.715}$$

$$= 48.94°C$$

This completes the simulation of this simple system. The utility of the IFD in reducing the number of equations to be solved simultaneously is evident.

[a] Since V_{56} is high, $\dot{m}_{12} = \dot{m}_{56}$ is high, and this results in P_2 becoming "too low" while P_5 becomes too high.
[b] Since the specific heats of the two streams are equal, the fluid enthalpy term in the energy equation can be substituted by temperature.

5.3 OFF-DESIGN PERFORMANCE PREDICTION

Like all engineering systems, the thermal systems too are designed for certain operating (rating) conditions. However, in actual practice these conditions may be encountered only for few hours in a year. Most of the time the thermal systems operate at off-design conditions. The ambient temperature and humidity would usually be different from the design conditions; and so would be the imposed load. System simulation procedures enable us to predict the "off-design" performance of thermal systems and thus help identify any "harmful" consequences (like freezing of water in a chiller or overheating of a surface) that might occur. This information is useful in designing suitable control strategies to ensure safe and optimal operation over a wide range of operating conditions.

We shall illustrate this through a few examples given below.

Example 5.7

A 1-TR window air conditioner using the refrigerant R-22 has following rating conditions:

$$\text{Cooling capacity} = 3.5 \text{ kW}$$
$$\text{Evaporator temperature} = 7°C$$
$$\text{Volumetric flow rate of air} = 15 \text{ m}^3/\text{min}$$
$$\text{Air entry temperature} = 25°C$$
$$\text{Condensation temperature} = 55°C$$

The air flow rate over the cooling coil can be changed by altering its fan speed. The manufacturer wants to specify the minimum permissible fan speed to ensure that there is no possibility of frosting over the coil. Check whether the fan speed can be safely reduced to one-third of the air flow rate over the coil.

Assume that the refrigerant evaporates inside the coil to dry saturated vapor state, and gets heated inside the compressor shell by 25°C before entering the compressor. Estimate the compressor volumetric efficiency (η_v) by the equation:

$$\eta_v = 1.05 - 0.05 \left(\frac{P_{cond}}{P_{evap}} \right)^{0.85}$$

Assume that there is no change in the room temperature and the condensation temperature, the expansion device adjusts to the new operating conditions, there is only sensible cooling in evaporator, and the overall heat-transfer coefficient of the coil is proportional to $\dot{m}_a^{0.8}$, where $\dot{m}_a$ is the mass flow rate of air flowing over the coil.

Take specific heat of moist air = 1.02 kJ/kg K and its density = 1.15 kg/m^3.

Solution

Since it is mentioned in the problem that in spite of change in the air flow rate over the cooling coil, there is no change in the condensation temperature, or expansion device behavior, the problem essentially reduces to balancing the performance of the cooling coil with that of the compressor (see Example 5.5) at the reduced air flow rate. Therefore, we first find the NTU of the cooling coil and the volumetric flow rate of the refrigerant through the compressor using the specified performance at the rated conditions.

Cooling Coil

$$\text{Air mass flow rate} = 15 \times 1.15 = 17.25 \text{ kg/min}$$

$$\text{or} \quad \dot{m}_a = 0.2875 \text{ kg/s}$$

Since there is only sensible cooling of air, we can write its energy balance as:

$$(3.5 \times 1000) = (0.2875)(1.02 \times 1000)\Delta T$$

$$\text{which gives} \quad \Delta T = 11.935°C$$

Hence, the temperature of air leaving the cooling coil is

$$25 - 11.935 = 13.065°C$$

$$\text{Effectiveness of coil} = \frac{11.935}{(25 - 7)} = 0.66306$$

Since for evaporator coil, effectiveness = $1 - \exp(-NTU)$

We get $1 - e^{-NTU} = 0.66306$

$$\Rightarrow NTU = 1.087837 = \frac{UA}{C_{min}} = \frac{UA}{(0.2875 \times (1.02 \times 1000))}$$

We get $UA = 319.0$ W/K

Compressor

To carry out compressor analysis, we first find the refrigerant mass flow rate by plotting the state points on a $p - h$ chart (Fig. 5.21). The values of the enthalpy at various state points can be found from the $p - h$ chart for refrigerant R-22, as:

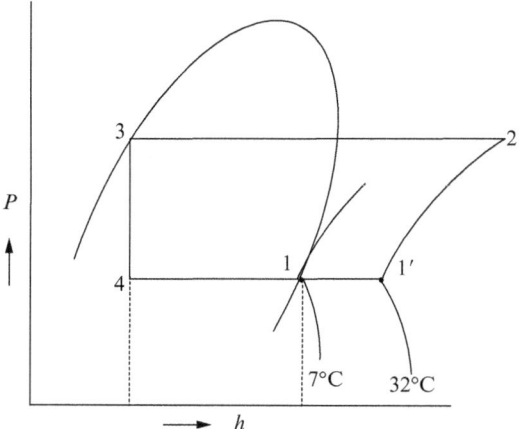

Fig. 5.21 P-h diagram for window air conditioner.

$$h_3 = h_4 = 271 \text{ kJ/kg enthalpy of liquid } R\text{-}22 \text{ at } 55° \text{ C.}$$

$$h_1 = 407.5 \text{ kJ/kg} \quad P_2 = 2.1754 \text{ MPa} \quad P_1 = P_{1'} = 0.6217 \text{ MPa}$$

Energy balance of the evaporating refrigerant gives

$$\dot{m}_r = \frac{3.5}{(407.5 - 271)} = 0.02564 \text{ kg/s}$$

Further, specific volume of the refrigerant at entry to the compressor, $v_{1'} = \frac{1}{25} = 0.04 \text{ m}^3/\text{kg}$. Its volumetric efficiency is

$$\eta_v = 1.05 - 0.05 \left(\frac{P_2}{P_{1'}} \right)^{0.85}$$

$$= 1.05 - 0.05 \left(\frac{2.1754}{0.6217} \right)^{0.85}$$

$$= 0.905$$

The volumetric displacement of the compressor can now be found from the equation:

$$\dot{m}_r = \frac{V_c \times \eta_v}{v_{1'}}$$

$$\Rightarrow V_c = \frac{\dot{m}_r v_{1'}}{\eta_v} = \frac{0.02564 \times 0.04}{0.905} = 0.001133 \text{ m}^3/\text{s}$$

To determine the evaporator temperature at which the system would work when the air flow rate over the coil is reduced, we need to make a suitable assumption for its value and then check whether the rate of evaporation of refrigerant in the coil matches with the new rate of mass flow rate through the compressor at this evaporator temperature.

Let us assume evaporator temperature = $1°C$.

From the thermodynamic property tables for refrigerant R-22, we get the new property values as:

$$h_3 = h_4 = 271 \text{ kJ/kg} \quad \text{(since condenser temperature does not change)}$$

$$h_1 = 405.4 \text{ kJ/kg}$$

$$v_{1'} (\text{at } 26°C) = \frac{1}{20} = 0.05 \text{ m}^3/\text{kg} \quad P_1 = P_{1'} = 0.5146 \text{ MPa}$$

Compressor volumetric efficiency:

$$\eta_v = 1.05 - 0.05 \left(\frac{2.1754}{0.5146} \right)^{0.85} = 0.8797$$

$$\text{Hence} \quad \dot{m}_r = \frac{0.001133 \times 0.8797}{0.05} = 0.019934 \text{ kg/s}$$

Now we find the rate of evaporation of refrigerant in the cooling coil. The NTU is altered directly due to the change in the mass flow rate of air and also indirectly by its influence on the U-value.

$$\text{The new value of UA} = 319.0 \times \left(\frac{1}{3} \right)^{0.8} = 132.46 \text{ W/K}$$

$$\text{and NTU} = \frac{UA}{C_{min}} = \frac{132.46}{(0.2875/3) \times 1020} = 1.35509$$

$$\text{Therefore} \quad \epsilon = 1 - \exp(-1.35509)$$

$$= 0.74208$$

The heat transfer in the cooling coil is

$$Q_{coil} = 0.74208 \times \left(\frac{0.2875}{3} \times 1020 \right) \times (25 - 1)$$

$$= 1740.9 \text{ W}$$

Hence the rate of evaporation of refrigerant in the cooling coil is

$$\dot{m}_{coil} = \frac{1704.9}{(h_1 - h_4)} = \frac{1740.9}{(405.4 - 271) \times 1000}$$

$$= 12.953 \times 10^{-3} \text{ kg/s}$$

This is smaller than the mass flow rate through the compressor (19.934×10^{-3} kg/s) calculated above. This implies that actual temperature in the evaporator is even lesser than $1°C$. Let us now assume a much lower value, say $-5°C$ and check whether this equalizes the two mass flow rates.

From the thermodynamic property tables, we get for $-5°C$ evaporator temperature:

$$h_3 = h_4 = 271 \text{ kJ/kg (as before)}$$
$$h_1 = 403.16 \text{ kJ/kg}$$
$$v_{1'} (\text{at } 20°C) = \frac{1}{16.5} = 0.0606 \text{ m}^3/\text{kg} \quad P_1 = P_{1'} = 0.422 \text{ MPa}$$
$$\eta_v = 1.05 - 0.05 \left(\frac{2.1754}{0.422}\right)^{0.85} = 0.8485$$

Hence $\quad \dot{m}_r = \dfrac{0.001133 \times 0.8485}{0.0606} = 0.01586 \text{ kg/s}$

The heat transfer in the cooling coil is

$$Q_{coil} = (0.74208) \times \left(\frac{0.2875}{3} \times 1020\right)(25+5)$$
$$= 2176.15 \text{ W}$$

Hence the rate of evaporation of refrigerant in the cooling coil

$$\dot{m}_{coil} = \frac{2176.15}{(403.16 - 271) \times 1000} = 0.016466 \text{ kg/s}$$

We now see that the rate of evaporation of refrigerant in the evaporator coil is slightly higher ($\sim$4%) than the pumping capacity of the compressor at $T_{evap} = -5°C$. This implies that actual evaporator temperature is slightly higher. Next estimate for T_{evap} could be found by using regula-falsi method. For this we define the "error" function as the difference of two mass flow rates, which is a function of the evaporator temperature, that is

$$f(T_{evap}) = \dot{m}_r - \dot{m}_{coil}$$

From the two iterations we have

$$f(1°C) = (19.934 - 12.953) = 6.981$$
$$f(-5°C) = (15.86 - 16.466) = -0.606$$

Thus the next estimate for T_{evap} would be (Section 2.2):

$$T_{evap} = \frac{6.981 \times (-5) + (0.606) \times 1}{6.981 + 0.606} = -4.52°C$$

The students are advised to verify that at this temperature, the refrigerant mass flow rate in the system, as calculated by two independent procedures are almost equal thus indicating the convergence of system-simulation procedure.

Note

It can be seen that under the reduced air flow rate in the evaporator, its temperature falls below $0°C$. There is thus a great likelihood of frosting occurring over the evaporator coil which will have deleterious consequences. Thus the system simulation reveals that fan speed control should be so designed as to ensure that air flow rate always remains higher than this value.

One could also find by simulation (Problem 5.3) what is the minimum permissible air flow rate over the evaporator coil to ensure that frost formation does not take place.

Example 5.8

In the indirect heating system of Example 5.4, the air exit temperature can be regulated by reducing the oil flow rate through the system by operating the value V. This increases the temperature of oil at the pump entrance. For safe operation of the pump it is necessary to keep the temperature at entry to be less than 130°C. Determine the minimum oil flow rate at which the system can be operated and the corresponding air exit temperature.

Solution

We will have to find the oil mass flow rate by trial and error. We got the steady-state oil mass flow rate in Example 5.4 as 5.799 kg/s and the oil temperature at pump entry, T_{II}, as 113.8°C. Since the permissible value of T_{II} is higher, we assume the reduced the oil flow rate as $\dot{m} = 3$ kg/min and simulate the system performance at this condition. We get

$$\epsilon_I = 1 - e^{-(m/10)^{0.8}} = 1 - e^{-0.3^{0.8}} = 0.3173$$

$$\epsilon_{II} = 1 - e^{-(m/10)^{0.7}} = 1 - e^{-0.3^{0.7}} = 0.3498$$

Since oil flow rate is being reduced, C_{oil} will continue to be the C_{min} in both the heat exchangers. Therefore, using the definition of effectiveness, Eq. (4.23), we get

$$\epsilon_I = \frac{T_I - T}{300 - T} = 0.3173$$

$$\epsilon_{II} = \frac{T_I - T_{II}}{T_I - 20} = 0.3498$$

Further the equation for temperature rise in the pump gives

$$T - T_{II} = 5 + \left(\frac{3}{5}\right)^{1.3} = 5.51°C$$

Solving the above three simultaneous linear equations we get

$$T = 134.38°C \quad T_I = 186.93°C \quad T_{II} = 128.87°C$$

Since T_{II} can be allowed to rise till 130°C, this implies the possibility of further reduction in the mass flow rate of oil.

Let us take $\dot{m} = 2.5$ kg/min and repeat the above calculations

$$\epsilon_I = 1 - e^{-0.25^{0.8}} = 0.281$$

$$\epsilon_{II} = 1 - e^{-0.25^{0.7}} = 0.3154$$

$$\epsilon_I = \frac{T_I - T}{300 - T} = 0.281$$

$$\epsilon_{II} = \frac{T_I - T_{II}}{T_I - 20} = 0.3154$$

$$T - T_{II} = 5 + \left(\frac{2.5}{5}\right)^{1.3} = 5.41°C$$

Solving the above three equations we get

$$T = 136.7°C \quad T_I = 182.587°C \quad T_{II} = 131.29°C$$

Since the temperature T_{II} is now greater than 130°C it follows that oil flow rate should be greater than 2.5 kg/min. To get a better estimate based on the above two iterations, we again use regula-falsi method. Defining an "error" function as the difference of T_{II} and 130°C, we can write

$$f(\dot{m}) = T_{II} - 130$$

For the two iterations we have

$$f(3) = 128.87 - 130 = -1.13$$

$$f(2.5) = 131.29 - 130 = +1.29$$

The next estimate for $\dot{m}$ can now be found using Eq. (3.16), as:

$$\dot{m} = \frac{3.0 \times 1.29 + 2.5 \times 1.13}{1.29 + 1.13} = 2.77$$

The detailed calculations for this new value of mass flow rate are given below.

$$\epsilon_I = 1 - e^{-0.277^{0.8}} = 0.301$$

$$\epsilon_{II} = 1 - e^{-0.277^{0.7}} = 0.3344$$

$$\epsilon_I = \frac{T_I - T}{300 - T} = 0.301$$

$$\epsilon_{II} = \frac{T_I - T_{II}}{T_I - 20} = 0.3344$$

$$T - T_{II} = 5 + \left(\frac{2.77}{5}\right)^{1.3} = 5.464°C$$

The solution of the above three equations is

$$T = 135.12°C \quad T_I = 184.75°C \quad T_{II} = 129.67°C$$

Since T_{II} is just less than 130°C, we could accept this as the converged solution.

The air exit temperature can be found from the analysis of HE II, by first finding the total heat transfer, Q_{II} as:

$$Q_{II} = 0.3344 \times (2.77 \times 2.5) \times (184.75.7 - 20)$$
$$= 381.5 \, kJ/min$$

Equating this to the heat gained by air, we get

$$15 \times 1.0 \times \Delta T = 381.5$$
$$\Rightarrow \Delta T = 25.4°C$$

The air exit temperature is therefore = 45.4°C.

At full load operation the air exit temperature must have been much higher. To find its value we use the results of Example 5.4. The rate of heat transfer at full load is

$$Q_{II} = 0.4948 \times (5.799 \times 2.5) \times (205.7 - 20)$$
$$= 1332.1 \, kJ/min$$

from where we get the air temperature rise as

$$\Delta T = \frac{1332.1}{15 \times 1.0} = 88.8°C$$

So the air exit temperature at full load operation is 108.8°C, which would reduce to 45.4°C at the lowest permissible oil flow rate.

PROBLEMS

5.1 Fig. 5.22 shows the schematic diagram of a modified gas–turbine cycle where the exhaust from the turbine is used to raise steam in a boiler. This steam is fed into an absorption refrigeration machine and the cold water obtained there is used to precool the air entering the compressor of the gas turbine plant. Draw IFDs for each of the components and for the complete cycle indicating clearly the recommended procedure for simulating its performance.

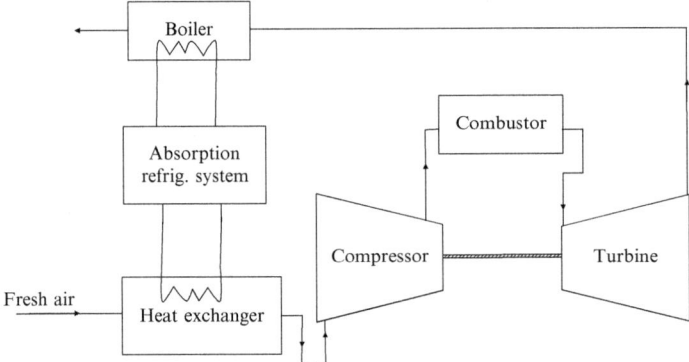

Fig. 5.22 Modified gas-turbine cycle.

5.2 Fig. 5.23 shows the schematic diagram of a regenerative cooling system.

Warm dry air at state 1 is cooled in the heat exchanger I with the help of cold water produced in the spray washer II. In this spray washer is admitted a part of the cold air leaving heat exchanger I. The water warmed in the heat exchanger I is then pumped back to spray chamber for removing this heat. Draw an IFD for each of the components and for the whole system and explain how can the states of air at 2 and 3 be determined from known state at 1 and the given designs of the HE I and the spray chamber.

5.3 Biogas is being produced in three different plants (Fig. 5.24) and is fed to two engines after mixing in suitable proportions to ensure the following:

Engine 1 Total mass flow rate : 1 kg/s

Mass fraction of methane: 0.4

Engine 2 Total mass flow rate: 2 kg/s

Mass fraction of methane: 0.5

Given, the mass fraction of methane produced in various biogas plants is

Plant 1: 0.25   Plant 2: 0.45   Plant 3: 0.55

Also $\dot{m}_4 = 2\,\dot{m}_3$ and $\dot{m}_8 = 0.2\,\dot{m}_2$

Show that the mass flow rates of the gases drawn from each plant ($\dot{m}_1, \dot{m}_2, \dot{m}_3$) (Fig. 5.24) can be determined by solving three simultaneous linear equations. Solve these equations by Gauss elimination method.

5.4 A refrigeration plant is operating at the following conditions in steady state.

Evaporator temperature = 4°C   Condenser temperature = 40°C

Condenser pressure: 1.016 MPa

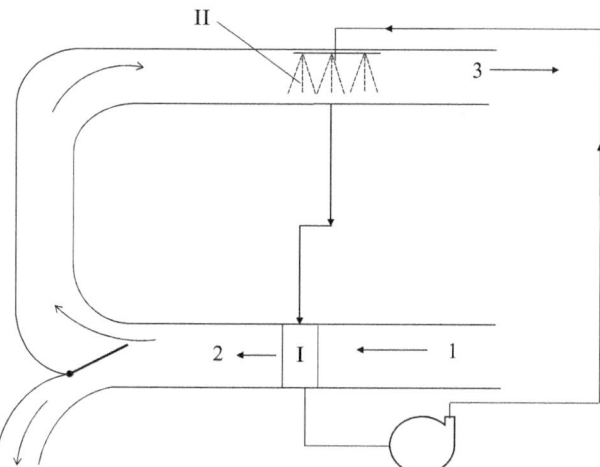

Fig. 5.23 Regenerative cooling system.

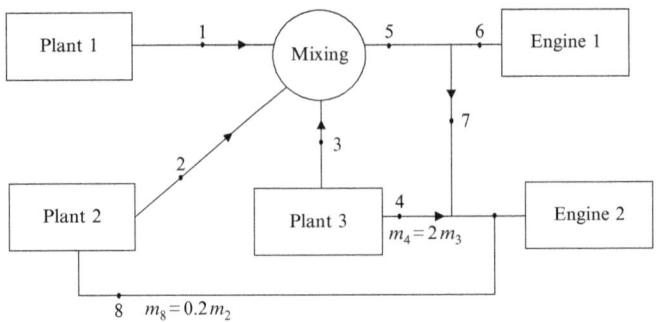

Fig. 5.24 Biogas-plant system.

Air temperature at inlet to the evaporator: 25°C
Air temperature at outlet of the evaporator: 12°C
Air mass flow rate: 0.3 kg/s
Estimate the new steady-state temperature after the air inlet temperature changes to 30°C, and the mass flow rate to 0.4 kg/s. Assume that the condenser is generously sized and therefore the condensation temperature does not change.
Given
(a) The mass flow rate through the compressor is given by the relation

$$m = k \cdot [1.05 - 0.05(P_{cond}/P_{evap})^{0.8}] \times \text{ vapor density at inlet}$$

where P_{cond} = condenser pressure, P_{evap} = evaporator pressure, and k is a constant.
(b) The effectiveness of the evaporator, ϵ, varies with the air mass flow rate according to the relation

$$\epsilon \alpha \text{ (mass flow rate)}^{-0.1}$$

(c) The vapors leaving the evaporator are always dry saturated.
(d) The thermodynamic properties of the refrigerant:

Condensate enthalpy: 106.19 kJ/kg

Temp. (0°C)	Pressure (MPa)	Spec. vol. of vapor (m³/kg)	Enthalpy of liquid (kJ/kg)	Enthalpy of vapor (kJ/kg)
0	0.2928	0.0689	50.02	247.23
4	0.3376	0.0600	55.35	249.53
8	0.3876	0.0525	60.73	251.80
12	0.4429	0.0460	66.18	254.03

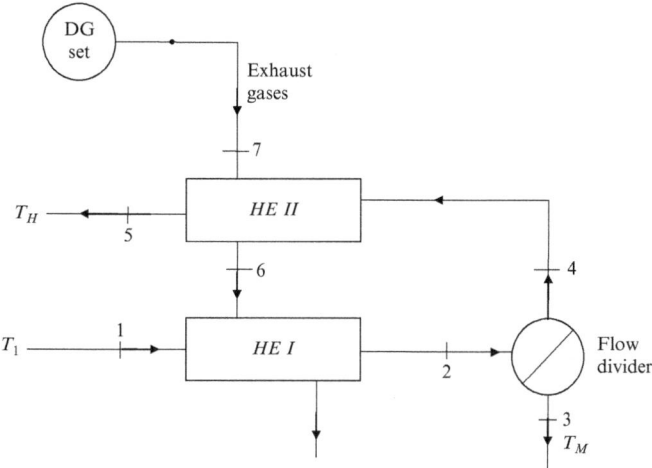

Fig. 5.25 Heat recovery system.

5.5 Fig. 5.25 shows a system to recover heat from the hot exhaust of a DG set. The system is designed to provide hot water at two temperatures, that is, a medium temperature T_M and a high temperature T_H. Determine the values of these temperatures under steady-state operation for following design conditions:

$$T_1 = 30°C \quad T_7 = 400°C \quad \dot{m}_1 = 2\,kg/s \quad \dot{m}_7 = 5\,kg/s$$
$$Cp_1 = 4200\,J/kg\,K \quad Cp_7 = 1100\,J/kg\,K \quad \text{Flow divider setting } m_3 = m_4$$
$$(UA)_I = 1000\,W/K \quad (UA)_{II} = 600\,W/K$$

5.6 Fig. 5.26 shows the arrangement of water flow in a fountain. Out of the total water pumped from the reservoir, one-third is sprayed out at opening 3 and another one-third at the opening 4.

   If the height of water sprayed at opening 5 is 2 m estimate the flow rate at each outlet and the height to which water would rise there. Given

$$\text{Pump characteristic: } \Delta P = 160 - 20Q - 4.0Q^2$$
$$\text{Each pipe element characteristic: } \Delta P = Q^2$$

where ΔP is in kPa and Q in m^3/min.

5.7 Fig. 5.27 shows the arrangement for flash cooling of liquid refrigerant R-134a by throttling a part of the liquid stream to a lower temperature.

   Determine the temperature T of liquid at state 2 and when the mass flow rate m is 0.2 kg/s using the following data:

$$h_l = 91.5\,kJ/kg \quad Cp_{liq} \text{ (average value)} = 1.37\,kJ/kg\,K$$
$$h_{fg} \text{ at } -12°C = 205.8\,kJ/kg \quad UA \text{ of heat exchanger} = 1.5\,kW/K$$

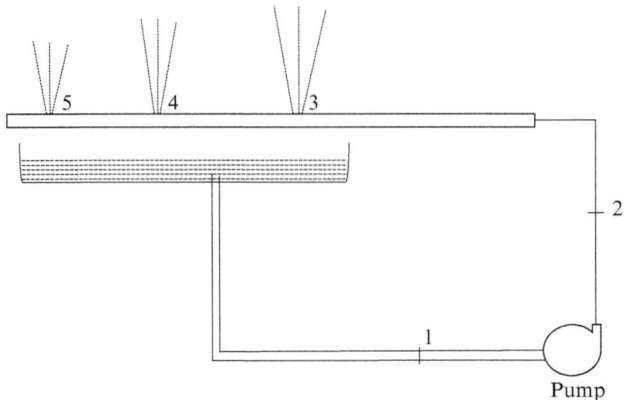

Fig. 5.26 Water fountain.

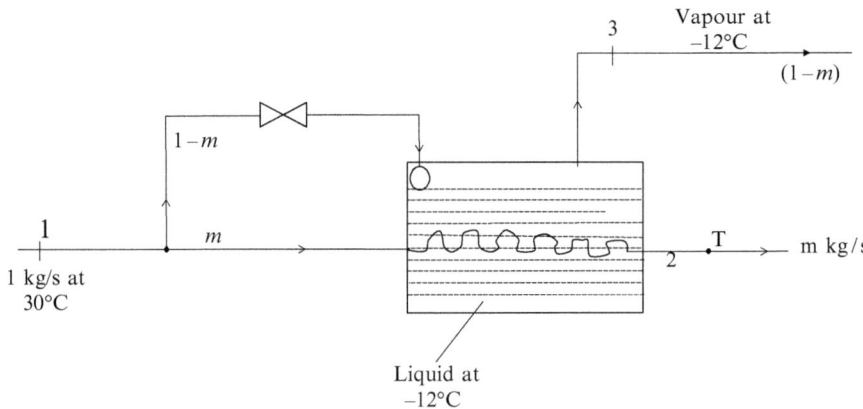

Fig. 5.27 Flash cooling of refrigerant.

5.8 Fig. 5.28 shows the arrangement made in a biogas plant to maintain its efficiency in varying climatic conditions. A portion of the gas produced is used to preheat water fed into the plant so that the gas production rate does not fall in cold weather. Draw an IFD to evolve a strategy to simulate this system. Given the gas production rate depends on the dung flow rate, water flow rate, and its temperature. The water flow controller and the gas distributor are manually adjusted.

5.9 Fig. 5.29 shows the turbine-condenser subsystem of a thermal power plant producing 500 MW power at the design conditions indicated in the figure. How would the power output be influenced by an increase of 5°C in the temperature of water entering the condenser and a simultaneous decrease of 10% in its mass flow rate.

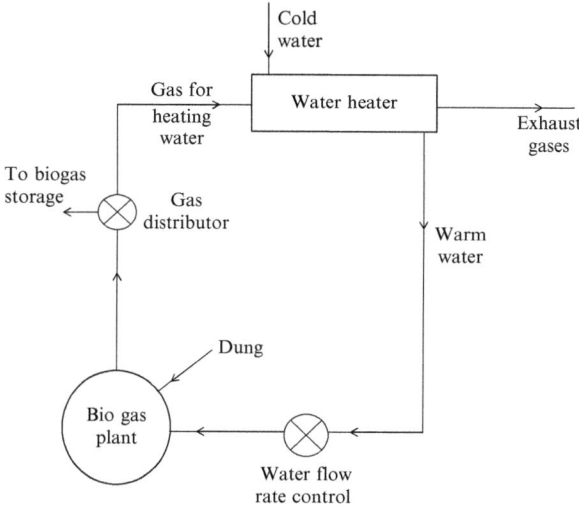

Fig. 5.28 Biogas plant with water heating.

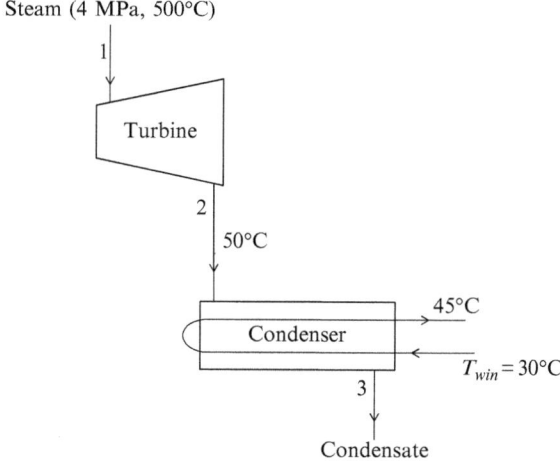

Fig. 5.29 Turbine condenser subsystem.

Assume expansion in the turbine is isentropic and that the mass flow rate through it remains unaltered. Given the overall heat-transfer coefficient in the condenser varies with change in water flow rate M_{water} as

$$U \propto M_{water}^{0.75}$$

Given the condensate remains saturated liquid under the altered conditions also.

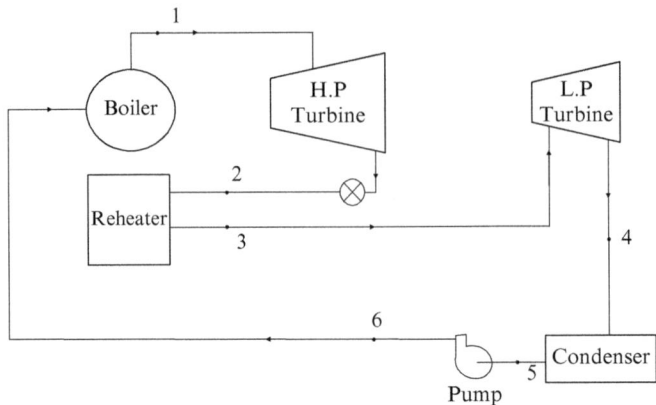

Fig. 5.30 Steam power plant with reheater.

5.10 Fig. 5.30 shows a steam power plant with reheater. Determine its steady-state operating conditions from following design/operations data:

State 1   Pressure = 100 bar   Temperature = 500°C

State 2   The valve at turbine 2 exit is so adjusted that the mass flow rate through the two turbines is equal

Reheater can be treated like a hot surface at a constant temperature of 500°C with $UA = 3000$ W/K

Condensate temperature $T_s = 40°C$

Determine the total power output and the pressure and temperature at states 2, 3, and 4. Equations for mass flow rate through turbines:

$$\dot{m}_{HP} = 5\Delta P^{0.5} + \Delta P^{0.6}$$
$$\dot{m}_{LP} = 6\Delta P^{0.5} + 1.5\Delta P^{0.6}$$

where ΔP is the total pressure drop in the turbine in bar and the mass flow rates are in kilogram per second. The isentropic efficiencies of the turbines are given by the following equations:

$$\eta_{HP} = 0.95 - 0.1(\Delta P/100)^{1.1}$$
$$\eta_{LP} = 0.90 - 0.1(\Delta P/20)^{1.1}$$

5.11 In Manikaran (HP) river water is available at 10°C and near the river bank hot saline water at 90°C is oozing out from a hot water spring. A dual-temperature water heating system (Fig. 5.31) has been designed to heat two river water streams to 50 and 30°C, respectively. The flow rate of hot water stream entering heat exchanger 1 is regulated so that the river water is heated to 50°C. In heat exchanger 2, the geothermal water leaving heat exchanger 1 is used to heat

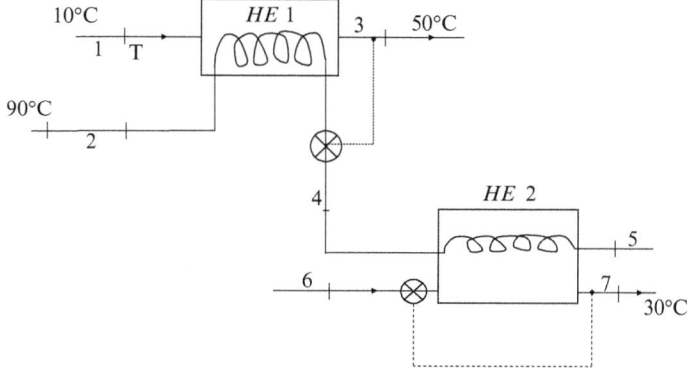

Fig. 5.31 Dual-temperature water heating system.

another stream of river water. Here the flow rate of this stream is adjusted suitably to maintain its temperature at heat exchanger exit at 30°C. Using the design data of the heat exchangers given below carry out simulation of the system and determine the flow rates and the temperatures of all the streams at entrance/exit of the heat exchangers.

HE1. Double pipe heat exchanger with inner pipe ID = 10 mm and OD = 11 mm

ID of outer pipe: 15 mm

Total length of each pipe: 60 m

Mass flow rate of stream 1: 0.2 kg/s

HE2. Double pipe heat exchanger with the two pipes being of the same diameters as above. Total length of each pipe: 50 m

Given the river water streams flow inside the inner pipe. Neglect the thermal resistance of the pipe wall.

5.12 In Example 5.7 determine the minimum permissible air flow rate over the cooling coil to ensure that the evaporator temperature does not fall below +1°C

CHAPTER 6

System Simulation: Case Studies

To bring out the validity, as also the utility of the system simulation procedures discussed in earlier chapters, we shall discuss in this chapter case studies of three thermal systems, that is, an industrial refrigeration system, an industrial combined cycle power plant (CCPP), and a desiccant-based cooling system whose technoeconomic feasibility was being investigated.

6.1 INDUSTRIAL REFRIGERATION PLANT

The refrigeration plant being presented as a case study is a 120-TR (nominal capacity) liquid chilling plant, manufactured by a prominent industrial house of India. The design details of the plant and the simulation procedure can be seen in the Dhar and Saraf [1] and Saraf [2]. It consists of a reciprocating compressor, a multipass water cooled condenser cum subcooler, a flooded chiller with a float-type expansion valve, and a liquid-vapor heat exchanger (LVHE) to superheat the saturated vapors leaving the chiller before these enter the compressor. Fig. 6.1 shows a schematic diagram of the plant.

Computer programs were prepared to simulate the performance of various individual components and validated against the experimented data. Thereafter these programs were suitably "assembled" into a comprehensive software to simulate the whole system. This "assembly" was done using the concept of information flow diagrams explained in the previous chapter.

We shall discuss, in brief, the characteristic features of the simulation of various components, the strategy for system simulation followed and a comparison between the rated and the predicted performance of the plant.

6.1.1 Component Simulation
Evaporator
The evaporator of this plant is a two-pass flooded liquid chiller, each pass having 114 integrally finned tubes each of 3.64 m effective length, 13.8 mm ID, 17.04 mm OD at fin root and 19.05 mm diameter over the fins. The fins are 0.379 mm thick and spaced 0.958 mm apart from each other (19 fins per inch of tube length). Fig. 4.15 shows the schematic diagram of the chiller. Since the refrigerant boils outside the tubes

Thermal System Design and Simulation
http://dx.doi.org/10.1016/B978-0-12-809449-5.00006-1

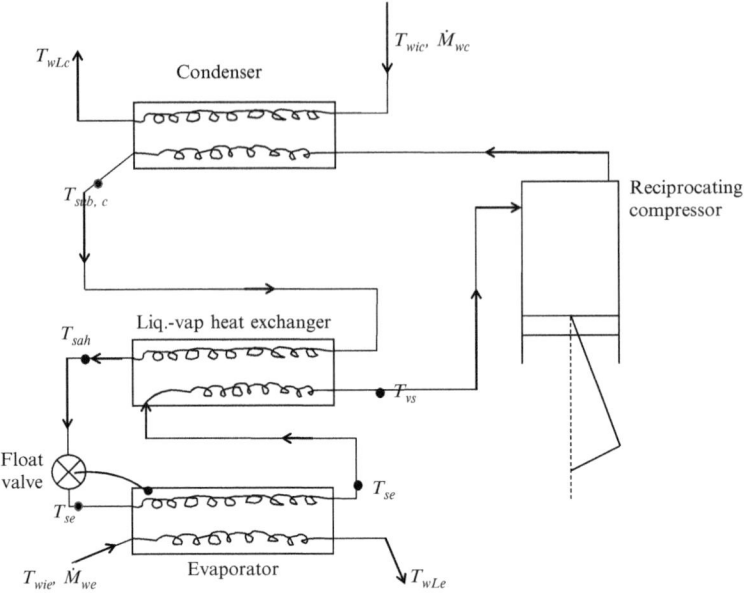

Fig. 6.1 Schematic diagram of the industrial liquid chiller.

carrying water, the refrigerant side heat-transfer coefficient varies along its length, and the procedure explained in the "Shell and tube heat exchanger with phase change in the shell-side fluid" in Section 4.1.4 has been adopted to predict its performance. The comparison between the predicted and the rated capacity of the chiller over a wide range of operating conditions has already been presented in Table 4.1. It can be seen that the prediction of cooling capacity is quite accurate (deviation less than 5%), but there is larger gap between the predicted and the rated values of water side pressure drop. It has been attributed to neglect of "return losses" within the chiller and inaccuracy in estimation of the entrance and exit losses due to lack of accurate data regarding area changes at the water inlet and exit from the chiller.

A flooded chiller is basically a heat exchanger where water is cooled by the boiling refrigerant. Thus its IFD is similar to the generic IFD of any heat exchanger (Fig. 5.2), but special characteristics of flooded chiller could be used to simplify it. The IFD finally adopted is shown in Fig. 6.2. Its distinguishing features are

(a) absence of refrigerant mass flow rate in input variables since the heat-transfer coefficient in pool boiling is not influenced by it;

(b) incorporation of h_i, the enthalpy of throttled refrigerant entering the chiller in input variables. This is necessary to predict the rate of evaporation of the refrigerant from the evaporator, that is, $\dot{m}_{re}$ which is given by:

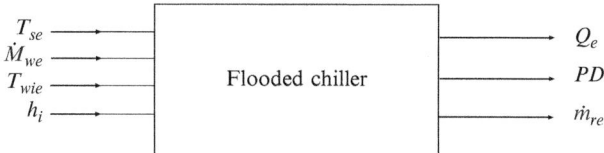

Fig. 6.2 Information flow diagram of the flooded chiller.

$$\dot{m}_{re} = \frac{Q}{h_{sat,e} - h_i} \tag{6.1}$$

where $h_{sat,e}$ is the enthalpy of saturated refrigerant vapors leaving the chiller. For a given refrigerant, $h_{sat,e}$ is a unique function of the saturation temperature T_{se}, an input variable.

Liquid-Vapor Heat Exchanger

The LVHE is a double-pipe heat exchanger 2.44 m long with the inner tube of 69.6 mm ID/76.0 mm OD, the ID of the outer tube being 127 mm. Since there is no phase change in any of the fluids, it is analyzed using the effectiveness approach discussed in Section 4.1.1. Its information flow diagram shown in Fig. 6.3 is basically of the form of generic IFD for heat exchangers, given in Fig. 5.2.

Its distinguishing features are discussed below:

(i) Since the mass flow rates of the fluids flowing inside and outside the inner tube are the same, the value of only one mass flow rate $\dot{m}_{rc}$ is needed as input information.

(ii) The actual temperature of the condensate entering the heat exchanger is $T_{subc} = T_{sc} - D_{sc}$ (Fig. 6.1), where T_{sc} is the temperature of the saturated liquid at condenser pressure and D_{sc} is the degree of subcooling achieved in the condenser. Therefore, two variables (i.e., T_{sc} and D_{sc}) are indicated as input information for specifying the actual temperature of the liquid entering heat exchanger (i.e., T_{subc} in Fig. 6.1). Since this heat exchanger has small length, the pressure drop in it has been neglected. Therefore, the pressure drop calculations are not indicated in the IFD.

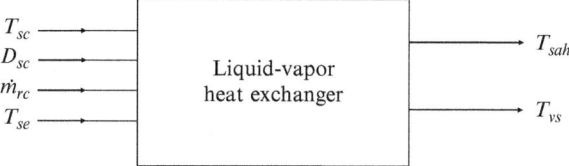

Fig. 6.3 Information flow diagram of liquid vapor heat exchanger.

Reciprocating Compressor

The superheated vapors leaving the LVHE at temperature T_{vs} are compressed in the reciprocating compressor to the condenser pressure. Detailed simulation of the compressor, analyzing the changes inside the piston with every 1°C rotation of the crank has been done following the method discussed in Section 4.3.1. As mentioned therein, the method gives a reasonably good prediction of the rated performance over a wide range of operating conditions. In the system simulation, rather than following this elaborate procedure, polynomial equations have been fitted into the rated performance data to get the volumetric efficiency (η_v), the polytropic index of compression n and the compressor power consumption (BHP) as a function of T_s (saturation temperature at suction pressure) and T_d (saturation temperature at condenser pressure). The IFD for the compressor is shown in Fig. 6.4.

While BHP is obtained directly from input information using the polynomial equation, the mass flow rate $\dot{m}_{rco}$ is calculated using the values of η_v and n (obtained through respective polynomial expressions involving T_s and T_d) along with the physical data of the compressor (i.e., its bore, stroke, no. of cylinders, clearance volume) and its RPM, using the equations for steady flow analysis of reciprocating devices, that is

$$\dot{m}_{rco} = \frac{\eta_v \cdot RPM \cdot \left(\frac{\Pi}{4}D^2 \times S\right) \times N_{cyl}}{\nu} \tag{6.2}$$

where D is the compressor bore, S the compressor stroke, N_{cyl} the number of cylinders, RPM the revolution per minute of crank, and ν the specific volume of the refrigerant at compressor suction conditions.

Similarly the temperature of the vapors leaving the compressor is estimated using the polytropic compression index n as (see Eq. 3.38)

$$T_v = (T_{vs} + 273.15)\left(\frac{P_d}{P_s}\right)^{\frac{n-1}{n}} - 273.15 \tag{6.3}$$

where the temperatures are in °C.

Thus all the output variables can be computed from given values of the input variables indicated in the IFD of Fig. 6.4.

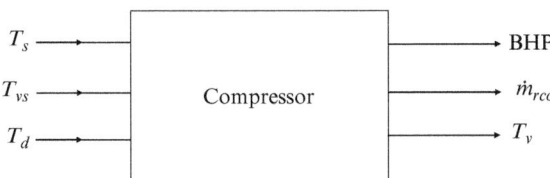

Fig. 6.4 Information flow diagram of a reciprocating compressor.

Condenser

Fig. 6.5 gives the line diagram of the shell and tube condenser used in this plant. The cooling water is bifurcated into two branches, each having three passes. Included in the lower branch B is a separate section to subcool the condensate.

The coolant in the very first pass of branch B is bifurcated into two parts: one flowing into the condensing region and another inside the subcooler. The subcooler is a small compartment at the bottom of the shell, separated from the rest of the condenser by a partition plate. A small opening in this plate allows the dripping condensate collecting over the plate to enter into the subcooler. In view of this asymmetry in the flow introduced by the subcooler, the water flow rate in the two branches of the first pass of section B (Fig. 6.5) can not be equal. It is calculated by trial and error to ensure that the overall pressure drop in both these branches is equal. The analysis of heat transfer is done by following an approach similar to that used in the case of flooded chiller since, here too, the heat–transfer coefficient is varying only on the outside of the tubes where the refrigerant vapors are condensing. In the subcooler, the condensate is being cooled with the help of incoming cold water, and this has been analyzed by treating it as a liquid–liquid heat exchanger with constant U value.

Considering the condenser on a whole, it is basically a heat exchanger and its IFD follows directly from the generic IFD of Fig. 5.2. Fig. 6.6 shows the IFD adopted for this particular condenser with subcooler. Its distinguishing features are

(i) The flow rate of the condensing refrigerant ($\dot{m}_{rc}$) is kept as an output variable, rather than an input variable since the heat-transfer coefficient for condensation outside tubes is generally not influenced by the mass flow rate (Eq. 3.268)

(ii) The variables influencing the heat transfer in the condensing region are the tube wall temperature and the saturation temperature of the refrigerant (T_{sc}); the tube

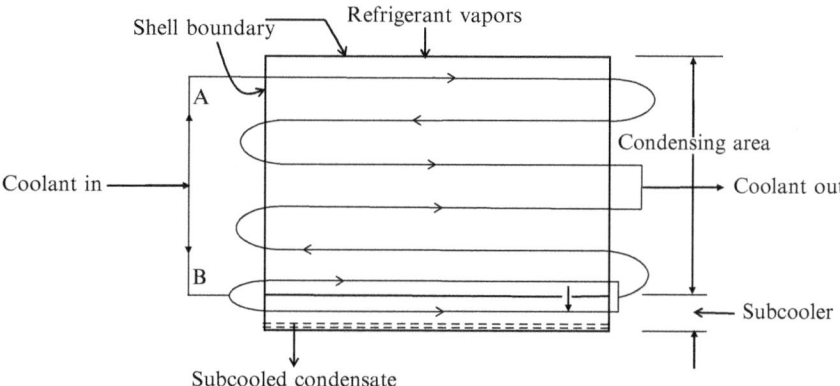

Fig. 6.5 Line diagram of shell and tube condenser.

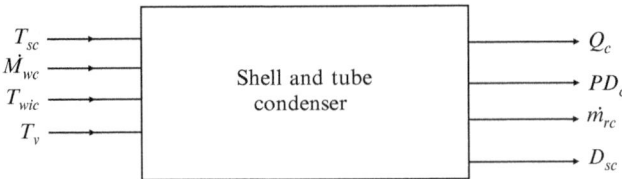

Fig. 6.6 Information flow diagram for the shell and tube condenser.

wall temperature being governed mainly by the cooling water temperature T_{wic}. Therefore, T_{sc} and T_{wic} are identified as input information needed for simulation.

(iii) The actual temperature of vapors entering the condenser, T_v, is needed to calculate the mass flow rate of condensing refrigerant on the basis of the heat–transfer rate in the condensing region. Thus T_v which is an output information in the IFD of reciprocating compressor (Fig. 6.4) is also identified as an input information here.

(iv) The degree of subcooling achieved in the subcooler, D_{sc}, is indicated as an output variable. This acts as an input information for the analysis of the LVHE where the condensate is further cooled (Fig. 6.3).

Table 6.1 shows a comparison between the predicted and the rated performance of the condenser at a few typical operating conditions.

The results indicate an excellent agreement between the predicted and the rated values of the total heat transfer, the maximum deviation being less than 4%. The predictions of the degree of subcooling and the pressure drop also show a good agreement, though the deviations are somewhat higher.

Float-Type Expansion Valve

The system is fitted with a pilot operated float valve to throttle the refrigerant from the condenser pressure to the evaporator pressure. As we know, the flow rate through this expansion device gets automatically adjusted to the load changes which it senses through changes in the liquid level in the flooded chiller. Consequently the presence

Table 6.1 Condenser performance: rated and predicted

Section no.	T_{sc} (°C)	T_{wic} (°C)	$\dot{M}_{wc}$ (kg/s)	D_{sc}(°C) Rated	Pred.	Q_c (TR) Rated	Pred.	Pres drop (kPa) Rated	Pred.
1	61.7	37.8	9.463	8.3	7.7	110	110.37	8.238	7.845
2	51.7	37.8	14.195	8.3	7.6	145	142.32	17.848	16.181
3	46.1	32.2	9.463	8.3	7.9	110	110.99	8.238	7.845
4	40.6	26.7	14.195	8.3	8.1	145	144.54	17.848	16.181
5	43.3	32.2	10.410	8.3	7.1	95.0	95.56	9.905	9.316
6	43.3	32.2	18.233	8.3	7.0	135	133.7	27.949	25.4
7	46.1	37.8	21.135	8.3	5.7	110	109.24	36.481	33.048

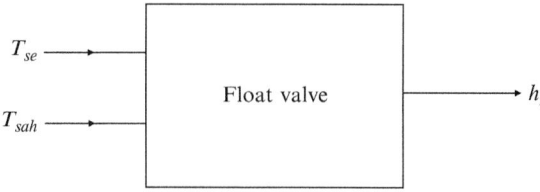

Fig. 6.7 Information flow diagram of the float valve.

of this valve does not introduce any additional constraint in so far as the steady-state performance is concerned. Accordingly, in its information flow diagram (Fig. 6.7), we only have variables indicating the enthalpy of the throttled refrigerant (h_i) to be the same as the enthalpy of the incoming subcooled refrigerant, which depends upon its saturation pressure (in IFD, the saturation temperature T_{sc}) and its actual temperature, T_{sah}, which is the temperature of the condensate leaving the LVHE (Fig. 6.1).

The enthalpy h_i is the input information in the IFD of the flooded chiller (Fig. 6.2).

6.1.2 System Simulation

The strategy for simulation of the liquid chilling plant has been developed by appropriately combining the IFDs of various components by interlinking the variables, which appear as an output from one component and an input in another.

Fig. 6.8 shows the IFD for the complete system obtained in this manner. Following the convention of Chapter 5, (see Fig. 5.9) each variable whose value is externally input is placed in a rectangular box with letter I inside. These are the mass flow rates and the temperatures of the coolants entering the condenser and the evaporator, that is, $\dot{M}_{wc}$, $\dot{T}_{wic}$ and $\dot{M}_{we}$, T_{wie}, respectively.

As is clear from Fig. 6.8, the system IFD forms a closed loop with multiple nests. An iterative strategy is therefore necessary for system simulation.

The best strategy, as identified by a scrutiny of the IFD is to assume the values of three variables (i.e., T_{sc}, T_{se}, and T_{vs}), that is, the saturation temperatures of the refrigerant in the condenser and the evaporator, and the temperature of the superheated refrigerant vapors at compressor suction. Each of these is indicated in Fig. 6.8 by an ellipse (Chapter 5). An additional 'component', that has been indicated in Fig. 6.8 is the box titled, piping. This has been introduced to account for the pressure drop in the long well insulated pipe line connecting the evaporator exit to the compressor suction. The company manufacturing this unit assumes it to cause a drop in the saturation temperature by 1°F (~0.55°C); the same has been indicated in Fig. 6.8, that is, $T_s = T_{se} - 0.55$, T_s being the saturation temperature of the refrigerant vapors at the compressor suction and T_{se} being the temperature at the evaporator exit.

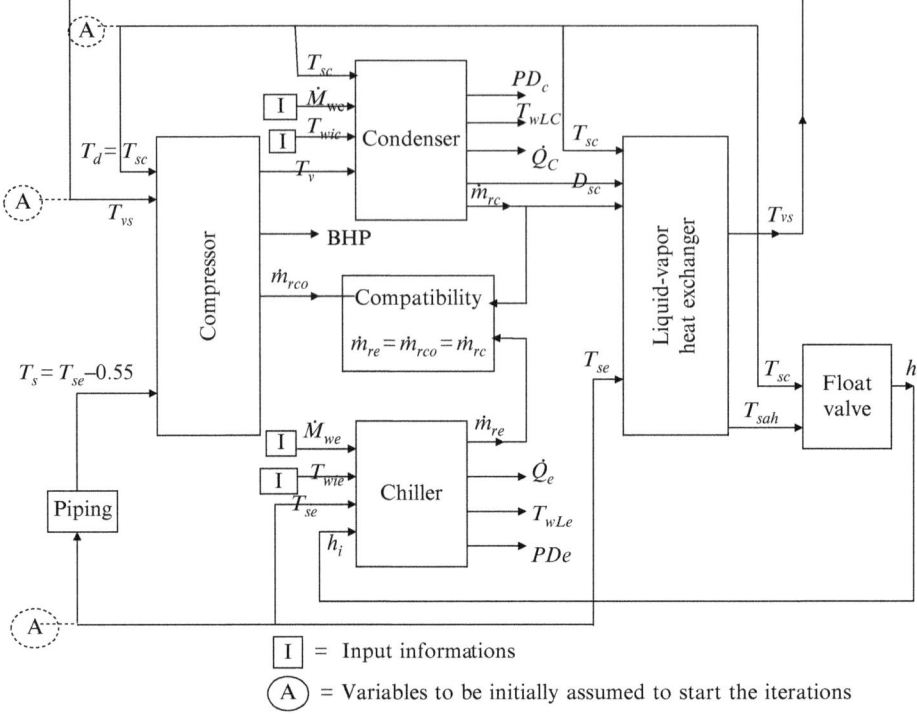

Fig. 6.8 Information flow diagram of the liquid chilling plant.

Once the values of the three variables T_{sc}, T_{se}, and T_{vs} have been assumed (say as T_{sc*}, T_{se*}, and T_{vs*}) the simulation of the components is initiated from the compressor. Since all the input information needed for compressor simulation is available, the values of the output variables $\dot{m}_{rco}$ (compressor mass flow rate), and T_v (the temperature of vapors leaving the compressor) can be calculated. Using the value of T_v thus obtained, along with the assumed value of T_{sc}, and the specified values of water flow rate ($\dot{M}_w$) and its temperature (T_{wic}), the condenser simulation can be done. Thus the condensate mass flow rate ($\dot{m}_{rc}$) and the degree of subcooling, D_{sc}, can be determined.

The LVHE can now be analyzed to predict the liquid temperature at the exit (T_{sah}) and the temperature of the superheated vapors leaving it (i.e., T_{vs}).

The enthalpy of the throttled liquid-vapor mixture leaving the expansion valve, h_i can now be calculated at assumed condenser saturation temperature T_{sc*} and the actual temperature of the subcooled condensate T_{sah}, from the thermodynamic properties of the refrigerant. Using this value of h_i in the chiller simulation, along with the values of the other variables (T_{se*} whose value has already been assumed, $\dot{M}_{we}$ and T_{wie}, externally input variables), the mass flow rate of refrigerant evaporating in the chiller $\dot{m}_{re}$ and the other output variables can be calculated.

Having thus completed a cycle of simulation of all the components we assess the correctness of our assumptions (of T_{sc*}, T_{se*}, and T_{vs*}) by checking for satisfaction of the following compatibility equations:

$$T_{vs} = T_{vs*} \tag{6.4}$$

$$\dot{M}_{rco} = \dot{M}_{re} \tag{6.5}$$

$$\dot{M}_{re} = \dot{M}_{rc} \tag{6.6}$$

Depending upon the nature and extent of deviations in these equations, another set of values of the assumed variables is identified and iterations carried out till convergence is obtained. Warner's method (Section 2.2.1, Example 5.5) has been employed to determine successive estimates of the correct values of the assumed variables. Had we followed a direct mathematical approach by writing down all the equations governing the performance of various components, it would have required solution of 18 nonlinear equations for 18 variables—clearly a much more elaborate task than required in above strategy with only 3 variables.

6.1.3 Validation

The validity of the system simulation procedure has been tested by comparing the predicted performance with the rated performance obtained from the manufacturer (Tables 6.2 and 6.3).

It is seen that the predictions of the refrigeration capacity and the compressor power input are within ±1% of the rated values. The predicted values of the cardinal operating conditions (Table 6.3), viz. T_{se} T_{sc} are mostly within ±0.5°C of the rated values. The degree of subcooling in the condenser, and the temperature of the refrigerant vapors at compressor suction are predicted within ±1°C, while the temperature at compressor exit is predicted within ±0.5°C. The refrigerant mass flow rate is also predicted with ±1%, of the rated value in most of the cases. This close correspondence between the rated and the predicted performance over a wide range of operating conditions establishes the validity of the complete system simulation method. This method can therefore be used for assessing the comparative performance of alternative designs of various components. This is discussed in detail in Chapter 9.

6.2 COMBINED CYCLE POWER PLANT

As the very name suggests, a CCPP uses a combination of two cycles—that is, the gas turbine cycle and a steam cycle—to produce power at a much higher efficiency than a cycle based only on either of the two. This combination is achieved by using the hot gases leaving the turbine of the gas-turbine power plant in a waste-heat recovery boiler

Table 6.2 Comparison between the rated and the predicted performance of the refrigeration plant for condenser water inlet temperature: 32.22°C

Section no.	Operating conditions			Refrig. cap. (TR)			Heat trans. in cond. (TR)			Comp. BHP		
	$\dot{M}_{we}$ (kg/h)	T_{wie} (°C)	$\dot{M}_{wc}$ (kg/h)	Rated	Pred.	%er	Rated	Pred.	%er	Rated	Pred.	%er
1	68,137	11.67	67,163	111.25	111.69	0.39	136.75	129.78	−5.1	118.5	117.92	−0.49
2	74,951	11.67	68,523	112.81	113.23	0.37	138.34	131.35	−5.05	118.96	118.39	−0.48
3	31,764	11.67	69,691	114.51	114.56	0.36	139.69	132.70	−5.0	119.34	118.80	−0.45
4	90,850	11.67	71,037	115.64	116.08	0.38	141.19	134.21	−4.94	119.78	119.25	−0.44
5	115,833	11.67	73,970	118.73	119.19	0.39	144.42	137.33	−4.91	120.66	120.14	−0.43
6	90,850	17.22	90,952	136.2	136.86	0.40	161.62	154.85	−4.19	124.76	124.85	0.07
7	90,850	20.56	105,529	149.0	150.07	0.72	174.16	167.71	−3.7	127.28	127.86	0.45
8	81,765	15.56	81,606	127.72	128.59	0.68	153.44	146.71	−4.39	122.93	122.92	0.01

Table 6.3 Comparison between the rated and the predicted operating conditions of the refrigeration plant

Section no.	Chiller sat. T (°C)		Conden. sat. T. (°C)		Vap. temp. at suc. (°C)		Vap. temp. at dis. (°C)		Deg. of subcool. (°C)		Refr. flow rate (kg/min)	
	Rated	Comp.	Rated	Comp.	Rated	Comp.	Rated	Comp.	Rated	Comp.	Rated	Comp.
1	2.22	2.48	43.33	42.78	5.21	5.99	74.02	73.80	8.33	7.27	142.97	143.84
2	2.58	2.84	43.33	42.80	5.52	6.31	73.61	73.43	8.33	7.25	144.88	145.76
3	2.89	3.16	43.33	42.82	5.79	6.59	73.25	73.11	8.33	7.24	146.53	147.42
4	3.24	3.51	43.33	42.83	6.10	6.89	72.88	72.76	8.33	7.23	148.34	149.31
5	3.96	4.22	43.33	42.87	6.73	7.53	72.32	72.03	8.33	7.21	152.11	153.17
6	7.66	8.12	43.33	43.15	9.97	10.99	68.20	68.31	8.33	7.12	173.36	175.19
7	10.29	10.86	43.33	43.37	12.30	13.43	65.86	65.83	8.33	7.07	188.81	191.62
8	5.93	6.36	43.33	43.09	8.46	9.42	70.29	70.06	8.33	7.19	163.06	164.95

to generate high-pressure steam, which in turn is used to operate a conventional steam power plant. The CCPP presented here as a case study is a 800 MW plant (consisting of 2 modules of 400 MW each) operating near New Delhi (~50 km away) and designed by a reputed public sector company of India. Fig. 6.9 shows the schematic diagram of one module of the plant, consisting of two gas turbines with two waste-heat recovery boilers feeding steam to a single set of HP and LP steam turbines. Each waste-heat recovery boiler consists of a series of heat exchangers arranged in a sequence to extract heat optimally from the hot exhaust coming out of the gas turbine. Thus the very first heat exchanger encountered by hot exhaust is the high-pressure superheater, followed by the high-pressure evaporator. The saturated high-pressure steam produced in the evaporator is superheated in the HP superheater and then sent to the high-pressure steam turbine. The other heat exchangers in the WHRB are, in sequence, the HP economizer 2; low-pressure steam superheater; LP evaporator; a section where the LP economizer and the HP economizer 1 are sharing the hot gases leaving the LP evaporator; and last of all the condensate preheater.

Other components of the plant are the air compressor, gas and steam turbines, liquid pumps, etc. We shall briefly discuss the characteristic features of the simulation of each component and their IFD, and the strategy for system simulation evolved by combining these IFDs.

6.2.1 Component Simulation

As mentioned in Section 4.4.1 to determine the performance of rotating machines like gas compressor, gas/steam turbine, centrifugal pump, etc., from the first principles, by developing detailed process models is an extremely complex task needing extensive CFD simulation of the flow and the temperature fields inside the equipment. Therefore, for their simulation usually the experimental performance data are regressed in the form of equations between pertinent parameters. In so far as the other equipment are concerned, namely the combustion chamber, the evaporators, superheaters, economizers, etc., their detailed process models can be developed using the basic principles of thermodynamics, heat and mass transfer, and fluid flow. This approach has been adopted in developing the strategy for simulation of various components of the CCPP, presented below. For a detailed account of these procedures, the reader may refer to the doctoral dissertation of Seyedan [3] and the paper by Seyedan et al. [4].

Air Compressor

The air compressor used in gas turbine power plant is an axial flow compressor whose characteristics are usually given in the form of graphical relationships between

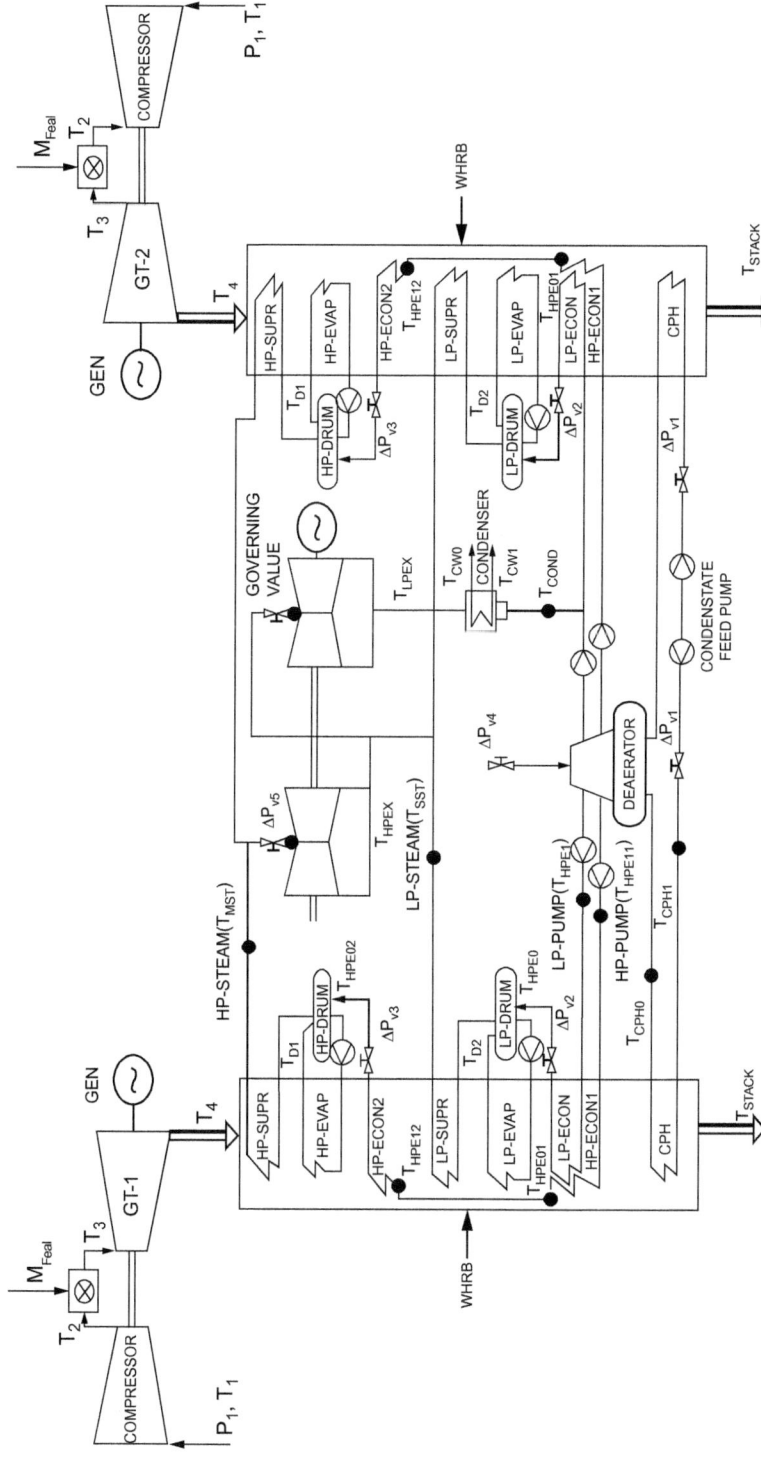

Fig. 6.9 Schematic diagram of one module of the combined cycle power plant.

dimensionless numbers, compressor mass flow parameter, compressor speed parameter, pressure ratio, and efficiency. The first two of these are defined as:

$$\text{Compressor mass flow parameter} = \frac{\dot{m}\sqrt{RT}}{PD^2}$$

$$\text{Compressor speed parameter} = \frac{ND}{\sqrt{RT}}$$

Since the universal gas constant R and the compressor impeller diameter D are constant, often these terms are omitted and dimensional mass flow parameter and speed parameter are defined as:

$$CMP = \frac{\dot{m}\sqrt{T}}{P}$$

$$\text{Speed parameter} = \frac{N}{\sqrt{T}} \tag{6.7}$$

and used to correlate the two performance indices, that is, the pressure ratio and the efficiency in the form of characteristic curves. Using these curves (regressed into polynomial equations) one can calculate, for any value of the mass flow rate and the inlet conditions (state 1) the outlet temperature of air, and the work input needed for compression, with the help of basic thermodynamic relationships (see Eqs. 3.38, 3.46). Thus the complete characteristics of the air compressor can be expressed as:

$$CPR = \frac{P_2}{P_1} = f_1\left(\frac{\dot{m}_a\sqrt{T_1}}{P_1}, \frac{N_c}{\sqrt{T_1}}\right) \tag{6.8}$$

$$\eta_c = f_2\left(\frac{\dot{m}_a\sqrt{T_1}}{P_1}, \frac{N_c}{\sqrt{T_1}}\right) \tag{6.9}$$

$$\Delta T_{21} = T_2 - T_1 = \frac{T_1}{\eta_c}\left[\left(\frac{P_2}{P_1}\right)^{\frac{\gamma-1}{\gamma}} - 1\right] \tag{6.10}$$

$$W_c = CMP \cdot Cpa \cdot \sqrt{T_1}P_1\left(CPR^{\frac{\gamma-1}{\gamma}} - 1\right) \tag{6.11}$$

It is evident from these equations that if the values of just three variables (i.e., P_1, T_1, and $\dot{m}_a$) are known, the compressor speed N_c being constant during steady-state operation, it would be possible to calculate all other parameters using Eqs. (6.8)–(6.11). This suggests the IFD for the air compressor shown in Fig. 6.10.

The Combustion Chamber
The high-pressure air coming out of the compressor (mass flow rate, $\dot{m}_a$) is heated in the combustion chamber by burning natural gas (mass flow rate, $\dot{m}_f$). The natural

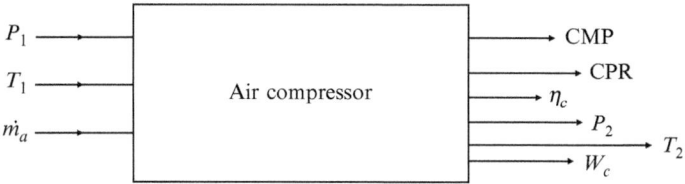

Fig. 6.10 Information flow diagram for the compressor.

gas has mostly methane (84.51%); the other constituents[1] being ethane (7.7%), carbon dioxide (5.2%), propane (2.45%), butane (0.13%), and nitrogen about (0.01%). The final temperature of the products of combustion depends on the air-fuel ratio in the combustor. The theoretical air-fuel ratio for complete combustion, as found by the rules of thermodynamics of reactive mixtures (Section 3.1.2), is 2.1, corresponding to the stoichiometric equation:

$$0.8451\,CH_4 + 0.077\,C_2H_6 + 0.0245\,C_3H_8 + 0.0013\,C_4H_{10}$$
$$+ 0.052\,CO_2 + 0.0001\,N_2 + 2.1\,O_2 + 2.1 \times 3.76\,N_2$$
$$\Rightarrow 1.1298\,CO_2 + 2.0257\,H_2O + [2.1 \times 3.76 + 0.0001]\,N_2 \qquad (6.12)$$

Usually, the air admitted in the combustor is much more than stoichiometric amount. If χ be the percentage of theoretical air supplied then:

$$\chi = \frac{(AF)\ actual}{(AF)} \times 100 = \frac{(\dot{m}_a/\dot{m}_f)}{2.1} \times 100 \qquad (6.13)$$

and the chemical equation for combustion of natural gas is (cf. Eq. 3.107):

$$0.8451\,CH_4 + 0.077\,C_2H_6 + 0.0245\,C_3H_8 + 0.0013\,C_4H_{10}$$
$$+ 0.052\,CO_2 + 0.0001\,N_2 + (\chi/100) \times 2.1\,O_2 + (\chi/100) \times 2.1 \times 3.76\,N_2$$
$$\Rightarrow 1.1298\,CO_2 + 2.0257\,H_2O + [(\chi/100) \times 2.1 \times 3.76 + 0.0001]\,N_2$$
$$+ [(\chi/100 - 1) \times 2.1]\,O_2 \qquad (6.14)$$

Assuming the combustion process to be adiabatic, the temperature of the products of above reaction can be found by using the SFEE for reactive mixtures (Eq. 3.115), which can be concisely written as:

$$\Sigma N_{pi}(h_i^*)_P = \Sigma N_{ri}(h_i^*)_R \qquad (6.15)$$

where

$$h_i^* = h_i(T, P) - h_i(25°C, 1\ atm) + \Delta h_{fi}$$

---

[1] As specified in the plant design.

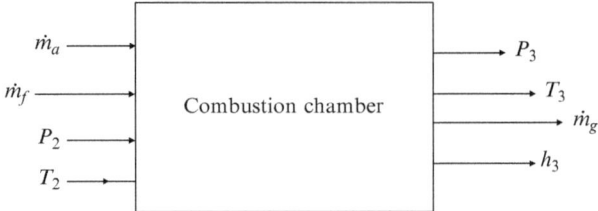

Fig. 6.11 Information flow diagram of combustion chamber.

Δh_{fi} being the enthalpy of formation of compound i. Since the condition of the reactants is known, the value of RHS of Eq. (6.15) can be calculated, and then the temperature of the products of combustion (T_3) is found by iteration so that Eq. (6.15) is satisfied. This is represented by the information flow diagram of Fig. 6.11.

As per the recommendations of the manufacturer, the pressure drop in the combustion chamber is taken as 2.5% of P_2, so that $P_3 = 0.975P_2$. The mass flow rate of gases leaving the combustion chamber $\dot{m}_g$ should clearly be equal the sum of $\dot{m}_a$ and $\dot{m}_f$.

Turbines

The performance of all turbines, irrespective of whether these are gas or steam turbines is represented by dimensionless numbers, that is, turbine mass parameter (TMP), turbine pressure ratio (TPR), turbine speed parameter, and the efficiency η_t. These parameters arising from dimensional analysis are akin to the corresponding parameters for the compressor discussed previously. In the case of turbines, the turbine pressure ratio is usually taken as the independent variable, along with the speed parameter, and curves relating other parameters to these are provided by the manufacturer. These curves can be converted into regression equations to get the following functional relationships for the gas turbines of the power plant, the state points being as indicated in Fig. 6.9 (cf. Eqs. 6.8–6.11):

$$\frac{\dot{M}_T\sqrt{T_3}}{P_3} = f_3\left(TPR, \frac{N_T}{\sqrt{T_3}}\right) \tag{6.16}$$

$$\eta_t = f_4\left(TPR, \frac{N_T}{\sqrt{T_3}}\right) \tag{6.17}$$

$$T_4 - T_3 = \eta_t T_3\left[1 - \left(\frac{1}{TPR}\right)^{\frac{\gamma-1}{\gamma}}\right] \tag{6.18}$$

$$W_T = \eta_t \; TMP \; P_3\sqrt{T_3} \; C_{PT}\left[1 - \left(\frac{1}{TPR}\right)^{\frac{\gamma-1}{\gamma}}\right] \tag{6.19}$$

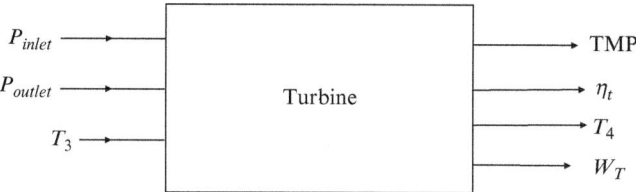

Fig. 6.12 Information flow diagram for turbine.

These calculations are represented by the IFD of Fig. 6.12, TPR being equal to the ratio of inlet and outlet pressures, that is, $TPR = P_{inlet}/P_{outlet} = P_3/P_4$ for the gas turbine of Fig. 6.9. Similar IFDs have been developed for the steam turbines, the input information again being the pressures at the inlet and the exit of the turbine, and the temperature of steam at turbine inlet.

Pumps

There are many liquid pumps in the CCPP. Their performance characteristics are usually available from the manufacturer in the form of head versus capacity and efficiency versus capacity curves. The two commonly employed IFDs for liquid pumps, as discussed in Section 5.1, are shown in Fig. 5.5. The IFD adopted in the present simulation is given in Fig. (6.13).

Here the pressure at its inlet (P_i) and outlet (P_o), and the temperature at inlet T_i are the input variables. Based on these the pump mass flow rate $\dot{m}_p$ and efficiency η_p are calculated using the pump characteristics given by the manufacturer. The other output variables are viz. W_p (pump work), h_o (enthalpy), and T_o the temperature at pump exit are calculated using basic thermodynamic equations:

$$\dot{m}_p = f_5(P_o, P_i) \tag{6.20}$$

$$\eta_p = f_6(P_o, P_i) \tag{6.21}$$

$$W_p = f_7(v_f, P_o, P_i) = \frac{v_f(P_o - P_i)}{\eta_p} \tag{6.22}$$

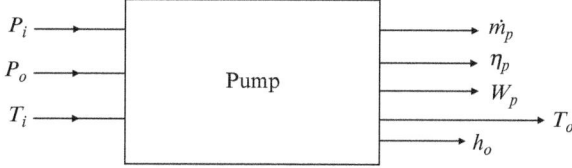

Fig. 6.13 Information flow diagram of a pump.

$$h_o = h_i + W_p \tag{6.23}$$

$$T_o = f_8(h_o, P_o) \tag{6.24}$$

Single-Phase Heat Exchangers

There are numerous single-phase heat exchangers in the power plant like the economizers, the condensate preheater and the superheaters. These are basically cross-flow heat exchangers with hot-flue gases flowing outside the finned tubes and water/steam flowing inside the tubes. The simulation of these heat exchangers is done following the effectiveness—NTU approach discussed at length in Section 4.1.2. The main task is thus that of determining the overall heat-transfer coefficient (Eq. 3.226).

Since the WHRB is recovering heat from a clean air-gas mixture, the fouling on the hot surface is insignificant and the fouling factor is incorporated only for liquid (water) and steam side. The values of heat-transfer coefficient inside the tubes are determined using Gnielinski correlation (Eq. 3.251) while the gas side heat-transfer coefficient is found using Zukauskas and Karni [5] and Zukauskas and Ulinskas [6] correlation, which is basically an equation of the type (Eq. 3.248), giving values of various coefficients/constants over a range of values of Re, and geometric configuration factors such as longitudinal and transverse pitch, tube diameters, fin outer diameter, etc.

The pressure drop of fluid flowing inside the tubes is determined using Darcy equation (3.128), while that of gases flowing through the finned tube bundle is determined by using Eq. (3.139) recommended by Zukauskas. The total pressure drop on the gas-side across a heat exchanger core is calculated by taking into account the acceleration pressure drop, friction pressure drop as also the entrance and exit losses (Eq. 4.127). Accordingly, the IFD of the single-phase heat exchangers is the standard (IFD) of Fig. 5.2, reproduced here (Fig. 6.14) for sake of completeness and to indicate appropriate symbols.

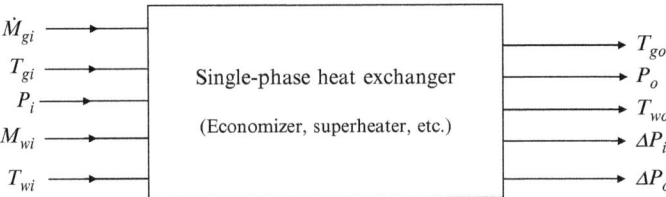

Fig. 6.14 Information flow diagram for single-phase heat exchangers of WHRB.

Evaporator

In the two evaporators of the CCPP, forced convection boiling of water drawn out from the drum takes place. Water flows inside the tubes and is heated by hot gases flowing outside. The water flow rate is kept sufficiently high so that the water is only partially evaporated (typical vapor fraction at exit ~0.2) when it leaves the evaporator and reenters the evaporator drum. The heat-transfer coefficient inside the evaporator changes considerably due to phase change, and its analysis should therefore be done as a heat exchanger with varying overall heat-transfer coefficient, Section 4.1.4. However, in practice it is seen that the variation in U value is not very large since it is strongly influenced by the gas side heat-transfer coefficient which is much lesser than the heat-transfer coefficient of boiling water and remains constant over the heat exchanger length. Accordingly, Seyedan [3] has assumed U to remain constant at its mean value, and analyzed the evaporator as a heat exchanger with one fluid remaining at a constant temperature. The calculations are therefore done using standard NTU-effectiveness approach (Eq. 4.47) as per the IFD of Fig. 6.15, which follows from the standard heat exchanger IFD of Fig. 5.2. Here, instead of the temperature of hot water entering the evaporator, the drum pressure is used as input variable (temperature = saturation temperature at this pressure). Again since all the water is not converted into steam, the dryness function of water-steam mixture leaving the evaporator, x_f (calculated on the basis of the total heat transfer) is kept as an output variable.

The heat-transfer coefficient of water boiling inside the tubes has been found using Chen's correlation (Eq. 3.280–3.286) while the heat-transfer coefficient of hot gases flowing over the finned tubes is found using Zukauskas correlation mentioned above.

Condenser

The condenser used in the power plant is a water-cooled shell and tube-type condenser. As in the case of evaporator, the condenser too is analyzed as a heat exchanger with constant U value and its effectiveness determined using Eq. (4.47). Its information flow diagram (Fig. 6.16) is similar to the standard IFD of Fig. 5.2.

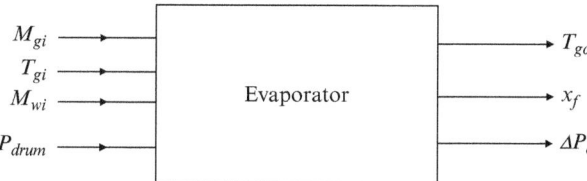

Fig. 6.15 Information flow diagram of the evaporator.

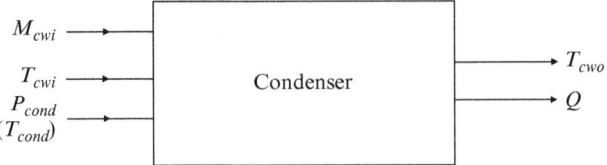

Fig. 6.16 Information flow diagram for the condenser.

Since the heat-transfer coefficient of steam condensing outside tubes is not in-fluenced by its mass flow rate, the input information needed are only three, that is, the mass flow rate and the temperature of the cooling water (M_{cwi}, T_{cwi}) and the temperature/saturation pressure (T_{cond}/P_{cond}) of the steam. The total heat transfer in the condenser Q is kept as an output variable, besides the temperature of the cooling water leaving it (i.e., T_{cwo}).

The heat-transfer coefficient of condensing steam is found using Nusselt's correlation (Eq. 3.268) and that of water flowing through the tubes is determined using Gnielinski correlation (Eq. 3.251).

6.2.2 System Simulation

It is evident that the IFD's of various components of the power plant could be combined in a variety of ways to evolve a strategy for simulation of the complete system. The final strategy which has been adopted after a number of trials is indicated in Fig. 6.17.

To start the simulation of the system, values of nine different variables are assumed. These are

   (i) $\dot{m}_a$: Mass flow rate of air flowing through the compressor
  (ii) P_{HPE1}: Pressure at the exit of high-pressure water pump
 (iii) P_{LPE1}: Pressure at the exit of low-pressure water pump
  (iv) P_{D1}: HP drum pressure
   (v) T_{HPEI2}: Temperature at the inlet to the HP economizer 2
  (vi) P_{D2}: LP drum pressure
 (vii) P_{cond}: Stream pressure in the condenser
(viii) ΔP_{v1}: Condensate control value pressure drop
  (ix) DPB: Total gas-side pressure drop in the WHRB
The sequence of calculations (i.e., simulation of various components) is as follows.

Air Compressor
Referring to the IFD of Fig. 6.10, it is clear that once a value of the mass flow rate of air, $\dot{m}_a$, is assumed, the simulation of the air-compressor can be done since the values of the other two input variables (i.e., ambient air temperature and pressure) are provided externally. The pressure and temperature of the air leaving the compressor (P_2 and T_2)

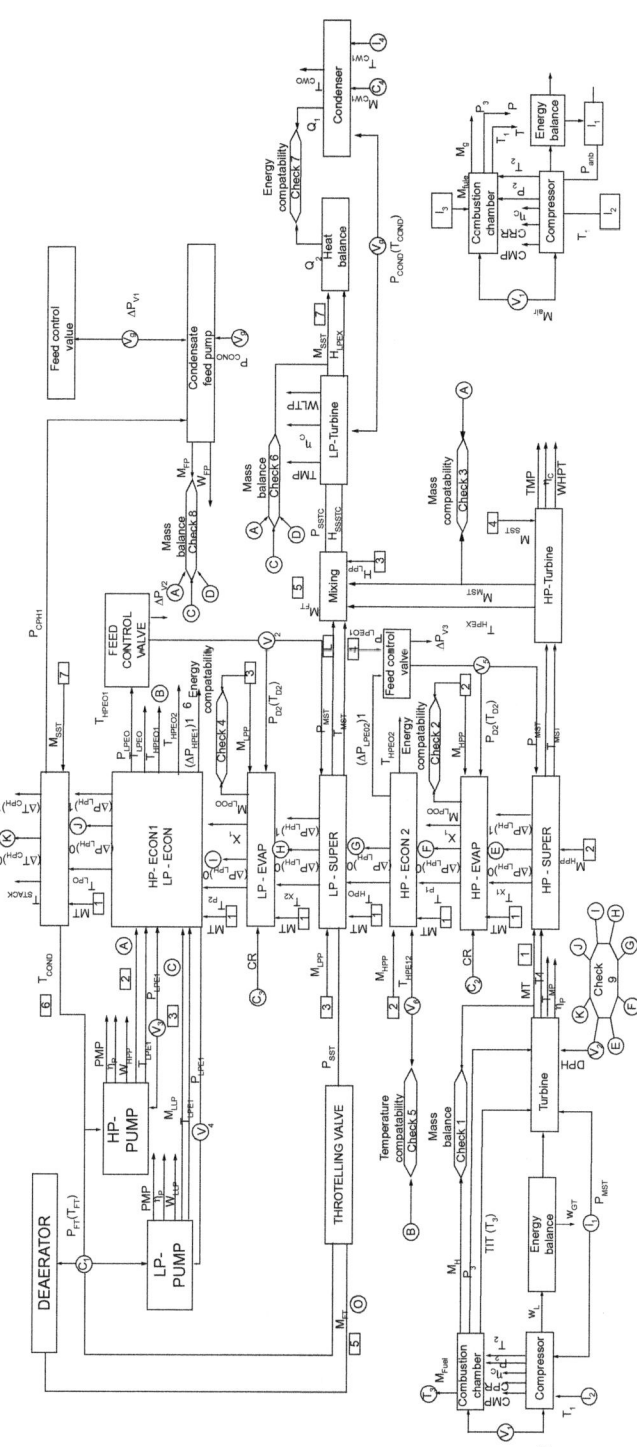

Fig. 6.17 Information flow diagram for a dual-pressure combined cycle power plant.

are two important outputs of compressor simulation, besides the other variables indicated in Fig. 6.10 and also in Fig. 6.17, "compressor" block.

The Combustion Chamber

The mass flow rate of fuel admitted into the combination chamber is an external input. The values of all other input variables in Fig. 6.11 are thus known and therefore the combustion chamber simulation can be done as explained in Section 6.2.1. Thus the temperature (T_3), the pressure (P_3), the mass flow rate ($\dot{m}_g$), and the temperature (TIT) of the hot gases leaving the combustion chamber can be determined (Fig. 6.17, "Combustion chamber" block).

The Gas Turbine

The temperature and pressure at the inlet to the gas turbine (T_3, P_3) are known as output of combustion chamber simulation. The exit pressure is $P_{outlet} = P_4 = P_{atmospheric} + DPB$. Since DPB, the total gas-side pressure drop in the WHRB has been assumed, P_4 is also known and thus the turbine simulation can be done in accordance with the IFD of Fig. 6.12, and all the outputs indicated therein can be obtained. These include its power output W_T, from which we could subtract W_c, the power input in the compressor obtained earlier (Fig. 6.10) to get the net power output of the gas–turbine cycle of the CCPP. Further, we also get the values of the turbine mass parameter TMP ($= \dot{m}_T \sqrt{T_3}/P_3$) from which mass flow rate through the turbine, $\dot{m}_T$, can be calculated. It should equal the mass flow rate leaving the combustion chamber, $\dot{m}_g$, calculated earlier. This provides us with the first check on the nine assumptions made to initiate the simulation process (Block check 1, in Fig. 6.17).

HP, LP Pumps

Before initiating the simulation of various heat exchangers in the WHRB, the simulation of the two water pumps is done. Since the inlet pressure to the pump is fixed (=deaerator pressure), and the exit pressures P_{HPE1} and P_{LPE1} have been assumed, the simulation of the pumps can be carried out as explained previously; IFD in Fig. 6.13. The exit temperature and the mass flow rate, and the work input needed for each pump can be determined using Eqs. (6.20)–(6.24); see blocks "LP pump" and "HP pump" in Fig. 6.17.

HP superheater, Evaporator, Economizer

The hot gases leaving the gas turbine first impinge on the HP superheater inside the WHRB (Fig. 6.9). Since the temperature and flow rate of both the hot gases and water entering the HP superheater are now known, its analysis can be done as explained in the sub-section "Single phase heat exchangers" in Section 6.2.1, and the state of the outgoing streams can be found (see Fig. 6.14). Next the hot gases flow across the evaporator tubes. Since the pressure of steam in the HP drum, P_{DI} has been already assumed, and the value of the circulation ratio ($CR = \dot{m}_{wevap}/\dot{m}_{HPP}$) is externally specified, all the input information needed for simulation of the evaporator is now available (Fig. 6.15).

The temperature of the hot gases leaving the evaporator (and flowing to the HP economizer 2), as also the drop in the pressure, can be determined from the evaporator simulation procedure. This forms the input information for the analysis of HP economizer 2, where HP water is heated by the hot gases. The flow rate of water is known, but its temperature at entry is not known a priori since the extent to which it is heated in economizer 1 will be estimated much later. That is the reason for assuming the temperature at the inlet to economizer 2 (variable T_{HPEI2}). Thus all the input information needed for its analysis is now available and its simulation can be carried out (Fig. 6.14). The temperature of water leaving this economizer and entering the HP evaporator can thus be found. This completes the information regarding the four streams entering/leaving the HP evaporator drum (Fig. 6.9). Its energy balance provides a second check (check 2 block, Fig. 6.17) on the assumptions made initially.

The pertinent input and output variables in the terminology of CCPP are shown in Fig. 6.17, blocks named, "HP-super," "HP-Evap," and "HP-Econ 2."

LP Super Heater, LP Evaporator

As can be seen from the schematic diagram of the CCPP Fig. 6.9, the hot gases leaving HP economizer 2 flow to the LP superheater and LP evaporator in sequence. Accordingly their simulation is also done in this sequence following an approach similar to that followed for HP superheater and HP evaporator. The mass flow rate of water-steam mixture in the LP evaporator is again found by multiplying the externally specified value of circulation rates CR, with the water mass flow rate through the LP pump, calculated earlier.

Once the pressure and temperature of the steam leaving LP superheater are known, the analysis of the turbines can be carried out.

HP, LP Turbines

The condition of steam at the inlet to the HP turbine is known from the output of HP superheater. The pressure at the exit of HP turbine equals that at the exit of LP superheater, which has now been determined. Thus all the input information needed for the analysis of HP turbine (Fig. 6.12) is available, and its performance can be determined. The temperature of the steam at HP turbine exit is now known and we can determine the conditions after mixing of this stream with that coming out of LP superheater. This gives us the conditions at the entry to the LP turbine. The pressure at the exit of LP turbine is the condenser pressure, P_{cond}, assumed in the beginning. Thus the analysis of the LP turbine can also be carried out and its work output, as also the state and the mass flow rate of steam leaving it (and entering the condenser), can be determined. The mass flow rate of steam as calculated from LP turbine characteristics should match with that obtained earlier from the mixing of steams leaving HP turbine and that leaving LP superheater. This provides us with another check on the initial assumptions, indicated as check 6 in Fig. 6.17. Similarly the mass flow rate through HP turbine, as calculated on the basis of its characteristics should be equal to the mass flow rate through HP pump. This compatibility requirement is indicated in Fig. 6.17 as the block "Check 3."

Condenser

Since the condenser pressure is assumed initially, the mass flow rate and the temperature of the steam entering it has been determined from LP turbine analysis, and the mass flow rate and the temperature of the cooling water entering the condenser are externally specified, the simulation of the condenser can be carried out (Fig. 6.16). The total rate of heat transfer thus calculated is checked against that needed to condense the incoming steam at the assumed pressure. This heat balance provides another check on the assumptions. It is indicated in the system simulation IFD, Fig. 6.17, by the box "Check 7."

LP Economizer and HP Economizer 1

In the next section of the WHRB, following the LP superheater and the evaporator, the hot gases leaving the LP evaporator flow through LP economizer and HP economizer 1 arranged in parallel (see Fig. 6.9). Therefore, these two heat exchangers are analyzed together, and the drop in the temperature of hot gases is due to the total heat transfer to both of them. Accordingly, in the IFD for system simulation, Fig. 6.17, these are shown together in a single block. Since both of these are single-phase heat exchangers, their analysis is done using the generic IFD of Fig. 6.14. The simulation provides us the values of water temperature at the exit of HP economizer 1 (T_{HPE01}) and the LP economizer (T_{LPE0}). Since the temperature at the exit of HP economizer 1 must equal the earlier assumed temperature at the inlet to HP economizer 2, this provides us with another check on the initial assumptions. This is indicated by the block "Check 5" in Fig. 6.17. Again, as in the case of the HP evaporator, since the water leaving the LP economizer, enters the drum of LP evaporator (Fig. 6.9), knowing its thermodynamic state enables us to check the energy balance of the drum, the state of other three streams being known. This provides us with yet another check on the assumptions, indicated as "Check 4" in the IFD of Fig. 6.17.

Condensate Preheater

The condensate preheater, placed at the rear of the WHRB, can now be analyzed since all the input information necessary for the simulation of this heat exchanger, that is, the temperature and mass flow rate of hot gases, and the temperature and mass flow rate of condensate entering it from the condenser, are known. Thus the temperature and pressure of hot gases entering the stack are determined as also the temperature of the condensate preheated therein. The pressure drop in it is also known.

Condensate Feed Pump and Deaerator

Since the pressure drop across the condensate control value (i.e., ΔP_{v1}) has already been assumed, the pressure drop in the condensate preheater has been calculated above, and the pressure in the deaerator is an externally specified constant value, the pressure at the exit of the condensate feed pump can be determined to be $= P_{deaerator} + \Delta P_{condensate\ preheater} + \Delta P_{v1}$. The pressure at its inlet is the condenser pressure. Thus knowing the required pressure head, the mass flow rate pumped by it can be found using pump characteristics (Fig. 6.13). Similarly the flow rate through the valve connecting the LP steam header to the deaerator can be found from the specified valve characteristics. Since the mass flow rates through the LP and HP pump have already been determined, one can check for the mass balance of the deaerator. This is indicated as "Check 8" in the IFD of Fig. 6.17.

The ninth and the final check on the assumptions is provided by comparison of the total gas–side pressure drop in the WHRB, which equals the sum of calculated gas–side pressure drops while flowing through all the eight heat exchangers, with the initially assumed value of the total pressure drop DPB. This is indicated as a hexagonal box marked "Check 9" near the bottom of Fig. 6.17.

These nine "checks" can be visualized as nine nonlinear equations in nine variables whose values were assumed to start the simulation of the CCPP. The task of system simulation thus reduces to finding appropriate values of these nine variables so that the compatibility conditions are satisfied. This has been done by using Warner's method (Section 3.2.1) and the predicted performance of the power plant obtained at various loads for which the rated values were made available by the manufacturer. Tables 6.4, 6.5 and 6.6 show a comparison of the rated power output and the operating conditions with the values predicted by the system simulation procedure.

Close agreement between the predicted and the rated performance validates the complete system simulation as also the component simulation procedures.

6.3 LIQUID DESICCANT-BASED AIR-CONDITIONING SYSTEM (LDAC)

The third case study presented below is quite different from the above two. The earlier studies involved simulation of industrial systems actually in use in the field, while in this section we shall present the procedure developed for predicting the performance of a proposed novel air-conditioning system based on liquid desiccants. Using this procedure, a detailed parametric analysis was carried out to identify the factors which would strongly influence its performance. The ultimate objective was to develop optimal design of the system to meet typical air-conditioning requirements and then

Table 6.4 Comparison of the predicted and the rated operating conditions at various loads

Fuel flow rate (kg/s)	Compressor pressure ratio		Turbine inlet temp. (°C)		Compressor air flow (kg/s)		Main steam flow (kg/s)	
	Comp.	Rated	Comp.	Rated	Comp.	Rated	Comp.	Rated
6.23	7.37	7.34	941	943	348	348.02	89.06	91.25
7.76	8.73	8.59	1003	1011.7	401.3	394.74	107.12	108.31
9.01	9.42	9.41	1104	1090	412.8	414.76	128.44	133.83
9.43	10.19	10.24	1055	1050	458.7	462.16	124.8	125.2
10.0	10.68	10.39	1080	1090	478.2	462.16	132.01	135.87

Table 6.5 Comparison of the predicted and the rated operating conditions at various loads (contd)

Secondary steam flow (kg/s)		Main steam pressure (bar)		Secondary steam pressure (bar)		Condenser pressure (bar)	
Comp.	Rated	Comp.	Rated	Comp.	Rated	Comp.	Rated
107.14	107.82	45.68	47.3	3.68	3.70	0.087	0.088
127.9	127.32	54.6	56.41	4.42	4.44	0.099	0.098
150.54	153.78	68.27	69.67	5.21	5.30	0.138	0.112
149.8	149.4	64.0	65	5.22	5.14	0.114	0.110
151.36	159.15	68.43	70.73	5.48	5.49	0.119	0.116

Table 6.6 Comparison of the rated and the predicted power outputs

Fuel flow rate (kg/s)	Compr. inlet temp. (°C)	Cond. coolant temp (°C)	Gas cycle		Steam cycle		Total	
			Comp.	Rated	Comp.	Rated	Comp.	Rated
6.23	27	32	139.06	139.24	109.0	108.3	248.06	247.97
7.76	27	32	199.8	199.2	128.5	131.1	328.3	330.3
9.01	50	36	236.9	235.6	156.5	158.83	393.4	392.0
9.43	27	32	259.9	260.4	150.9	151.0	410.8	410.0
10.0	27	32	281.0	279.0	158.6	162.4	439.6	441.4

compare its initial and operating costs with those of a conventional air-conditioning system (Section 9.3.3). This case study demonstrates the utility of system simulation in development of innovative thermal systems.

In desiccant-based air-conditioning systems, evaporative cooling is used to achieve drop in temperature of air, and desiccants are used to remove moisture. Further, air-to-air heat exchangers are used to improve the energy efficiency of the system. Jain et al. [7] carried out an extensive study of various configurations of the desiccant cooling systems, and identified the cycle shown on a psychometric chart in Fig. 6.18 as most suited for Indian climatic condition. Fig. 6.19 shows the schematic diagram of the system based on this cycle.

The supply air at state 4 picks up heat and moisture in the room to maintain it at state 5. A part of room air is exhausted, but the bulk is taken out for recirculation after processing. This air is first evaporatively cooled to state 6, and then this "cooled" air is used to precool the dehumidified air leaving the absorber, in air-to-air heat exchanger. The air, now at state 7, is mixed with fresh outdoor air needed for ventilation and the mixture sent to the water-cooled liquid-desiccant-based moisture absorber. The dehumidified air at state 2 is precooled to state 3 in the heat exchanger 1 (Fig. 6.19) where it transfers heat to the evaporatively cooled room air (process 6–7) as mentioned earlier. The dehumidified air at state 3 is further cooled to state 4 in an evaporative cooler. Here it also gets humidified, process 3–4, Fig. 6.18, but since the moisture content at state 3 is very low, even after humidification in the evaporative cooler the supply air at state 4 has moisture content lower than in the room. It can therefore pickup both moisture and heat generated in the room before reaching the room air condition, state 5. Thus the air cycle is completed.

For continuous operation of the system it is necessary to regenerate the liquid desiccant, which picks up moisture from the air while dehumidifying it from state 1 to 2. This is indicated in the schematic diagram shown in Fig. 6.19. The "weak"

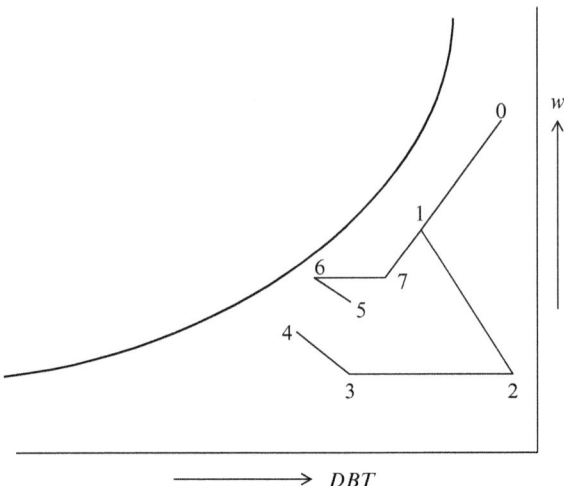

Fig. 6.18 Representation of the LDAC system on the psychrometric chart.

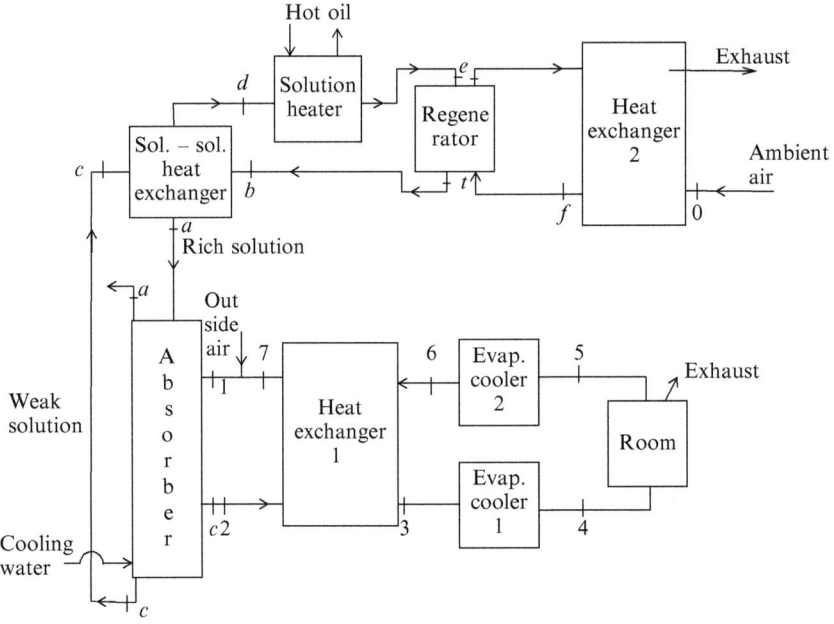

Fig. 6.19 Schematic diagram of the proposed LDAC system.

solution leaving the absorber[2] is first heated in the heat exchanger (SSHE) with the help of hot "rich" solution coming out of the regenerator. The "weak" solution is further heated in the solution heater (SH) using any heat source like burning natural/biogas, solar energy, etc. This hot weak solution is then "concentrated" in the regenerator by bringing it in contact with ambient air which picks up moisture from the hot solution. The concentration of the desiccant in the solution increases and it is taken out for reuse in the absorber. Before being sent to the absorber, this rich solution is cooled down in the SSHE, in the process heating the weak solution entering the heat exchanger from the absorber. The hot humid air leaving the regenerator is used to preheat the ambient air (in the air-to-air heat exchanger 2; Fig. 6.19) before it enters the regenerator.

The absorber and the regenerator have conventional falling film design. In the absorber, we have a shell and tube type of configuration, with air and the desiccant solution flowing downward inside the tubes. The cooling water flows in the shell, outside the tubes, to remove the heat of absorption of moisture in the desiccant. The regenerator has a plate-type configuration with the hot "weak" solution flowing downward and the air, preheated in heat exchanger 2 (Fig. 6.19), flowing upward. As the moisture is picked up from the desiccant solution, the concentration of the desiccant in the solution increases, and at the bottom of the regenerator we have "rich solution" ready to be reused in the absorber for dehumidification.

6.3.1 Component Simulation
The Absorber and the Regenerator

The absorber and the regenerator are essentially heat and mass exchangers. In both, heat and mass transfer occur between the falling desiccant film and air coming in contact with it. In the absorber we have, in addition, water circulating around the tubes through which air and desiccant are flowing, and it carries away a part of the heat of adsorption of water in the desiccant. Their simulation needs a step-by-step solution of the governing heat and mass transfer equations as explained in Chapter 4, Section 4.2.3.

In the absorber, both the desiccant solution and air enter at the top (point a) but cooling water enters at the bottom (location c Fig. 6.19). So the step-by-step solution is facilitated if an assumption is made regarding the temperature of the cooling water leaving the absorber (T_{Wa}). Its validity could be checked by comparing the predicted value of water temperature at location c with the given value. Accordingly, the IFD for the absorber is formed, as shown in Fig. 6.20.

The input conditions (all at location a, i.e., top of the absorber) are the mass flow rate, the temperature, and the concentration of desiccant ($\dot{m}_{sa}$, TS_a, C_a); mass flow rate

---

[2] Liquid-desiccant solution is said to be "weak" when the concentration of desiccant falls down after passing through the moisture absorber, and is said to be "rich" after some moisture is driven off in the regenerator.

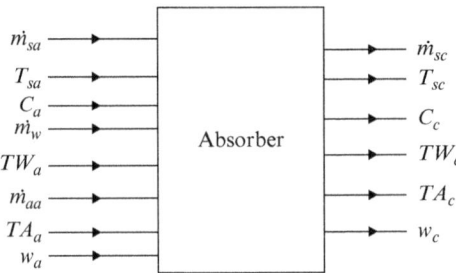

Fig. 6.20 Information flow diagram of the absorber.

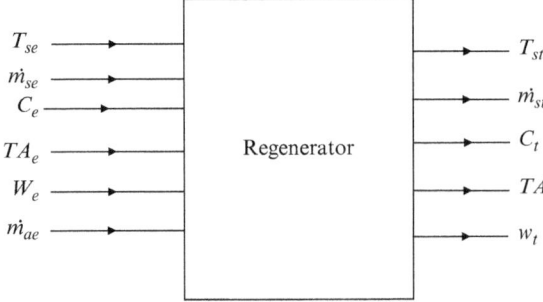

Fig. 6.21 Information flow diagram of the regenerator.

and temperature of cooling water ($\dot{m}_w$, TW_a); and the mass flow rate, the temperature, and the humidity ratio of air ($\dot{m}_{aa}$, TA_a, W_a), and the output conditions are the values of these properties (except $\dot{m}_w$, which remains unchanged) at location c, which are obtained by step-by-step solution of the governing differential equations (see the "Mass balance of moisture content" in Section 4.2.3).

In a similar manner the IFD of regenerator can be constructed. Here we have only two streams—the desiccant solution falling down from the top (location e in Fig. 6.19) and the air entering at the bottom (location t in Fig. 6.19).

Fig. 6.21 shows the IFD. Here too, the input information is the complete set of properties of the two fluids at the top location of the regenerator, and the output information is the set of properties at the bottom of the regenerator.

The Heat Exchangers
The system uses four heat exchangers. Two of these are air-to-air heat exchangers, viz. one in the supply air circuit between the return air (state change 6–7) and the process air (state change 2–3); and another in desiccant solution circuit between the ambient air before entry to the regenerator (state change 0–f) and the humid air

leaving the regenerator at state e. The remaining two heat exchangers, viz. SSHE and SH in Fig. 6.19, involve heat transfer between liquids. All these heat exchangers can be analyzed using the standard NTU-effectiveness approach outlined in Section 4.1, and their information flow diagrams can be chosen from the alternative IFDs of heat exchanger discussed in Chapter 5 (Figs. 5.2 and 5.3), depending on the requirements of overall system simulation and the availability of values of various variables. To determine the NTU of the heat exchangers, the values of heat-transfer coefficients on both sides of the heat-transfer surface are to be found. This is done using Gnielinski's correlation (3.251) for all internal/confined flows and Eq. (3.248) for flow outside tube arrays in air-to-air heat exchangers.

Evaporative Coolers

The system uses two evaporative coolers to reduce air temperature with the help of evaporating water. The detailed simulation of the process coolers has not been done, instead a certain humidifying efficiency η_H exogenously input. This enables direct calculation of the outlet temperature from the known inlet conditions:

$$T_{outlet} = T_{inlet} + \eta_h(WBT_{inlet} - T_{inlet}) \tag{6.25}$$

Since WBT remains constant during the process of adiabatic saturation encountered in these coolers, $WBT_{outlet} = WBT_{inlet}$ and the outlet moisture content w_{outlet} can be found using the values of T_{outlet} and WBT_{outlet} in the psychrometric chart. Eq. (6.25) is also used to estimate the temperature of water leaving the cooling tower, TW_{cTO}, on the basis of the specified efficiency of the tower (I_{10}) and the condition of the inlet air (I_8, I_9) (see Fig. 6.22).

6.3.2 System simulation

As in the case studies presented earlier, the strategy for the simulation of the LDAC system has been devised by combining the information flow diagrams of individual components judiciously. The IFD for the system finally adopted after a number of trials is shown in Fig. 6.22. As indicated in the figure, the values of seven variables are assumed suitably to carry out the simulation of all the components and seven compatibility conditions have been obtained which check for the correctness of these assumptions. The variables assumed (indicated as A_1, A_7 in Fig. 6.22) are

A_1: Solution temperature at absorber inlet (top), TS_a
A_2: Solution concentration at absorber inlet (top), C_a
A_3: Temperature of water at absorber top, TW_a
A_4: Temperature of air entering absorber, TA_a
A_5: Humidity ratio of air entering absorber, W_a
A_6: Temperature of air leaving regenerator, TA_e
A_7: Humidity ratio of air leaving regenerator, W_e

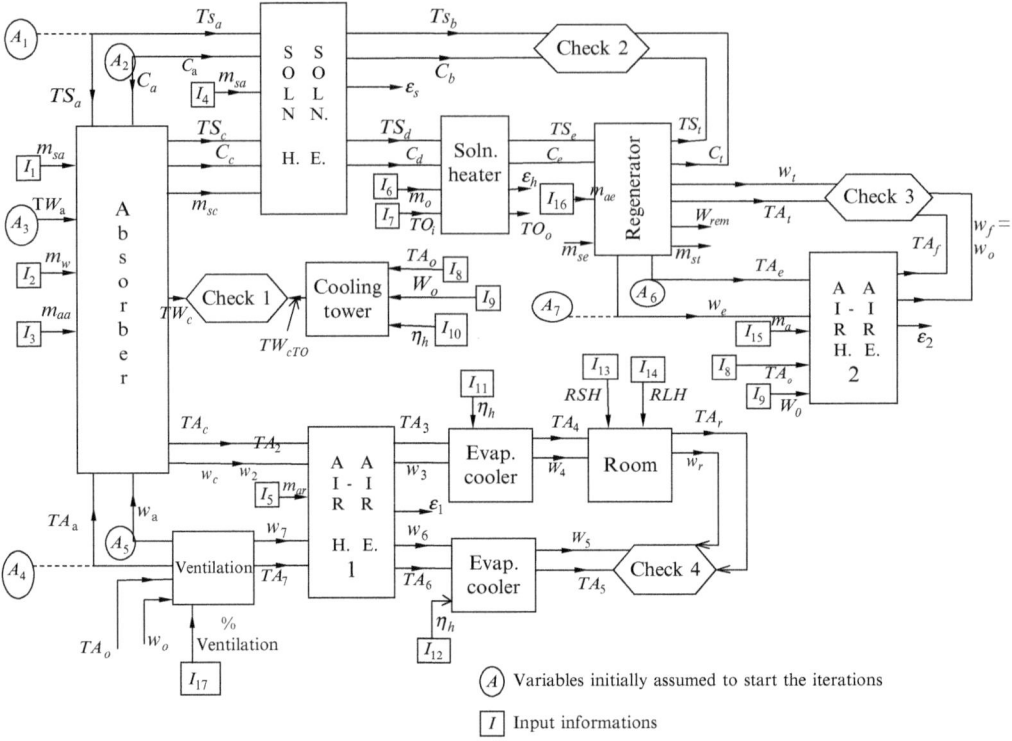

Fig. 6.22 Information flow diagram of the complete LDAC system.

The values of a large number of variables (17 in all, $I_1 \rightarrow I_{17}$) have been externally input during the simulation. Since we have not incorporated pressure drop calculations in our analysis (for the sake of simplicity, and for other reason that will be explained later), the mass flow rates of all the streams—the rich solution, the weak solution, air, water, heating oil in SH—are external inputs ($I_1 \rightarrow I_6, I_{15}, I_{16},$). The other inputs include temperature of heating oil in SH (I_7), the temperature and humidity of outdoor air (I_8, I_9), the humidifying efficiencies of the cooling tower and the evaporative coolers ($I_{10} \rightarrow I_{12}$), the room sensible and latent heat loads (I_{13}, I_{14}), and the ratio of ventilation air to return air (I_{17}). Knowing the values of all these variables the calculations are done in the following sequence.

The absorber is the first component to be analyzed. As can be seen from Fig. 6.22, all the input information needed for its analysis is available (either assumed $A_1 \rightarrow A_5$, or input externally $I_1 \rightarrow I_3$). Thus absorber simulation can be carried out and the values of variables indicated as output information in the "absorber" block (Fig. 6.22) can be determined. The mass flow rate, the temperature, and the concentration of the weak solution leaving the absorber at location c (i.e., $\dot{m}_{sc}, TS_c, C_c$) and the temperature of water at section c (TW_c) are thus known. The simulation of the SSHE can now be carried out since all the properties of the cold weak-solution stream entering it (state c) and that of the rich solution stream leaving it (state a, Fig. 6.19) are known. Through this simulation the properties of these two streams at the other end of the heat exchanger (i.e., TS_b, C_b and TS_d, C_d) can be determined.

The weak solution thus preheated in the SSHE enters the solution heater (SH) at state d. Here it is heated with the help of the heating medium (taken as hot oil) entering at prescribed temperature TO_i and mass flow rate $\dot{m}_o$. Simulation of this heat exchanger can thus be carried out to get the temperature of the weak solution at entry to the regenerator, TS_e. Its concentration of course remains unchanged and so $C_e = C_d$.

The weak solution leaving the solution heater at state e enters the regenerator. Since the properties of air leaving it at location e are assumed (A_6, A_7), the simulation of the regenerator can be carried out by step-by-step solution of the governing differential equations starting from the top and proceeding downwards. Thus the states of air and the desiccant solution at regenerator bottom, location t (Fig. 6.19), can be found out.

To complete the "solution circuit" of the LDAC system, the analysis of the air-air heat exchanger coupled to the regenerator is carried out. The state of hot, humid air entering it (at e) is known from the initial assumptions and that of the ambient air is external input (I_8, I_9, I_{15}). Thus all the data needed for its simulation are known and the state of air leaving it and entering the regenerator (state f) can be found. This is checked against the values obtained through regenerator analysis, as discussed later.

The simulation of the process air circuit is tied up with that of the absorber. The condition of air entering the absorber TA_a, w_a having been initially assumed (A_4, A_5), is actually the state of air at point 1 (Fig. 6.19) on the air side, resulting from the mixing of the ambient air admitted for ventilation and the air leaving the air-to-air heat exchanger 1, at state 7. As the percentage of ventilation air admitted is an external input (I_{17}) using the basic equations for analyzing mixing of two streams (Eqs. 3.86, 3.87), the state of air at location 7 be found from the known temperature and humidity of the ambient air and at state 1.

The analysis of the air-to-air heat exchanger 1 can now be carried out using the known states of the two streams, that is, state 2 and state 7 (Fig. 6.19), by using the standard effectiveness—NTU approach. Thus we get the temperature and humidity ratio of air at locations 3 and 6. The analysis of the two evaporative coolers is carried out using the specified values of their humidifying efficiencies. Thus the state of air supplied to the room (state 4) and that leaving the room (state 5) are known. Next the analysis of the room loads is done. Since both the room sensible heat again (RHS) and the room latent heat gain (RLH) are specified inputs, and the state of supply air (state 4) is known, we can calculate the resulting room conditions, that is, the room air temperature TA_r, and its humidity W_r. These can be checked against the values obtained from analysis of the evaporative cooler 2 (state 5) as indicated in the IFD of Fig. 6.22. This completes

Table 6.7 Compatibility conditions for simulating LDAC

Check no.	Equation	Remarks
1	$TW_c = TW_{cTO}$	TW_{cTO} calculated using Eq. (6.25)
2	$TS_b = TS_t; C_t = C_b$	The states of the rich solution leaving regenerator and that entering the SSHE obtained independently from their analysis
3	$TA_t = TA_f; w_t = w_f$	The states of air leaving HE2 and that at entry to the regenerator obtained independently from their analysis
4	$TA_s = TA_r; w_5 = w_r$	The states of room air as obtained from the analysis of the supply air and the room loads, and that obtained from the analysis of the evaporative cooler EC_2

the one cycle of simulation of the complete system. The seven conditions which check for the correctness of the initial assumptions are indicated in Table 6.7.

Thus the task of system simulation is, from a mathematical perspective, reduced to finding the correct values of the seven assumed variables which satisfy the these seven equations. As in the earlier two case studies this has also been done using Warner's method (Section 3.2.1)

In order to validate the model an experimental set up was built in the IIT Delhi laboratory [8]. During the experimentation it became evident that the absorber tubes and the regenerator plates were not getting properly "wetted" by the falling film of the desiccant solution. As a result not all the surface area is available for heat and mass transfer. The comparison between the experimental results and the simulation predictions was used to estimate the fraction of the actual area available for heat and mass transfer in both the absorber and the regenerator. These two factors F_w and F_h (w: wet h: heat transfer) were computed for various operating conditions. It was noted that the typical values of F_w and F_h for the regenerator ($F_w = 0.6$ and $F_h = 1.0$) were higher than those for the absorber ($F_w = 0.2$ and $F_h = 0.7$). This was attributed to differences in the design of the liquid spray mechanisms and led to development of innovative designs of liquid spray devices and subsequently the regenerators [9].

6.4 EPILOG

The utility of system simulation in engineering practice must be evident from the above three case studies. It enables us to predict the off-design performance of complex thermal systems, which is very helpful in developing control systems for efficient operation under part load and also for identifying their safe operating range.

It can also help in identifying the components which are contributing the most to underperformance of the system and thus point out the direction in which developmental effort should be focused to get the maximum improvement in performance.

The most significant use of such system simulation programs is in developing better designs of various components of complex thermal systems like a thermal power plant or a refrigeration plant. As brought out in Chapter 1 through the simple example of design of a heat exchanger, we can develop numerous feasible designs of any thermal system, meeting with all the design requirements, by changing the values of variables like tube sizes, their number, fluid velocities, configurations of heat exchangers, sizes of pumps or compressor, etc. We could then choose among them based on a suitable optimization criterion. Since the number of feasible designs is quite large, it is necessary to devise suitable strategy to narrow down the search domain in consonance with the optimization criterion. This can be most efficiently done with the help of mathematical/numerical optimization techniques. These help us to generate progressively "better" designs of various components of a system whose performance is simulated, at each step of the search, by using system simulation programs of the kind presented in this chapter. We shall discuss this whole approach in detail in the following chapters.

REFERENCES

[1] P.L. Dhar, G.R. Saras, Computer Simulation, and Design of Refrigeration System, Khanna Publishers, New Delhi, 1987.
[2] G.R. Saraf, Computer simulation and optimization of a refrigeration system, PhD thesis, Department of Mechanical Engineering, IIT Delhi, 1979.
[3] B. Seyedan, Simulation and optimization studies of a combined cycle power plant, PhD thesis, IIT Delhi, 1995.
[4] B. Seyedan, P.L. Dhar, R.R. Gaur, G.S. Bindra, Computer simulation of a combined cycle power plant, Heat Recov. Syst. CHP 15 (7) (1995) 619–630.
[5] A. Zukauskas, J. Karni, High Performance Single Phase Heat Exchangers, Hemisphere Publishing Co., New York, 1989.
[6] A. Zukauskas, R. Ulinskas, Heat Transfer in the Tube Banks in Cross Flow, Hemisphere Publishing Co., New York, 1988.
[7] S. Jain, P.L. Dhar, S.C. Kaushik, Evaluation of liquid desiccant based evaporative cooling cycles for typical hot and humid climates, Heat Recov. Syst. CHP 14 (1994) 621–632.
[8] S. Jain, Simulation, design and fabrication of a liquid desiccant augmented evaporative cooling system, MTech thesis, Mechanical Engineering Department, IIT Delhi, 1990.
[9] A.K. Asati, Performance studies of falling film absorber and regenerator, PhD thesis, IIT Delhi, 2007.

CHAPTER 7

Introduction to Optimum Design

While designing a complex thermal system—or any of its individual components—a designer is faced with the task of selecting from among numerous alternative designs which fulfill the main technical requirements like the tonnage of a refrigeration plant or the MWs of power to be generated by a power plant. As illustrated in Chapter 1 through the example of design of a heat exchanger to recover heat from the flue gases leaving a DG set, we have considerable choice regarding its configuration, the number, the length, and the diameter of the tubes used, the longitudinal and lateral pitches of the tube pattern, etc. Thus we could literally have thousands of feasible designs all recovering the heat at the desired rate. Which of these should the designer recommend?

The answer to this question would depend upon the "other" requirement/preferences of the client. Often the client is interested in a product with the lowest initial cost; then this forms the basis of selection of design among the set of a feasible designs. There could clearly be many other objectives like minimal running cost, minimal life cycle cost, minimal carbon footprint, minimal floor area requirement for installation, etc. If there are more than one objectives, usually one of these is chosen as the objective function to be optimized (minimized or maximized as the case may be) and the rest are converted into constraints to be satisfied by the "optimum" design. Thus while designing a power plant, the overall cost of the plant could be the objective function to be minimized, while the net power requirement could be kept as a constraint to be satisfied. Similarly while designing the fins for cooling a VLSI circuit, minimizing the weight of fins could be the objective function and the given cooling duty could be a constraint to be satisfied by any feasible design.

The constraints on the design can also arise from other intrinsic requirements. Thus in heat exchangers involving flow over a tube bank we have geometrical constraints. In in-line tube configuration the tube pitches have to be greater than the tube diameter, that is, $P_T > D$, $P_L > D$ (Fig. 3.7A). In staggered tube configuration the diagonal pitch P_D also becomes important, $P_T > D$ and $P_D > D$ (Fig. 3.7B). The conservation laws also introduce some constraints. While designing the reciprocating devices, there may be constraints on the stroke/bore ratio. Constraints may also have to be imposed for other "nonthermal" reasons like avoiding excessive noise, erosion, or even vibrations. While optimizing a component of a complex thermal system, additional constraints may have to be put to ensure that introduction of the new component would not adversely

Thermal System Design and Simulation
http://dx.doi.org/10.1016/B978-0-12-809449-5.00007-3

affect the overall system performance. We shall illustrate it later with the example of optimizing only the evaporator of a refrigeration plant.

Once the objective function and the constraints are identified, the variables which determine the equipment design are finalized. Not all these variables may influence the objective function; some may appear only in the constraints.

7.1 GENERAL FORMULATION OF AN OPTIMUM SYSTEM DESIGN PROBLEM

In the light of the above brief discussion it should be clear that formulation of any optimum design problem has three aspects; that is, identification of
(a) the objective function;
(b) the constraints; and
(c) the variables to be optimized.
The objective function would be a mathematical function of all (or some) of the variables. Similarly the constraints would be either directly on the values of the variables or on certain relationships between them. The type of optimization method that would be suitable for a problem depends primarily on the nature of the objective function and the constraints (Chapter 8). Thus if both are linear functions, very well-established analytical methods are available for finding the optimum values of the variables. If the objective function is a polynomial function of the variables, again, in some cases, exact analytical solutions can be obtained. However in the most commonly encountered situations both the objective function and the constraints are complex nonlinear functions of the variables necessitating the use of numerical methods.

The constraints are further classified into two categories, namely equality constraints and inequality constraints. Thus the requirement that the power output of a plant be 500 MW can be expressed as an equation:

$$Power = 500 \, MW$$

On the other hand the requirement that gas velocities anywhere in a heat exchanger should not exceed 14 m/s would be expressed as an inequality constraint:

$$Velocity \leq 14 \, m/s$$

The variables which we come across in thermal system design can be classified into three categories, namely continuous, discrete, and logical variables. The variables, which can have real numbers as their value, are called continuous variables in contrast to discrete variables which can have only certain discrete values, but not all values within the range. Thus the length of a heat exchanger is a continuous variable while the number of its passes is a discrete variable (we may have 2-pass or 3-pass heat exchanger; a 2.5-pass heat exchanger makes no sense). Some other variables pertinent to thermal systems are

Continuous variables: Tube pitch, baffle pitch, mass flow rate, and temperature.

Discrete variables: Number of baffles, number of tubes in various passes, and number of stages in multistage compression.

The third category of variables—the "logical" variables—are those which do not have any numerical value but indicate configuration choice; for example, the relative directions of the two fluids flowing in a heat exchanger, the tube arrangement (in line or staggered tube configuration), type of tubes (bare or finned), etc. Such variables can not be directly admitted into a mathematical optimization formulation and have to be treated one-by-one through scenario building approach.

We shall illustrate the methodology of problem formulation through a few examples.

Example 7.1

Consider the problem of designing the indirect water heating system of Fig. 5.14 where process air (flow rate 15 kg/min) is to be heated by 50°C in heat exchanger II, using the heat-transfer oil heated by flue gases in heat exchanger I. The flue gas flow rate is 20 kg/min and the other data are (NB: Some of the equations are different from that in Example 5.4)

Average specific heat of air and flue gas: 1.0 kJ/kg K

Average specific heat of heat-transfer oil: 2.5 kJ/kg K

Oil Pressure Drops

$$HE\ I = 1.2\,m^{1.8} \cdot A_I\ \text{kPa}$$
$$HE\ II = 1.3\,m^{1.7} \cdot A_{II}\ \text{kPa}$$

where $\dot{m}$ is the mass flow rate of oil in kilogram per minute and A_I and A_{II} are the surface areas of the two heat exchangers in square meter.

Pump Head-Flow Characteristics

1. ΔP (in kPa) $= 100 - 5\dot{m} - 0.5\dot{m}^2$
2. Oil temperature rise in pump (°C) $= 5 + (\dot{m}/5)^{1.3}$

HE Performance

$$\text{Effectiveness of } HE\ I, \epsilon_I = 1 - e^{-(\dot{m}\ A_I/20)^{0.8}}$$
$$\text{Effectiveness of } HE\ II, \epsilon_{II} = 1 - e^{-(\dot{m}\ A_{II}/15)^{0.7}}$$

Formulate the problem of optimum design of the two heat exchangers of this system, for minimum life cycle cost. Given the cost data in Rupees:

$$\text{Present worth of life cycle cost of pump power} = 10^6 \Delta P$$

$$\text{Cost of heat exchangers} = 10^4 \cdot \text{Surface area}$$

The cost of the pump can be neglected.

Solution

The objective function to be minimized is clearly:

$$F = 10^6 \Delta P + 10^4 (A_I + A_{II})$$
$$= 10^6 (100 - 5\dot{m} - 0.5\dot{m}^2) + 10^4 (A_I + A_{II})$$

The Constraints

The process air has to be heated by 50°C. This implies

$$Q_{II} = 15 \times 50 \times 1.0 \, \text{kJ/min}$$
$$= 750 \, \text{kJ/min}$$

So the heat exchanger II should transfer this much heat to the process air. To estimate the rate of heat transfer in the heat exchangers, we must know their effectiveness, and the heat capacity of the two fluids. Since the mass flow rate of oil is not known a priori, we assume that its mass flow rate will be small enough so that C_{oil} is the smaller than C_{air} as well as C_{flue}. After solving the problem, if this assumption turns out to be incorrect, we would have to solve the problem again taking the correct value of C_{min}.

$$\text{Now} \quad C_{oil} = \dot{m} \times 2.5 \, \text{kJ/min K}$$
$$C_{air} = 15 \times 1.0 = 15 \, \text{kJ/min K}$$
$$C_{flue} = 20 \times 1.0 = 20 \, \text{kJ/min K}$$

Heat transfer in HE_{II} is (see Fig. 5.14)

$$Q_{II} = \epsilon_{II} \, C_{min}(T_I - 20°C)$$
$$= [1 - e^{-(\dot{m} \, A_{II}/15)^{0.7}}][2.5\dot{m}][T_I - 20°C] \, \text{kJ/min}$$

The first constraint thus becomes

$$Q_{II} = [1 - e^{-(\dot{m}A_{II}/15)^{0.7}}][2.5\dot{m}][T_I - 20] = 750$$

Another constraint would arise from the analysis of heat transfer in *HE I*. We first need to find the temperature of oil entering it from the pump. This is turn needs the temperature of oil entering the pump, that is, T_{II}. This can be related to T_I through heat balance of HE_{II}:

$$Q_{II} = 750 = \dot{m} \times 2.5 \times (T_I - T_{II})$$

which gives

$$T_{II} = T_I - \frac{300}{\dot{m}}$$

The temperature rise in the pump is given as:

$$\Delta T = T - T_{II} = 5 + (\dot{m}/5)^{1.3}$$

$$\therefore \quad T = T_{II} + 5 + (\dot{m}/5)^{1.3} = T_I + 5 + (\dot{m}/5)^{1.3} - \frac{300}{\dot{m}}$$

The heat transfer needed in *HE I* is therefore

$$Q_I = \dot{m} \times 2.5 \times (T_I - T)$$

$$= 2.5\dot{m}\left[-5 - (\dot{m}/5)^{1.3} + \frac{300}{\dot{m}}\right]$$

From the effectiveness relationship we get

$$Q_I = 2.5 \times \dot{m} \times \epsilon_I \times (300 - T)$$

$$= 2.5\dot{m}\left[1 - e^{-(\dot{m}\, A_I/20)^{0.8}}\right](300 - T)$$

Thus the heat-transfer requirements of *HE I* introduces a constraint:

$$2.5\dot{m}\left[1 - e^{-(\dot{m}A_I/20)^{0.8}}\right]\left(300 - T_I - 5 - \left(\frac{\dot{m}}{5}\right)^{1.3} + \frac{300}{\dot{m}}\right)$$

$$= 2.5\dot{m}\left[-5 - \left(\frac{\dot{m}}{5}\right)^{1.3} + \frac{300}{\dot{m}}\right]$$

which can be simplified to

$$\left[1 - e^{-(\dot{m}A_I/20)^{0.8}}\right]\left(300 - T_I - 5 - \left(\frac{\dot{m}}{5}\right)^{1.3} + \frac{300}{\dot{m}}\right)$$

$$= -5 - \left(\frac{\dot{m}}{5}\right)^{1.3} + \frac{300}{\dot{m}}$$

Another constraint arises from balancing the pressure head developed by the pump with the pressure drop in the two heat exchangers (assuming the pressure drop in the connecting pipes is insignificant). This gives

$$\Delta P = 100 - 5\dot{m} - 0.5\dot{m}^2 = 1.2\dot{m}^{1.8}A_I + 1.3\dot{m}^{1.7}A_{II}$$

Thus the complete formulation of the optimum design problem would be

$$\text{Minimize} \quad F = 10^6(100 - 5\dot{m} - 0.5\dot{m}^2) + 10^4(A_I + A_{II})$$

subject to the constraints:

(a) $\left[1 - e^{-(\dot{m}A_{II}/15)^{0.7}}\right](\dot{m})[T_I - 20] = 300$

(b) $\left[1 - e^{-(\dot{m}A_I/20)^{0.8}}\right]\left(300 - T_I - 5 - \left(\frac{\dot{m}}{5}\right)^{1.3} + \frac{300}{\dot{m}}\right) = -5 - \left(\frac{\dot{m}}{5}\right)^{1.3} + \frac{300}{\dot{m}}$

(c) $100 - 5\dot{m} - 0.5\dot{m}^2 = 1.2\dot{m}^{1.8}A_I + 1.3\dot{m}^{1.7}A_{II}$

The variables are $\dot{m}, A_I, A_{II}$, and T_I.

In the above formulation, T_I may appear as an extraneous variable not directly related to the heat exchanger design. A little reflection would reveal that it is directly dependent on the other three variables and can, in fact, be eliminated from the optimization problem by using the constraint (a) to express it in terms of the other variables. If this is done, algebraic expression becomes a bit messy, but nonetheless we get an alternate formulation of the optimization problem with only two constraints, that is

$$\left[1 - e^{-(\dot{m}A_I/20)^{0.8}}\right]\left[275 - \frac{300}{\dot{m}\left(1 - e^{-(\dot{m}A_{II}/15)^{0.7}}\right)} - \left(\frac{\dot{m}}{5}\right)^{1.3} + \frac{300}{\dot{m}}\right]$$

$$= -5 - \left(\frac{\dot{m}}{5}\right)^{1.3} + \frac{300}{\dot{m}}$$

and

$$100 - 5\dot{m} - 0.5\dot{m}^2 = 1.2\dot{m}^{1.8}A_I + 1.3\dot{m}^{1.7}A_{II}$$

In this formulation the variables are $\dot{m}, A_I, A_{II}$, all directly related to the equipment design.

Example 7.2

Consider the problem of designing the three heat exchangers of the refrigeration system of Fig. 5.15, for the following rating conditions.

Evaporator

   Coolant flow rate: 2.0 kg/s
   Inlet temperature: 10°C
   Outlet temperature: 5°C
   Specific heat: 4 kJ/kg K
   Effectiveness $(\epsilon_e) = 1 - e^{-(A_e/5)^{0.9}}$

Condenser

   Coolant flow rate: 3.0 kg/s
   Inlet temperature: 30°C
   Specific heat: 4.2 kJ/kg K
   Effectiveness $(\epsilon_c) = 1 - e^{-(A_c/4)^{0.9}}$

Subcooler

$$\text{Effectiveness } (\epsilon_s) = 1 - e^{-(A_s/6)^{0.8}}$$

The compressor characteristic equations are

Refrigeration capacity (kW): $Q_{comp} = 20.0 + 5.0\,T_e - 0.02\,T_e\,T_s - 0.3\,T_c - 0.02\,T_e\,T_c$

Compressor power (kW): $P = 1.2\,T_c - 3\,T_e - 0.004\,T_c^2 - 0.01\,T_c\,T_s$

where T_e is the evaporator saturation temperature, T_c is the condenser saturation temperature, and T_s is the temperature of superheated vapors leaving the subcooler and entering the compressor; A_e, A_c, and A_s being the surface areas of the evaporator, the condenser, and the subcooler, respectively.

Formulate the optimization problem for minimal life cycle cost of the system. Take the system life as 15 years. Given the cost data:

Electric power rate: Rs. 6/kW h

Average annual operating hours: 6000 h

Assume electric power rate will increase every year by 5%; and the inflation rate and the interest rate over next 15 years would be around 4% and 8%, respectively.

Cost of Heat Exchangers Based on Surface Area

Evaporator: Rs. 8000 per m^2
Condenser: Rs. 10,000 per m^2
Subcooler: Rs. 5000 per m^2

Solution

We first formulate the objective function. It has two components, that is, the running cost spread over 15 years of the operating life of the system, and the initial cost of the heat exchangers. We first find present worth of the total running cost, considering both the escalation in power rates and the inflation and interest rates.

The effective interest rate, accounting for inflation can be found using Eq. (2.79) as:

$$i_{eff} = \frac{i - f}{1 + f} = \frac{0.08 - 0.04}{1 + 0.04} = 0.03846$$

Therefore the present worth of total running cost would be (assuming year-end payment):

$$\frac{P(6 \times 6000)}{1 + 0.03846} + \frac{P(6 \times 6000) \times 1.05}{(1 + 0.03846)^2} + \frac{P(6 \times 6000) \times (1.05)^2}{(1 + 0.03846)^3} + \cdots + \frac{P(6 \times 6000) \times (1.05)^{14}}{(1 + 0.03846)^{15}}$$

$$= \frac{(36,000)P}{1.03846}\left\{1 + \frac{1.05}{1.03846} + \left(\frac{1.05}{1.03846}\right)^2 + \left(\frac{1.05}{1.03846}\right)^3 + \cdots + \left(\frac{1.05}{1.03846}\right)^{14}\right\}$$

$$= \frac{(36,000)P}{1.03846}\{1 + 1.0111126 + 1.022349 + 1.03371 + \cdots + 1.16733\}$$

$$= \frac{(36,000)P \times 16.22493}{1.03846} = (36,000 \times 15.624)P = (562,465)P\,\text{Rs}$$

where P kW is the average power consumption of the plant under steady-state operation.

The objective function to be minimized is therefore (in Rupees):

F = Initial cost + Present worth of running cost

$\quad$ = $8000\ A_e + 10{,}000\ A_c + 5000\ A_s + 562{,}465\ P$

$\quad$ = $8000\ A_e + 10{,}000\ A_c + 5000\ A_s + 562{,}465(1.2\ T_c - 3\ T_e - 0.004\ T_c^2 - 0.01\ T_c\ T_s)$

We now identify the constraints to be satisfied by any feasible design. The main requirement from the system is that it should cool 2.0 kg/s of the coolant from 10 to 5°C. This demands a refrigerating effect of Q_e, where

$$Q_e = 2.0 \times 4 \times 5 = 40.0\ \text{kW}$$

This is the first constraint to be satisfied. To enable this heat transfer to take place in the evaporator its surface area A_e should be sufficient to give the needed effectiveness ϵ_e. This implies

$$\epsilon_e = \frac{Q_e}{(2 \times 4)(10 - T_e)} = \left(1 - e^{-(A_e/5)^{0.9}}\right)$$

Substituting $Q_e = 40\ \text{kW}$, we get the constraint as:

$$5 = (10 - T_e)\left(1 - e^{-(A_e/5)^{0.9}}\right)$$

Similar constraints arise from the consideration of the effectiveness of the condenser and the subcooler:

$$\epsilon_c = \frac{Q_c}{(3 \times 4.2)(T_c - 30)} = \left(1 - e^{-(A_c/4)^{0.9}}\right)$$

$$\epsilon_s = \frac{T_s - T_e}{T_c - T_e} = \left(1 - e^{-(A_s/6)^{0.8}}\right)$$

where subcooler effectiveness has been expressed in terms of the difference in the temperatures of the two streams (see Fig. 5.15, Example 5.5).

We also need to match the performance of the compressor with the requirements. Thus considering the refrigeration requirements we get the constraint:

$$Q_e = Q_{comp} = 20 + 5\ T_e - 0.02\ T_e\ T_s - 0.3\ T_c - 0.02\ T_e\ T_c$$
$$= 40\ \text{kW}$$

The overall energy balance of the complete system demands that

$$Q_c = Q_e + P = 40 + 1.2\ T_c - 3\ T_e - 0.004\ T_c^2 - 0.01\ T_c\ T_s$$

Substituting the above expression for Q_c in the condenser heat-transfer constraint mentioned previously, we can eliminate Q_c as a variable in the optimization problem. The final complete formulation of the optimization problem is

Minimize $F = 8000\ A_e + 10{,}000\ A_c + 5000\ A_s + 562{,}465(1.2\ T_c - 3\ T_e - 0.004\ T_c^2$

$-0.01\ T_c\ T_s)$

Subject to constraints:

$$(10 - T_e)\left(1 - e^{-(A_e/5)^{0.9}}\right) = 5$$

$$12.6(T_c - 30)\left(1 - e^{-(A_c/4)^{0.9}}\right) = 40 + 1.2\ T_c - 3\ T_e - 0.004\ T_c^2 - 0.01\ T_c\ T_s$$

$$(T_s - T_e) = (T_c - T_e)\left(1 - e^{-(A_s/6)^{0.8}}\right)$$

$$5\ T_e - 0.02\ T_e\ T_s - 0.3\ T_c - 0.2\ T_e\ T_c = 20$$

The variables to be optimized are T_c, T_e, T_s, A_c, A_e, A_s, all of them being continuous variables with nonnegative values.

An important aspect of optimum design of systems is to be able to identify the kind of data that are required to formulate the problem. This is often not self-evident especially in novel systems like biogas and biomanure plants encountered in rural settings. Allocation of various local energy resources like cattle dung and biomass among alternative devices like biogas, biomanure, briquetting, or gasifier plants is another crucial factor. Example 7.3 illustrates the complexity of formulating the problem of optimal design of a typical rural energy system.

Example 7.3

A village *panchayat*[a] has been sanctioned Rs. 10 crores to set up an integrated energy system to meet all the energy needs of the village and its 200 families. It consults an energy expert who suggests the following system based on the three energy resources, namely biomass, cattle dung, and solar energy (Fig. 7.1).

Each of these resources could be utilized in different ways. Thus biomass could be used in a gasifier which generates producer gas for an engine-generator set, which produces electricity. It could be pelletized to make briquettes for meeting the cooking energy requirements, or alternatively it could also be fed into biogas plants to enhance their biogas output.

The cattle dung could be used in biogas plants or converted into biomanure along with biomass. The use of solar energy is suggested to produce electricity through photovoltaic cells.

Further, the biogas produced in the system could have twin uses. First, direct use as fuel for cooking; and secondly to be converted into bio-CNG after cleaning and compression and sold to earn revenue for the village. The excess electricity generated, if

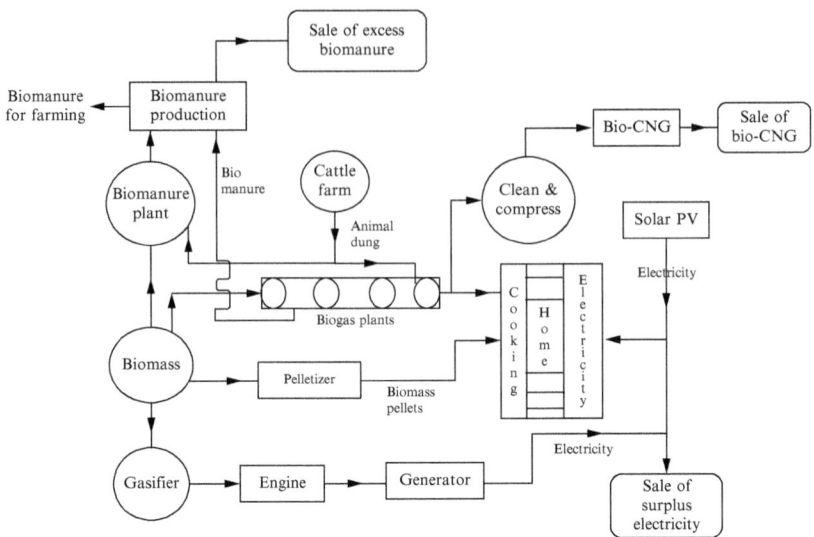

Fig. 7.1 Integrated energy system.

any, could also be sold back to grid. Similarly, the excess biomanure produced (i.e., after meeting the farming requirements of the village) could also be sold to earn revenue.

The village *panchayat* wants to choose the best possible strategy of utilizing the available resources so as to maximize the yearly earnings of the village and spend the entire sanctioned amount to buy and install various gadgets.

Convert this requirement of the village *panchayat* into an optimization problem, identifying clearly the objection function, the constraints, and the design variables. Identify the cost and fuel consumption data that you would need for solving this problem.

Solution

The objective function is quite clear from the problem statement, that is, to maximize the annual earning of the village. One financial constraint is also quite explicitly stated: The total cost of the system should be 10 crores (or lesser), but to convert these statements into mathematical expressions we need data on the initial cost per unit capacity of various equipment as also their yearly running cost arising from manpower and maintenance requirements. Thus the data needed are

Gasifier

Initial cost per unit capacity (in m^3/h) = ICG
Running cost per hour per unit capacity = RCG
Biomass required (in kg/h) per unit capacity = BMG

Engine Generator Set

Initial cost per unit power output (in kW): ICEG

Running cost per hour per unit power output: RCEG
Gas required (in m^3/h) per unit power output: GEG

Pelletizer

Initial cost per unit output (in kg/h): ICP
Running cost per hour per unit output: RCP
Biomass required (kg/h) per unit output: BMP

Biogas Plants

Initial cost per unit biogas output (m^3/h): ICBG
Running cost per hour per unit output: RCBG
Biomass (kg/h) and dung required per unit output: BMBG/DBG
Biomanure produced (kg/h) per unit biogas output: BMRBG

Biomanure Plants

Initial cost per unit output (kg/h): ICBMR
Running cost per hour per unit output: RCBMR
Biomass and dung required per unit output: BMBMR/DBMR
Biomanure produced (kg/h) per unit biogas output: BMRBG

Biogas Cleaning and Compression

Initial cost per unit bio-CNG (m^3/h) output: ICBCNG
Running cost per hour per unit bio-CNG output: RCBCNG
Biomass required per unit bio-CNG: BGBCNG

Solar Photovoltaic System

Initial cost per unit output (kW) output: ICSPV
Running cost per unit output : RCSPV

We would also need data on the total energy resources available in the village, that is
Average biomass availability per day: BMA
Average cattle dung available per day: DA
Similarly, the pertinent requirements of the village also need to be quantified:
Biomanure needed per day (average) in the village: BMRDAILY
Daily energy requirement for cooking (average) in the village: ECOOK DAILY
Daily energy requirement (average) other than cooking (i.e., lighting, fans, pumps, etc.) in kW h: EE DAILY
We also need to know the number of operating hours per day of various plants. We indicate these as, HMR, HBG, HSPV, HEG, HP, HBCNG, HG, being the daily effective operating hours of the biomanure plant, biogas plant, solar photovoltaic plant, electric generator, Pelletizer, bio-CNG plant, and gasifier, respectively.

Once all these data are collected, the optimum design of the integrated energy system can be evolved. The objective function to be maximized is

F = Annual earning of the village

= Earning from the sale of excess biomanure, bio-CNG, and surplus electricity

To find the amount of these surpluses, we need to find how much of each is produced and how much is consumed annually. This in turn depends on the "resources" available and the in-house demand of "products" in the village. Thus considering their "balance" we get the following equations in terms of the new variables defined as follows.

XBMP: Fraction of total biomass uses in the Pelletizer

XBMBG: Fraction of total biomass used in the biogas plant

XBMBMR: Fraction of total biomass used in the biomanure plant

XBMG: Fraction of total biomass used in the gasifier plant = $1 - XBMP - XBMBG - XBMBMR$

XBGCOOK: Fraction of biogas used for cooking

XBGCNG: Fraction of biogas used for producing bio-CNG = $1 - XBCOOK$

XDBG: Fraction of total dung used in the biogas plant

XDBM: Fraction of total dung used in the biomanure plant = $1 - XDBG$

QP: Rating of the Pelletizer plant in kg/h = $BMA \star XBMP/(BMP \star HP)$

QBG: Rating of the biogas plant in m^3/h = $[XDBG \star DA/DBG + BMA \star XBMBG/BMBG]/HBG$

QG: Rating of gasifier plant in m^3/h = $BMA \star XBMG/(BMG \star HG)$

QEG: Rating of engine generator plant (kW)

QMR: Rating of biomanure plant in kg/h = $[BMA \star XBMBMR/BMBMR + DA \star XDBM/DBMR]/HMR$

QBCNG: Rating of the Bio-CNG plant in m^3/h = $XBGBCNG \star QBG \star HBG/(BGBCNG \star HBCNG)$

QSPV: Rating of solar photovoltaic plant (kW)

We can now write an expression for the objective function in terms of these variables:

$$
\begin{aligned}
F = {}& (QMR * HMR - BMRDAILY) * PBMR + QBCNG * HBCNG * PBCNG \\
& + (QSPV * HSPV + QEG * HEG - EEDAILY) * PEE \\
& - QMR * RCBMR * HMR - QBG * RCBG * HBG \\
& - QSPV * RCSPV * HSPV - QEG * RCEG * HEG - QP * RCP * HP \\
& - QBCNG * RCBCNG * HBCNG - QG * RCG * HG
\end{aligned}
$$

where PBMR, PBCNG, and PEE are the selling price of 1 kg of biomanure, 1 m^3 of bio-CNG, and 1 kW h of electric energy, respectively.

Substituting for QMR and QBG from the above equations we can write another expression for the objective function in terms of the design variables as:

$$
\begin{aligned}
F = {}& \left(\frac{BMA * XBMBMR}{BMBMR} + \frac{DA * XDBM}{DBMR} \right) * (PBMR - RCBMR) \\
& - BMRDAILY * PBMR + \frac{XBGBCNG}{BGBCNG} * \left(\frac{XDBG * DA}{DBG} + \frac{BMA * XBMBG}{BMBG} \right) \\
& * (PBCNG - RCBCNG) + (QSPV * HSPV * (PEE - RCSPV) + QEG * HEG
\end{aligned}
$$

$$* (PEE - RCEG) - EEDAILY * PEE) - \left(\frac{XDBG * DA}{DBG} + \frac{BMA * XMBMG}{BMBG} \right)$$

$$* RCBG - \frac{BMA * XBMP}{BMP} * RCP - \frac{BMA * XBMG}{BMG} * RCG$$

In this expression, the design variables are $XBMBMR$, $XDBM$, $XBGBCNG$, $XDBG$, $XBMBG$, $XBMP$, $XBMG$, $QSPV$, QEG, all other terms are constants, being the input information needed for design.

We can now delineate the constraints on the system design. The first of these is the financial constraint, viz. the total cost of all the equipment should be less than, at the most equal to, the allocated sum of Rs. 10 crores. This gives

$$ICG * QG + ICEG * QEG + ICP * QP + ICBG * QBG + ICBMR * QMR$$
$$+ ICBCNG * QBCNG + ICSPV * QSPV \leq 10^8$$

Expressing various capacities in terms of the design variable, the above constraint can be written as

$$\frac{ICG * BMA * XBMG}{BMG * HG} + ICEG * QEG + ICP * \frac{BMA * XBMP}{BMP * HP}$$
$$+ ICBG * \left(\frac{XDBG * DA}{DBG} + \frac{BMA * XBMBG}{BMBG} \right) \div HBG$$
$$+ ICBMR * \left(\frac{BMA * XBMBMR}{BMBMR} + \frac{DA * XDBM}{DBMR} \right) \div HMR$$
$$+ ICBCNG * \left(\frac{XBGBCNG}{BGBCNG} * \left[\frac{XDBG * DA}{DBG} + \frac{BMA * XBMBG}{BMBG} \right] \right) \div HBCNG$$
$$+ ICSPV * QSPV \leq 10^8$$

The variables involved here are $XBMG$, $XBMP$, $XDBG$, $XBMBG$, $XBMBMR$, $XDBM$, $XBGBCNG$, QEG, and $QSPV$.

We also need to put constraints to ensure that all the energy requirements of the entire village are certainly met. This gives

$$QSPU * HSPV + QEG * HEG \geq EEDAILY$$
$$QMR = \left(\frac{BMA * XBMBMR}{BMBMR} + \frac{DA * XDBM}{DBMR} \right) \geq BMRDAILY$$
$$QBG * HBG * XBGCOOK * ECVBG + QP * HP * ECVP \geq ECOOKDAILY$$

The above condition can further be written in terms of the design variables as:

$$\frac{XDBG * DA}{DBG} * XBGCOOK * ECVBG + \frac{BMA * XBMP}{BMP} * ECVP \geq ECOOKDAILY$$

where $ECVBG$ and $ECVP$ are the effective calorific values of the biogas and the biomass pellets, respectively.

We can eliminate a few of the design variables by using the mass conservation principle at the splitters in Fig. 7.1. This gives the relations:

$$XBGCOOK = 1 - XBGBCNG$$

$$XBMG = 1 - XBMP - XBMBG$$

$$XDBM = 1 - XDBG$$

Thus the complete optimization problem can be written as:

Maximize

$$F = \left[\frac{BMA}{BMBMR} \cdot XBMBMR + \frac{DA}{DBMR} \cdot (1 - XDBG) \right] * (PBMR - RCBMR)$$

$$- BMRDAILY * PBMR + XBGBCNG * \left[\frac{DA}{DBG * BGBCNG} \cdot XDBG \right.$$

$$+ \frac{BMA}{BMBG * BGBCNG} \cdot XBMBG \right] * (PBCNG - RCBCNG)$$

$$+ (QSPV * HSPV * (PEE - RCSPV)) + QEG * HEG * (PEE - RCEG)$$

$$- EEDAILY * PEE - \left(\frac{DA}{DBG} \cdot XDBG + \frac{BMA}{BMBG} \cdot XBMBG \right) \cdot RCBG$$

$$- \frac{BMA}{BMP} \cdot XBMP \cdot RCP - \frac{BMA}{BMG} \cdot (1 - XBMP - XBMBG) \cdot RCG$$

subject to the constraints:

(a) $\dfrac{ICG * BMA}{BMG * HG}(1 - XBMP - XBMBG) + ICEG * QEG + \dfrac{ICP * BMA}{BMP * HP} \cdot XBMP$

$$+ \frac{ICBG * DA}{DBG * HBG} \cdot XDBG + \frac{ICBG * BMA}{BMBG * HBG} \cdot XBMBG + \frac{ICBMR * BMA}{BMBMR * HMR} \cdot XBMBMR$$

$$+ \frac{ICBMR * DA}{DBMR * HMR} \cdot (1 - XDBG) + \frac{ICBCNG}{BGBCNG * HBCNG} * \frac{DA}{DBG} \cdot (XBGBCNG * XDBG)$$

$$+ \frac{ICBCNG}{BGBCNG * HBCNG} \cdot \frac{BMA}{BMBG} \cdot XBGBCNG \cdot (1 - XBMP - XBMBG)$$

$$+ ICSPV * QSPV \leq 10^8$$

(b) $\dfrac{BMA}{BMBMR} \cdot XBMBMR + \dfrac{DA}{DBMR} \cdot (1 - XDBG) \geq BMRDAILY$

(c) $\dfrac{DA * ECVBG}{DBG} \cdot XDBG * XBGCOOK + \dfrac{BMA * ECVP}{BMP} \cdot XBMP \geq ECOOKDAILY$

(d) $QSPV * HSPV + QEG * HEG \geq EEDAILY$

The variables to be optimized are

$$XBMBMR, XDBG, XBGBCNG, XBMBG, XBMP, XBGCOOK, QSPV, QEG$$

The rest of the terms in above expressions are the constants pertaining to the cost and the performance of various equipment which are data inputs. All the variables are continuous variables. To reduce the region of search for the values of the variables we may also introduce additional constraints on the values of various X-variables, all of which being positive fractions should be less than 1 and nonnegative, that is

$$1 \geq XBMBMR \geq 0$$
$$1 \geq XDBG \geq 0$$
$$1 \geq XBGBCNG \geq 0$$
$$1 \geq XBMBG \geq 0$$
$$1 \geq XBMP \geq 0$$
$$1 \geq XBGCOOK \geq 0$$

The values of the other two variables have also to be nonnegative; that is, $QSPV \geq 0$ and $QEG \geq 0$.

[a] Local self-government of a village in India.

7.2 OPTIMUM DESIGN OF A COMPONENT

Sometimes we come across situations where only one component of a complex system is to be optimized. This could be due to financial constraints, or because the overall system performance is highly sensitive to the design of one of the components. It may also happen that in an industrial set up the designs of other components can not be changed due to on-site considerations or nonavailability of many choices in these components.

Since most thermal systems operate in a cycle, any single component would influence the performance of others. It is therefore necessary to put additional constraints to ensure that the overall system performance is not altered, and no significant changes are necessary in the design of other components. To illustrate this, let us consider the case of optimizing the design of a DX-chiller of an industrial refrigeration plant. Since the four main components of the system, that is, the chiller, the compressor, the condenser, and the expansion valve, are interconnected, any change in the chiller design would influence the performance of all other components, change the operating conditions as also the overall cooling effect. It is therefore necessary to ensure that both the refrigerating effect in the chiller, as also the operating conditions of the other components are not altered, that is, the thermodynamic cycle on which the system is operating (1–2–3–4, Fig. 7.2) remains unaltered. This implies that the new design of the chiller should give the same cooling capacity, and have the same pressure drop in the refrigerant, as also in the coolant, as in the original design.

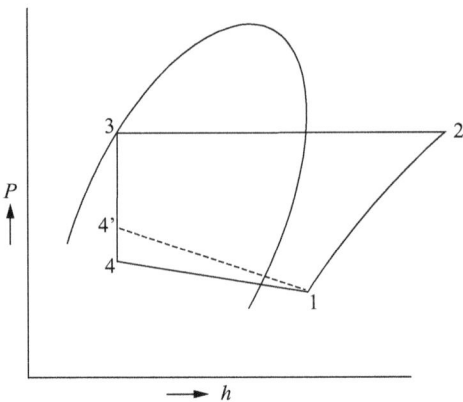

Fig. 7.2 $p - h$ diagram of the refrigeration system.

Since these constraints drastically reduce the possibilities of alternative designs, Dhar [1] has suggested a modification by relaxing the condition of the refrigerant-side pressure drop being the same as in the original design. Instead, the condition of the vapors leaving the evaporator, state 1 in Fig. 7.2, has been constrained to the same as in the original design. This ensures that the performance of the compressor, and the condenser are not influenced by the changes in the design of the evaporator. The only change needed is in the expansion valve. The pressure drop across this valve has to be adjusted in tune with the change in the pressure drop in the evaporator (i.e., expansion process in 3–4', instead of 3–4) to ensure the constancy of the evaporator exit conditions (i.e., state 1, Fig. 7.2). This is not likely to alter the cost of the expansion valve. The detailed problem formulation is discussed in Example 7.4.

As another example of optimization of only a few components of a complex system, let us consider the case of the combined cycle power plant whose detailed simulation has been discussed in Section 6.2. The company supplying this plant wanted to optimize only the design of the waste heat recovery boilers (WHRB) each of which consists of eight heat exchangers, since this was the main component being designed and manufactured in house. All other components (except its condenser) were being imported. Here too, as is evident from Fig. 6.2, changes in the design of the WHRB would influence both the downstream steam cycle, as also the upstream gas cycle performance. We need to put suitable constraints to ensure that the overall performance of the system, that is, the power produced is not altered. If we want to restrict that the operating conditions are also not changed, then additional constraints have to be put on the overall gas-side pressure drop in the WHRB, as also on the steam state leaving each heat exchanger. This, of course, would severely limit the possibilities of improvement in the WHRB design. A better strategy would be to allow the operating conditions on the gas side, as also on the steam side to float, and only ensure that the overall power output of the complete power plant equals the rated value. This is discussed further in Chapter 9, Section 9.3.2.

Example 7.4

Formulate the problem of optimum design of a DX-chiller (for minimum initial cost) with the following rated performance.

Capacity: 40TR   Water-side pressure drop = 71.7 kPa
Refrigerant-side pressure drop = 13.8 kPa
Configuration: shell and tube type configuration with two refrigerant circuits and eight passes. Twenty-one baffles on the shell side.

Solution

As per problem specification, the objective function should be the total initial cost of the chiller. This depends on the total number of tubes, their diameter, length and thickness, and the diameter and thickness of the shell. The tubes are usually of copper (in view of its high thermal conductivity) and the shell is of steel, in view of its high mechanical strength. Their cost is directly proportional to the weight and thus the objective function can be written as:

$$F = \sum_{i=1}^{Np} N_t(i) \cdot \frac{\pi}{4}(d_o^2 - d_i^2) \cdot L \cdot \rho_c \cdot C_c + \frac{\pi}{4}(D_{so}^2 - D_{si}^2) \cdot L \cdot \rho_s \cdot C_s$$

where d_0, d_i are the outside and inside diameters of the copper tubing; $N_t(i)$ the number of tubes in i_{th} pass; N_p the number of tube passes in the chiller; ρ_c the density of copper; L the length of the tubes; C_c the cost of copper tubes in rupees per kilogram; D_{so}, D_{si} are the outer and inner diameters of the shell; ρ_s the density of steel; and C_s the cost of steel shell in rupees per kilogram.

All the variables occurring in the expression for F are obviously the design variables, that is, $N_t(i)$, d_o, d_i, L, D_{so}, D_{si}, and N_p. Besides these there are other variables also which have not been included in F since their contribution to the cost is not significant, but they influence the chiller performance significantly. These are the variables which influence the shell-side flow field and hence the pressure gradient and the heat-transfer coefficient on the shell side. These variables are the number of baffles (NB), baffle pitch (BP), baffle cut (BC), and tube pitch (P_T).

It can be seen that these variables are of two kinds, namely continuous and discrete valued variables. The "continuous" design variables, which can have any real number as their value, are the tube length L, shell diameters D_{so}, D_{si}, baffle pitch BP, baffle cut BC, and the tube pitch P_T. The remaining variables can have only certain discrete values. These are the number of passes N_p, number of tubes in various passes $N_t(i)$, and the number of baffles NB, all of which can have only integer values; and the tube diameters d_o/d_i which are constrained to be chosen among certain specified values of tube diameters commonly manufactured by the industry. Besides these one could also include certain "logical" variables, for example, the tube arrangement (in line square, staggered square, or triangular), the direction of flow of the refrigerant relative to that of coolant, and type of the tube (i.e., bare, internally and/or externally finned, etc.).

Now considering the constraints, to ensure that the optimized chiller can be seamlessly fitted into the existing refrigeration system without any major change in any other component it is obvious that its refrigerating effect (Q) should be 40TR, the refrigerant exit conditions should be the same as in the original design, and the water-side pressure drop (ΔP_w) should be 71.7 kPa. The refrigerant-side pressure drop need not be constrained since, as explained previously, changes in its value can be taken care of by adjustments in the expansion device.

We also need to introduce additional constraints to ensure geometric compatibility of the design configuration. Thus to ensure that NB baffles, spaced a distance BP apart from each other can be accommodated within the shell length L it is necessary that

$$L \geq BP(N_B - 1) + N_B * BT$$

where BT is the thickness of each baffle.

Again it is necessary to ensure that the total number of tubes in all the passes can be accommodated within the shell diameter. This needs a comprehensive procedure based on an analysis of the geometry of the shell cross section.

To summarize, the optimization problem can be written as:

$$\text{Minimize} \quad F = \sum_{i=1}^{Np} N_t(i) \cdot \frac{\pi}{4}(d_o^2 - d_i^2) \cdot L \cdot \rho_c \cdot C_c + \frac{\pi}{4}(D_{so}^2 - D_{si}^2) \cdot L \cdot \rho_s \cdot C_s$$

Subject to Constraints

$Q = 40\text{TR}$

$\Delta P_{water} = 71.7\,\text{kPa}$

$L \geq BP(N_B - 1) + N_B * BT$

$\sum_{i=1}^{Np} N_t(i) \leq$ Max. no of tubes of outer diameter d_0 that can be accommodated

in a shell of diameter

D_{si}, with a tube pitch P_T

Design Variables

Continuous: $L, D_{so}, D_{si}, BP, BC, P_T$

Discrete: $d_o, d_i, N_p, [N_t(i), i = 1, 2, \ldots, N_p], N_B$

Logical: tube arrangement, relative direction of flow of the two fluids (refrigerant and water), type of tube

7.3 EPILOG

In the highly competitive modern market economy, optimizing the design is of paramount importance. It is not only linked to the system performance and cost, but also to marketing strategies. In the past, manufacturers of domestic items like refrigerators and air conditioners used to design for minimum cost to attract more customers. However, with increasing energy costs and the need for reducing energy consumption, energy efficiency has become very important. Thus the concept of star rating of domestic appliance has been introduced so that the customers are assured that higher "stared" appliances are more energy efficient[1] and are therefore willing to pay a higher price for the same. Manufacturers are thus encouraged to design for specified energy efficiency of the appliances, and not opt for the traditional default design of lowest cost. Besides energy efficiency, the noise level and air cleaning features can be other important criteria which could govern the sale of domestic equipment. Environmental regulations are also demanding that the effluents of power plants and even automobile engines meet certain standards. These clearly will also have a bearing on the equipment design and need to be empirically factored in the design process.

PROBLEMS

7.1 A simple gas turbine plant is being modified by introducing a boiler in the turbine exhaust which recovers a part of the heat and feeds it into an absorption refrigeration system (Fig. 5.22). The cooling effect produced in the absorption system is used to precool the fresh air entering the compressor. This increases the mass flow rate of air through the gas turbine cycle, thus increasing it power output. This increase in power output is to be balanced against the increase in initial cost of the additional boiler, absorption refrigeration system, and the heat exchanger. Formulate the problem of optimizing these additional equipments (considered as a single module) identifying the cost data that would be needed.

7.2 The air-cooled condenser of the refrigeration system of a cold storage has got spoiled due to erosion caused by dust. It is proposed to replace it by a new air cooled condenser. Formulate the problem of optimizing the design of the condenser so that the performance of the system is not impaired.

7.3 It has been decided to put a finned surface of size $L \times W$ over a power electronics VLSI circuit to dissipate 1000 W of heat (Fig. 7.3). The two design variables are the fin spacing and the height of the fins. Formulate the problem of optimizing the fin design to ensure that the volume of the fins is minimum.

---

[1] Thus in India a 5-star split A/C has a minimum COP of 3.3 while an air conditioner with a COP of 2.5 would qualify for 1 star.

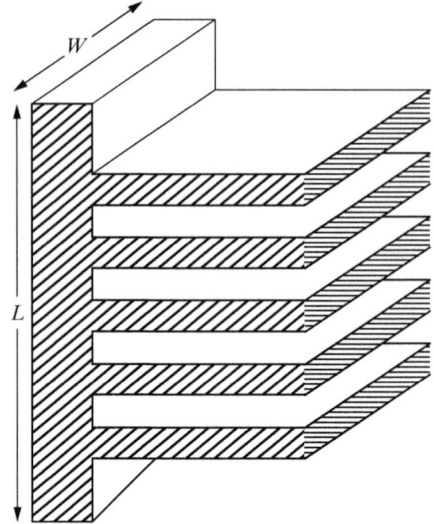

Fig. 7.3 Heat dissipation by fins.

7.4 Formulate the problem of maximizing the total earning of a simple combined cycle power plant (Fig. 7.4) over its life using the given data.

Present worth of power generated during the life time:	$100 \star$ Power in kW
Efficiency of gas turbine plant:	30%
Present worth of fuel cost during the plant's life time:	$7.5 \star (Q \text{ in kW})$

First cost of combustor: $0.1 \ Q^{0.2}$

First cost of evaporator: $0.15 \star$ Surface area in m^2

First cost of superheater: $0.2 \star$ Surface area in m^2

First cost of economizer: $0.18 \star$ Surface area in m^2

First cost of condenser: $0.25 \star$ Surface area in m^2

Enthalpy of superheated steam at state 8: $2340 + 2.3 \ T$ (in $^\circ$C)

U_{evap}: 300 W/m^2 K

U_{sup}: 100 W/m^2 K

U_{econ}: 200 W/m^2 K

U_{cond}: 600 W/m^2 K

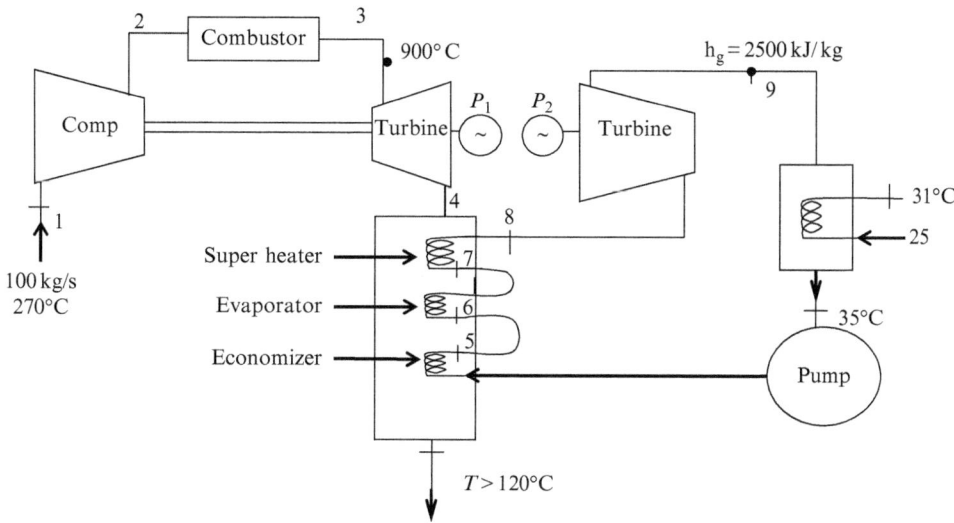

Fig. 7.4 Combined cycle power plant.

The design variables are the surface areas of the condenser, economizer, evaporator, and superheater, and the mass flow rate of water in the condenser. The mass flow rate of air entering the compressor is 100 kg/s, the temperature of the gas turbine exhaust is 500°C and the temperature of the hot gases entering the gas turbine is 900°C. The saturation temperature of steam in the evaporator is 200°C and the condensation temperature is 35°C. The cold water enters the condenser at 25°C and leaves at 31°C. The temperature of the gases leaving the stack should not be less than 120°C.

Express all the constraints in terms of the design variables. Make appropriate assumptions to simplify the analysis.

7.5 Attached to a big cattle shed is a large biogas producing 100 m³ of biogas daily. The biogas can be used in three different applications indicated in Fig. 7.5. Formulate the problem of optimizing the distribution of the gas among the three alternatives so as to maximize the net earnings over the 15–year life span of the whole system. Identify the data that would be needed for optimization.

7.6 Water is being pumped from a river to irrigate three fields of a farmer (Fig. 7.6). Since the electricity costs are high, the farmer wants to optimize the diameters of various pipes so that the total cost of the system taken over its complete life span of 25 years is minimal. Formulate the optimization problem. Assume the cost of water pipes in rupees per meter length is given as

$$Cost = 40 \, D^{1.2} \quad \text{where } D \text{ is the diameter in mm}$$

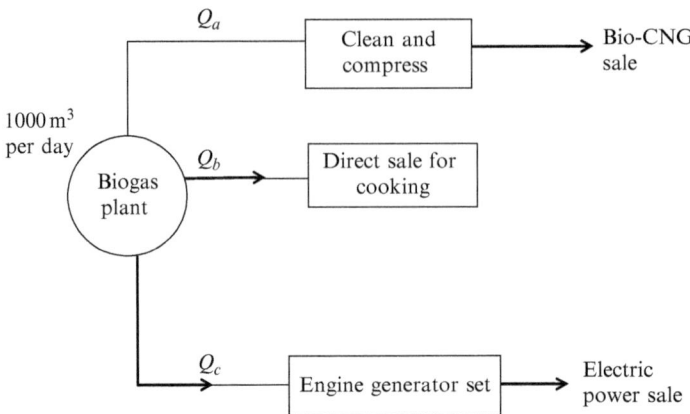

Fig. 7.5 Biogas plant.

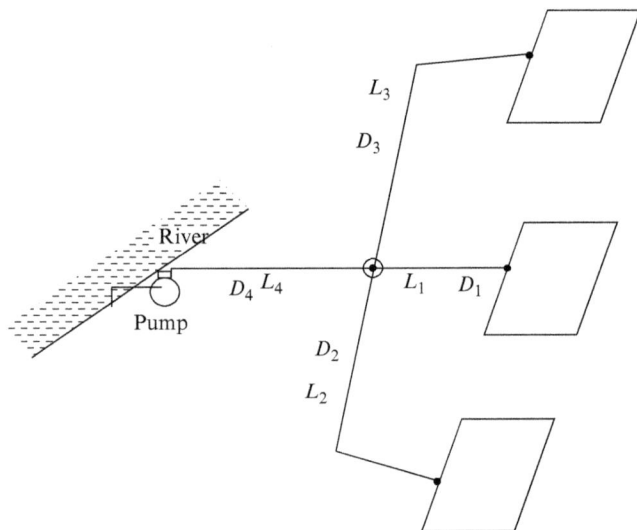

Fig. 7.6 Pumping water from a river.

7.7 The steam leaving the turbine of a thermal power plant is condensed in a water cooled condenser. The hot water produced here is used for keeping the plant control room warm. To reduce the size of the condenser, the steam temperature (and therefore its pressure) should be kept high. However, this would reduce the net power output in the turbine. Formulate the problem of optimizing the design of the condenser, without altering the design of the rest of the components of the power plant.

7.8 A water pipe line of 200 mm diameter is carrying 80 kg/s of cold water at 5°C for distribution to the fan-coil units air conditioning various rooms in a building. It is necessary to insulate it to prevent loss of cooling. The external diameter of the insulated pipe is constrained to be 250 mm. A variety of insulating materials are available—those with better thermal insulating property costing considerably larger. The loss of cooling effect at this temperature effectively costs the air-conditioning system about Rs. 5000 per TR. Formulate the problem of choosing the thermal insulation for this 500 m long pipe.

Take design environment temperature as 40°C, and the heat-transfer coefficient on the insulated pipe as 20 W/m^2 K.

REFERENCE

[1] P.L. Dhar, Optimization in refrigeration systems, PhD thesis, Department of Mechanical Engineering, IIT Delhi, 1974.

CHAPTER 8

Optimization Techniques

Over the past few decades, optimization of engineering equipment has attracted the attention of both engineers and mathematicians, and so a large number of optimization techniques have been developed. In this chapter we shall present a brief review of a select few methods which are especially suited to the optimization of thermal systems. These methods are usually classified into two broad categories, viz. analytical methods and numerical methods. The analytical methods are rarely employed in engineering practice since these are quite awkward to use for the complex problems generally encountered in system design (Chapter 7). However, these are helpful in understanding the theoretical basis of numerical methods; and some analytical methods can be used to solve very simple practical problems too. We shall therefore study both analytical and numerical methods often used in optimizing thermal systems.

8.1 ANALYTICAL METHODS

The analytical methods—also sometimes called the calculus methods—originated from the contributions of Newton, Liebnitz, Lagrange, and others. Considering the unconstrained optimization (i.e., minimization or maximization) of a multivariable function $f(x_1, x_2, \ldots, x_n)$, a function of n variables $x_1, x_2, \ldots, x_n$, we can prove from differential calculus and Taylor's series expansion that the optimum values $x_1^*, x_2^*, \ldots, x_n^*$ are such that

$$f'_{xi}(x_1^*, x_2^*, \ldots, x_n^*) = 0 \tag{8.1}$$

This result is often written in a more compact vectorial form as

$$f'_{xi}(\vec{X}^*) = 0 \quad \text{or} \quad \nabla f = f'_{x1}\hat{i}_1 + f'_{x2}\hat{i}_2 + \cdots + f'_{xn}\hat{i}_n \tag{8.2}$$

$$\text{where} \quad \vec{X} = \begin{bmatrix} x_1 \\ x_2 \\ \ldots \\ x_n \end{bmatrix} \tag{8.3}$$

Here the n-dimensional vector $\vec{X}$ of design variables is called the design vector, and the vector ∇f is called the gradient vector. To find whether the stationery point $\vec{X}^*$ is a

Thermal System Design and Simulation
http://dx.doi.org/10.1016/B978-0-12-809449-5.00008-5

minima or a maxima, we determine the values of the second derivatives and check for the positive definiteness of $n \times n$ J-matrix defined as

$$J = \left(\frac{\partial^2 f}{\partial x_i \partial x_j} \right)_{X=X^*} \quad \begin{matrix} i = 1, 2, \ldots, n \\ j = 1, 2, \ldots, n \end{matrix} \tag{8.4}$$

at the optimal "point." If the matrix is positive definite, then the optimal vector X^* is a minima. If it is negative definite, then X^* is a maxima. A matrix J is said to be positive definite if the quadratic from $|A^T J A|$ is always positive for all values of the n-dimensional vector A. There are standard methods to establish positive definiteness of a matrix, but usually we do not need to apply them. In thermal system optimization it is possible, in most cases, to find from physical considerations whether the stationary point is a minima or a maxima.

Example 8.1

A heat exchanger is being designed with 50 tubes of diameter D and length L. The total net cost of the heat exchanger including both the initial cost, the life time cost of its operation and excluding the revenue from heat transfer is given by the equation:

$$F = 10{,}000 + 1.5 \times 10^6 D^{1.2} \cdot L^{1.2} + \frac{0.00002 \cdot L}{D^{4.5}} - 16 \times 10^6 (1 - e^{-0.026 L D^{0.8}})$$

Determine the values of the diameter and the length of tubes for minimum cost.

Solution

Following the notation introduced above, the design vector is $\vec{X} = \begin{bmatrix} D \\ L \end{bmatrix}$ and the minimum value of F will occurs when Eq. (8.2) is satisfied, that is,

$$\frac{\partial F}{\partial D} = 0; \quad \frac{\partial F}{\partial L} = 0$$

This gives

$$1.5 \times 10^6 (1.2 D^{0.2}) L^{1.2} + \frac{0.00002 L (-4.5)}{D^{5.5}} + 16 \times 10^6 (-0.026 L \times 0.8 D^{-0.2}) e^{-0.026 L D^{0.8}} = 0$$

$$1.5 \times 10^6 D^{1.2} (1.2 L^{0.2}) + \frac{0.00002}{D^{4.5}} + 16 \times 10^6 (-0.026 D^{0.8}) e^{-0.026 L D^{0.8}} = 0$$

Simplifying and dividing these equations we get

$$\frac{1.8 \times 10^6 D^{0.2} L^{1.2} - \frac{0.00009 L}{D^{5.5}}}{1.8 \times 10^6 D^{1.2} L^{0.2} + \frac{0.00002}{D^{4.5}}} = \frac{0.8 L}{D}$$

Cross-multiplying and simplifying this equation gives

$$L^{0.2} = \frac{0.29444 \times 10^{-9}}{D^{5.70}}$$

Using this result to express L in terms of D in the first equation obtained above we get a nonlinear equation for optimum D (after dividing both sides by L) as

$$1.8 \times 10^6 D^{0.2} L^{0.2} - \frac{0.00009}{D^{5.5}} - \frac{0.3328 \times 10^6}{D^{0.2}} \quad \exp[-0.026 L D^{0.8}] = 0$$

$$\Rightarrow \quad \frac{0.00044}{D^{5.5}} - \frac{0.3328 \times 10^6}{D^{0.2}} \quad \exp\left[-0.026\left(\frac{0.2944 \times 10^{-9}}{D^{5.7}}\right) D^{0.8}\right]$$

Solving this equation by any of the procedure outlines in Chapter 2 we get $D = 0.2113\,\mathrm{m}$ and $L = 1.2188\,\mathrm{m}$

8.1.1 Constrained Optimization

In engineering optimization we often come across situations where the design variables are not free to take any value but are constrained to be within certain limits, or by certain interrelationship among them. Such problems are termed as constrained optimization problems. A general formulation can be written as

$$\text{Minimize} \quad f(\vec{X}) \qquad \vec{X} = [x_1, x_2, \ldots, x_n]^T \tag{8.5}$$

subject to constraints

$$g_k(\vec{X}) = 0 \qquad 1 \leq k \leq m \quad (m < n) \tag{8.6}$$

There exist a number of approaches to solve such problems but the easiest, and the most popular technique is through the use of, what are called, the Lagrange Multipliers. The first step in all these procedures is to eliminate as many constraints as possible by expressing some of the n-variables in terms of the rest. Thereafter, the remaining constraints are incorporated with the objective function through the use of unknown multipliers λ_k, to define an augmented function Y:

$$Y = f(\vec{X}) + \sum_{k=1}^{m} \lambda_k g_k(\vec{X}) \tag{8.7}$$

It can be showed that the minima of this augmented function, considered as an unconstrained function, is the same as the minima of $f(\vec{X})$. Using Eq. (8.2) we get n equations, which along with the m constraint equations (i.e., Eq. 8.6), provide us with $n + m$ equations for $n + m$ unknowns, viz. the optimal design vector $\{x_1, x_2, \ldots, x_n\}^T$ and the m Lagrange multipliers, $\lambda_k (k = 1, 2, \ldots, m)$.

Example 8.2

A finned surface having numerous pin fins each of diameter d and length L is to be designed for minimum weight W and a heat transfer of a given amount, Q. The simplified expressions for W and Q are:

$$W = C_1 + C_2 d^2 L + C_3 d$$
$$Q = C_4 dL + C_5 d$$

Determine optimal values of the design variables d and L.

Solution

The augmented function using the Lagrange multiplier λ to incorporate the constraint in the objective function is:

$$Y = C_1 + C_2 d^2 L + C_3 d + \lambda (Q - C_4 dL - C_5 d)$$

Equating the gradient vector ∇Y to zero gives:

$$\frac{\partial Y}{\partial d} = 2C_2 dL + C_3 + \lambda(-C_4 L - C_5) = 0$$
$$\frac{\partial Y}{\partial L} = C_2 d^2 + \lambda(-C_4 d) = 0$$

The second equation can be simplified to get the value of Lagrange multiplier λ in terms of the design variables as:

$$\lambda = \frac{C_2 d}{C_4}$$

Substituting this in the first equation we get:

$$2C_2 dL + C_3 + \frac{C_2 d}{C_4}(-C_4 L - C_5) = 0$$

which on collecting terms gives:

$$L = \frac{C_5}{C_4} - \frac{C_3}{C_2 d}$$

Substituting this result in the constraint for heat transfer Q we get:

$$Q = C_4 d \left(\frac{C_5}{C_4} - \frac{C_3}{C_2 d} \right) + C_5 d$$

which can be simplified to give the optimal value of d as:

$$d = \frac{Q + C_3 C_4 / C_2}{2C_5} = d^*$$

This, when substituted in the equation for L, gives:

$$L = \frac{C_5}{C_4} - \frac{2C_3 C_5}{C_2(Q + C_3 C_4 / C_2)} = L^*$$

The value of the optimal objective function (i.e., weight, W) is:

$$W^* = C_1 + C_2 d^{*2} L^* + C_3 d^*$$

$$= C_1 + C_2 \left[\frac{Q + C_3 C_4 / C_2}{2C_5} \right]^2 \star \left[\frac{C_5}{C_4} - \frac{2C_3 C_5}{C_2(Q + C_3 C_4 / C_2)} \right] + C_3 \left[\frac{Q + C_3 C_4 / C_2}{2C_5} \right]$$

$$= C_1 + \frac{C_2 C_5}{C_4} \left(\frac{Q + C_3 C_4 / C_2}{2C_5} \right)^2 - 2C_3 C_5 \frac{(Q + C_3 C_4 / C_2)}{(2C_5)^2} + C_3 \left[\frac{Q + C_3 C_4 / C_2}{2C_5} \right]$$

$$= C_1 + \frac{C_2}{C_4 C_5} \left(\frac{Q + C_3 C_4 / C_2}{2} \right)^2$$

and the value of the Lagrange multiplier is:

$$\lambda = \frac{C_2 d^*}{C_4} = \frac{C_2}{2C_4 C_5} \left(Q + \frac{C_3 C_4}{C_2} \right)$$

Comment: The optimal value of W is clearly dependent on the value of the heat transfer (Q) acting as a constraint on the optimum design. The change in the value of W with any change in Q is given by:

$$\left(\frac{\partial W}{\partial Q} \right) = \frac{C_2}{C_4 C_5} \cdot \frac{1}{4} \cdot 2 \left(Q + \frac{C_3 C_4}{C_2} \right)$$

$$= \frac{C_2}{2C_4 C_5} \left(Q + \frac{C_3 C_4}{C_2} \right)$$

which is exactly equal to λ, the Lagrange multiplier used in the formulation. It means that the sensitivity of optimal value to the value of the constraint—sometimes called the sensitivity coefficient (SG)—equals the value of the Lagrange multiplier. This is true not just in this example, but is a general result. If there are more than one constraints, each Lagrange coefficient gives the sensitivity of the optimal value to that constraint. This is usually expressed as:

$$SC_1 = \lambda_1; \quad SC_2 = \lambda_2; \quad \ldots; \quad SC_m = \lambda_m \tag{8.8}$$

Example 8.3

Flue gas at 600°C is being used to heat water in two stage process (Fig. 8.1). HE1 is a high quality heat exchanger whose effectiveness is approximately equal to $0.2\sqrt{A_1}$ (A_1 being its surface area) and the cost is $25A_1 \times 10^4$ Rs. HE_2 is a cheaper heat exchanger

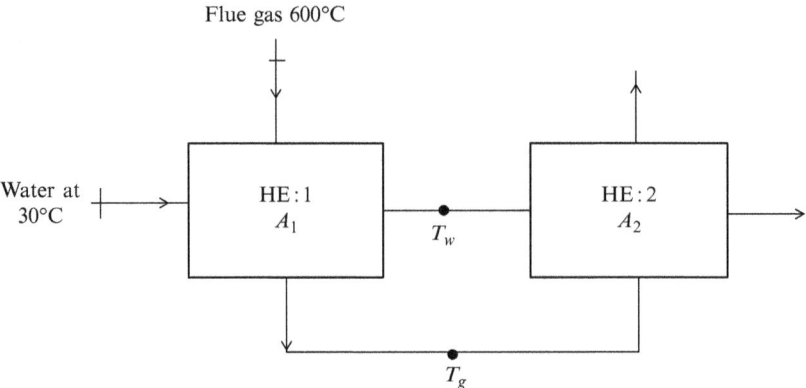

Flue gas 600°C

Water at
30°C

HE:1
A_1

T_w

HE:2
A_2

T_g

Fig. 8.1 Two stage water heater system.

costing $10A_2 \times 10^4$ Rs. and giving an effectiveness of $0.1\sqrt{A_2}$, A_2 being its surface area in m². The other data are:

Flue gas flow rate: 60 kg/min   Specific heat: 1 kJ/kg°C
Water flow rate: 20 kg/min   Specific heat: 4.2 kJ/kg°C

If the total amount available to buy the heat exchangers is 45×10^4 Rs., determine the surface areas of the two heat exchangers which will maximize the total heat recovered in the heat exchangers.

Solution

Clearly the objective function to be maximized is the total heat transfer (Q) in the two heat exchangers and the constraint is that their total cost should be 45×10^4 Rs. The constraint can be easily written in terms of the design variables A_1 and A_2 as

$$25A_1 \times 10^4 + 10A_2 \times 10^4 = 45 \times 10^4$$
$$\text{or} \quad 2.5A_1 + A_2 = 4.5$$

We now express the total heat transfer Q in the two heat exchangers in terms of the design variables. The heat transfer in the first heat exchanger Q_1, can be directly calculated using the effectiveness, the temperatures of the two streams and the minimum of their heat capacity rates.

$$\text{Now} \quad C_{fluegas} = \frac{60}{60} \times 1000 = 1000 \text{ W/°C}$$

$$C_{water} = \frac{20}{60} \times 4200 = 1400 \text{ W/°C}$$

$$\text{Clearly} \quad C_{min} = 1000 \text{ W/°C}$$

$$\text{and} \quad Q_1 = 1000 \times (0.2A_1^{1/2})(600 - 30) \text{ W}$$
$$= 1.14 \times 10^5 A_1^{1/2} \text{ W}$$

Further using this expression for Q_1, we calculate by heat balance the exit temperatures of the two streams. This gives:

$$\frac{60}{60} \times 1000 \times (600 - T_g) = 1.14 \times 10^5 A_1^{1/2}$$

$$\frac{20}{60} \times 4200 \times (T_w - 30) = 1.14 \times 10^5 A_1^{1/2}$$

from, which we get:

$$T_g = 600 - 114 A_1^{1/2}$$

$$T_w = 30 + 81.43 A_1^{1/2}$$

Now, using the given expression for the effectiveness of the second heat exchanger, we can determine the heat transfer taking place in it, Q_2, as:

$$Q_2 = (0.1 A_2^{1/2})(1000)(T_g - T_w)$$

$$= 100 A_2^{1/2}(570 - 195.43 A_1^{1/2})$$

Thus the total heat transfer in the two heat exchangers is:

$$Q = Q_1 + Q_2 = 1.14 \times 10^5 A_1^{1/2} + 100 A_2^{1/2}(570 - 195.43 A_1)$$

The optimization problem can now be summarized as:

$$\text{Maximize} \quad Q = 1.14 \times 10^5 A_1^{1/2} + 100 A_2^{1/2}(570 - 195.43 A_1)$$

subject to constraint $2.5 A_1 + A_2 = 4.5$

Though we could solve it by eliminating one of the two design variables by using the constraint, here we will solve it by using Lagrange multiplier λ. The augmented function becomes:

$$Y = 1.14 \times 10^5 A_1^{1/2} + 100 A_2^{1/2}(570 - 195.43 A_1) + \lambda(-2.5 A_1 - A_2 + 4.5)$$

Equating ∇Y to zero, the design vector being $\begin{bmatrix} A_1 \\ A_2 \end{bmatrix}$, we get:

$$\frac{\partial Y}{\partial A_1} = \left(\frac{1.14 \times 10^5}{2}\right) A_1^{-1/2} - 100 \times 195.43 A_2^{1/2} - \lambda(2.5) = 0$$

$$\frac{\partial Y}{\partial A_2} = \frac{100}{2} A_2^{-1/2}(570 - 195.43) A_1 - \lambda = 0$$

From the above two equations we get:

$$\frac{1.14 \times 10^5}{2} A_1^{-1/2} - 100 \times 195.43 A_2^{1/2} = \frac{250}{2}(570 - 195.43 A_1) A_2^{-1/2}$$

Simplifying this equation we get:

$$5.7A_1^{-1/2} - 1.9543A_2^{1/2} = 1.25(5.7 - 1.9543A_1)A_2^{-1/2}$$
$$= 7.125A_2^{-1/2} - 2.44288A_1A_2^{-1/2}$$

$$\Longrightarrow 5.7\left(\frac{A_2}{A_1}\right)^{1/2} - 1.9543A_2 = 7.125 - 2.44288A_1$$

Substituting from the constraint, $A_2 = 4.5 - 2.5A_1$ we get an equation for optimal A_1:

$$5.7\left(\frac{4.5 - 2.5A_1}{A_1}\right)^{1/2} - 1.9543(4.5 - 2.5A_1) = 7.125 - 2.44288A_1$$

This nonlinear equation in A_1 can be solved by using any of the methods outlined in Chapter 2 to get:

$$A_1^* = 0.805\,\text{m}^2$$
$$\therefore A_2^* = 4.5 - 2.5A_1^* = 2.4875\,\text{m}^2$$

which gives the value of maximum heat transfer Q as:

$$Q = 1.14 \times 10^5 (0.805)^{1/2} + 100(2.4875)^{1/2}(570 - 195.43 \times 0.805)$$
$$= 16.737 \times 10^4\,\text{W} \quad \text{or} \quad 167.37\,\text{kW}$$

The students are advised to solve this problem directly by reducing it to optimization of a function of single variable and verify the above result.

In some cases the constraints are not in the form of equations (like Eq. 8.6), but are inequalities, viz.

$$g_k(\vec{X}) \leq 0 \quad k = 1, 2, \ldots, m \tag{8.9}$$

where m is no longer necessarily less than n, the number of independent variables.

To use the Lagrange multiplier method for such problems, the inequalities are converted into equations by introducing additional variables called the *slack variables*. Thus constraints (8.9) are modified as:

$$g_k(\vec{X}) + x_{n+k}^2 = 0 \quad k = 1, 2, \ldots, m \tag{8.10}$$

where $x_{n+1}, x_{n+2}, \ldots, x_{n+m}$ are the m slack variables. Optimization of the function (8.5) subject to constraints (8.10) can now be carried out as before by defining an augmented function with the help of m Lagrange multipliers $\lambda_1, \lambda_2, \ldots, \lambda_m$ and treating the objective function as dependent on $n + m$ variables $x_1, x_2, \ldots, s_n, x_{n+1}, x_{n+2}, \ldots, x_{n+m}$.

Besides these general approaches, powerful techniques have been developed for special types of objective functions and constraints. Thus, when both the objective function and the constraints are linear functions of the variables, the technique of linear programming is very efficient. Similarly if both the objective function and the constraints can be expressed in the form of sums of polynomials, geometric programming (GP) is the method of choice. Since in thermal system optimization we rarely come across linear functions, while polynomial functions are often encountered, we shall study in some detail the basic principles of GP.

8.1.2 Geometric Programming

A typical optimization problem amenable to use of GP is:

$$\text{Minimize:} f = 100 + \frac{20}{x_2} + 30x_2^{0.8} - 4x_1^2 \tag{8.11}$$

$$\text{subject to:} x_1 x_2^2 + x_1\sqrt{x_2} = 100 \tag{8.12}$$

The concept of degree of difficulty (DD) is very helpful in identifying the "extent" to which GP algorithm would help in solving the problem. It is defined as:

$$DD = T_o + T_c - (N + 1)$$

where T_o and T_c indicate the number of terms involving the variables in the objective function and those in the constraints respectively, and N is the number of variables. Thus the value of DD for the example problem given above would be

$$DD = 3 + 2 - (2 + 1) = 2$$

While the basic algorithm of GP is applicable to all problems involving only polynomial functions, it is especially convenient when, $DD = 0$. For $DD > 0$, the algorithm would involve solution of nonlinear equations and this may not be less time consuming as compared to Lagrange multipliers. We shall therefore focus only on the solution of problems with $DD = 0$, beginning with the general unconstrained minimization problem:

$$\text{Minimize} \quad f(\vec{X}) = \sum_{j=1}^{N}(c_j \quad x_1^{a_{1j}} \quad x_2^{a_{2j}} \quad \ldots \quad x_n^{a_{nj}}) \tag{8.13}$$

where $c_j > 0, x_i > 0$ and $a_{1j}, a_{2j}, \ldots, a_{nj}$ are real constants.[1] In order that the DD of this problem is zero, clearly $N = n + 1$.

According to the GP algorithm, the optimal value of the function, f^*, can be represented in a product form as:

$$f^* = \left(\frac{c_1 x_1^{a_{11}} x_2^{a_{21}} \ldots x_n^{a_{n1}}}{w_1}\right)^{w_1} \left(\frac{c_2 x_1^{a_{12}} x_2^{a_{22}} \ldots x_n^{a_{n2}}}{w_2}\right)^{w_2} \ldots \left(\frac{c_N x_1^{a_{1N}} x_2^{a_{2N}} \ldots x_n^{a_{nN}}}{w_N}\right)^{w_N} \tag{8.14}$$

---

[1] In view of these constraints, some authors prefer use of term posynomials instead of polynomials.

where the values of the exponents are chosen so as to satisfy the following equations:

$$a_{11}w_1 + a_{12}w_2 + \cdots + a_{1N}w_N = 0$$
$$a_{21}w_1 + a_{22}w_2 + \cdots + a_{2N}w_N = 0$$
$$\vdots$$
$$a_{n1}w_1 + a_{n2}w_2 + \cdots + a_{nN}w_N = 0$$
$$\text{and} \quad w_1 + w_2 + w_3 + \cdots + w_N = 1 \tag{8.15}$$

For zero DD, $N = n + 1$ and therefore in Eq. (8.15) we have N linear equations for N variables which can be easily solved to get the values of the exponents $w_1, w_2, \ldots, w_N$. These values when substituted back in Eq. (8.14) give a very simple expression for the optimal values, viz.

$$f^* = \left(\frac{c_1}{w_1}\right)^{w_1} \left(\frac{c_2}{w_2}\right)^{w_2} \cdots \left(\frac{c_N}{w_N}\right)^{w_N} \tag{8.16}$$

It can be proved that the exponents also indicate the contribution of the individual terms of the objective function to the total optimal value, that is,

$$w_1 = \frac{(c_1 x_1^{a_{11}} x_2^{a_{21}} \ldots x_n^{a_{n1}})^*}{f^*}$$

$$w_2 = \frac{(c_2 x_1^{a_{12}} x_2^{a_{22}} \ldots x_n^{a_{n2}})^*}{f^*}$$

$$\ldots$$

$$w_N = \frac{(c_N x_1^{a_{1N}} x_2^{a_{2N}} \ldots x_n^{a_{nN}})^*}{f^*} \tag{8.17}$$

This is helpful in determining the optimal values of the variables $x_1, x_2, \ldots, x_n$. We shall illustrate the method by a few examples.

Example 8.4

A rectangular steam tank has $1.5\,\text{m}^2$ of surface exposed to ambient air. These exposed surfaces are to be insulated with glass wool, $k = 0.4\,\text{W/m K}$. Determine the optimal value of the thickness of the glass wool, using the following data:

Cost of glass wool insulation blankets of thickness $x\,\text{cm}$: $500x^{0.8}$ Rs. per m^2
Cost of heat loss through the tank: C_E/kWh where C_E is the cost of 1 kWh of electric energy.
Life time operating hours $= 10^5$
Take the temperature drop across the insulation as 200°C.
Solve the problem for two values of C_E, that is, $C_E = 10\,\text{Rs./kWh}$ and $C_E = 5\,\text{Rs./kWh}$

Solution

The rate of heat transfer through the insulation of x cm thickness is:

$$Q = \frac{0.04 \times 1.5 \times 200}{x/100} \, W = \frac{1200}{x} \, W$$

$$\text{Life time cost of heat lost} = \frac{1200}{x} \times \frac{10^5 \cdot C_E}{1000}$$

$$= \frac{12 \times 10^4}{x} C_E$$

Thus the objective function to be minimized is:

$$F = 500 \times 1.5 x^{0.8} + \frac{12 \times 10^4}{x} C_E$$

Using the GP algorithm, the optimal value of F can be directly written as:

$$F^* = \left(\frac{750}{w_1}\right)^{w1} \left(\frac{12 \times 10^4 C_E}{w_2}\right)^{w2}$$

$$\text{where } 0.8 w_1 + w_2(-1) = 0$$

$$w_1 + w_2 = 1$$

These equations give $w_1 = 1/1.8$ $w_2 = 0.8/1.8$ and the optimal value of objective function:

$$F^* = (750 \times 1.8)^{1/1.8} \left(\frac{12 \times 10^4 C_E \times 1.8}{0.8}\right)^{0.8/1.8}$$

which for $C_E = 10$ Rs./kWh gives:

$$F^* = 39578.4 \, \text{Rs.}$$

To find the optimal value of x, we use Eq. (8.17). This gives:

$$\frac{1}{1.8} = \frac{(750 x^{\star 0.8})}{39578.4}$$

$$\rightarrow x^* = \left(\frac{39578.4}{1.8 \times 750}\right)^{1.25} = 68.22 \, \text{cm}$$

For $C_E = 5$ Rs./kWh, we get:

$$F^* = 29084.9 \, \text{Rs.}$$

and to find x^*, we use this time:

$$\frac{0.8}{1.8} = \frac{(\frac{12 \times 10^4 \times 5}{x^*})}{29084.9}$$

$$\text{which gives } x^* = 46.42 \, \text{cm}$$

Comment

As expected, with decrease in the energy costs, the insulation requirement decreases. The values of the exponents w_1 and w_2 also indicate the relative contribution of the two terms to the total objective function. Here, the first term—the fixed cost—contributes about 55.55% to the total while the second term—the running cost—contributes the rest 44.45%. This ratio does not changes with change in the energy cost C_E. It would change if, for example, the fixed cost was influenced differently by the change in the thickness of the insulation. Thus, if the cost of glass wool varied with x as $500x$, then the new values of the exponents would be $w_1, w_2 = 1/2$, that is, each term would contribute equally to the total cost. The students are advised to check how this would affect the optimal values.

We thus see that GP provides valuable insights into the optimization problem.

Example 8.5

River water is to be used to cool a high-power electronic system E (Fig. 8.2). The total cost of the electronic circuit, C_E depends on its temperature, which in turn depends on the rate of heat dissipation from it:

$$C_E = 12,000\,T_E$$

where T_E is the temperature in °C. The total heat to be dissipated is 2 MW and the overall heat transfer coefficient in the cooling circuit (of surface area 20 m²) can be estimated using the equation:

$$h = 1000\,Q^{0.8} \text{ W/m}^2 \text{ K}$$

where Q is the water flow rate in liters/second.

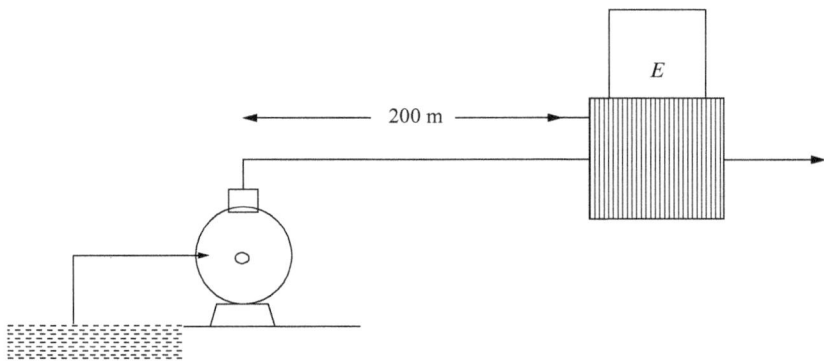

Fig. 8.2 Cooling power electronic system.

The river bank is located 200 m from the electronic system and the cost of the piping system per meter length, C_p, is given by:

$$C_P = 150\,D^{1.2}$$

where D is the pipe diameter in centimeter.

The life time cost of electricity needed to run the water pump is given by the equation

$$C_r = 10^3 \cdot Q \cdot \Delta P$$

where Q is flow rate in L/s and ΔP is the total pressure drop in N/m^2.

Determine the optimal values of the flow rate Q, and the pipe diameter D, which will minimize the life cycle cost of the system. Take environment temperature as 30°C, and pipe friction factor as 0.03.

Solution

The objective function has three components, viz. the initial cost of the electronic system and the piping system, and the running cost of the pump. We shall developed expressions for each of these in terms of the two design variables Q and D.

$$C_E = 12{,}000\, T_E$$

where T_E is governed by the requirement of dissipating 2 MW of heat. This gives

$$2 \times 10^6 = h \cdot A \cdot \Delta T = (1000 Q^{0.8})(20)(T_E - 30)$$

This gives $T_E = \dfrac{100}{Q^{0.8}} + 30$

Therefore $C_E = 36 \times 10^4 + \dfrac{12 \times 10^5}{Q^{0.8}}$

To convert C_r in terms of D and Q, we have to determine the pressure drop in pipeline. This can be done using Darcy equation (taking water density as 1000 kg m^3):

$$\Delta P = \frac{(1000)(0.03) \times 200}{(D \times 10^{-2})} \cdot \frac{1}{2} \left[\frac{Q \times 10^{-3}}{\frac{\pi}{4} \times D^2 \times 10^{-4}} \right]^2 \frac{N}{m^2} = 4.8634 \times 10^7 Q^2 / D^5$$

The objective function can now be written as:

$$f = 36 \times 10^4 + \frac{12 \times 10^5}{Q^{0.8}} + 150 D^{1.2} \times 200 + 4.8634 \times 10^{10} \frac{Q^3}{D^5}$$

which has a DD $= 3 - (2 + 1) = 0$.

Leaving aside the constant value of 36×10^4, we can get f^*, the optimum value of the remaining terms, from GP formulation as: $f^* = \left(\dfrac{12 \times 10^5}{w_1} \right)^{w_1} \left(\dfrac{3 \times 10^4}{w_2} \right)^{w_2} \left(\dfrac{4.8634 \times 10^{10}}{w_3} \right)^{w_3}$

where the coefficients w_1, w_2, and w_3 are given by the equations:

$$Q: \quad w_1(-0.8) + 0 + w_3 \times 3 = 0$$
$$D: \quad w_2 \times 1.2 + w_3 \times (-5) = 0$$
$$w_1 + w_2 + w_3 = 1$$

The three equations can be easily solved to get:

$$w_1 = 0.42056, \quad w_2 = 0.46729, \quad w_3 = 0.11215$$

Substituting these values we can calculate the optimal value of the objective function, f^*, as:

$$f^* = \left(\frac{12 \times 10^5}{0.42056}\right)^{0.42056} \left(\frac{3 \times 10^4}{0.46729}\right)^{0.46729} \left(\frac{4.8634 \times 10^{10}}{0.11215}\right)^{0.11215}$$

$$= 1{,}847{,}025$$

To find the values of optimal pipe diameter D^*, and the optimal flow rate Q^*, we make use of Eq. (8.17). This gives:

$$\frac{12 \times 10^5}{Q^{*0.8}} = w_1 f^* = 0.42056 \times 1847025$$

$$\Rightarrow Q^* = 1.722 \, \text{L/s}$$

$$\text{and} \quad 150 D^{*1.2} \times 200 = w_2 f^* = 0.46729 \times 1847025$$

$$\Rightarrow D^* = 16.44 \, \text{cm} \tag{8.18}$$

The total optimal cost of the entire system is:

$$f = 36 \times 10^4 + f^* = 2207025 \, \text{Rs.}$$

The constrained optimization problems can also be solved using GP in a similar manner. The only distinction is that the exponents of the polynomial expressions in the constraints have additional unknown multipliers in Eq. (8.14). These are determined by using the additional relationships arising from equating the sums of the exponents of the objective function polynomials and the polynomials of the constraints *separately* equal to 1 as explained below.

Indicating each of the terms of Eq. (8.13) by p_i ($p \equiv$ polynomial expression, that is, $c_i \, x_1^{a_{1i}} \, x_2^{a_{2i}} \ldots x_n^{a_{ni}}$), we can write the constrained optimization problem as:

$$\text{Minimize} \quad f = p_1 + p_2 + p_3 + \cdots + p_N \tag{8.19}$$

subject to constraints (say):

$$p_{N+1} + p_{N+2} = 1$$
$$p_{N+3} + p_{N+4} + p_{N+5} = 1 \tag{8.20}$$

According to the GP algorithm, the optimal value of the objective function is:

$$f^* = \left(\frac{p_1}{w_1}\right)^{w_1} \left(\frac{p_2}{w_2}\right)^{w_2} \left(\frac{p_3}{w_3}\right)^{w_3} \cdots \left(\frac{p_N}{w_N}\right)^{w_N} \left(\frac{p_{N+1}}{w_{N+1}}\right)^{M_1 w_{N+1}} \left(\frac{p_{N+2}}{w_{N+2}}\right)^{M_1 w_{N+2}}$$

$$\left(\frac{p_{N+3}}{w_{N+3}}\right)^{M_2 w_{N+3}} \left(\frac{p_{N+4}}{w_{N+4}}\right)^{M_2 w_{N+4}} \left(\frac{p_{N+5}}{w_{N+5}}\right)^{M_2 w_{N+5}} \tag{8.21}$$

where the exponents $w_1, w_2, \ldots, w_{N+5}$ and the multipliers M_1, M_2 are the solution of the linear equations:

$$a_{11}w_1 + a_{12}w_2 + \cdots + a_{1N}w_N + M_1 a_{1,N+1}w_{N+1} + M_1 a_{1,N+2}w_{N+2}$$
$$+ M_2 a_{1,N+3}w_{N+3} + M_2 a_{1,N+4}w_{N+4} + M_2 a_{1,N+5}w_{N+5} = 0$$
$$a_{21}w_1 + a_{22}w_2 + \cdots + a_{2N}w_N + M_1 a_{2,N+1}w_{N+1} + M_1 a_{2,N+2}w_{N+2}$$
$$+ M_2 a_{2,N+3}w_{N+3} + M_2 a_{2,N+4}w_{N+4} + M_2 a_{2,N+5}w_{N+5} = 0$$
$$\cdots$$
$$a_{n1}w_1 + a_{n2}w_2 + \cdots + a_{nN}w_N + M_1 a_{n,N+1}w_{N+1} + M_1 a_{n,N+2}w_{N+2}$$
$$+ M_2 a_{n,N+3}w_{N+3} + M_2 a_{n,N+4}w_{N+4} + M_2 a_{n,N+5}w_{N+5} = 0$$

along with the conditions:

$$w_1 + w_2 + \cdots + w_n = 0$$
$$M_1 w_{N+1} + M_1 w_{N+2} = M_1$$
$$M_2 w_{N+3} + M_2 w_{N+4} + M_2 w_{N+5} = M_2 \tag{8.22}$$

Thus we have $n+3$ linear equations for $N+7$ variables, viz. $w_1, w_2, \ldots, w_{N+5}, M_1 + M_2$. If the DD of the problem zero, then:

$$0 = N + 5 - (n + 1)$$
$$\Rightarrow \quad n = N + 4$$

Thus $n + 3 = N + 7 \Rightarrow$ Above linear equations are sufficient to determine the values of all the variables, and thus the optimal solution can be found. We shall illustrate the procedure with the help of a few examples.

Example 8.6

Solve Example 8.5 treating it as "constrained" problem with variables Q, D and ΔP.

Solution

Here instead of substituting for ΔP in terms of Q and D from Darcy equation, we keep ΔP also as a variable, and treat the Darcy equation as a constraint relating $Q, \Delta P$ and D. The objective function can therefore be written as:

$$f = 36 \times 10^4 + \frac{12 \times 10^5}{Q^{0.8}} + 10^3 Q \cdot \Delta P + 150 \times 200 D^{1.2}$$

and the constraint is:

$$\Delta P = 4.8634 \times 10^7 \frac{Q^2}{D^5}$$

which can be written in the standard form of GP formulation as:

$$10^{-7} \times 0.2056 \frac{\Delta P D^5}{Q^2} = 1$$

According to the GP algorithm, the optimal solution for $f_1 = (f - 36 \times 10^4)$ is:

$$f_1^* = \left(\frac{12 \times 10^5}{w_1}\right)^{w_1} \left(\frac{10^3}{w_2}\right)^{w_2} \left(\frac{3 \times 10^4}{w_3}\right)^{w_3} \left(\frac{10^{-7} \times 0.2056}{w_4}\right)^{Mw_4}$$

and the equation relating these coefficients are:

$$
\begin{aligned}
Q: &\; -0.8w_1 + w_2 - 2Mw_4 = 0 \\
D: &\; \quad 1.2\, w_3 + 5\, Mw_4 = 0 \\
\Delta P: &\; \qquad w_2 + Mw_4 = 0 \\
&\; w_1 + w_2 + w_3 = 1 \\
&\; Mw_4 = M \;\Rightarrow\; w_4 = 1
\end{aligned}
$$

Solving the above four equations we get:

$$w_1 = 0.42056; \quad w_2 = 0.11215; \quad w_1 = 0.46729; \quad M = -0.11215$$

The optimal value of f_1 is:

$$f_1^* = \left(\frac{12 \times 10^5}{0.42056}\right)^{0.42056} \left(\frac{10^3}{0.11215}\right)^{0.11215} \left(\frac{30000}{0.46729}\right)^{0.46729} \left(\frac{10^{-7} \times 0.2056}{1}\right)^{-0.11215}$$

$$= 1,847,082$$

which matches closely with the value found in Example 8.5. Since the coefficients w_1, w_2, w_3 are also matching (note: w_1, w_3 got interchanged in this example since we inadvertently changed the order of terms in the objective function), the optimal values of Q^* and D^* would also match with those obtained earlier.

Example 8.7

Solve Example 8.5 taking both T_E and ΔP as variables besides Q and D.

Solution

The expression for the objective function becomes quite simple:

$$f = 12{,}000 T_E + 10^3 Q \cdot \Delta P + 30{,}000 D^{1.2}$$

The constraints are:

$$10^{-8} \times 2.056 \frac{\Delta P D^5}{Q^2} = 1$$

$$\text{and} \quad T_E = \frac{100}{Q^{0.8}} + 30$$

which can be written in the standard from as:

$$\frac{T_E}{30} - \frac{100}{30 \cdot Q^{0.8}} = 1$$

This constraint cannot be directly accommodated in the GP formulation since all the coefficients have to be positive. We can overcome this difficulty by defining a new variable $T'_E = T_E - 30$ and convert this constraint into:

$$T_{E'} = \frac{100}{Q^{0.8}}$$

$$\text{or} \quad \frac{T'_E Q^{0.8}}{100} = 1$$

which is an admissible constraint, the coefficient of the expression now being positive. Substituting for T_E in terms of T'_E we get the objective function as:

$$f_1 = f - 36 \times 10^4 = 12{,}000 T_{E'} + 10^3 Q \cdot \Delta P + 30{,}000 D^{1.2}$$

and the constraints:

$$10^{-8} \times 2.056 \frac{\Delta P D^5}{Q^2} = 1$$

$$\text{and} \quad \frac{T_{E'} Q^{0.8}}{100} = 1$$

According to the GP algorithm for constrained problems the optimal value of f_1 is given by expression (Eq. 8.21):

$$f_1^* = \left(\frac{12000}{w_1}\right)^{w_1} \left(\frac{10^3}{w_2}\right)^{w_2} \left(\frac{30000}{w_3}\right)^{w_3} \left(\frac{10^{-8} \times 2.056}{w_4}\right)^{M_1 w_4} \left(\frac{0.01}{w_5}\right)^{M_2 w_5}$$

and the relationship between various coefficients are (equal 8.20, 8.21):

$$T_{E'} : w_1 + M_2 w_5 = 0$$

$$Q : w_2 - 2M_1 w_4 + 0.8 M_2 w_5 = 0 \qquad (8.23)$$

$$\Delta P : w_2 + M_1 w_4 = 0$$

$$D : 1.2 w_3 + 5 M_1 w_4 = 0$$

$$w_1 + w_2 + w_3 = 1$$

$$w_4 = 1; \quad w_5 = 1$$

Solving the above equations we get:

$$w_1 = 0.42056; \quad w_2 = 0.11215; \quad w_3 = 0.46229; \quad w_4 = 1; \quad w_5 = 1$$
$$M_2 = -0.42056 \quad M_1 = -0.11215$$

Substituting these values in the above expression for f_1^* we get:

$$f_1^* = \left(\frac{12000}{0.42056}\right)^{0.42056} \left(\frac{1000}{0.11215}\right)^{0.11215} \left(\frac{30000}{0.46729}\right)^{0.46729} \left(\frac{10^{-8} \times 0.2056}{1}\right)^{-0.11215}$$
$$\left(\frac{0.01}{1}\right)^{-0.42056} = 1,847,074$$

which matches closely with the solution obtained above. Further, since the exponents also match, the optimal values of various variables will also match with those obtained earlier.

8.1.3 Calculus of Variations

In the analytical methods considered hitherto, the objective was to find the value(s) of the variable(s) which optimized (i.e., either minimized or maximized) a function of these variable(s). However, we also come across situations like finding the shape of an object for minimum drag or the profile of a fin to maximize the heat transfer—where we need to determine a function which optimizes a function of that function, usually termed as a *functional*. This problem is addressed by the calculus of variations (COVs). The most commonly encountered functional is a definite integral whose integrand involves the function to be determined. Thus the simplest statement of the basic problem of COV is:

$$\text{Optimize} \quad I = \int_{x_1}^{x_2} F(x, y, y') dx \tag{8.24}$$

where $y(x)$ is a smooth (differentiable) function of x and can be viewed as family of curves satisfying the specified boundary conditions. The simplest version of these conditions is:

$$y(x_1) = y_1; \quad y(x_2) = y_2 \tag{8.25}$$

By following a procedure similar to that used in ordinary differential calculus, it can be proved that the optimal $y(x)$ is a solution of the differential equation, called the Euler–Lagrange (E-L) equation:

$$\frac{\partial F}{\partial y} - \frac{d}{dx}\left(\frac{\partial F}{\partial y'}\right) = 0 \tag{8.26}$$

By expressing the full derivative expression in terms of the partial derivatives, E-L equations can also written in the form:

$$\left(\frac{\partial^2 F}{\partial y'^2}\right) \frac{d^2 y}{dx^2} + \left(\frac{\partial^2 F}{\partial y' \partial y}\right) \frac{dy}{dx} + \left(\frac{\partial^2 F}{\partial y' \partial x} - \frac{\partial F}{\partial y}\right) = 0 \tag{8.27}$$

which is a second-order differential equation for y(x). This equation could be linear or nonlinear, depending upon the nature of the integrand $F(x, y, y')$.

As a simple illustration of COV we would consider the problem of finding the curve connecting two points (x_1, y_1) and (x_2, y_2) which has the shortest length. We already know its solution intuitively and the same can be obtained by first developing the functional I and then applying the E-L equations.

$$I = \int \frac{ds}{dx} \cdot dx$$

$$= \int \sqrt{1 + \left(\frac{dy}{dx}\right)^2} \, dx$$

$$\text{Thus } F \equiv \sqrt{1 + \left(\frac{dy}{dx}\right)^2} \equiv F(y', x)$$

Substituting in the E-L equation, Eq. (8.26) we get:

$$\frac{d}{dx}\left(\frac{\partial F}{\partial y'}\right) = 0$$

$$\text{or} \quad \frac{d}{dx}\left[\left(\frac{dy}{dx}\right) \cdot \left\{1 + \left(\frac{dy}{dx}\right)^2\right\}^{-1/2}\right] = 0$$

Integrating we get:

$$\left(\frac{dy}{dx}\right)\left[1 + \left(\frac{dy}{dx}\right)^2\right]^{-1/2} = c, \quad \text{a constant}$$

$$\Rightarrow \quad \left(\frac{dy}{dx}\right)^2 = c^2\left[1 + \left(\frac{dy}{dx}\right)^2\right]$$

$$\Rightarrow \quad \left(\frac{dy}{dx}\right)^2 = \frac{c^2}{1 - c^2} = c_1^2, \quad \text{say}$$

$$\Rightarrow \quad \frac{dy}{dx} = c_1$$

which on integration gives:

$$y = c_1 x + c_2$$

showing that the curve of minimum length is a straight line.

The problems generally encountered in thermal systems need optimization of a functional subject to constraints. Such problems can be tackled by defining a Lagrange multiplier λ, incorporating the constraint with the functional and determining the extremum of the augmented functional (cf. Section 8.1.1),

Thus if the constraint is:

$$\int_{x_1}^{x_2} G(x, y, y')dx = k \tag{8.28}$$

where k is a constant, the augmented functional is (cf. Eq. 8.7):

$$F^* = F(x, y, y') + \lambda G(x, y, y') \tag{8.29}$$

and the E–L equations are:

$$\frac{\partial F^*}{\partial y} - \frac{d}{dx}\left[\frac{\partial F^*}{\partial y'}\right] = 0 \tag{8.30}$$

The optimum function y is found as before by solving Eq. (8.30) and using the constraint Eq. (8.28), and the boundary conditions to determine λ, and the constants of integration.

We shall illustrate the method below by an example.

Example 8.8

Find the optimum profile of a pin fin of specified total volume (or mass) and length which will give maximum heat transfer from the hot surface to the surrounding fluid.

Solution

Let the profile be represented by the function $y(x)$, as indicated in Fig. 8.3. The equation governing the heat transfer through an infinitesimal section of the fin (assumed to have unit width) can be written as:

$$\left|-k \cdot 2y\frac{dT}{dx}\right|_x - \left|-k \cdot 2y\frac{dT}{dx}\right|_{x+dx} = h(2ds)(T - T_\infty)$$

where T is the fin temperature at position x, h is the heat transfer coefficient at fin surface, and k its thermal conductivity. This on simplification gives:

$$\frac{d}{dx}\left(k \cdot y\frac{dT}{dx}\right) = h \cdot (T - T_\infty)\left[1 + \left(\frac{dy}{dx}\right)^2\right]$$

Assuming for the sake of simplicity that it is a thin fin, $\frac{dy}{dx} \ll 1$, and the equation becomes:

$$\frac{d}{dx}\left(k \cdot y\frac{dT}{dx}\right) = h(T - T_\infty) \tag{8.31}$$

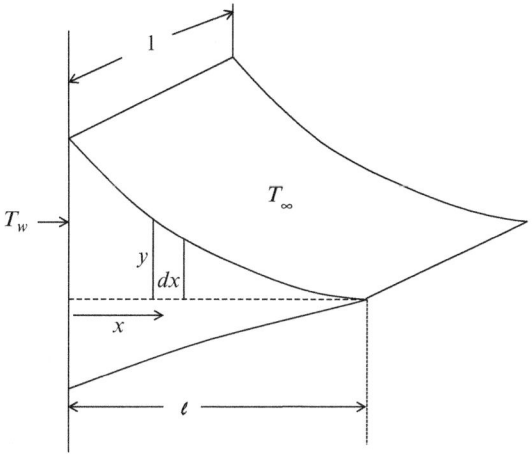

Fig. 8.3 Optimum profile of a pin fin.

We nondimensionalize the various variables by defining:

$$\theta = \frac{T - T_\infty}{T_o - T_\infty}; \quad X = \frac{x \cdot h}{k}; \quad Y = \frac{y \cdot h}{k}; \quad L = \frac{hl}{k} \qquad (8.32)$$

The Eq. (8.31) can be transformed in terms of these nondimensional variables (assuming k to be constant)as:

$$\frac{d}{dX}\left(Y\frac{d\theta}{dX}\right) = \theta \qquad (8.33)$$

The boundary conditions are (assuming negligible heat transfer at fin tip):

$$\text{at } x = 0, \quad T = T_o \Rightarrow X = 0; \quad \theta = 1 \qquad (8.34)$$

$$\text{at } x = l, \quad y\frac{dT}{dx} = 0 \Rightarrow X = L; \quad Y\frac{d\theta}{dX} = 0 \qquad (8.35)$$

To determine the total heat transfer through the fin, we multiply Eq. (8.33) by θ on both sides and then integrate to get:

$$\int_o^L \theta \frac{d}{dx}\left(Y\frac{d\theta}{dX}\right) = \int_o^L \theta^2 \, dX$$

$$\Rightarrow \left|\theta . Y . \frac{d\theta}{dX}\right|_o^L - \int_o^L \left(\frac{d\theta}{dX}\right)\left(Y\frac{d\theta}{dX}\right) dX = \int_o^L \theta^2 \, dX$$

Using the boundary conditions (8.34) and (8.35) the above equation can be simplified to give:

$$-\left[Y\theta \frac{d\theta}{dX}\right]_{x=0} = \int_o^L \left[Y\left(\frac{d\theta}{dX}\right)^2 + \theta^2\right] dX \qquad (8.36)$$

Now, applying Fourier's law, the heat transfer through the fin can be calculated as:

$$q = -2k \left| y \frac{\partial T}{\partial x} \right|_{x=0}$$

which can also be written, using Eqs. (8.32) and (8.36) as:

$$Q = \frac{q}{2k(T_o - T_\infty)} = -\left[Y \frac{d\theta}{dX} \right]_{x=0} = \int_o^L \left[Y \left(\frac{d\theta}{dX} \right)^2 + \theta^2 \right] dX$$

The fin volume V is given by the equation:

$$V = \int_o^L 2y \cdot 1 \, dx$$

which can also be transformed in terms of the dimensionless variables as:

$$V^* = \frac{V(h/k)^2}{2} = \int_o^L Y dX \tag{8.37}$$

Thus the optimization problem can be succinctly written as:

$$\text{Maximize} \quad Q = \int_o^L \left[Y \left(\frac{d\theta}{dX} \right)^2 + \theta^2 \right] dX \tag{8.38}$$

subject to the constraint:

$$V^* = \int_o^L Y dX \tag{8.39}$$

This can be solved using Lagrange multiplier, invoking Eqs. (8.29) and (8.30).

The augmented function is:

$$F^* = \int_o^L \left[Y \left(\frac{d\theta}{dX} \right)^2 + \theta^2 + \lambda Y \right] dX \tag{8.40}$$

and the E-L equation becomes:

$$\frac{\partial F^*}{\partial Y} - \frac{d}{dX} \left[\frac{\partial F^*}{\partial Y'} \right] = 0 \tag{8.41}$$

Here F^* does not involve $Y' \left(= \frac{dY}{dX} \right)$ term, θ is a function of X only and so the E-L equation can be written as:

$$\left(\frac{d\theta}{dX} \right)^2 + \lambda = 0$$

$$\left(\frac{d\theta}{dX} \right) = \pm(-\lambda)^{1/2}$$

Since the temperature gradient cannot be an imaginary quantity, λ must be negative, and writing $\lambda^- = -\lambda$, we get:

$$\frac{d\theta}{dX} = \pm(\lambda^-)^{1/2}$$

Further, since θ is decreasing with X (for $T_o > T_\infty$), we adopt the $-$ve sign:

$$\Rightarrow \quad \theta = -(\lambda^-)^{1/2}X + C \tag{8.42}$$

where C is the constant of integration. Above equation shows that the temperature profile in the optimal fin is linear. The value of constant C can be found from the boundary condition Eq. (8.34):

$$\text{at} \quad X = 0, \theta = 1$$

$$\Rightarrow \quad C = 1 \tag{8.43}$$

Thus the temperature profile in the optimal fin is:

$$\theta = 1 - \sqrt{\lambda^-}X \tag{8.44}$$

The optimal fin profile can be obtained by substituting this temperature profile in Eq. (8.33), which is:

$$\frac{d}{dX}\left(Y\frac{d\theta}{dX}\right) = \theta$$

$$\Rightarrow \quad \frac{d}{dX}(-Y \cdot \sqrt{\bar{\lambda}}) = 1 - \sqrt{\bar{\lambda}}X$$

$$\Rightarrow \quad Y = \frac{X^2}{2} - \frac{X}{\sqrt{\bar{\lambda}}} - \frac{C_1}{\sqrt{\bar{\lambda}}} \tag{8.45}$$

The value of constant C_1 is found using the boundary condition (Eq. 8.35) as follows:

$$\text{at} \quad X = L; \quad Y\frac{d\theta}{dX} = 0$$

$$\Rightarrow \quad \left[\frac{L^2}{2} - \frac{L}{\sqrt{\lambda^-}} - \frac{C_1}{\sqrt{\lambda^-}}\right]\left(-\lambda^{-1/2}\right) = 0$$

$$\Rightarrow \quad C_1 = +\left[\left(\frac{L^2}{2}\sqrt{\lambda^-} - L\right)\right]$$

Substituting it back in Eq. (8.45) we get the fin profile as:

$$Y = \frac{X^2 - L^2}{2} - \frac{X - L}{\sqrt{\lambda^-}} \tag{8.46}$$

The value of the Lagrange multiplier λ^- is found using the constraint, Eq. (8.39):

$$V^* = \int_0^L Y dX$$

$$= \int_0^L \left[\frac{X^2 - L^2}{2} - \frac{X - L}{\sqrt{\lambda^-}} \right] dX$$

$$= \frac{-L^3}{3} + \frac{-L^2}{2\sqrt{\lambda^-}}$$

$$\Rightarrow \quad \sqrt{\lambda^-} = \frac{L^2}{2\left[V^* + \frac{L^3}{3} \right]}$$

$$\text{or} \quad \lambda^- = \frac{L^4}{4(V^* + L^3/3)^2}$$

The results can also be expressed in terms of the primary variables by substituting for them in the above expressions using Eqs. (8.32) and (8.44), This gives:

$$\sqrt{\bar{\lambda}} = \frac{l^2}{V + \frac{2}{3}\frac{hl^3}{k}}$$

$$\theta = 1 - \sqrt{\lambda^-}\frac{h \cdot x}{k}$$

$$y = \frac{h}{2k}(x^2 - l^2) - \frac{x - l}{\sqrt{\lambda^-}}$$

Above equation gives us the optimum profile of a pin fin of specified volume which will maximize the heat transfer to the surroundings.

8.1.4 Pontryagin's Maximum Principle

Pontryagin's maximum principle, which can be seen as an extension of the COV, is widely used to obtain the strategy for optimal control of continuous processes. We shall restrict our discussion to processes whose performance equations are ordinary differential equations, viz.

$$\frac{dx_1}{dt} = f_1(x_1, x_2, \ldots, x_n; \theta)$$

$$\frac{dx_2}{dt} = f_2(x_1, x_2, \ldots, x_n; \theta)$$

$$\vdots$$

$$\frac{dx_n}{dt} = f_n(x_1, x_2, \ldots, x_n; \theta) \tag{8.47}$$

with the initial values $x_j(0) = x_j^0$; $for\ j = 1, 2, \ldots, n$.

Here $x_1, x_2, \ldots, x_n$ ($\equiv \vec{X}$) are the state variables and θ is the control variable, all functions of t.

The objective is to find an optimal control $\theta^* = \theta^*(t)$ and the corresponding path $\vec{X}^* = \vec{X}(t)$ $0 \leq t \leq T$ which maximizes a general functional of the final values of the state, viz.

$$S(\theta) = c_1 x_1(T) + c_2 x_2(T) + \cdots + c_n x_n(T) + \int_0^T f_o(x_1(t), x_2(t), \ldots, x_n(t); \theta(t)) dt \quad (8.48)$$

among all possible controls $\theta = \theta(t)$.

Pontryagin's Maximum Principle gives a necessary condition for $\theta = \theta^*(t)$ and the corresponding $\vec{x} = \vec{X}^*(t)$ to be optimal as:

$$S(\theta^*) = \max S(\theta)$$

This is determined through introduction of adjoint variables $z_1, z_2, \ldots, z_n$ and a "Hamiltonian" H defined as:

$$H(z_1, z_2, \ldots, z_n; x_1, x_2, \ldots, x_n; \theta) = z_1 f_1 + z_2 f_2 + \cdots + z_n f_n + f_o \qquad (8.49)$$

$$\text{with } \frac{dz_j}{dt} = -\frac{\partial H}{\partial x_j}, \quad j = 1, 2, \ldots, n \qquad (8.50)$$

These conditions complement the fact that:

$$\frac{dx_j}{dt} = \frac{\partial H}{\partial z_j}, \quad j = 1, 2, \ldots, n \qquad (8.51)$$

For each fixed time $t(0 \leq t < T)$, we choose θ^* to be the value of θ which *maximizes* the *Hamiltonian* H among all admissible values of θ's (keeping Z and X fixed;). This gives:

$$\frac{\partial H}{\partial \theta} = 0 \qquad (8.52)$$

The boundary conditions for the adjoint system are set in accordance with those for the given system. Thus for fixed terminal time T, free terminal state $x(T)$, the terminal conditions for the adjoint variables are: $z_j(T) = c_j, j = 1, 2, \ldots, n$. For time optimal problem, with fixed terminal state $X(T)$ and free terminal time T, $c_j = 0$; $j = 1, 2, \ldots, n; f_o = -1$. We solve the adjoint system with $H(z(T), X(T), \theta(T)) = 0$, with no terminal condition for the adjoint system.

The application of this principle, is illustrated below with the help of an example from thermal systems.

Example 8.9

A cold storage containing 600 tonnes of potatoes at 35°C is to be cooled to 0°C in 4 h using a refrigeration plant whose specifications are as under.

$$\text{Effective surface area of all the cooling coils} = 800\,\text{m}^2$$

Average heat transfer coefficient between

$$\text{air and the evaporator coils} = 100\,\text{W m}^2\,\text{K}$$

$$\frac{\text{Compressor power}}{\text{Refrigeration effect}} = 0.5 + 0.0001t - 0.02\theta$$

where θ is the evaporator temperature and t the total time of operation in seconds.[a] Given the specific heat of potatoes = 3.43 kJ/kg K. Neglect the thermal capacity of the cold store walls and determine the optimal control strategy to minimize the total power consumption during pull down. Assume the potatoes are cooled uniformly during the process.

Solution

If T indicates the temperature of the potatoes, its rate of change with time can be written using overall heat balance as:

$$600 \times 1000 \times 3.43 \times \frac{dT}{dt} = \frac{800 \times 100 \times (\theta - T)}{1000}$$

$$\Rightarrow \quad \frac{dT}{dt} = 3.8873 \times 10^{-5}(\theta - T)$$

Total power consumption during the pull-down, P, is:

$$P = \int_0^T \frac{800 \times 100 \times (T - \theta)}{1000} \cdot (0.5 + 0.0001t - 0.020\theta)dt$$

$$= \int_0^T 80(T - \theta)(0.5 + 0.0001t - 0.020\theta)dt$$

Here T is given to be 4 h, that is, $4 \times 3600 = 14{,}400$ s. Thus the functional to be *maximized* through Pontryagin's principle is:

$$S = -P = -\int_0^T 80(T - \theta)(0.5 + 0.0001t - 0.020\theta)dt$$

To find the optimal strategy $\theta(t)$, we first define the Hamiltonian H (Eqs. 8.49 and 8.50) as:

$$H = z_1 \cdot (3.8873 \times 10^{-5}(\theta - T)) - 80(T - \theta)(0.5 + 0.0001t - 0.020\theta)$$

This gives, using Eq. (8.50)

$$\frac{dz_1}{dt} = -\frac{\partial H}{\partial T} = -\left[-3.8873 \times 10^{-5}z_1 - 80(0.5 + 0.0001t - 0.020\theta)\right]$$

$$= 3.8873 \times 10^{-5}z_1 + 80(0.5 + 0.0001t - 0.020\theta)$$

The optimal strategy $\theta(t)$ is given by:

$$\frac{\partial H}{\partial \theta} = 0 = 3.8873 \times 10^{-5} z_1 - 80(-0.02T - 0.5 - 0.0001t + 0.040\theta)$$

$$\Rightarrow \quad z_1 = -20.5798 \times 10^5 [0.02T + 0.5 + 0.0001t - 0.040\theta)]$$

Substituting this in the above equation for $\frac{dz_1}{dt}$ gives for optimal θ:

$$\frac{dz_1}{dt} = 80(-0.02T + 0.02\theta) = 1.6(\theta - T)$$

We can get another expression for $\frac{dz_1}{dt}$ by differentiating the expression for z_1 obtained above. This gives:

$$\frac{dz_1}{dt} = -20.5798 \times 10^5 \left[0.02\frac{dT}{dt} + 0.0001 - 0.04\frac{d\theta}{dt} \right]$$

Combining these two equations we can get a relationship between $\frac{dT}{dt}$ and $\frac{d\theta}{dt}$:

$$-20.5798 \times 10^5 \left[0.02\frac{dT}{dt} + 0.0001 - 0.04\frac{d\theta}{dt} \right] = 1.6(\theta - T)$$

Substituting the performance equation to express $(\theta - T)$ in terms of $\frac{dT}{dt}$, and simplifying we get:

$$-20.5798 \times 10^5 \left[\frac{dT}{dt} + 0.005 - 2\frac{d\theta}{dt} \right] = 80(\theta - T)$$

$$= \frac{80}{3.8873 \times 10^{-5}} \frac{dT}{dt}$$

$$= 20.5798 \times 10^5 \frac{dT}{dt}$$

$$\text{or} \quad \frac{dT}{dt} + 0.005 - 2\frac{d\theta}{dt} = -\frac{dT}{dt}$$

$$\Rightarrow \quad \frac{d\theta}{dt} = \frac{dT}{dt} + 0.00025$$

Differentiating the performance equation we get:

$$\frac{d^2 T}{dt^2} = 3.8873 \times 10^{-5} \left(\frac{d\theta}{dt} - \frac{dT}{dt} \right) = 3.8873 \times 10^{-5} \times 0.00025$$

$$= 0.9718 \times 10^{-8}$$

Solving this equation we get:

$$T = 0.9718 \times 10^{-8} \times \frac{t^2}{2} + c_1 t + c_2$$

Since at $t = 0$ $T = 35$, this gives $c_2 = 35$. Also at $t = 14,400$, $T = 0$. This gives:

$$35 + c_1 \times 14,400 + \frac{0.9718 \times 10^{-8} \times (14,400)^2}{2} = 0$$

$$\therefore c_1 = -36.0075/14,400 = -0.0025$$

This gives optimal $T(t)$ as:

$$T(t) = 35 - 0.0025t + 0.4859 \times 10^{-8}t^2$$

To find the optimal strategy for regulating the evaporator temperature, we use the performance equation:

$$\frac{dT}{dt} = 3.8873 \times 10^{-5}(\theta - T)$$

$$\text{Now optimal } \frac{dT}{dt} = 0.9718 \times 10^{-8}t - 0.0025$$

$$\therefore \theta = \frac{0.9718 \times 10^{-8}t - 0.0025}{3.8873 \times 10^{-5}} + T$$

$$= 0.25 \times 10^{-3}t - 64.312 + T$$

$$= 0.25 \times 10^{-3}t - 64.312 + 0.4859 \times 10^{-8}t^2 - 0.0025t + 35$$

$$= 0.4859 \times 10^{-8}t^2 - 0.00225t - 29.312$$

The above equation gives the optimal control strategy for evaporator temperature to minimize the power consumption.

[a] Compressor power requirement increases with time due to frost formation.

8.1.5 Discrete Maximum Principle

In thermal systems we often come across situations where processes occur in stages, for example, in multistage compressors and multistage refrigeration systems. These processes can be visualized as a discretized version of continuous process with continuously changing control variables.

Accordingly, to optimize such systems a discretized version of Pontryagin's maximum principle has been developed by Chang [1] and Katz [2]. We shall present the discrete maximum principle for a multistage system with feedback (Fig. 8.4) following Fan and Wang [3].

Fig. 8.4 shows the schematic diagram of such a system consisting of N stages connected in a series with a portion of the output from the last stage fed back to the first stage. The transformation of the process stream at stage n is given by a set of performance equations (cf. Eq. 8.47):

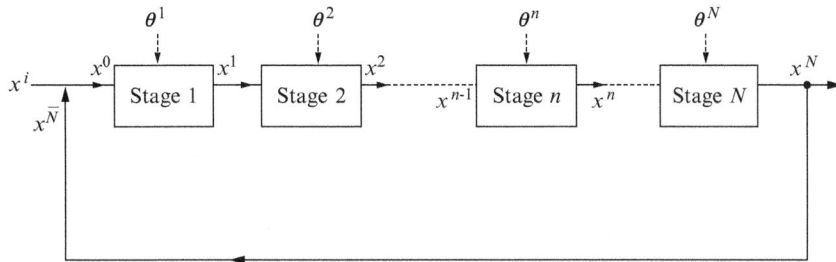

Fig. 8.4 Multistage process with feedback.

$$x_1^n = T_1^n(x_1^{n-1}, x_2^{n-1}, \ldots, x_s^{n-1}; \quad \theta_1^n, \theta_2^n, \ldots, \theta_t^n)$$
$$x_2^n = T_2^n(x_1^{n-1}, x_2^{n-1}, \ldots, x_s^{n-1}; \quad \theta_1^n, \theta_2^n, \ldots, \theta_t^n)$$
$$\vdots$$
$$x_s^n = T_s^n(x_1^{n-1}, x_2^{n-1}, \ldots, x_s^{n-1}; \quad \theta_1^n, \theta_2^n, \ldots, \theta_t^n) \tag{8.53}$$

where T_n is the transformation function, $(\vec{X})$ is the s-dimensional vector denoting the state of the process stream, and $\vec{\theta}$, a t-dimensional vector, is the control variable.

The initial feed enters the stage at rate $\dot{m}_i$ whereas the feedback rate is $\dot{m}_f$. The feed stream and the recycle stream mix to produce the state $\vec{X}^o$ at which the feed enters stage 1. This state $\vec{X}^o$ is obtained using the mixing rules:

$$\vec{X}^o = M(\vec{X}^i, \vec{X}^n; \quad \dot{m}_i, \dot{m}_f) \tag{8.54}$$

where M is the mixing operator.

Considering the case when the flow rates $\dot{m}_i$ and $\dot{m}_f$ are constant, the above equation can be written as:

$$\vec{X}^o = M(\vec{X}^i, \vec{X}^n)$$

Further, the flow rate of the stream entering the stage 1 is:

$$\dot{m}_o = \dot{m}_i + \dot{m}_f \tag{8.55}$$

Typical optimization problem for such a system is to find the sequence of control variables $\vec{\theta}$ to maximize $\sum_{i=1}^{s} c_i x_i^N$ for a specified input state $\vec{x}^i$, where $c_i, i = 1, 2, \ldots, s$ are some specified constants.

The procedure for solving this optimization problem is analogous to Pontryagin's method. First we introduce adjoint variables $z_1, z_2, \ldots, z_s$, and a Hamiltonion H^n satisfying:

$$H^n = \sum_{i=1}^{s} z_i^n T_i^n(x^{n-1}; \theta^n) \ldots \quad n = 1, 2, \ldots, N \tag{8.56}$$

or, in vector form:

$$\vec{H} = \vec{Z}^n \cdot \vec{X}^n$$

with the adjoint variables given by equations:

$$z_i^{n-1} = \frac{\partial H^n}{\partial x_i^{n-1}} \quad i = 1, 2, 3, \ldots, s; \quad n = 1, 2, 3, \ldots, N \qquad (8.57)$$

or in the vector notation:

$$\vec{Z}^{n-1} = \frac{\partial H^n}{\partial \vec{X}^{n-1}}$$

$$\text{and} \quad z_i^N - \Sigma z_j^0 \cdot \frac{\partial M_j(x_i, x^N)}{\partial x_i^N} = c_i \quad i = 1, 2, \ldots, s \qquad (8.58)$$

The optimal sequence of decisions $\vec{\theta}^n$ is then found from the conditions

$$\frac{\partial H^n}{\partial \theta^n} = 0 \text{ or } H^n = \text{maximum} \quad n = 1, 2, \ldots, N \qquad (8.59)$$

Both x and z are considered fixed while maximizing the Hamiltonian. For optimization problem in which some of x_i^N, say x_a^N and x_b^N, are preassigned, and therefore not included in the objective function, the basic algorithm remains the same and Eq. (8.58) is applied for values of $i = 1, 2, \ldots, s; i \neq a, b$

We shall illustrate this method by considering a simple example of optimizing a multistage compressor.

Example 8.10
In a multistage compressor, air is being compressed from pressure P_i to a very high pressure P_f. After each stage the compressed air is cooled back to the initial temperature T_i. Estimate the pressure ratio in various stages so as to minimize the total work input during compression.

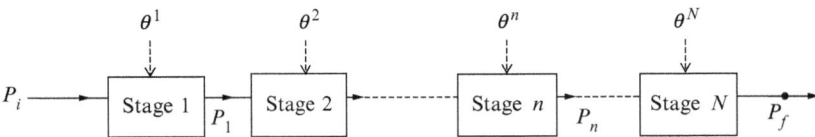

Fig. 8.5 Multistage compression.

Solution
Fig. 8.5 shows the schematic diagram of the multistage compressor. Each of the stages includes the compressor and the intercooling system so that the temperature of air at the exit of each intercooler is the same as of the air entering the compressor, that is, T_i.

To express the problem in the formalism of the discrete maximum principle, we define the state variables and the control variables as follows:

x_1^n : The pressure at the exit of stage n

θ^n : The pressure ratio in the nth stage

x_2^n : The total work input in all the stages from 1 to n

Thus the state equations can be written as:

$$x_1^n = x_1^{n-1} \cdot \theta^n$$

$$x_2^n = x_2^{n-1} + RT_i \frac{\gamma}{\gamma - 1} \left[(\theta^n)^{\frac{\gamma-1}{\gamma}} - 1 \right]$$

where $x_i^o = P_i; x_2^o = 0; R$ is the universal gas constant, T_i the temperature of air at the beginning of compression, and γ the ratio of specific heats of air. This expression for work input in the nth stage presumes the compression is isentropic. In case it is polytropic, we replace γ by r the index of compression (i.e., compression process being $pv^r = \text{constant}$).

The optimization problem becomes:

$$\text{Minimize} : x_2^N$$

$$\text{or Maximize} = (-x_2^N)$$

To apply the discrete maximum principle we define the Hamiltonian as:

$$H^N = z_1^n(x_1^{n-1}\theta^n) + z_2^n(x_2^{n-1} + C(\theta^n)^{\frac{\gamma-1}{\gamma}} - C) \quad n = 1, 2, \ldots, N$$

where the constant $C = RT_i\gamma/(\gamma - 1)$. The adjoint variables z_n^1, z_2^n are given by Eq. (8.57) as:

$$z_1^{n-1} = \frac{\partial H^n}{\partial x_1^{n-1}} = z_1^n\theta^n \quad n = 1, 2, \ldots, N$$

$$z_2^{n-1} = \frac{\partial H^n}{\partial x_2^{n-1}} = z_2^n \quad n = 1, 2, \ldots, N$$

Further Eq. (8.58) becomes [$M \equiv 1$, since there is no feedback]:

$$z_2^N = c_2 = -1$$

The value of the other adjoint variable z_1^N is not set equal to c_1, since x_1^N is prespecified as P_f, the final pressure.

The optimal values of control variable are then found using Eq. (8.59):

$$\frac{\partial H^n}{\partial \theta^n} = 0 \Rightarrow z_1^n x_1^{n-1} + z_2^n \cdot C \frac{\gamma - 1}{\gamma} \cdot (\theta^n)^{-1/\gamma} = 0$$

Now since $z_2^{n-1} = z_2^n$ and $z_2^N = -1, \Rightarrow z_2^n = -1$ for $n = 1, 2, \ldots, N$

Thus above equation becomes:

$$z_1^n x_1^{n-1} = C \cdot \frac{\gamma - 1}{\gamma} \cdot (\theta^n)^{-1/\gamma}$$

Combining this with above equations, viz.

$$z_1^{n-1} = z_1^n \theta^n$$
$$\text{and} \quad x_1^n = x_1^{n-1} \theta^n$$

we get:

$$z_1^{n-1} = z_1^n \theta^n = \left[\frac{C}{x_1^{n-1}} \cdot \frac{\gamma - 1}{\gamma} \cdot (\theta^n)^{-1/\gamma} \right] \theta^n$$

Using this expression for $n + 1$, instead of n gives:

$$z_1^n = \left[\frac{C}{x_1^n} \cdot \frac{\gamma - 1}{\gamma} \cdot (\theta^{n+1})^{-1/\gamma} \right] \theta^{n+1}$$

Dividing these two we get:

$$\frac{z_1^{n-1}}{z_1^n} = \frac{(\theta^n)^{1-\frac{1}{\gamma}}}{x_1^{n-1}} \cdot \frac{x_1^n}{(\theta^{n+1})^{1-\frac{1}{\gamma}}} = \theta^n$$

Further since $\frac{x_1^n}{x_1^{n-1}} = \theta^n$, we get:

$$(\theta^n)^{1-\frac{1}{\gamma}} = (\theta^{n+1})^{1-\frac{1}{\gamma}}$$

or

$$\theta^n = \theta^{n+1} \quad \text{for} \quad n = 1, 2, \dots, N - 1$$

that is, the pressure ratio in each stage is equal. Since the total pressure change is from P_i to P_f, it follows that:

$$\theta^n = \left(\frac{P_f}{P_i} \right)^{\frac{1}{N}} \quad \text{for} \quad n = 1, 2, \dots, N$$

8.2 NUMERICAL METHODS

For most practical problems of optimizing the design of thermal systems, numerical methods are needed since its becomes almost impossible to apply the analytical techniques. Thus, for example, while optimizing the design of a refrigeration system we often need to design for a specified refrigeration capacity. This is converted into a constraint on the design variables. It requires the simulation of the complete system and checking whether the design being considered gives the requisite capacity. Since the simulation of the system may entail solution of differential equations (see Section 6.1) it is not possible to express the capacity of the refrigeration plant as an explicit (or even implicit) function of the design variables. This rules out the possibility of using the analytical techniques.

The general approach of numerical techniques can be summed up in one sentence (considering, say, the case of minimization of the objective function): Starting from a feasible design vector (usually termed as a "point" in multidimensional space) the "direction" (in multidimensional space) in which the objective function value decreases is determined by a "suitable procedure," a new "point" obtained and the procedure is repeated till no more reduction in the value of objective function is possible.[2] The numerical techniques differ in the "suitable procedure" adopted to move from the initial design to a better design. This in turn depends upon the number of design variables and the number and the type of constraints. We shall review a few of these methods.

8.2.1 Single Variable Functions

Most practical problems of optimization of thermal systems involve more than one variable. Still, it is imperative to learn about the techniques for optimizing the functions of a single variable since many techniques for multivariable function optimization involve finding sequentially the optimum value of each variable without altering the values of other variables. The numerical methods involve logical search for the optimum which presumes that the objective function $f(x)$ is unimodal (Fig. 8.6), that is, if it has (say) a minimum at $x = x^*$, then

$$f(x_1) > f(x_2) > f(x^*) \quad \text{if} \quad x_1 < x_2 < x^*$$
$$f(x^*) < f(x_3) > f(x_4) \quad \text{if} \quad x^* < x_3 < x_4 \tag{8.60}$$

In most physical problems, the objective function is unimodal and little difficulty is encountered in arriving at the optimum value. However, even for multimodal functions the local optimum can be located for each local unimodal region (see Fig. 8.7) and then the global optimum found by comparing various local optima.

In the simplest method of finding the optimum of a single variable function, we start from a base point x_0 (Fig. 8.6) and find the value of the objective function $f(x_0)$. We assume a step size h and find the objective function value at $x_1 = x_1 + h$. If $f(x_1) < f(x_0)$ while the minima is being sought, then our direction of search is correct and we proceed further with x values of $x_2 = x_1 + h$; $x_3 = x_2 + h, \dots$ and so on till we arrive at the value x_k such that $f(x_k) > f(x_{k-1})$. Then it follows from the unimodality condition (8.24),

---

[2] This approach can also be used in non-linear regression (Sec 2.3.2) – treating the unknown coefficients as the 'design variables' for minimizing the function given by Eq. 2.33.

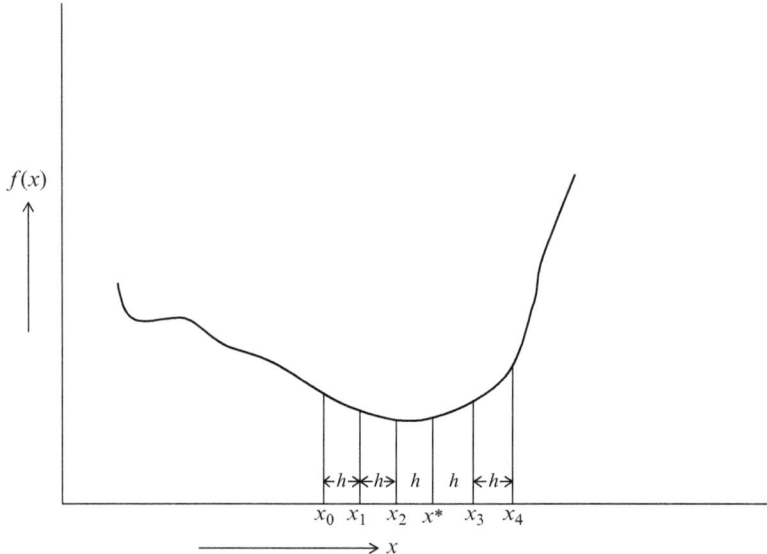

Fig. 8.6 A unimodal function.

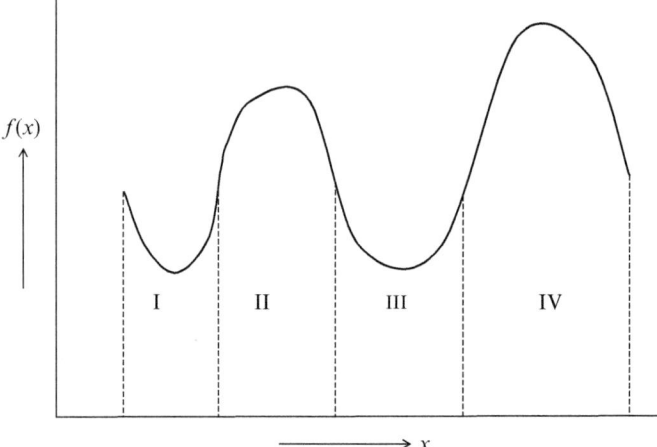

Fig. 8.7 A multimodal function with local unimodal regions.

that the required minima lies between x_k and x_{k-2} (between x_3 and x_2, in Fig. 8.6). Had the first trial given a higher value, then the search would have been carried out in the opposite direction (i.e., taking step size as $-h$) and the minima bracketed as above.

Using a fixed step size to search for the optima is highly inefficient, especially when the base point x_0 is far away from the optimum. Therefore, numerous accelerated search

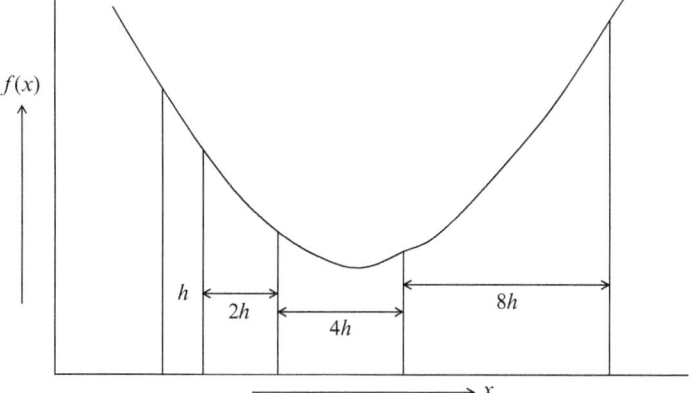

Fig. 8.8 Accelerated search.

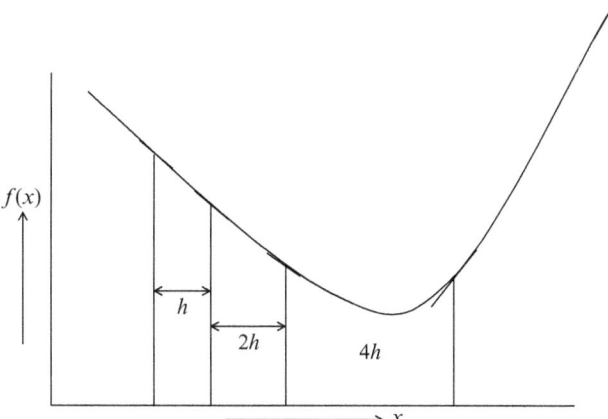

Fig. 8.9 Accelerated search with gradient evaluation.

algorithms have been devised. One such algorithm progressively doubles the step size, once the direction of search is correctly identified (Fig. 8.8).

In case the values of the derivatives can be easily calculated, we can use the criterion of "change of sign of derivative" to locate the optimum within two consecutive points (Fig. 8.9). In the later case, one could also directly find the root of the equation $f'(x) = 0$ using any of the methods discussed in Section 2.2.

There exist many other techniques of optimizing functions of a single variable, especially those for which upper and lower bounds of the variable are specified. The basic approach, of course, remains the same, but there are some differences in the selection of the step size. The details can be seen in books on optimization techniques [4].

Example 8.11

Solve Example 8.4 by using the logical search procedures discussed above for a C_E value of Rs. 10/kWh.

Solution

The objective function for $C_E = 10$ becomes

$$f(x) = 750x^{0.8} + \frac{12 \times 10^5}{x}$$

Let us start with a base value of $x = 0.5$ and the first step size $h = 1$. We shall illustrate both the methods, viz. using only the objective function, and that using the gradient. The value of the gradient is

$$f'(x) = 600x^{-0.2} - \frac{12 \times 10^5}{x^2}$$

The calculations are shown in below tabular using accelerated search: As is clear from the table, if we use only $f(x)$ calculations, the optimum x is bracketed within 31.5 and 127.5. However, according to the gradient calculation, we can say that optimum x lies between 63.5 and 127.5.

x	f(x)	f'(x)
0.5	2,400,431	−4,799,311
1.5	801,037.4	−532,780
3.5	344,900.4	−97,492
7.5	163,759.4	−20,932
15.5	84,138.7	−4,648
31.5	49,945	−908.4
63.5	39,660	−36.03
127.5	45,675	+153.7

x	f(x)	f'(x)
63.5	39,660.17	−36.03
64.5	39,628.35	−27.7
66.5	39,588.73	−12.1838
70.5	39,595.5	14.7247

We could reduce the range of values within which optimum x lies by starting from $x = 63.5$ and proceeding with same step size and accelerating it as before. The calculations are summarized above: Using the change of sign of derivative as the guiding principle, we can say x^*, the value of optimal x, lies between 66.5 and 70.5.

Pinpointing x^*, could be done by using a very small step size while starting the search from $x = 66.5$. Thus the value of x^* can be obtained as 68.22 cm, and the optimal value of the objective function as 39,578 as in Example 8.4.

Comment
It is evident that if we had adopted a fixed step size of $h = 1$, starting from the same base point of $x = 0.5$, much larger number of calculations would be needed to arrive at the optimum point than in the above case with increasing step size.

8.2.2 Multivariable Functions

Numerical methods for optimizing multivariable functions are generally categorized as "direct search methods"—those requiring only the evaluation of objective function values at various "points," and "gradient methods"—those which require in addition the evaluation of gradients of the objective function. Since in most practical cases of thermal systems, the objective functions are complex functions (often implicit) of the design variables, it is not convenient to determine the gradients and the gradient methods become too unwieldy. Consequently here we shall discuss only the "direct search" methods of optimizing multivariable functions.

We consider first the case of optimizing functions without any constraining conditions. The simplest method called *univariant search* is a direct extension of the one-dimensional search methods discussed above. Starting from the base point search is carried out sequentially for locating the best value of each of the variables. Fig. 8.10 illustrates the methods for minimization of function of two variables. Point A ($x_1 = x_{1A}; x_2 = x_{2A}$) is the starting point of search. First, the value of the variable $x_2(= x_{2A})$ is kept constant and the search carried for optimal value of variable $x_1(= x_{1B})$ at which the function has the smallest value (of f_2). This is shown as point B in Fig. 8.10. Next, the value of the variable $x_1(= x_{1B})$ is kept constant and the value of $x_2(= x_{2C})$ at which the function has the lowest value is found. Further search is continued from point C in similar manner till no further reduction in the value of the objective function is possible.

This simple method has two limitations, viz. it tends to oscillate near the optima and secondly it fails at sharp valleys, that is, in regions with sharp changes in the values of the objective function. This is illustrated in Fig. 8.11 for a function with sharp edged contours. It can be seen that the search process would stop at point A since search in the neighborhood along all the directions would only increase the function values. This point is clearly far away from the true optimum X.

Many heuristic algorithms have been devised to overcome this limitations, and one such more effective method is the *sequential simplex method* [5]. Unlike univariant search, in this method the search directions are not fixed, but altered keeping in view the previous trials. The method starts with evaluation of the objective function at the

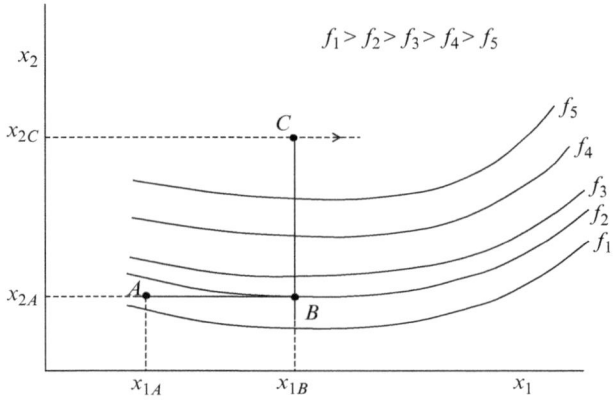

Fig. 8.10 Univariant search.

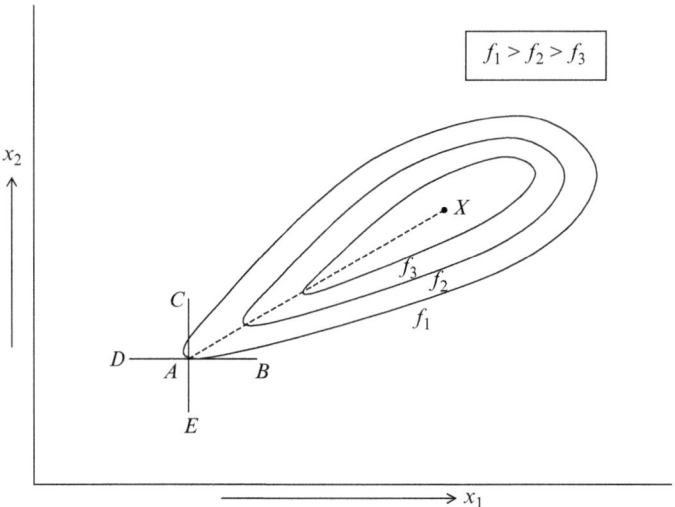

Fig. 8.11 Failure of univariant search.

vertices of a regular geometrical figure in the n–dimensional space (n being the number of variables), called the "Simplex." For $n = 2$ the figure is a triangle, and for $n = 3$ it is a tetrahedron. The "vertex" at which the objective function value is the highest (in a minimization problem) is termed as the "worst" vertex and replaced by a new point. This new point is found along the direction away from the "worst" point and in alignment with the "center of gravity" of the rest of the vertices, such that the geometric shape of the simplex is preserved. This new simplex is again examined for

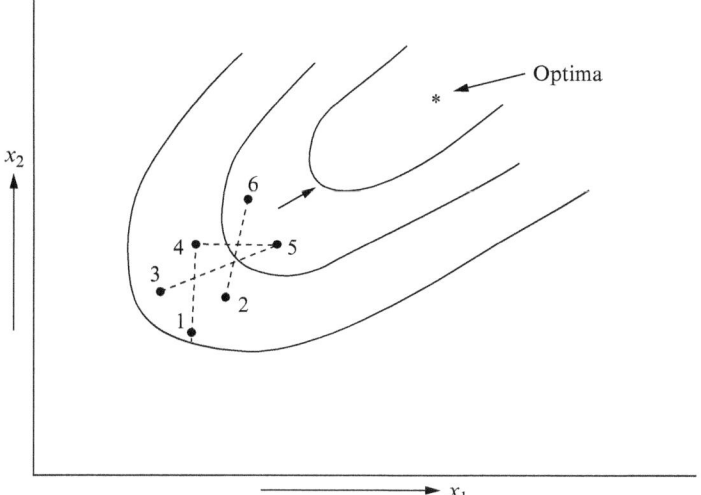

Fig. 8.12 Sequential simplex method.

the worst vertex and the process repeated till the simplex straddles the optima. Fig. 8.12 shows a few stages of a two-dimensional sequential simplex method.

Thus starting from the simplex 1-2-3 (equilateral triangle for 2-D search), the objective function values are evaluated at every vertex. The "worst" point, corresponding to the largest value of the objective function, is clearly point 1. This point is then replaced by the point 4 located along the line joining 1 and the "center of gravity" of the remaining points, viz. points 2 and 3, and extending it away from 1 such that 3-2-4 forms the new simplex. Among its three vertices the "worst" vertex now is point 3. This is replaced by point 5 found by the same procedure to get the new simplex 2-4-5. Thus the simplex rolls on, changing its direction of search sequentially as dictated by the nature of the function, so that it slowly approaches the true optima. In order to avoid oscillations due to repetitions of simplices, additional rules have been prescribed to be followed while rejecting and generating vertices.

For optimizing functions subject to certain constraints different strategies have been developed. If these are equality constraints $[g_k(x) = 0, k = 1, \ldots, m]$ the best strategy is to use these constraints to express some of the variables in terms of the rest, and incorporate these expressions in the objective function, thus eliminating some of the variables. In case this is not possible, the constraints $[g_k(x)]$ are incorporated with the objective function (f) by using a sequence of very large number P_k and the augmented objective function (F) (also called penalty function) thus obtained is then optimized as an unconstrained function:

$$F = f + \sum_{k=1}^{m} P_k \{g_k(x)\}^2 \tag{8.61}$$

It can be shown (as is intuitively obvious) that as P_k values are increased to infinity, the minimum of function F is obtained at the same point as that of constrained function f since this minimum will occur only when $[g_k(x)]$ is as small as possible, that is, zero.

In case the constraints are in the form of inequalities like $g_k(x) \leq 0$ for $k = 1, 2, 3, \ldots, m$, then either of the two kinds of approaches to the solution can be adopted. In the first approach the inequality constraints are converted into equality constraints by using additional (slack) variables $x_{n+1}, \ldots, x_{n+m}$ as indicated below.

$$g_k(\vec{X}) + X_{n+k}^2 = 0 \quad k = 1, 2, \ldots, m \tag{8.62}$$

The optimization of the objective function subject to the above constraints can then be done using the method mentioned above.

In the second approach, an unrestricted search is carried out following any of the methods discussed above. However, after each search the new points are checked for their "feasibility" (i.e., satisfaction of the constraints), and whenever any constraint is violated, immediate return into the feasible region is made following certain heuristic rules. The most important criteria for the success of these rules is that the search should not get stranded at local optima, for away from the true optima. One of the most successful strategies, which has been extensively used for optimization of thermal systems is the "complex" method given by Box [6].

In this method, which is an extension of the simplex method described above, starting from the first feasible "point," $(k-1)$more "points" are generated by using the equation

$$X_{ij} = X_{Li} + r_{ij}(X_{Hi} - X_{Li}) \quad 1 \leq i \leq n \quad j = 1, 2, \ldots, k \tag{8.63}$$

where r_{ij} are the pseudorandom numbers between 0 and 1; and X_{Li} and X_{Hi} are respectively the lower and upper bounds for the variables. Box [6] recommends a value $k = 2n$, where n is the number of design variables. These "points" are then checked for feasibility, and if any of the constraints is violated by any "point," it is repeatedly moved halfway in towards the "centroid" of the figure formed by the feasible points, until it is brought inside the feasible region. These k-feasible "points" thus obtained can be viewed as the vertices of a figure in n-dimensional design variable space. Box calls it a "complex" to distinguish it from the "simplex" mentioned above which preserved its regular shape while rolling along towards the optima.

The search algorithm is, in essence, quite similar to that of the simplex method. The objective function is evaluated at all the vertices of the "complex" and the vertex giving the "poorest value" (i.e., maximum value during minimization) is located. This "worst" point is now replaced by another point which is, as in the simplex method, along the line joining the rejected point and the centroid of the remaining figure. But it is at a distance from this centroid which is equal to $\alpha(\alpha > 1)$ times the distance of the "reflection" of the rejected point in the "centroid."

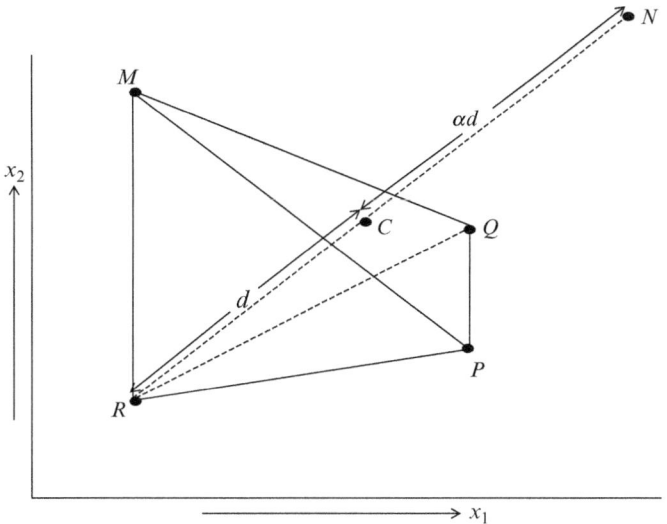

Fig. 8.13 Box-method: reflection of a rejected point.

Fig. 8.13 brings out this crucial step clearly, for a two dimensional problem. Here R is the rejected point, C is the centroid of the remaining complex (MPQ), and N is the new point. The variable values for this "new point" are given by the relation:

$$x_{iN} = x_{iC} + \alpha(x_{iC} - x_{iR}) \tag{8.64}$$

where the variable values for the centroid are:

$$X_{iC} = \frac{1}{k-1}\left(\sum_{j=1}^{k-1} x_{ij}\right) \tag{8.65}$$

This new point is checked for feasibility and if infeasible, it is repeatedly moved halfway towards the centroid till the new point falls within the feasible region.

The objective function value at this valid new point is then compared with those at other vertices of the "complex." If this new point is not the worst point again, the above process of rejection of the vertices is continued.

However, if the valid new point turns out to be the worst point of the new "complex" it is rejected and another new point obtained by moving halfway towards the centroid of the remaining "complex." This process is repeated till the new "point" is no longer the worst "point." This check ensures that the algorithm does not oscillate at steep "ridges" and "valleys" in the objective function contour.

This process of "rejection" and "regeneration" of vertices is continued till it is found that five successive evaluations of the objective function give "equal" value within the required accuracy. This means the "complex" has almost shrunk to a "point" and the optimum has been reached.

We shall illustrate this method with an example.

Example 8.12

Solve Example 8.1 numerically starting from an initial estimate of the variables as $D = 0.04$ cm and $L = 1.2$ m.

Solution

The objective function is:

$$f = 10{,}000 + 1.5 \times 10^6 D^{1.2} L^{1.2} + \frac{0.00002L}{D^{4.5}} - 16 \times 10^6 (1 - e^{-0.026LD^{0.8}})$$

We first use univariant search starting from the point of $[0.04; 1.2]$. Since the two variables D and L have widely different values we use different step sizes for search in the two direction. We start with step size of 0.005 m for D and 0.1 m for L. The calculation are summarized in the table below:

D	L	Function (f)	Remarks
0.04	1.2	11,306.44	
0.045	1.2	11,495.8	
0.035	1.2	9379.819	Direction of search changed to reduce the objective function
0.03	1.2	7777.017	
0.025	1.2	6627.74	
0.02	1.2	6317.67	
0.015	1.2	8626.039	

Search for better D values stopped since f increased with further reduction in its value from 0.02 to 0.015. Now we begin search in "L" direction, keeping D fixed at 0.02.

D	L	Function (f)	Remarks
0.02	1.3	6310.646	
0.02	1.4	6330.543	Objective function increased; start search along the "D" direction
0.025	1.3	6738.789	Objective function increased both by increasing
0.015	1.3	8723.745	and reducing D indicating proximity to the optima. So step size for D reduced
0.021	1.3	6277.442	
0.020	1.3	6316.31	Accepting $D = 0.021$ as the optimum value, we now search for optimum L
0.021	1.25	6270.19	step size of search for L reduced
0.021	1.20	6270.3	f increased slightly
0.021	1.23	6269.341	
0.021	1.22	6269.361	very small increase in "f" so previous value can be taken as optimum

Thus the search method arrives at $D = 0.021$ m and $L = 1.23$ m as the optima. This matches quite well with the result obtained in Example 8.1.

Let us also see how simplex method could be used to arrive at the solution, using the "half way to centroid" rule for improving the reflected point, should it turn out to be "worse" than the point being rejected. A few steps of detailed calculations are shown below.

Starting from the same base point of $D = 0.04$ m and $L = 1.2$ m we identify two more points in the proximity to form the starting simplex:

Vertex 1	[0.04, 1.2]
Vertex 2	[0.02, 1.3]
Vertex 3	[0.03, 1.25]

The values of the objective function are evaluated at these three points as indicated below:

	D	L	f
Vertex 1	0.04	1.2	11,306.44 ←
Vertex 2	0.02	1.3	6310.646
Vertex 3	0.03	1.25	7922.809

Clearly vertex 1 is the "worst" point since our aim is to minimize the objective function. To find a "better" point we reflect it in the centroid of the remaining simplex, which is $(0.025, 1.275)\{$i.e., $\frac{0.02+0.03}{2}, \frac{1.3+1.25}{2}\}$. Reflecting vertex 1 in this centroid gives the new point as:

$$D = 2 \times 0.025 - 0.04 = 0.01$$
$$L = 2 \times 1.275 - 1.2 = 1.35$$

The objective function value for this point turns out to be 31,459.8 which is higher than that at the vertex 1 sought to be rejected. Clearly this is not acceptable, and we try to get a "better" point by moving halfway towards the centroid. The new point thus obtained is:

$$D = \frac{0.01 + 0.025}{2} = 0.0175$$
$$L = \frac{1.35 + 1.275}{2} = 1.313$$

The objective function value corresponding to this point is 6868.28 which is "better" than that of the rejected point. We can thus accept it as a new vertex of the simplex which now has following vertices and objective function values;

0.0175	1.313	6868.28
0.02	1.3	6310.646
0.03	1.25	7922.809 ←

We again identify the worst vertex (shown by an arrow) and proceed as before in following steps.

$$\text{Centroid: } D = \frac{0.0175 + 0.02}{2} = 0.018875$$

$$L = \frac{1.3 + 1.313}{2} = 1.307$$

and get the new reflected "point" as:

$$D = 2 \times 0.01875 - 0.03 = 0.0075$$

$$L = 2 \times 1.307 - 1.25 = 1.363$$

The objective function value at this new point is 104,304.1, which is much higher than that of the point being rejected. So by repeatedly moving halfway we find a better reflected point as:

$$D = 0.01805 \quad L = 1.311 \quad f = 6668.8$$

The new simplex is:

0.0175	1.313	6868.28 ⟵
0.02	1.3	6310.646
0.01805	1.311	6668.8

The worst vertex is again identified and the procedure repeated.

$$\text{Centroid}: (0.01903, 1.306)$$

$$\text{Reflected point}: (0.02055, 1.298) \quad f = 6282.125$$

New simplex is:

	D	L	f
	0.02055	1.298	6282.125
	0.02	1.3	6310.646
"Worst" Point	0.01805	1.311	6668.8 ⟵

$$\text{Centroid}: (0.02028, 1.299)$$

$$\text{Reflected point}: (0.0225, 1.287) \quad f = 6351.2$$

Though this "point" has smaller objective function than the "rejected" vertex, in the new simplex it would still have the highest value, and would need to be rejected again. So we move halfway towards the centroid to generate a new point:

$$D = \frac{0.0225 + 0.02028}{2} = 0.02139$$

$$L = \frac{1.287 + 1.299}{2} = 1.293$$

Corresponding objective function value is 6282.97 which is still higher. So we make another move to get:

$$D = \frac{0.02139 + 0.02028}{2} = 0.02084$$

$$L = \frac{1.293 + 1.299}{2} = 1.296$$

which gives $f = 6276.729$ and is a "better" point. Further calculations are given below:

The new simplex:

	D	L	f
	0.02084	1.296	6276.729
	0.02055	1.298	6282.125
"Worst" point	0.02	1.3	6310.646 ⟵

Centroid : 0.02070   1.297

Reflected point : 0.0214   1.294 $f = 6383.55$ (not better)

Moving half way : 0.0211   1.295 $f = 6277.24$

New simplex

	D	L	f
	0.0211	1.295	6277.24
	0.02084	1.296	6276.729
Worst point	0.02055	1.298	6282.125 ⟵

Centroid : 0.02097   1.2955

Reflected point : 0.02139   1.293   6282.967   (not better)

Moving half way : 0.02118   1.2942   6278.2   (not better)

Again moving half way : 0.02107   1.2949   6276.879   (better)

Since the vertices are quite close to each other we accept this as the final solution. It is seen that while the optimum D value matches quite well with the results obtained in Example 8.1, the optimum length is about 5% more. This is due to very mild influence of L on the objective function value in contrast to extremely strong influence of diameter D.

Example 8.13

Solve Example 8.5 using Box's complex method, putting additional constraints $1 \le Q \le 2$ and $10 \le D \le 20$.

Solution

Problem statement:

$$\text{Minimize} \quad f = 36 \times 10^4 + \frac{12 \times 10^5}{Q^{0.8}} + 150D^{1.2} \times 200 + 4.8634 \times 10^{10} \frac{Q^3}{D^5}$$

we could divide it by 10^4 to get:

$$f' = 36 + \frac{120}{Q^{0.8}} + 3D^{1.2} + 4.8634 \times 10^6 \frac{Q^3}{D^5}$$

whose minimum will occur for the same values of D and Q as of f. The constraints are:

$$1 \le Q \le 2 \quad 10 \le D \le 20$$

Since we have two variables Q and D; $n = 2$, and so we generate four feasible points to form the starting complex. We take these as:

D	L	Function (f')
1.2	10	271.3
1.2	18	240.42
1.8	10	442.16 ⟵
1.8	18	222.25

Clearly the third "point" is the worst point. The centroid of the remaining three points is:

$$\frac{1.2 + 1.2 + 1.8}{3}, \quad \frac{10 + 18 + 18}{3}$$

$$\text{or} \quad (1.4, 15.333)$$

The new point obtained by reflection of the "worst" point in the centroid is (Eq. 8.64, taking $\alpha = 1.3$)

$$\text{New point:} \quad 1.4 \times 2.3 - 1.3 \times 1.8 = 0.88$$

$$15.333 \times 2.3 - 1.3 \times 10 = 22.266$$

Since this new reflected point violates both the constraints (viz. $2 \le Q \le 1$; $10 \le D \le 20$). We move it half-way toward the centroid in search for a feasible reflected point. This gives:

$$D = \frac{0.88 + 1.4}{2} = 1.14$$

$$L = \frac{22.266 + 15.333}{2} = 18.8$$

This point satisfies both the constraints, so we compute the objective function value at this point as $f = 248.5432$. Since this is lesser than the objective function value of the rejected point, and is not the "worst" point in the new "complex" obtained by replacing the rejected "worst" point by this point, we accept this as a feasible new point.

The new complex thus formed is:

Q	D	f
1.2	10	271.3 ←
1.2	18	240.42
1.8	18	222.25
1.14	18.8	248.54

The new worst point is clearly the very first point (indicated by an arrow sign) and we repeat the above procedure summarized in a tabular form below:

$$\text{Centroid} = (1.38, 18.266)$$

$$\text{Feasible reflected point} = (1.40925, 19.61) \quad f = 238.57$$

	Q	D	f	
Complex	1.40925	19.61	238.57	
	1.2	18	240.42	
	1.8	18	222.25	
	1.14	18.8	248.54	← Worst point

Centroid (1.46975, 18.536)

Reflected point (1.8984, 18.1943) $f = 222.0549$

	Q	D	f	
New complex	1.40925	19.61	238.57	
	1.2	18	240.42	← Worst point
	1.8	18	222.25	
	1.8984	18.1943	222.0549	

Centroid (1.702, 18.601)

Feasible reflected point (1.8659, 18.7969) $f = 223.718$

	Q	D	f	
New "complex"	1.40925	19.61	238.57	← Worst point
	1.8659	18.7969	223.718	
	1.8	18	222.25	
	1.8984	18.1943	222.0549	

Centroid (1.8547, 18.330)

  Feasible reflected point (1.9996, 17.915) $f = 221.719$

	Q	D	f	
New "complex"	1.9996	17.915	221.719	
	1.8659	18.7969	223.718	← Worst point
	1.8	18	222.25	
	1.8984	18.1943	222.0549	

Centroid (1.89933, 18.0363)

  Reflected point (1.9428, 17.0475) $f = 221.4911$

	Q	D	f	
New "complex"	1.9996	17.915	221.719	
	1.9428	17.0475	221.49	
	1.8	18	222.25	← Worst point
	1.8984	18.194	222.055	

Centroid (1.9469, 17.718)

  Feasible reflected point (1.9946, 17.627) $f = 221.622$

	Q	D	f	
New "complex"	1.9996	17.915	221.719	
	1.9428	17.0475	221.49	
	1.9946	17.627	221.622	
	1.8984	18.194	222.055	← Worst point

Centroid (1.979, 17.5298)

  Feasible and better reflected point (1.9921, 17.4218) $f = 221.658$

	Q	D	f	
New "complex"	1.9996	17.915	221.79	← Worst point
	1.9428	17.0475	221.49	
	1.9946	17.627	221.622	
	1.9921	17.4218	221.658	

Centroid (1.9765, 17.366)

  Feasible and better reflected point (1.96905, 17.187) $f = 221.6118$

	Q	D	f	
New "complex"	1.96905	17.187	221.6118	
	1.9428	17.0475	221.49	
	1.9946	17.627	221.622	
	1.9921	17.4218	221.658	← Worst point

Centroid (1.9688, 17.2873)

Reflected point (1.938596, 17.11252) $f = 221.4019$

	Q	D	f	
New "complex"	1.96905	17.187	221.6118	
	1.9428	17.0475	221.49	
	1.9946	17.627	221.622	← Worst point
	1.9386	17.1125	221.4019	

Centroid (1.95015, 17.1157)

Better reflected point (1.9212, 16.783) $f = 221.579$

	Q	D	f	
New "complex"	1.96905	17.187	221.6118	← Worst point
	1.9428	17.0475	221.49	
	1.9212	16.783	221.579	
	1.9386	17.1125	221.4019	

Centroid (1.9342, 16.981)

Better reflected point (1.8889, 16.7131) $f = 221.34$

Since the objective function values are quite close to each other, even through the complex has not "shrunk" to a point, we can be sure that the best value obtained so far, that is, $Q = 1.889$ L/s $D = 16.71$ cm with $f = 221.34$ is quite near the optimum. This is borne out by comparison with the result of Example 8.5 ($Q = 1.722$ L/s, $D = 16.44$ cm, $f = 220.7 \times 10^4$). The students are advised to carry out a few more steps and verify that the algorithm would lead to the correct optima.

8.2.3 Mixed Discrete-Continuous Variables

In many optimization problem, some of the variables can have only discrete values, while others can have continuous values. For example, in a heat exchanger design problem, the length of the heat exchanger is a continuous variable, while the tube diameter is a discrete variable since tubes of only certain diameters are manufactured in the industry.

Some other variables like number of tubes and the number of passes can have only integer values since a heat exchanger can have either 57 or 58 tubes, but not 57.6 tubes.

The optimization methods discussed above cannot be directly applied to such problems and modifications in the algorithms are necessary. A general approach is to treat "discreteness" of variables as an additional constraint and device suitable procedures to bring back the search in the feasible domain when this constraint is violated. Modifications in Box's complex method to enable its application in such problems have been suggested by Dhar [7]. A method for efficiently handling such problems, which has become quite popular in recent times, is the genetic algorithm (GA). A brief description of its salient features is presented in the next section.

8.2.4 Genetic Algorithms

GAs are basically probabilistic search algorithms which use the Darwinian principle of "survival of the fittest" to move the search towards the desired optima. Thus these algorithms can be seen as modifications of the Box's Complex method, where the generation of a new "point" to replace the "worst" point of the starting "complex" of feasible points is done by a stochastic procedure.

The main steps in a typical GA for optimizing a multivariable function are:
1. Generating an initial "population" of possible solutions randomly (Eq. 8.63). The population is chosen depending on the number of variables and is typically $4 \times n$ or larger.
2. The objective function value is calculated at each of these points and based on these a "fitness" value is assigned to each point which indicates it relative fitness for further propagation.

For maximization problems, a normalized value of the objective function is a good indicator of the relative fitness. Often the normalization is done on the basis of the average value of the objective function of all members of the starting population.

For minimization problems to generate nonnegative values of the fitness coefficient, suitable mapping functions have been devised to transform the objective function value to the fitness coefficient. Typical mappings are:

$$F(x_i) = \frac{1}{1 + f(x_i)} \tag{8.66}$$

or

$$F(x_i) = V - \frac{P.f(x_i)}{\sum_{i=1}^{P} f(x_i)} \tag{8.67}$$

where P is the population size, $f(x_i)$ is the value of objective function of the ith member of the population, and V is a large value to ensure nonnegative fitness values.

3. The next step is to generate new members of the population which hopefully, would be nearer the optimum (i.e., have higher "fitness" values) than those of the first set of members. It is this step that distinguishes GA from traditional search methods like Box-complex method. Here this is done by applying genetic operators based on the principle of natural selection which suggests propagation of the "fittest" members of a population. This involves four steps, viz.

(a) coding,

(b) reproduction or selection,

(c) crossover,

(d) mutation.

The first step involves coding the variable-vector in binary (1^s) and (0^s) string structures. Though in some algorithms the variables are directly used, these are exceptions; here we shall discuss only the conventional GA. The length of the string is usually determined according to the desired accuracy of representing floating point values. Thus if five bits are used to code a variable, then the entire search space would be mapped from (00000) to (11111) wherein we can have a maximum of 32 points $(0 \rightarrow 2^0 + 2^1 + 2^2 + 2^3 + 2^4)$. Using a linear rule, the value of the variable corresponding to a binary code $[r_0, r_1, r_2, r_3, r_4]$ would be

$$x_i = x_i^L + \frac{x_i^U - x_i^L}{2^5 - 1} \sum_{j=0}^{4} r_j 2^j \tag{8.68}$$

where x_i^L and x_i^V represent the lower and the upper limiting values of the variable.

Clearly the accuracy of representation is $1/31$ of the search space. If it is insufficient, the number of elements in the string needs to be increased.

If we have more than one variable, the total string length would include the substrings of each of the variables. Thus for an optimization problem involving three variables, each coded by a string of five elements the points would be represented by a long string of fifteen elements, for example (*01011   00101   10111*).

It is not necessary to have the same substring lengths for each variable. These can vary depending upon the search space and the degree of accuracy needed. Thus, some variables may even have integer values. This method is thus ideally suited for mixed-integer programming problems. After coding the new design vectors are generated through three genetic operators whose main objective is to maximize their fitness function values.

Reproduction or selection is the first operator applied on the population to select the members with high fitness values from which new solution vectors could be generated. This is in tune with the evolutionary theory which suggests that the process of natural selection favors the survival (through reproduction) of the fittest individuals of a population. Numerous reproduction operators have been suggested in GA literature, however their essential idea is the same, viz. to identify above average strings (i.e., solution vectors with high values of the fitness function) from the correct population and use their multiple copies for further reproduction. This is usually done on the basis of normalized values of the fitness function (i.e., $f(x_i)/(\Sigma f(x_i)/P)$, where P is the population size). A solution vector with a normalized value of 0.5 may thus be dropped, while that with a value of 2.5 should have its two copies used for further reproduction. These selected designs become the "parents" to produce new designs (offsprings) by "crossover."

The operator *crossover* is used to generate new solutions by combining the strings in the pool thus selected from the first population. This is done by crossing over the strings of the "parent" designs at randomly chosen crossover points to produce new "offspring" designs. As indicated in Fig. 8.14, this involves exchanging all the bits on the right side of the crossing site.

The crossover can also be done at more than one sites. Two site crossover is illustrated in Fig. 8.15. Here the bits between the two sites are exchanged to produce the two offsprings.

The third genetic operator, *"mutation"* is a tactical move to widen the search horizon when after repeated reproduction and crossover the population tends to become homogeneous, that is, the difference in the designs are not much. It mimics the process of "mutation" in genetics and involves randomly disturbing the genetic information, so that the zero's in the string become ones; and the ones become zeros. The new solution thus generated is not necessarily better; this depends upon the selection of the bits being exchanged, and also on whether the global optima has been reached or not. It has been seen that this operator "mutation" is very helpful in preventing stranding of the search at local optima.

The various operations involved in GA are not performed at every step. Practical guidelines have been established for the values of probability for each step. Typically for

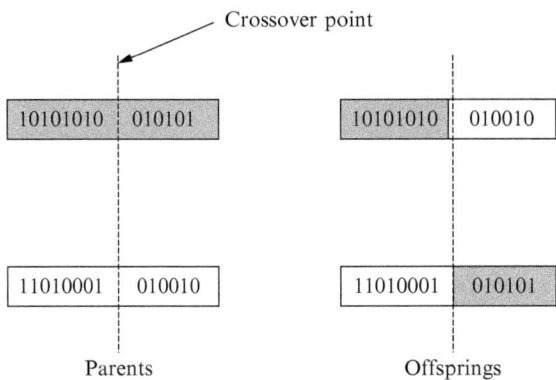

Fig. 8.14 Genetic operator "crossover."

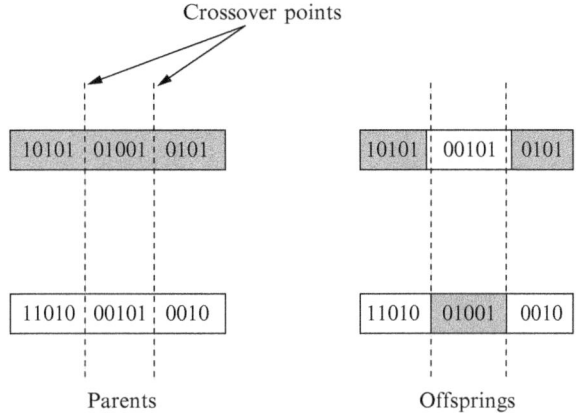

Fig. 8.15 Crossover at two sites.

a population with less than 30 members, the crossover probability is taken as 0.9 and the mutation probability as 0.01. For large size population (>100 members) the crossover probability may be 0.6 and the mutation probability 0.001. Again the crossover sites (and both their location number) are randomly selected. For details the readers are advised to refer to specialist, books, for example, Goldberd [8]. We shall illustrate the basic principles of GA in Example 8.8.

Example 8.14
Solve Example 8.4 by using a GA. Given the solution lies between 1 and 100.

Solution
The objective function to be *minimized* is

$$f(x) = 750x^{0.8} + \frac{12 \times 10^5}{x}$$

The search space is indicated as $0 \le x \le 100$.

This implies, in Eq. (8.68)

$$x^L = 0 \quad x^U = 100$$

Since the range is large, we choose a 8 bit string to represent the variable, and start with a population of four points. We generate randomly four binary codes for these points and calculate the actual value of the variable using Eq. (8.68). Thereafter the objective function value at each of these points is calculated using the above equation. The 'fitness' function is calculated using (8.66). Thereafter the genetic operators are employed sequentially to generate a new set of variable values. The selection of variables for further propagation is done on the basis of the normalized values of the fitness function. The selected strings are then crossed over at randomly selected sites with randomly selected strings and new strings generated. All these calculations have been tabulated below.

Strings								x	f(x)	Fitness	fit/favg	Copies
1	0	1	0	0	0	0	1	63.137	39,673.8	2.520×10^{-5}	1.347	2
0	1	0	1	0	1	1	0	33.725	48,096.3	2.079×10^{-5}	1.111	1
1	0	0	1	0	0	1	1	57.647	40,033.2	2.498×10^{-5}	1.334	1
0	0	0	0	1	1	0	0	4.706	257,589.3	3.882×10^{-6}	0.207	0
										favg 2.87×10^{-5}		

Strings								Copartner	Crossover Site	New Strings							
1	0	1	0	0	0	0	1	4	3	1	0	1	1	0	0	1	1
1	0	1	0	0	0	0	1	3	5	1	0	1	0	0	1	1	0
0	1	0	1	0	1	1	0	2	5	0	1	0	1	0	0	0	1
1	0	0	1	0	0	1	1	1	3	1	0	0	0	0	0	0	1

Evaluating New Strings								x	f(x)	Fitness	fit/favg	Copies
1	0	1	1	0	0	1	1	70.196	39,591.3	2.526×10^{-5}	1.0636	2
1	0	1	0	0	1	1	0	65.098	39,613.2	2.524×10^{-5}	1.063	1
0	1	0	1	0	0	0	1	31.765	49,707.1	2.012×10^{-5}	0.847	0
1	0	0	0	0	0	0	1	50.588	41,031.1	2.437×10^{-5}	1.026	1
									favg $= 2.374 \times 10^{-5}$			

Following similar procedure we carry out one more crossover procedure. The results are tabulated below:

Strings								Copartner	Crossover Site	New Strings							
1	0	1	1	0	0	1	1	3	4	1	0	1	1	0	1	1	0
1	0	1	1	0	0	1	1	4	6	1	0	1	1	0	0	0	1
1	0	1	0	0	1	1	0	1	4	1	0	1	0	0	0	1	1
1	0	0	0	0	0	0	1	2	6	1	0	0	0	0	0	1	1

Evaluating New Strings								x	f(x)	Fitness	fit/favg	Copies
1	0	1	1	0	1	1	0	71.37	39,610.6	2.524×10^{-5}	1.0078	1
1	0	1	1	0	0	0	1	69.41	39,583.2	2.526×10^{-5}	10086	1
1	0	1	0	0	0	1	1	63.92	39645.7	2.522×10^{-5}	1.0070	1
1	0	0	0	0	0	1	1	51.37	40,833.3	2.446×10^{-5}	0.9765	1
									favg $= 2.5045 \times 10^{-5}$			

Since the fitness values are approximately equal, we now introduce mutation to check whether we can improve the fitness by moving far away from the current x values. We randomly mutate two bits from each string and repeat all the calculations.

Mutated Strings								x	$f(x)$	Fitness
0	1	1	1	0	1	1	0	46.27	42,051.1	2.378×10^{-5}
1	0	1	0	1	0	0	1	66.27	39,591.7	2.526×10^{-5}
0	1	1	0	0	0	1	1	38.82	44,915.8	2.226×10^{-5}
1	0	0	0	1	1	1	1	56.08	40,195.1	2.488×10^{-5}

We notice that none of the x values away from the earlier values is a better solution. It suggests that optimum x is around $66.27 - 69.41$. Further crossover, between these strings could lead to the true optima. Thus crossing over between the first two points at site 5 would give a new string as

$$[1\ 0\ 1\ 0\ 1\ 1\ 1\ 0] \text{ for which } x = 68.24$$

$$f(x) = 39578.4 \text{ and the fitness function} = 2.53 \times 10^{-5}$$

which is the highest value obtained hitherto. Comparing it with solution given in the original Example 8.4 we note that this indeed in the global optima.

In case the optimization problem imposes some constraints on the variables, we follow a strategy similar to that discussed in Section 8.2.2 and incorporate the constraints with the objective function using appropriate penalty function, for example, Eq. (8.61). The augmented function is then treated as unconstrained and optimization carried out as discussed above. Repeating this process for different values of the penalty function, the constrained optima can be approached.

PROBLEMS

8.1 A single pass shell-and-tube heat exchanger is to be designed to transfer 100 kW of heat from a hot stream to cold water flowing through its tubes due to gravitational head. Determine the optimal shell diameter (D) and length (L) [both in meters] from following data.

Number of tubes per unit shell cross-section area : 150

Diameter of tubes : 3 cm

$$\text{Rate of heat transfer in Watts} = \frac{5000}{D^{0.8}} \times [\text{Surface area of tubes in m}^2]$$

$$\text{Cost of pumping hot water through the shell} = \frac{500\,L}{D^5}$$

$$\text{Cost of shell} = 10,000 \times \text{Total length in m}^2$$
$$\text{Cost of tubes} = 2000 \times \text{Total length in m}$$

Formulate the optimization problem as minimization of total cost expressed as a function of single variable.

8.2 Solve the above problem treating the specific heat transfer rate as a constraint and use Lagrange multiplier to obtain the optimal values of the design variables D and L.

8.3 Solve Problem 8.1 numerically starting from the first guess of $D = 0.3$ m and treating it as an unconstrained optimization problem. Use the univariant search technique.

8.4 Solve Problem 8.1 numerically treating it as a constrained optimization problem of two variables. Take the heat transfer requirement as an inequality constraint $Q \geq 100$ kW. Use Box complex method.

8.5 Treating Problem 8.1 as an unconstrained optimization problem of one variable, solve it numerically using GA.

8.6 Can we solve Problem 8.1 conveniently using GP?

8.7 Solve Example 8.3 by using GA, expressing it as unconstrained optimization problem in one variable.

8.8 Solve Example 8.9 by using COV assuming that the temperature of the cold store falls linearly with time.

8.9 In a 1-m long cylindrical vessel an exothermic reaction is taking place. The heat transfer coefficient for heat loss to the surroundings can be approximated by the equation:

$$h = 0.5D^{-1.5} + 2\Delta T^{0.25}D^{-0.25}$$

where ΔT is the temperature difference between the surface and the ambient air. The chemical reaction imposes a condition that:

$$\Delta TD^{0.25} = 10$$

Determine the optimum vessel diameter at which the heat loss is the minimum, treating it as a GP problem of two variables.

8.10 A 20-mm diameter cable is covered by an insulating material of thermal conductivity $k = 0.1$ W/m K, having very low surface emittance. Heat loss from its surface is by natural convection for which the average heat transfer coefficient is given as $h_{avg} = 2\left(\frac{\Delta T}{D}\right)^{0.25}$, where ΔT is the temperature difference between the external atmosphere (which is at 300 K) and the external surface temperature of the insulation. D is in meters. Determine the critical radius of insulation and the corresponding heat loss for a cable surface temperature of 400 K.

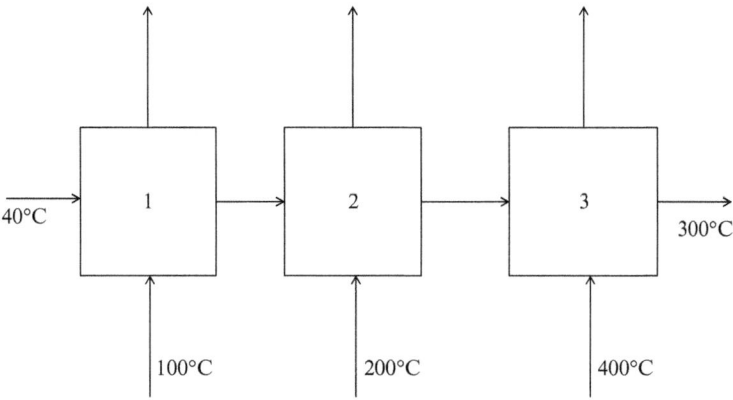

40°C

300°C

1   2   3

100°C   200°C   400°C

Fig. 8.16 Heat exchanger train.

8.11 Solve Problem 8.10 if besides convection radiative heat transfer also occurs from the surface of the insulated cable. Take surface emittance of the insulating sheath as 0.8.

8.12 Consider a train of three heat exchangers shown in Fig. 8.16, where cold water flowing at the rate of 20 kg/min is to be heated from 40 to 300°C, by using three different hot streams. The U values for the three heat exchangers are 100, 80, and 60 W/m² K, respectively. The three hot steams are flue gases with following properties:

Stream	$\dot{m}$	C_p
1	60 kg/min	1.1 kJ/kg K
2	80 kg/min	1.2 kJ/kg K
3	100 kg/min	1.3 kJ/kg K

Estimate the area required for each heat exchanger to ensure that their total surface area is the minimum. Use discrete maximum principle.

8.13 Solve Problem 8.11 numerically using Box Complex method.

8.14 The temperature inside a test room is to be raised from an ambient temperature of 0°C to a temperature of 40°C in 1 hour by using a heat pump of following specifications:

Condenser surface area: 5 m²

Heat transfer coefficient on air side: 80 W/m²

$$\frac{\text{Compressor power}}{\text{Condenser heat rejection rate}} = 0.4 + 0.01\,T_c$$

where T_c is the condenser surface temperature.

The thermal capacity of air and all other equipment in the room is 578 MJ/K. Determine the optimal strategy of controlling the condenser temperature to minimize the total power consumption during the warm up.

8.15 A car driver suddenly notices after taking a break in a long journey that the petrol quantity left is quite small. The next petrol pump is far away, D kilometers from this place, and he must reach there in time T seconds. How should he plan the journey so that he can minimize the fuel consumption?

Given the fuel consumption rate is a function of car acceleration:

$$\frac{dM_f}{dt} = \left(M \frac{d^2x}{dt^2} \right)^2$$

where M is the total mass of the car, including the passengers. Determine using COV, the speed with which the car should be driven over this distance and the minimum fuel consumption.

8.16 Solve Problem 8.15 by using Pontryagin's maximum principle.

REFERENCES

[1] S.S.L. Chang, Synthesis of Optimum Control Systems, McGraw Hill, New York, 1961.
[2] S. Katz, Best operating points for staged systems, Ind. Eng. Chem. Fundamental 1 (1962) 226.
[3] L.T. Fan, C.S. Wang, The Discrete Maximum Principle: A Study of Multistage Systems Optimization, John Wiley, New York, 1964.
[4] S.S. Rao, Engineering Optimization: Theory and Practice, New Age International Ltd., New Delhi, 2002.
[5] W. Spendley, G.R. Hext, F.R. Himsworth, Sequential applications of simplex designs in optimization and evolutionary operation, Technometrics 4 (1962) 441–461.
[6] M.J. Box, A new method of constrained optimization and comparison with other methods, Comput. J. 8 (1) (1965) 42–53.
[7] P.L. Dhar, Optimization in refrigeration systems, PhD thesis, Department of Mechanical Engineering, IIT, Delhi, 1974.
[8] D.E. Goldberd, Genetic Algorithms, Pearson Education, Delhi, India, 2006.

CHAPTER 9

Case Studies in Optimum Design

In this chapter we shall discuss a number of case studies of optimum design related to thermal systems and equipment used in them. First of all a few simple examples of thermodynamic optimization are presented. Thereafter the methodology of optimum design of some typical equipment used in thermal systems like the refrigerant condenser, flooded and DX chillers, various types of finned surfaces is discussed. Finally three comprehensive case studies on optimum design of a refrigeration system, a combined cycle power plant, and a liquid desiccant-based air-conditioning system are presented.

9.1 THERMODYNAMIC OPTIMIZATION

Thermodynamic optimization usually entails finding the optimum operating conditions of a system, usually to maximize its efficiency/coefficient of performance (COP). As an illustration of how this can be done, we shall present, in brief three examples pertaining to refrigeration systems.

9.1.1 Optimal Suction State in Vapor Compression Refrigeration Cycle

As the first example we consider the simple problem of optimum suction state for vapor compression cycle. Fig. 9.1 shows the $T - s$ diagram for a typical ideal cycle. For dry saturated state leaving the evaporator the cycle is $1^s - 2^s - 3 - 4$. It can be easily seen that for given evaporator and condenser temperatures the COP would depend upon the state of refrigerant leaving the evaporator and entering the compressor. Thus if we consider the limiting case of the suction state being the same as at the end of the throttling process (i.e., point 4 in Fig. 9.1), the COP is clearly zero since no refrigeration effect is obtained but there is a finite work input ($= h_{4'} - h_4$). If we consider a drier state, such as a, the COP is larger and it goes on increasing as we proceed toward point 1 in Fig. 9.1, the suction state corresponding to which the isentropic compression gives dry saturated vapors (state 2) at discharge. Up to this point, equal increments in the refrigeration effect are achieved at the cost of equal increments in work input, and hence the COP

Thermal System Design and Simulation
http://dx.doi.org/10.1016/B978-0-12-809449-5.00009-7

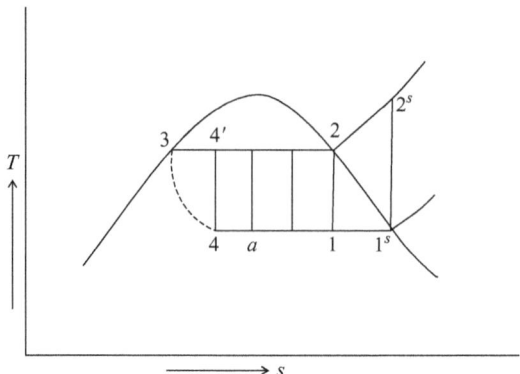

Fig. 9.1 Vapor compression cycle.

goes on increasing. Beyond this point, however, the work increases rapidly, and in some refrigerants rapidly enough to cause a fall in COP after attaining a maximum. In some other refrigerants this maximum is attained before the saturation point 1^s, while for others it continues to increase even beyond this point.

We could formulate the problem of locating this optimal state as an optimization problem of maximizing the COP which is a function of this state. For given condenser and evaporator temperatures, this suction state can be indicated by the enthalpy value h_1. Now COP can be written as:

$$\epsilon = \frac{h_1 - h_3}{h_2 - h_1} \tag{9.1}$$

Now, for given condenser temperature, h_3 is fixed, and h_2 is a function of h_1 alone. Therefore, COP is a function of h_1 alone and will be maximum if:

$$\epsilon = \frac{d\epsilon}{dh_1} = 0 \tag{9.2}$$

Using Eq. (9.1) in Eq. (9.2) we get upon simplification:

$$\frac{dh_2}{dh_1} = \frac{h_2 - h_3}{h_1 - h_3} \tag{9.3}$$

Eq. (9.3) is the condition from which the value of optimal h_1 can be found. A direct approach would be to get h_2 versus h_1 data for a refrigerant, fit an equation into it, find (dh_2/dh_1) and substitute in Eq. (9.3) to get a nonlinear equation for h_1. It was, however, found that due to inadequacy of superheated vapor data in immediate vicinity of the saturation state, even small difference between the regressed values and the actual values caused large errors in locating the optimal state.

This difficulty has been overcome by an ingenious approach [1]. Eq. (9.3) is transformed by subtracting 1 from its both sides to get:

$$\frac{d(h_2 - h_1)}{dh_1} = \frac{h_2 - h_1}{h_1 - h_3} \tag{9.4}$$

Its finite difference approximation is:

$$\frac{\Delta h_1}{\Delta (h_2 - h_1)} = \frac{h_1 - h_3}{h_2 - h_1} = \epsilon \tag{9.5}$$

It is very convenient to use above equation to locate the optima. All that we need to do is to calculate the values of left-hand side (LHS) and right-hand side (RHS) of Eq. (9.5) at the saturation state 1^s. Knowing the nature of variation of these two expressions with h_i (LHS would remain constant till point 1 and then decrease; RHS would increase with h_1 till point 1, its variation thereafter depending on the refrigerant) we can easily see that if the RHS of Eq. (9.5) is larger than the LHS at 1^s (see Fig. 9.2) the two curves would necessarily intersect at some point between 1 and 1^s, and the optimum lies in the wet region.

On the other hand, when the derivative value is larger even at the saturation state, it implies that optima, if it exists, lies in the superheated region Fig. 9.3.

Thus the search region is considerably narrowed and the optimum can be located accurately on the basis of Eq. (9.5).

Using this method, analysis of a number of refrigerants for two sets of operating temperatures, viz. (I) $-15°C/+30°C$ and (II) $-5°C/+40°C$, has been carried out. The results are summarized in Table 9.1.

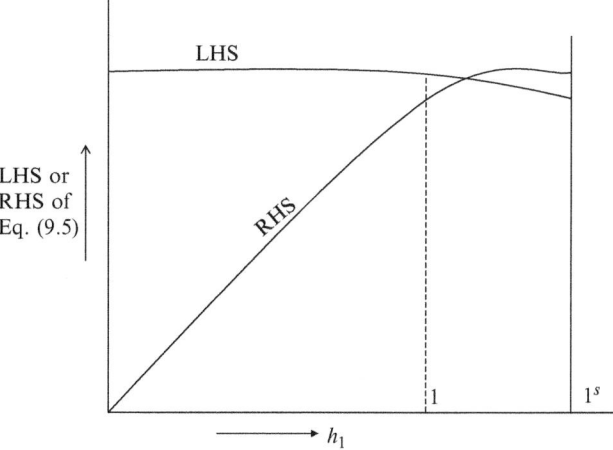

Fig. 9.2 Optimum suction state in wet region.

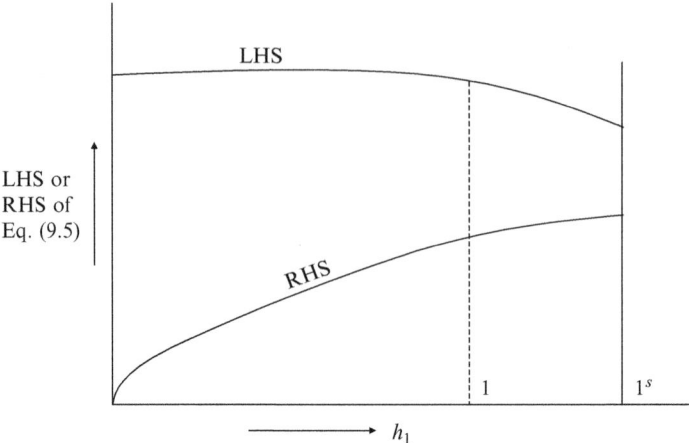

Fig. 9.3 Optimum suction state in the superheated region.

Table 9.1 Optimum suction state for a refrigeration cycle

	Cycle I		Cycle II		
Refrigerant	RHS	LHS	RHS	LHS	Remarks
12	4.7	4.98	6.66	7.29	Optimum beyond 100°C superheat
22	4.64	2.08	6.56	3.94	Optimum dryness fractions I: 0.97; II: 0.97
114	4.65	5.55	6.67	7.98	Optimum beyond 100°C superheat
717	4.76	2.27	6.86	3.66	Optimum dryness fractions I: 0.9; II: 0.92

9.1.2 Optimization of Multistage Refrigeration Systems

In refrigeration systems when the evaporator pressure is low, the compression of vapors up to the condenser pressure is often done in a number of stages with intercooling in between to reduce the work of compression, Fig. 9.4. For specified values of the two limiting pressures the intermediate pressures have therefore to be properly selected to minimize the total work input.

We have already seen in Example 8.10 how the discrete maximum principle can be used to predict the optimal interstage pressures in air compressors with perfect intercooling where the temperature of air entering each compressor is the same. In refrigeration compressor, however, perfect intercooling is not possible (see Fig. 9.4). The maximum amount of intercooling which can be done is up to the saturation temperature at the prevailing pressure. Since the intermediate temperatures in various stages are therefore different, we need to modify the analysis with perfect inter cooling discussed in Example 8.10.

Using the same symbols as in Fig. 8.5 (repeated here for sake of convenience as Fig. 9.5), x_1^n denoting the exit pressure, θ^n the pressure-ratio in the nth stage, we have the relationship:

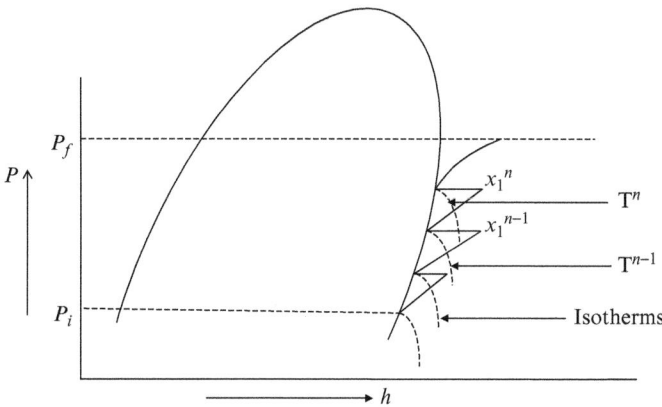

Fig. 9.4 Multistage refrigerant compressor.

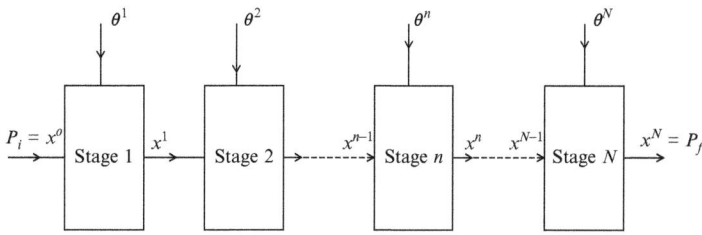

Fig. 9.5 Multistage compression.

$$x_1^n = x_1^{x-1}\theta^n \tag{9.6}$$

Here the temperature of the refrigerant entering each compressor is the saturation temperature at that pressure. Using symbol T^n to denote the temperature in stage n, we can write

$$T^n = \phi(x_1^n) \tag{9.7}$$

where ϕ is the saturation temperature-pressure equation for the refrigerant.

The work input in the nth stage, per unit mass flow (W^n) is:

$$W^n = R\frac{\gamma}{\gamma - 1}T^{n-1}\left[(\theta^n)^{\frac{\gamma-1}{\gamma}} - 1\right] \tag{9.8}$$

where γ is the index of compression (assumed to be the same in all the stages).

Defining another state variable x_2^n by equation:

$$x_2^n = x_2^{n-1} + \phi(x_1^{n-1})\left[(\theta^n)^{\frac{\gamma-1}{\gamma}} - 1\right] \tag{9.9}$$

with $x_2^0 = 0$, it follows that:

$$x_2^N = \sum \phi(x_1^{n-1}) \left([\theta^n]^{\frac{\gamma-1}{\gamma}} - 1 \right) = \frac{\Sigma W^n}{R \cdot \gamma / (\gamma - 1)} \qquad (9.10)$$

Thus x_2^n can be viewed as the total work done up to (and including) the nth stage. Clearly, minimization of x_2^N will ensure minimization of the total work input in all the compressors.

We can now use the discrete maximum principle to find the optimal values of the control variable θ (Section 8.1.5). Here the state vector x is two-dimensional and the vector θ is one dimensional. The Hamiltonian H^n becomes (cf. Eq. 8.56):

$$H^n = z_1^n (x_1^{n-1} \theta^n) + z_2^n \left(x_2^{n-1} + \phi(x_1^{n-1}) \left\{ (\theta^n)^{\frac{\gamma-1}{\gamma}} - 1 \right\} \right) \qquad (9.11)$$

Using Eq. (8.57) with H^n defined by Eq. (9.11) we get:

$$z_1^{n-1} = \frac{\partial H^n}{\partial x_1^{n-1}} = z_1^n \theta^n + z_2^n \left[\phi' \{\theta^n\}^{\frac{\gamma-1}{\gamma}} - 1 \right] \quad n = 1, 2, \ldots, N \qquad (9.12)$$

$$z_2^{n-1} = \frac{\partial H^n}{\partial x_2^{n-1}} = z_2^n \quad n = 1, 2, \ldots, N \qquad (9.13)$$

where

$$\phi' = \frac{d\phi(x_1^{x-1})}{dx_1^{n-1}}$$

Further using Eq. (8.58) here (with $M = 1$, since there is no feedback) gives:

$$z_2^N = c_2 = -1 \qquad (9.14)$$

Since our objective function to be maximized is $(-x_2^N)$. Eq. (8.58) cannot be applied for $i = 1$ since the final state corresponding to variable 1 (i.e., final pressure P_f) is specified.

Now since $z_2^{n-1} = z_2^n$ and $z_2^N = -1, \Rightarrow z_2^n = -1$ for $n = 1, 2, \ldots, N$ Thus, the Hamiltonian H^n becomes:

$$H^n = z_1^n x_1^{n-1} \theta^n - x_2^{n-1} - \phi(x_1^{n-1}) \left([\theta^n]^{\frac{\gamma-1}{\gamma}} - 1 \right) \qquad (9.15)$$

To get the optimal values of the decision variable we use Eq. (8.59). This gives:

$$\frac{\partial H^n}{\partial \theta^n} = z_1^n x_1^{n-1} - \phi(x_1^{n-1}) \cdot \frac{\gamma-1}{\gamma} (\theta^n)^{-\frac{1}{\gamma}} = 0$$

$$\Rightarrow \quad z_1^n = \frac{\gamma-1}{\gamma} \frac{[\theta^n]^{-\frac{1}{\gamma}} \phi(x_1^{n-1})}{x_1^{n-1}} \quad n = 1, 2, \ldots, N \qquad (9.16)$$

Using Eq. (9.16) in Eq. (9.12) gives, in addition to an equation for z_1^o, a set of $N-1$ recurrence relationships:

$$\frac{(\theta^n)^{-1/\gamma}\phi(x_1^{n-1})}{x_1^{n-1}} = \frac{(\theta_1^{n+1})^{\frac{\gamma-1}{\gamma}}\phi(x_1^n)}{x_1^n}$$

$$-\frac{\gamma}{\gamma-1}\left[(\theta_1^{n+1})^{\frac{\gamma-1}{\gamma}}-1\right]\frac{\partial\phi(x_1^n)}{\partial x_1^n} \quad n=1,2,\ldots,N \tag{9.17}$$

Eq. (9.17) together with Eq. (9.6) constitutes a set of $(2N-1)$ independent equations in $(2N-1)$ variables, viz. N decision variable θ^n and $(N-1)$ unknown state variables x_1 (recall $x_1^o = P_i, x_1^N = P_f$ are specified). Simultaneous solution of these equations would give us the optimum pressure ratio in various stages.

It is interesting to see that if we consider the case of perfect intercooling, that is, $\phi(x_1^o) = \phi(x_1^n) = T_1^o$ (a constant), $n = 1, 2, \ldots, N$ then Eq. (9.17) reduces to

$$\theta^n = \theta^{n+1}$$

that is, the pressure ratios in each stage are equal (Example 8.10).

As an illustration, these equations have been solved to get the optimum intermediate pressure for two stage ammonia refrigeration system.

Table 9.2 Optimum interstage pressures for refrigeration systems with interstage cooling

P_o	P_f	Optimal P_1	$\sqrt{P_oP_f}$
2.068 bar (30 psi)	13.79 bar (200 psi)	5.364 bar (77.8 psi)	5.343 bar (77.5 psi)
2.758 bar (40 psi)	13.79 bar (200 psi)	6.188 bar (89.75 psi)	6.164 bar (89.4 psi)
3.447 bar (50 psi)	13.79 bar (200 psi)	6.964 bar (101.0 psi)	6.895 bar (100.0 psi)

The results are shown in Table 9.2. It is seen that the optimal interstage pressure is only slightly higher than the geometric pressure which is the optimal pressure with perfect intercooling.

A more practical multistage refrigeration system is the one with *flash inter cooling*. Fig. 9.6 shows its schematic diagram and Fig. 9.7 the thermodynamic representation of various processes on a p-h diagram. We consider a compressor and its flash intercooler as one subsystem indicated by dashed line box in Fig. 9.6. Compressed vapors of enthalpy h_{n-1}^s and liquid-vapor mixture of enthalpy h_n^L enter the flash intercooler while saturated liquid of enthalpy h_{n-1}^L, and saturated vapors of enthalpy h_n flow out of it. Since $h_{n-1}^L < h_n^L$ and $h_n < h_{n-1}^s$, it follows that the mass flow rate of the liquid leaving the flash intercooler must be lesser than the mass flow rate of liquid vapor mixture entering it.

We denote by a_i the mass flow rate of the liquid that evaporates in the ith flash intercooler and do the analysis for a unit mass flow rate of refrigerant in the evaporator.

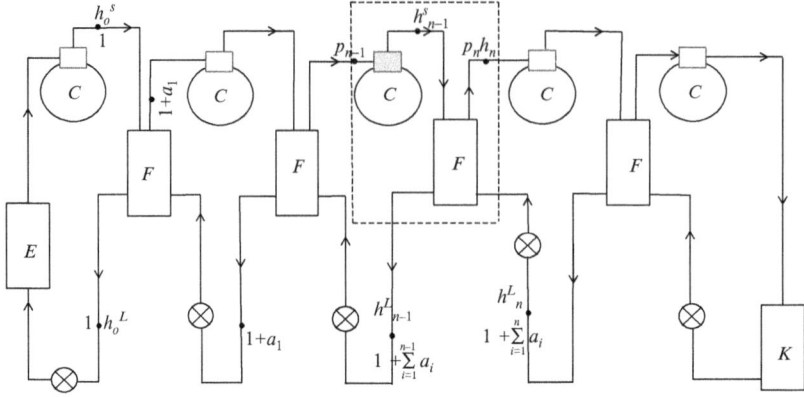

Fig. 9.6 Schematic diagram of a multistage refrigeration system with flash intercooling.

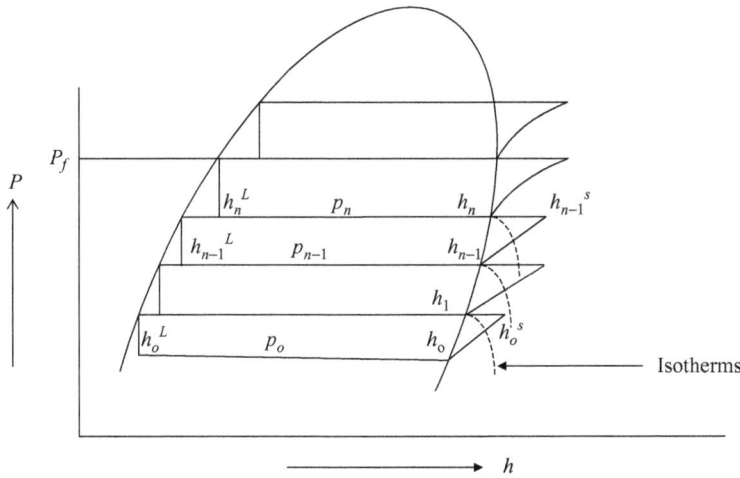

Fig. 9.7 p–h diagram of a multistage refrigeration system with flash intercooling.

The symbol h with superscript L denotes liquid enthalpy, with superscript s denotes superheated vapor enthalpy, and no superscript corresponds to saturated vapor enthalpy. Various processes are indicated in the p–h diagram also (Fig. 9.7).

Energy balance for the flash cooler gives a relationship for a_i^s as:

$$
h_{n-1}^s \left(1 + \sum_{i=1}^{n-1} a_i\right) + h_n^L \left(1 + \sum_{i=1}^{n} a_i\right) = h_n \left(1 + \sum_{i=1}^{n} a_i\right)
$$

$$
+ h_{n-1}^L \left(1 + \sum_{i=1}^{n-1} a_i\right) \quad n = 2, 3, \ldots, N-1 \qquad (9.18)
$$

Rearranging Eq. (9.18) we get:

$$1 + \sum_{i=1}^{n} a_i = \frac{h_{n-1}^s - h_{n-1}^L}{h_n - h_n^L} \left(1 + \sum_{i=1}^{n-1} a_i\right) \quad n = 2, 3, \ldots, N-1 \qquad (9.19)$$

From the energy balance for the first subsystem we get:

$$1 + a_1 = \frac{h_o^s - h_o^L}{h_1 - h_1^L} \qquad (9.20)$$

Combining Eqs. (9.19) and (9.20) we can get:

$$1 + \sum_{i=1}^{n} a_i = \prod_{i=1}^{n} \frac{(h_{i-1}^s - h_{i-1}^L)}{h_i - h_i^L} \quad n = 1, 2, \ldots, N-1 \qquad (9.21)$$

We have to determine the intermediate pressures so as to get minimum work per unit of refrigeration (η). Taking the compression processes to be isentropic, the work input would be equal to the enthalpy rise, and we can write the following expression for η:

$$\eta = \frac{\begin{array}{l}(h_o^s - h_o) + (1 + a_i)(h_1^s - h_1) + (1 + a_1 + a_2)(h_2^s - h_2) + \\ \cdots + \left(1 + \sum_{i=1}^{N-1} a_i\right)(h_{N-1}^S - h_{N-1})\end{array}}{(h_o - h_o^L)} \qquad (9.22)$$

In the view of the occurrence of h_o^L in the denominator of η, and its dependence on the next stage pressure, it is not possible to use the discrete maximum principle to get the conditions for minimum η. So direct minimization of η as a function of the interstage pressures has been done. To facilitate differentiation we first define a new variable x_i as:

$$x_i = \frac{h_{i-1}^s - h_{i-1}^L}{h_i - h_i^L} = x_i(p_{i-1}, p_i, p_{i+1}) \qquad (9.23)$$

Thus, while differentiating with respect to p_j (intermediate pressure) only terms with x_{j-1}, x_j or x_{j+1} as factors will be important.

Now for minimal η, the condition would be:

$$\frac{\partial \eta}{\partial p_j} = 0 \quad j = 1, 2, \ldots, N-1 \qquad (9.24)$$

Combining Eqs. (9.21)–(9.24) we get, on simplification, for $(N-1) \geq j \geq 2$:

$$0 = \frac{\partial x_{j-1}}{\partial p_j}(h^s_{j-1} - h_{j-1}) + x_{j-1}\frac{\partial h^s_{j-1}}{\partial p_j} + x_{j-1}x_j\left(\frac{\partial h^s_j}{\partial p_j} - \frac{\partial h_j}{\partial p_j}\right)$$

$$+ \left(x_{j-1}\frac{\partial x_j}{\partial p_j} + x_i\frac{\partial x_{j-1}}{\partial p_j}\right)(h^s_j - h_j) + \left[x_{j-1}x_j\frac{\partial x_{j+1}}{\partial p_j} + x_{j-1}\frac{\partial x_j}{\partial p_j}x_{j+1}\right.$$

$$\left. + x_j\frac{\partial x_{j-1}}{\partial p_j}x_{j+1}\right]X(j) \tag{9.25}$$

where

$$X(j) = h^s_{j+1} - h_{j+1} + x_{j+2}(h^s_{j+2} - h_{j+2}) + x_{j+2}x_{j+3}(h^s_{j+3} - h_{j+3})$$

$$+ \cdots + x_{j+2}x_{j+3}\ldots x_{N-1}(h^s_{N-1} - h_{N-1}) \tag{9.26}$$

and for $j = 1$

$$0 = (h_o - h^L_o)\left[\frac{\partial h^s_o}{\partial p_1} + \frac{\partial x_1}{\partial p_1}(h^s_1 - h_1) + x_1\left(\frac{\partial h^s_1}{\partial p_1} - \frac{\partial h_1}{\partial p_1}\right) + \left(x_1\frac{\partial x_2}{\partial p_1} + x_2\frac{\partial x_1}{\partial p_1}\right)X(1)\right]$$

$$+ \left(\frac{\partial h^L_o}{\partial p_1}\right)\left((h^s_o - h_o) + x_1(h^s_1 - h_1) + x_1x_2X(1)\right) \tag{9.27}$$

where $X(1)$ is obtained by putting $j = 1$ in Eq. (9.26).

Eqs. (9.25) and (9.27) give us $N - 1$ equations in $N - 1$ intermediate pressures and by solving these simultaneous equations optimal pressures can be obtained. In particular we get for two stage system, by putting $N = 2$ in Eq. (9.27), an equation for the intermediate pressure p_1 as:

$$(h_o - h^L_o)\left[\frac{\partial h^s_o}{\partial p_1} + \frac{\partial x_1}{\partial p_1}(h^s_i - h_i) + x_1\left(\frac{\partial h^s_1}{\partial p_1} - \frac{\partial h_1}{\partial p_1}\right)\right]$$

$$+ \left(\frac{\partial h^L_o}{\partial p_1}\right)\left[(h^s_o - h_o) + x_1(h^s_1 - h_1)\right] = 0 \tag{9.28}$$

where $x_1 = f(p_1)$, is a function of only one variable, p_1. Since p_o and p_2 are given, we have:

$$h^L_o = \phi(p_1)$$
$$h^S_o = \psi(p_o, p_1) = \psi(p_1)$$
$$h^S_1 = \theta(p_1, p_2) = \theta(p_1) \tag{9.29}$$

where ϕ_1, ψ, and θ are appropriate functions relating the enthalpy values to the pertinent pressures. Thus for a two-stage system the values of optimal p_1, for given values of p_o and p_2 can be iterated from Eq. (9.28) by using thermodynamic property tables, approximating the derivatives by their finite difference equivalents. Table 9.3 gives the results for the same conditions as in Table 9.2.

Table 9.3 Optimum interstage pressures with flash intercooling

p_o (bar/psi)	p_2 (bar/psi)	p_1 (bar/psi)	Geometric mean pressure (bar/psi)
2.068 (30)	13.79 (200)	5.66 (82.1)	5.343 (77.5)
2.758 (40)	13.79 (200)	6.66 (96.6)	6.164 (89.4)
3.447 (50)	13.79 (200)	7.756 (112.5)	6.895 (100)

It can be seen that the optimal pressures differ considerably from the geometric mean pressure.

9.1.3 Optimum Interstage Temperature for Cascade Refrigeration

Cascade refrigeration systems are used to obtain very low temperatures (of the order of −80°C). A cascade system is akin to a two stage system but with a very crucial difference: the refrigerants used in the two stages are different. The coupling of the two systems is achieved by using the evaporating refrigerant of the "high temperature stage" to condense the refrigerant leaving the compressor of the "low temperature stage" as shown in Fig. 9.8.

The thermodynamic diagram of the process is shown in Fig. 9.9, the two domes corresponding to two refrigerants. It is clear from the diagram that the overall COP of the system, for given values of the condenser and evaporator temperatures, will depend upon the temperatures T_{k1}, the condensing temperature of "low temperature stage" refrigerant, and T_{02}, the temperature of the evaporator of the "high temperature stage." These two temperatures are not independent, but related to each other through the heat

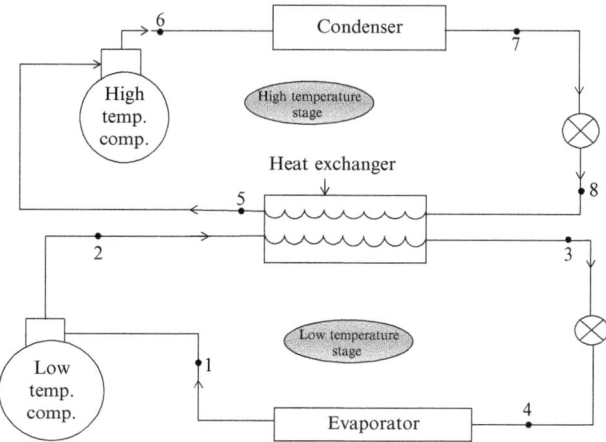

Fig. 9.8 Schematic diagram of a cascade system.

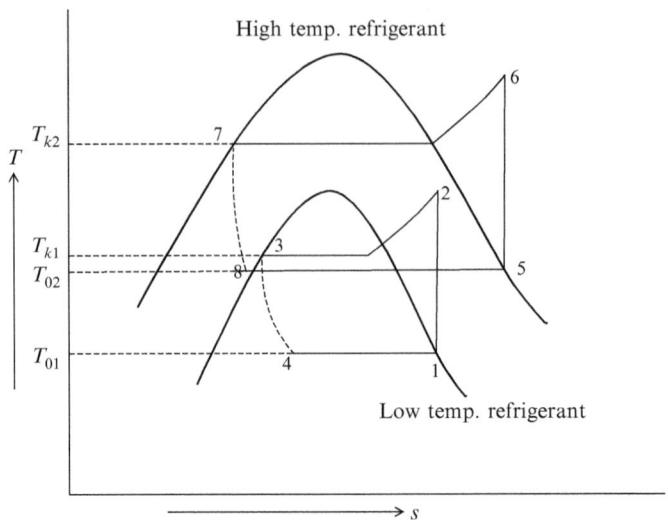

Fig. 9.9 Thermodynamic diagram of cascade system processes.

exchanger characteristics. Thus we can write:

$$T_{02} = f(T_{k1}) \tag{9.30}$$

The overall COP of the system, ϵ is obviously given by the following equation:

$$
\begin{aligned}
\epsilon &= \frac{\text{Refrigerating effect in the evaporator}}{\text{Sum of work input in the two compressors}} \\
&= \frac{\dot{m}(h_1 - h_4)}{\dot{m}_1(h_2 - h_1) + \dot{m}_2(h_6 - h_5)}
\end{aligned} \tag{9.31}
$$

where $\dot{m}_1$ is the mass flow rate of refrigerant in the "low temperature stage" and $\dot{m}_2$ the mass flow rate in the "high temperature stage." These two mass flow rates are further related through the energy balance in the heat exchanger which demands:

$$\dot{m}_1(h_2 - h_3) = \dot{m}_2(h_5 - h_8) \tag{9.32}$$

Combining Eqs. (9.31) and (9.32) we get

$$\epsilon = \frac{\epsilon_1 \epsilon_2}{\epsilon_1 + \epsilon_2 + 1} \tag{9.33}$$

$$\text{where } \epsilon_1 = \frac{h_1 - h_4}{h_2 - h_1} = \text{COP of the "low temperature cycle"} \tag{9.34}$$

$$\epsilon_2 = \frac{h_5 - h_8}{h_6 - h_5} = \text{COP of the "high temperature cycle"} \tag{9.35}$$

It is clear from Eqs. (9.33)–(9.35) that, for prescribed values of the low temperature cycle evaporator temperature (T_{01}) and the high temperature cycle condenser temperature (T_{k2}), the overall COP is a function of the T_{02} and T_{k1}. Further, in view of Eq. (9.30) we can write:

$$\epsilon = \epsilon(T_{k1}) \tag{9.36}$$

and the condition for maximum ϵ is:

$$\frac{d\epsilon}{dT_{k1}} = 0 \tag{9.37}$$

which on using Eq. (9.33) gives:

$$\epsilon_2(\epsilon_2 + 1)\frac{d\epsilon_1}{dT_{k1}} + \epsilon_1(\epsilon_1 + 1)\frac{d\epsilon_2}{dT_{k1}} = 0 \tag{9.38}$$

Since ϵ_1 and ϵ_2 are both function of T_{k1}, Eq. (9.38) becomes a nonlinear equation in T_{k1} and it can be solved by any of the standard methods enumerated in Section 2.2 to get the desired optimum coupling temperature T_{k1}. The corresponding T_{02} can then be determined using Eq. (9.30).

Using this method the optimum coupling temperature has been found [2] for a number of cascade systems. The design conditions were assumed to be:

The highest condensing temperature: $40°C$

The lowest evaporating temperature: $-80°C$

R–13 was assumed to be the refrigerant in the low temperature stage. R–12, R–22, and R–717 were assumed to be the refrigerants in the high temperature stage for the three cascade systems considered.

First of all stage COPs ϵ_1 and ϵ_2 were found out for various values of T_{k1} and T_{02}, respectively, by using the thermodynamic property data in Eqs. (9.34) and (9.35). Their functional relationships with T_{k1} and T_{02}, respectively, were then obtained by fitting polynomial equations in the data.

The relationship between T_{02} and T_{k1}, which depends upon the design of the coupling heat exchanger was assumed, for the sake of simplicity, to be a linear one, that is,

$$T_{02} = T_{k1} - \Delta T \tag{9.39}$$

The analysis was done for three different values of ΔT to study its effect on the optimal operating conditions.

Substituting the functional relationships for ϵ_1 and ϵ_2 as functions of the temperatures T_{k1} and T_{k2}, along with Eq. (9.39) in Eq. (9.38) we get the nonlinear equation for T_{k1}. This was solved by using Newton-Raphson technique (Section 2.2). The

Table 9.4 Optimal operating conditions for cascade systems

Refrig.		R − 12	R − 22	R − 717
$\Delta T = 5°C$	T_{k1}	−29.36	−29.1	−16.73
	ϵ	1.0599	1.0445	1.0239
	r_1	7.86	7.905	11.48
	r_2	11.49	11.21	8.82
$\Delta T = 10°C$	T_{k1}	−25.81	−25.72	−12.46
	ϵ	0.9830	0.9691	0.9390
	r_1	8.75	8.78	12.9
	r_2	12.30	12.03	9.12
$\Delta T = 15°C$	T_{k1}	−22.12	−22.09	−8.27
	ϵ	0.9111	0.8986	0.8596
	r_1	9.8	9.8	14.48
	r_2	13.08	12.80	9.46

optimal operating conditions thus obtained for the three cascade systems are indicated in Table 9.4. The terms r_1 and r_2 indicate the compression ratios in the two stages.

It can be seen that the optimal operating conditions depend strongly on the nature of the refrigerant as also on the heat exchanger performance. Unlike the usual multistage compressor, the compression ratios in the two stages are not equal. While in halo–carbon refrigerants the compression ratio of the compressor in the low temperature stage is lesser than that of the compressor in the high temperature stage, in ammonia the reverse is true.

9.2 OPTIMUM DESIGN OF COMPONENTS

In this section we shall present a few case studies on optimization of some components used in thermal systems. From the perspective of optimization the components can be divided into two categories, viz. those which form a part of a system operating in a cycle, and those which can be considered in a stand-alone mode, as any change in their design does not affect the thermal performance of other parts of the system. In the former category come the components like the evaporator/condenser of a refrigeration plant or a power plant, while in the latter come equipment like the finned surface of power electronic device. We shall first study the methodology of optimum design of a few components of the second category.

9.2.1 Finned Surfaces

Finned surfaces are often used these days to transfer heat from electronic devices like VLSI circuits, thermoelectric modules, power electronic devices etc. Usually, the total heat to be transferred out of the device is known and the challenge is to design a finned surface of minimum volume to enable this. We have already considered in Example 8.8 the conjugate problem of finding the optimum profile of a single fin of specified total volume, which will yield maximum heat transfer rate. However, the optimal profile of fins in a finned surface is not likely to be same in view of additional constraint imposed by the fact that the fins (of unknown base thickness) have to be accommodated within the height available (Fig. 9.10), while maintaining a certain specified distance between the fins. We shall present the general method suggested by Dhar and Arora [3] for optimizing such finned surfaces of various geometries.

We consider first a flat finned surface Fig. 9.10 and use the coordinate system shown therein, with $x = 0, y = 0$ being at the fin tip, y being half the thickness of the fin at any distance x from the fin tip. We can write the following equation for the total fin volume:

$$V = NW \int_{o}^{L} 2y dx \qquad (9.40)$$

where N is the total number of fins on the surface and W, their width. Since all these fins have to be accommodated within the total height of the surface we get a relationship between $N, 2c$ the fin spacing, and $y(L)$—half the thickness of the fin at the root:

$$H = 2cN + 2Ny(L) \qquad (9.41)$$

The total heat dissipation from the finned surface is clearly the sum of heat transfers from the fins and the base surface. This is,

$$Q = N \left[2ch(T_L - T_\infty) + \int_{0}^{L} 2h(T - T_\infty) dx \right] W \qquad (9.42)$$

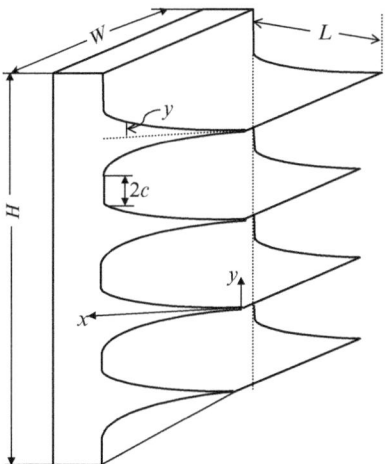

Fig. 9.10 Flat finned surface.

where we have assumed that the fins are thin and therefore fin surface area $\approx dx \cdot W$; and the heat transfer coefficient over the fin base equals that over the rest of the fin height. $T(x)$ is the fin temperature at distance x and $T\infty$ the ambient temperature.

Following Example 8.8 (Eq. 8.31) the basic energy equation for the fin can be written as:

$$\frac{d}{dx}\left(y\frac{dT}{dx}\right) = \frac{h}{k}(T - T_\infty) \tag{9.43}$$

The optimization problem can now be formulated as:

$$\text{Minimize } V = 2NW \int_0^L y\,dx \tag{9.44}$$

subject to the constraints:

$$H = 2cN + 2Ny(L) \tag{9.45}$$

$$Q = 2NW \left[ch_L(T_L - T_\infty) + \int_0^L h(T - T_\infty)\,dx \right] \tag{9.46}$$

$$\frac{d}{dx}\left(y\frac{dT}{dx}\right) = \frac{h}{k}(T - T_\infty) \tag{9.47}$$

$$\left(y\frac{dT}{dx}\right)_{x=0} = 0 \quad T(L) = T_L \tag{9.48}$$

where Eq. (9.48) represents the boundary condition that the heat loss through the tip is zero.

The objective function, Eq. (9.44), involves a variable N, besides the functional of $y(x)$. We could replace N in terms of other variables using Eq. (9.45), but it is seen that the resulting expression involves $y(L)$ which is a boundary value of the unknown function $y(x)$. It is therefore not possible to solve this problem directly using calculus of variations.

Dhar and Arora [3] suggest assuming the temperature profile to belong to a class of functions:

$$T = T_\infty + (T_L - T_\infty)\left(\frac{x}{L}\right)^\alpha \tag{9.49}$$

In Example 8.8 we saw while optimizing a single fin that the temperature profile is linear, that is, $\alpha = 1$. We can thus hope that it should be possible to capture the temperature profile in the optimal fins of finned surface with above equation with $\alpha \neq 1$.

Substituting Eq. (9.49) in Eq. (9.47), together with the boundary conditions of Eq. (9.48), we get after its integration:

$$y = \frac{hx^2}{k\alpha(\alpha + 1)} \tag{9.50}$$

Using this expression for y, the optimization problem can be formulated as:

$$\text{Minimize } V = \frac{2hW}{k\alpha(\alpha+1)} \cdot \frac{NL^3}{3} \tag{9.51}$$

$$\text{where } L = \frac{a_1\alpha \pm \sqrt{a_1^2\alpha^2 - 4\alpha(\alpha+1)ca_2}}{2} \tag{9.52}$$

$$a_1 = \frac{WH\Delta Tk}{Q} \qquad a_2 = \frac{k}{h} - \frac{WH\Delta Tk}{Q} \tag{9.53}$$

$$\Delta T = T_L - T_\infty \qquad N = \frac{Q/(2hW\Delta T)}{[L/(\alpha+1)] + c} \tag{9.54}$$

Substituting Eqs. (9.52) and (9.54), in Eq. (9.51) we can express V as a function of α alone, and by using the search methods (Section 8.2.1) the optimal α can be found out.

Often the space available for accommodating finned surface inside the equipment is limited and this can restrict the fin height. The optimal fin profile with restricted height would differ from that obtained above since L is now specified. We therefore assume a new temperature profile with a new parameter b:

$$T - T_\infty = \Delta T\left(1 - \frac{L-x}{b}\right)^\alpha; \quad \alpha > 0, \quad b > L \tag{9.55}$$

Eq. (9.55) reduces to Eq. (9.49) if $b = L$. Using Eq. (9.55) as the new temperature profile, in Eq. (9.47), we can proceed as above and get the fin profile as:

$$y = \frac{h}{k\alpha(\alpha+1)} \frac{(b-L+x)^{\alpha+1} - (b-L)^{\alpha+1}}{(b-L+x)^{\alpha-1}} \tag{9.56}$$

The optimization problem thus becomes:

Minimize:

$$V = \frac{2hN}{k\alpha(\alpha+1)}\left[\frac{b^3 - (b-L)^3}{3} - \frac{(b-L)^{\alpha+1}}{b^{\alpha-2}} \cdot \frac{F_1}{2-\alpha}\right.$$
$$\left. + \frac{(b-L)^3 F_1}{2-\alpha} - (b-L)^3 F_2 ln\left(\frac{b}{b-L}\right)\right] \tag{9.57}$$

subject to constraints $\tag{9.58}$

$$Q = 2Nh\Delta T\left(\frac{b^{\alpha+1} - (b-L)^{\alpha+1}}{b^\alpha(\alpha+1)} + c\right) \tag{9.59}$$

$$H = 2N\left(c + \frac{h}{k\alpha(\alpha+1)} \cdot \frac{b^{\alpha+1} - (b-L)^{\alpha+1}}{b^{\alpha-1}}\right) \tag{9.60}$$

$$\alpha > 0 \tag{9.61}$$

where the two functions F_1 and F_2 are defined as:

$$F_1 = \begin{bmatrix} 0 & \text{when} & \alpha = 2 \\ 1 & \text{otherwise} \end{bmatrix} \text{ and } F_2 = \begin{bmatrix} 1 & \text{when} & \alpha = 2 \\ 0 & \text{Otherwise} \end{bmatrix} \qquad (9.62)$$

As before, the constraints of Eqs. (9.59) and (9.60) can be used to express the variables b and N in terms of α and then these can be substituted in the objective function (Eq. 9.57) to express V as a function of α alone. Optimal value of α corresponding to the minimum fin material can then be determined as before using search procedure. Dhar and Arora [3] give the results of applying this procedure to solve a number of typical cases for two different fin materials (cast iron (CI) and aluminum (AL)) and seven different fin spacings (4–40 mm) for values of $Q/\Delta T$ ranging from 40 to 70.[1] Fig. 9.11 shows the values of fin profile index α versus fin spacing for both AL and CI fins for various values of heat flux from the surface. It can be seen that values of α differ considerably from 1, the value corresponding to optimal single fin. The fin volume/unit length for AL fins for various in spacings is shown in Fig. 9.12 (V series). It is evident

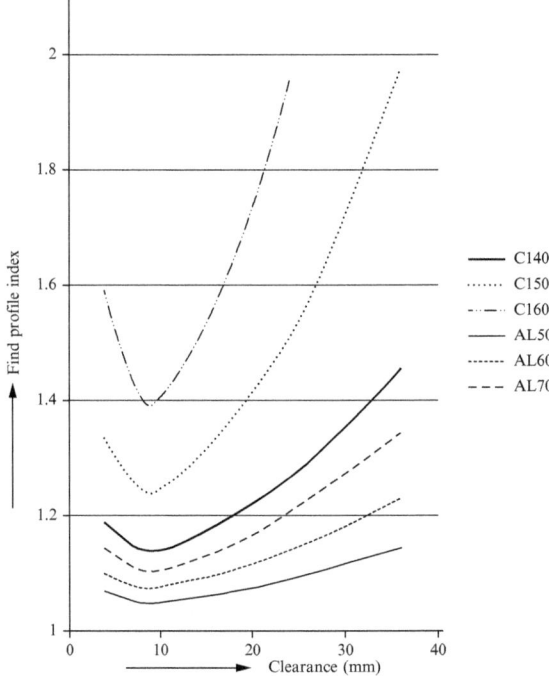

Fig. 9.11 Fin profile index as a function of fin spacing.

[1] Units are kcal/h/m^2/K.

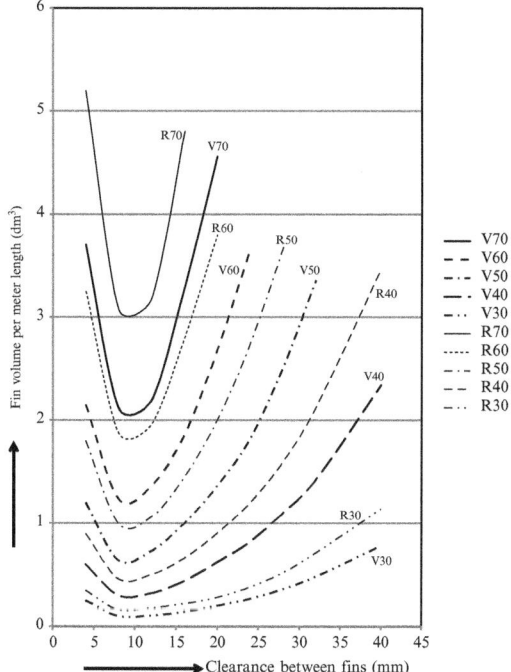

Fig. 9.12 Fin volume per unit length as a function of fin spacing.

that for every value of heat flux, there exists a fin spacing for which the fin volume is the minimum.

In order to ascertain whether the use of these fins of nonuniform cross–section was really useful, Dhar and Arora [3] have also carried out the optimization of finned surface with rectangular fins, and compared its results with those reported above.

With rectangular fins the whole analysis becomes quite simplified since the fin profile is already specified as:

$$y(x) = y_o, \quad \text{a constant} \tag{9.63}$$

substituting this equation in the basic energy conservation Eq. (9.43) we get

$$y_o \frac{d^2 T}{dx^2} = \frac{h}{k}(T - T_\infty) \tag{9.64}$$

This is a standard second-order differential equation whose closed form solution is well known. Substituting that solution in Eq. (9.46) and simplifying we get the final expression for V as a function of y_o as:

$$V = \frac{H\gamma_o}{c + \gamma_o} \cdot \sqrt{\frac{k\gamma_o}{h}} tanh^{-1}\left(\frac{Qc + Q\gamma_o - WchH\Delta T}{WH\sqrt{hk\gamma_o}\Delta T}\right) \tag{9.65}$$

$$\text{and} \quad L = \frac{V(\gamma_o + c)}{WH\gamma_o} \tag{9.66}$$

The optimum semi root thickness, γ_o, which minimizes V given by Eq. (9.65), can be found as before, and the result compared with variable thickness fins. Fig. 9.12 shows a summary of all the results. It can be seen that a substantial reduction in fin volume is obtained by using optimal variable thickness fins (V series), instead of optimal rectangular fins dissipating the same heat flux (R series), the saving being as high as 45% in some cases.

The same approach can be extended to optimization of fins on a cylindrical surface. Fig. 9.13 shows a cylindrical surface of length H, having transverse fins spaced a distance $2c$ apart. Using symbol δ to indicate half the thickness of fins, and invoking the thin fin assumption the basic minimization problem can be written as:

$$\text{Minimize } V^* = \frac{V}{4\pi} = N\int_{r_i}^{r_o} \delta r dr \tag{9.67}$$

subject to constraints:

$$Q^* = \frac{Q}{4\pi h(T_i - T_\infty)} = N\int_{r_i}^{r_o}\left(\frac{T - T_\infty}{T_i - T_\infty}r dr + cr_i\right) \tag{9.68}$$

$$\frac{d}{dr}\left(\delta \cdot r \cdot \frac{dT}{dr}\right) = \frac{rh}{k}(T - T_\infty) \tag{9.69}$$

$$H = 2N(c + \delta_i) \tag{9.70}$$

$$T(r_i) = T_i \quad \left(\delta\frac{dT}{dr}\right) = 0 \text{ at } r = r_o \tag{9.71}$$

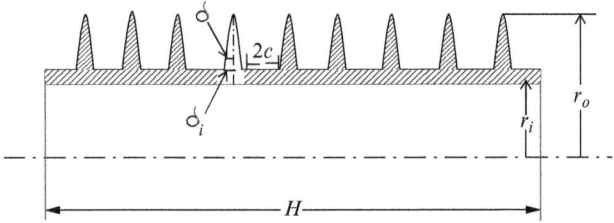

Fig. 9.13 Cylindrical surface with fins.

Following the same procedure as with a flat surface we assume following parametric temperature profile in the fins:

$$\frac{T - T_\infty}{T_i - T_\infty} = \left(\frac{r_o^2 - r^2}{r_o^2 - r_i^2}\right)^\alpha$$

$$\text{or} \quad T = T_\infty + \left(\frac{r_o^2 - r^2}{r_o^2 - r_i^2}\right)^\alpha (T_i - T_\infty) \tag{9.72}$$

Using this in the energy equation and combining terms, as in the analysis of a flat surface, the problem is reduced to:

$$\text{Minimize } V = \frac{Nh}{4k\alpha(\alpha + 1)} \cdot \left(r_o^4 \ln\frac{r_o}{r_i} + \frac{r_o^4 - r_i^4}{4} - r_o^2(r_o^2 - r_i^2)\right) \tag{9.73}$$

$$\text{where } N = \frac{A_2 \pm \sqrt{(A_2^2 - 4A_1A_2)}}{2A_1} \tag{9.74}$$

$$r_o^2 = 2(\alpha + 1)\left(\frac{Q}{N} - Cr_i\right) + r_i^2 \tag{9.75}$$

$$A_1 = 2c + \frac{2h(\alpha + 1)c^2}{k\alpha} \tag{9.76}$$

$$A_2 = H + \frac{4h(\alpha + 1)Q^*c}{k\alpha r_i} \tag{9.77}$$

$$A_3 = \frac{2h(\alpha + 1)Q^{*2}}{k\alpha r_i^2} \tag{9.78}$$

Optimum value of α giving the minimum fin volume can then be determined as in the case of a flat surface. Typical results of this analysis for CI fins for different amounts of heat flux from the surface are shown in Fig. 9.14. Once again it is seen that there exists an optimum fin spacing for which the fin volume is the minimum. Further, as before, carrying out analysis for the case when fin height is restricted it is seen that the fin volume, [curves [R] in Fig. 9.14] is slightly greater than the corresponding case when the fin height is unrestricted [U].

9.2.2 DX-Chiller

DX or dry-expansion chillers are commonly used in mid-sized refrigeration plants. Their distinguishing feature is complete evaporation, with a little superheating, of the refrigerant flowing inside the tubes of a heat exchanger. The most common heat exchanger configuration is the multipass shell-and-tube type heat exchanger with the refrigerant evaporating inside tubes and in the process cooling the coolant (usually water) circulating around the tubes inside the shell. We have already seen in Section 4.1.4 the method of simulating the performance of such chillers. Here we shall discuss the approach followed by Dhar [4] to optimize its design.

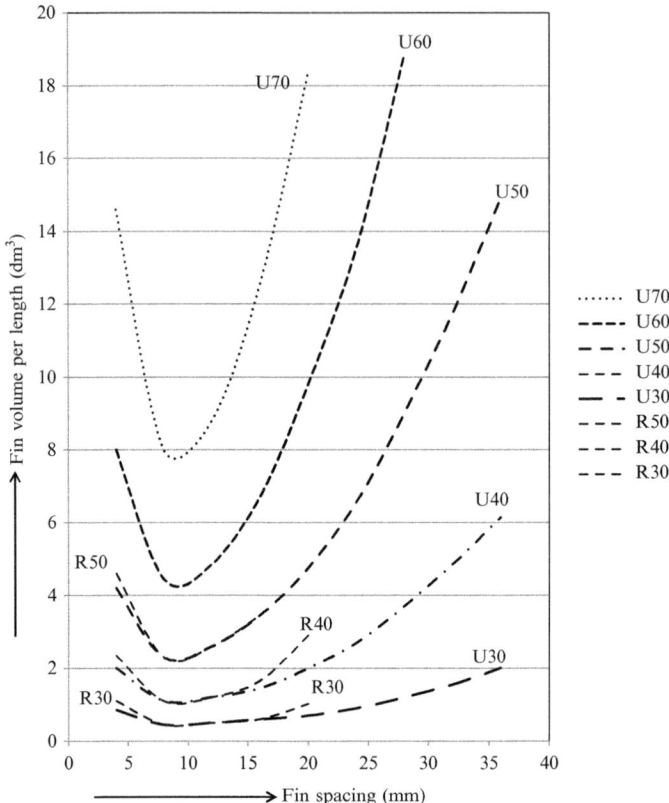

Fig. 9.14 Optimal fin volume for fins on cylindrical surface.

The first step in any optimization problem is to identify the variables to be optimized. Here all the three types of variables mentioned in Section 7.1 are present. Thus the continuous variables are: the shell diameter (d_s), the shell length (H), the Baffle pitch (BP), the tube pitch (P_T), and the baffle cut (BC); the discrete variables are the number of passes, n; the tube diameter, d_t, the number of baffles N_B, and the number of tubes in various passes, Tubes (*i*); and the logical variables are the tube arrangement (in-line square, staggered square or triangular), direction of flow of the refrigerant relative to that of the coolant, and the type of tubes (bare, internally finned, etc.)

All these variables are not independent, and for many of them "optimal" interrelationships have been derived from best engineering practices. Further there is the issue of minimizing the inventory in the manufacturing unit. Keeping in view all these, the tube diameter d_t and the tube pitch P_T as also the tube arrangement have been kept unaltered. Further since any change in the number of passes would drastically alter the complete simulation program (see Section 4.1.4), the number of passes n has also been kept as a logical variable, to be changed outside the main algorithm. Further, it is an established industrial practice that to achieve the superheating of the vapors in the minimum length, the relative direction of flow of the refrigerant and the coolant should be such that the refrigerant vapors leave the chiller at the "warmest" place, that is, the coolant entry position. Thus the refrigerant flow direction is fixed in accordance with the number of passes. We are thus left with only one logical variable, namely the type of tubes—finned or unfinned, besides the variables d_s, H, BP, BC, N_B, $Tubes(i)$.

The second step in formulating the optimization problem is to identify the constraints on various variables. Since the chiller is just one component of the complete refrigeration system which operates in a closed cycle, any change in its design will directly influence the performance of the whole system. Therefore, while optimizing the chiller it has to be ensured that the changes in its design are carried out subject to suitable constraints which ensure that not only does its refrigeration capacity remain unaltered, but the performance of the other parts of the system is not affected.

In order to achieve this with minimum changes in rest of the system, the evaporator exit conditions have been constrained to be the same as in the original design (see Fig. 9.15). This ensures that there are no changes in the performance of the components downstream. The only change needed is in the expansion valve whose setting has to be changed in tune with the changes in the pressure drop in the optimized chiller. If the pressure drop in the optimized chiller is more than that in the original design the expansion valve setting needs to be altered to ensure that the refrigerant flow rate through the valve remains unaltered (or it has to be replaced with another valve) in spite of reduced pressure drop through it (i.e., $p_3 - p_4'$ instead of $p_3 - p_4$). Any such change would not affect the performance of the system and also not alter the cost of the expansion valve significantly.

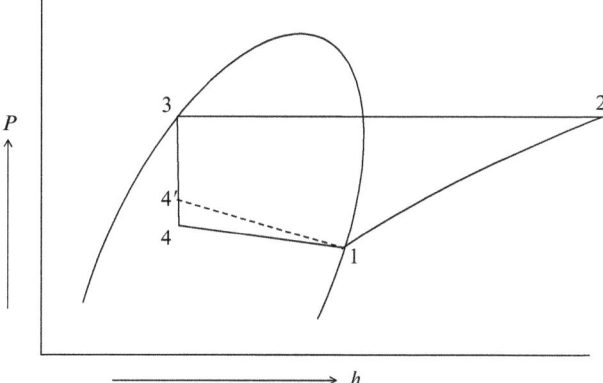

Fig. 9.15 Operating cycle.

These constraints ensure that the performance of the refrigeration system is not changed due to change in the design of the chiller. However, the cost of pumping coolant in the chiller can still vary with chiller design. To avoid this an additional constraint has been put: the shell-side pressure drop remains the same as in the original design.

Another constraint arises from the geometric compatibility, viz. to ensure that "N_B" baffles spaced a distance "BP" apart can be accommodated within the shell of length "H." This demands:

$$H \geq BP * (N_B - 1) + N_B * BT \tag{9.79}$$

where BT is the thickness of each baffle. Since this condition will be satisfied even by a ridiculously small value of N_B which would result in nonuniform distribution of coolant in the shell-side this inequality has been replaced by a more practical condition:

$$N_B = Int(H/BP) - K \tag{9.80}$$

where the function Int denotes integer value, and K is a constant which allows for the length of the unbaffled portion of the chiller and the thickness of the baffles.

Another "geometric" constraint arises from the requirement that the total number of tubes in all the passes should be less than or equal to the maximum number of tubes which can be accommodated within the shell diameter.

Again, in conformity with the industrial practice it has been assumed that the number of tubes in each pass is less than or, at the most, equal to the number of tubes in the succeeding pass,[2] that is,

$$\text{Tubes}(i - 1) - \text{Tubes}(i) \leq 0 \quad i = 2, \dots, n \tag{9.81}$$

Lastly we have the usual constraints that various variables have nonnegative values. Suitable values have also been put for the upper and lower bounds of various variables.

Thus, to sum up the constraints are:

(i) The refrigerant exit conditions are the same as in the original design.
(ii) The shell side coolant pressure drop is the same as in the original design.
(iii) The chiller cooling capacity is also unaltered.
(iv) Eq. (9.80).
(v) Inequality (9.81).
(vi) The total number of tubes in the chiller is less than (or equal to) the maximum number of tubes that can be accommodated within the shell.
(vii) The values of all the variables are within the specified limiting values.

---

[2] This is necessitated by the increasing refrigerant volumetric flow rate as it evaporates inside the chiller tubes.

The third step in formulating an optimization problem is to identify the objective function. It is apparent from the constraints put on the optimum design that the operating cost of the system as also the chiller would not be influenced by the changes in the chiller design. Therefore, the objective function has been taken as the initial cost of the chiller, which needs to be minimized. To express this in terms of the design variables it has been assumed that the total cost of the chiller is proportional to the total surface area of the copper tubes used in it. This approximation is justified since the cost of copper tubes is a major component of the chiller cost and the cost of the steel shell and other accessories is also indirectly dependent upon the number, length and the diameter of the tubes.

Thus the objective function can be written as:

$$F = \left\{ \sum_{i=1}^{n} \text{Tubes}(i) \right\} d_t \cdot H \qquad (9.82)$$

The complete optimization problem can now be stated as:

Minimize F (Eq. 9.82).

Subject to constraints (i)–(vii) above.

Design variables: d_s, H, BP, BC, N_B, Tubes (i).

Now since four of the constraints are equality constraints these have been directly incorporated so as to reduce the number of variables.

Thus the first constraint has been incorporated by using this as a boundary condition in the chiller simulation procedure (Section 4.1.4). The constraints (ii) and (iv) have been satisfied by iterating for BP (using Eq. (9.80) to find the number of baffles) so that the shell side pressure drop equals the rated value. Similarly, the constraint (iii) has been taken care of by iterating every time for the chiller length H so that the chiller capacity as calculated by the simulation procedure discussed in Section 4.1.4 equals the rated value.

Thus all the equality constraints are eliminated and the optimization problem reduces to finding the optimal values of variables d_s, BC, Tubes (i) so as to minimize the function F (Eq. 9.82) subject to the inequality constraints (v)–(vii). Since some of the variables can have only integer values, this is a mixed–integer nonlinear programming problem.

This method has been used to optimize the design of the industrial 40 *TR* chiller with eight refrigerant passes. The optimization was done using Box's complex method modified suitably to enable it to handle mixed integer programming problems [4]. Table 9.5 compares some of the alternative designs obtained during the search [Dhar and Arora [5]] with the existing design (S.No. 1). It can be seen that some of the new designs are better (i.e., have lower F values) than the existing design. The new design at S.No. 3 above has a F values is 15% less than that for the existing design.

Table 9.5 Optimum designs of 40 TR industrial DX-Chiller

	Variables											Pressure at entry (bar)
	Tubes (*i*)								d_s(m)	BC(m)	F (m²)	
S.no.	1	2	3	4	5	6	7	8				
1	12	16	20	24	30	32	32	34	0.391	0.073	8.437	5.35
2	12	16	20	21	23	24	24	25	0.391	0.073	7.362	5.395
3	12	19	20	21	22	23	23	23	0.374	0.078	7.112	5.385
4	13	18	20	22	23	24	24	24	0.376	0.0753	7.257	5.373
5	14	17	21	22	23	24	24	24	0.374	0.079	7.2	5.366
6	13	18	21	22	22	23	23	23	0.3785	0.0759	7.175	5.375

Table 9.6 Optimum 4-pass designs of 40 TR industrial DX-Chiller

	Variables						Objective function (m²)	Refrigerant pressure at entry (bar)
	Tubes (*i*)				d_s(m)	BC (m)		
S.no.	1	2	3	4				
1	30	40	40	48	0.391	0.073	8.687	5.225
2	31	35	35	38	0.386	0.0583	8.075	5.225
3	15	20	30	35	0.366	0.0507	7.281	5.325
4	12	18	25	30	0.305	0.1219	6.581	5.385
5	17	19	19	21	0.294	0.1045	5.937	5.375
6	17	19	19	21	0.327	0.1045	5.831	5.371
7	18	18	18	21	0.318	0.1051	5.8	5.383
8	14	16	16	22	0.327	0.099	5.74	5.451

Optimization of the design of this chiller has also been done for a different value of the number of passes, *n*. While studying the detailed simulation results of the 8-pass designs it was noted that even before entering the sixth pass the refrigerant is completely vaporized and that in part of subsequent passes the superheated vapor is actually being cooled. Therefore, optimum chiller designs with a smaller number of passes, $n = 4$, have also been obtained. Table 9.6 shows some of the results.

It can be seen that the best 4-pass chiller design has the objective function value of $F = 5.74/m^2$ as against the value of $F = 8.437$ of the original design, which implies a saving of $\approx 32\%$ in the cost of chiller.

Dhar [4] also reports the effect of another logical variable—the type of tubing, that is, plain or finned—on the optimum design of the chiller. Four different types of internally finned tubes, shown in Fig. 9.16, have been considered. In order to assess the savings that could be obtained by employing these finned tubes, the base chiller design has been taken as the best bare tube design obtained earlier (viz. S.No. 8 of Table 9.6). The results obtained are given in Table 9.7.

It can be seen that the chiller designs employing finned tubes require much less surface area than the base tube design, the decrease being as high as 52.8% with fin

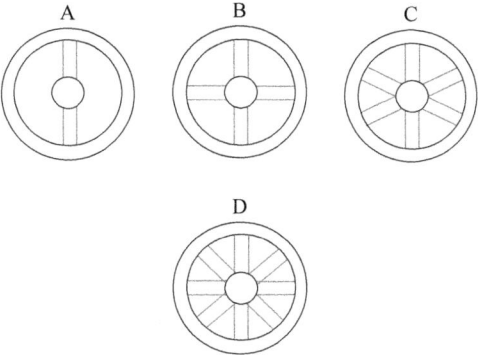

Fig. 9.16 Internally finned tubes for DX-chiller.

Table 9.7 Optimum 40TR designs with finned tubes

Type of fin	Objective function F (m²)	Refrigerant pressure at entry (bar)	Pressure drop across valve (bar)	Sat. temp at chiller entry (°C)	Temp. diff. (coolant-ref.) at entry (°C)
No fin	5.74	5.451	11.168	2.827	8.284
A	3.6567	5.687	10.932	4.167	6.944
B	3.0064	5.825	10.794	4.924	6.187
C	2.712	6.047	10.572	6.123	4.988
D	3.0152	6.475	10.144	8.349	2.762

design C (Fig. 9.16). This is also expected since, besides increasing the heat transfer coefficient, these fins also provide larger heat transfer area per unit tube length. These two factors completely overwhelm the detrimental influence of increasing pressure drop and consequent drop in the temperature difference between the refrigerant and the coolant as indicated in the last column of Table 9.7.

However, this detrimental influence becomes dominant in the fin design D as the temperature difference at chiller entry drops to 2.762°C from the value of 4.988°C with fin design C. These values are in contrast to the chiller–entry temperature difference of 8.284°C for the optimum chiller using plain tubes. As a result the objective function value with fins "D" is higher than with fins "C", which is the best amongst the finned tubes investigated.

9.2.3 Flooded Chiller

Flooded chillers are widely used in refrigeration industry, especially for large refrigeration capacity applications. Following the general approach outlined above for DX-Chillers, Saraf and Dhar [6] have carried out the optimum design of an industrial flooded liquid cooler. As before, to decouple the chiller from the rest of the components of the refrigeration plant, the chiller exit conditions and its cooling capacity have been kept the same as in the original design. Since there is no pressure drop on the refrigerant side of the flooded chiller, this constraint also maintains the rated pressure difference across the float value. Thus the mass flow rate of the refrigerant entering the chiller and that leaving it remain unaltered. The other constraints, as also the variables and the objective function to be optimized, are discussed below.

The primary design variables are the tube length and the number of tubes in various passes. All other variables like the number of passes, the tube diameter, type of tubes (finned or bare), and tube arrangement have been considered as "logical" variables. The optimum values of these logical variables are found through the case study approach, that is, finding out for each value of a logical variable, the values of the other design variables giving the optimum design and then choosing the best amongst all these optimum designs.

Unlike in the case of DX–Chiller, Saraf and Dhar have not put any constraint on the coolant pressure drop. To account for the impact of varying pressure drop on the system performance, the objective function has been taken as the sum of the first cost and the operating cost over the life of the equipment. The first cost is taken proportional to the total cost of copper tubes. The operating cost is calculated on the basis of the cost of pumping the coolant through the chiller. This gives the total cost T as:

$$T = \left\{ \sum_{i=1}^{Np} \text{Tubes } (i) \right\} HC_t + K \cdot L_H \dot{m}_w \cdot \Delta P C_k \tag{9.83}$$

where C_t is the cost of copper tubes (and other accessories) per unit tube length; C_k is the cost of 1 kWh of power consumption; K the unit conversion factor to get the power consumption in kilowatt, L_H the number of actual operating hours during the entire expected life of chiller; $\dot{m}_w$ the mass flow rate of the coolant; and ΔP the pressure drop in coolant while flowing thorough the chiller. To combine the cost parameters, the objective function has been defined as:

$$F = \frac{T}{C_t} = \left\{ \sum_{i=1}^{Np} \text{Tubes } (i) \right\} H + (8760 \cdot K)\theta \dot{m}_w \cdot \Delta P \tag{9.84}$$

$$\text{where } \theta = Ly\frac{C_k}{C_t} \text{ and } L_H = 8760 \cdot L_y \tag{9.85}$$

θ is thus the factor which takes into account the actual operational life of the chiller in years, L_y, and the relative cost of unit power consumption to the cost of copper tubing per unit length. Since the velocity of the coolant flowing through the chiller tubes can vary, a constraint has been put on its value to limit erosion/corrosion as:

$$V \leq 3 \, \text{m/s} \tag{9.86}$$

Besides this, we have the usual constraints on the upper and lower bounds for various variables (x_i). Thus, to sum up, the constraints are:

$$h_i = h_{iR}$$
$$Q \geq Q_R$$
$$V \leq 3$$
$$L(i) \leq x(i) \leq H(i) \quad L = 1, \ldots, n \tag{9.87}$$

where subscript R indicates rated values, symbol h_i stands for refrigerant enthalpy at inlet to the chiller, Q the chiller capacity in TR, $L(i)$ is the lower limiting value, and $H(i)$ is the upper limiting value of the ith variable. The constraint on chiller capacity is converted into an inequality constraint since, as discussed above in subsection, the optimization method developed can be used only with inequality constraints. It is obvious from the nature of the problem that the lowest cost will occur at the lowest value of Q which equals Q_R.

   This procedure has been used to find the optimum design of an industrial $120\,TR$ chiller manufactured by an Indian industry. The capacity of the chiller designs evolved during optimization has been found using the simulation procedure presented in Section 4.1.4. Table 9.8 shows the optimum designs for various values of the relative cost factor θ.

   It can be seen that, as expected, the optimum design is strongly dependent on the value of the relative cost factor θ. Thus, for value of $\theta \simeq 0.0305$, the existing design is also the optimum design, but as θ increases (say, due to increase in the energy costs), the optimum design has larger number of shorter tubes so as to reduce the coolant side pressure drop and thus restrict the increase in the pumping cost. Thus for $\theta = 0.1524$,

Table 9.8 Optimum flooded chiller designs

		Design variables			Objective function representing			Coolant
		No. of tubes		Length	First	Pumping	Total	ΔP
S.no.	θ^a	Pass 1	Pass 2	(m)	cost[b]	cost	cost	(kN/m^2)
1	0.0244	102	105	3.876	802.29	116.67	918.95	21.81
2	0.0305	114[c]	114	3.639	829.59	116.09	945.68	17.36
3	0.0366	116	119	3.567	838.36	129.81	968.17	16.17
4	0.0488	128	132	3.345	869.6	136.68	1006.28	12.78
5	0.0610	137	139	3.220	889.24	148.62	1037.86	11.11
6	0.0762	147	152	3.066	916.78	154.47	1071.25	9.24
7	0.0914	155	160	2.971	935.70	164.28	1099.98	8.190
8	0.1524	185	190	2.678	1004.39	183.56	1187.95	5.49

[a] θ in: m year/kWh.
[b] Cost in: Rs./(Rs./m).
[c] Existing design of the chiller.

Table 9.9 Effect of number of passes on optimum flooded chiller designs

S.no.	θ	Type of design	Optimum design variables						Obj. function			Coolant
			No. of tubes in pass				Σ	Length (m)	First cost	Pump. cost	Total cost	ΔP (kN/m²)
			1	2	3	4						
1	0.0305	2 pass	114	114	–	–	228	3.639	829.59	116.09	945.68	17.36
2	0.0305	4 pass	113	114	115	118	460	1.81	831.97	127.1	959.07	19.02
3	0.0610	2 pass	137	139	–	–	276	3.22	889.24	148.62	1037.86	11.1
4	0.0610	4 pass	139	140	143	144	566	1.59	897.82	157.98	1055.6	11.77
5	0.0914	2 pass	155	160	–	–	315	2.97	935.7	164.28	1099.98	8.20
6	0.0914	4 pass	161	162	164	167	654	1.45	948.85	170.65	1119.5	8.53
7	0.1524	2 pass	185	190	–	–	375	2.68	1004.39	183.56	1187.95	5.49
8	0.1524	4 pass	191	193	196	198	778	1.35	1017.52	194.08	1211.60	5.79

the optimum design has 375 tubes of 2.678 m length as against the existing design which has 228 tubes of 3.639 m length.

The effect of various logical variables like the fin design, number of passes, tube diameter, etc., has also been studied by Dhar and Saraf [7]. Table 9.9 shows the influence of number of tube passes on the optimum design.

Two values of the tube passes—2 and 4—have been considered for all values of θ. It is seen that the optimum designs for four pass chillers have, in comparison with the corresponding two pass designs, nearly double the number of tubes of half of the length. Thus the first cost of the chiller does not change appreciably with change in the number of passes. The slight increase for 4-pass designs can be attributed to the influence of increase in the coolant side pressure drop (due to expansion, contraction, and change in flow directions at more places). Nevertheless, the increase in the total cost with 4-pass designs is marginal, the maximum difference being of the order of 2% for $\theta = 0.1524$.

To study the effect of another "logical" variable, viz. the design of finned tubing, some changes are needed in the formulation of the objective function. To appreciate this, let us consider the cross-sectional view of the integrally finned tubes used in flooded chillers, shown in Fig. 9.17.

Various dimensions of the finned tube, viz. d_i, d_o, d_f, t_1, t_2 shown in this figure are bound to influence the chiller performance through effect on the heat transfer coefficient and the external surface area. But these would also influence the cost per unit length of the tube. If we assume that this cost is proportional to the weight (or the volume) of the per unit length, we get the following relationship between the cost per unit length of the existing tube (parameters indicated by subscript e) and the tube with new dimensions (subscript n):

$$\frac{C_{tn}}{C_{te}} = R = \frac{d_f^2 t_1 + d_o^2 t_2 - d_i^2 f_p}{d_{fe}^2 t_{1e} + d_{oe}^2 t_{2e} - d_{ie}^2 f_{pe}} \tag{9.88}$$

where

$$f_p = t_1 + t_2 \tag{9.89}$$

$$f_{pe} = t_{1e} + t_{2e} \tag{9.90}$$

In view of this, the objective function for optimum design, defined by Eq. (9.84), needs to be modified to account for the change in the cost per unit length, C_t as given by Eq. (9.88). Since for any set of changed dimensions the cost ratio R (Eq. 9.88) remains constant (being independent of the design variables) the only change required is that in the objective function definition (Eq. 9.84), the relative cost parameter θ should be replaced by θ_n defined as:

$$\theta_n = \frac{\theta}{R} \tag{9.91}$$

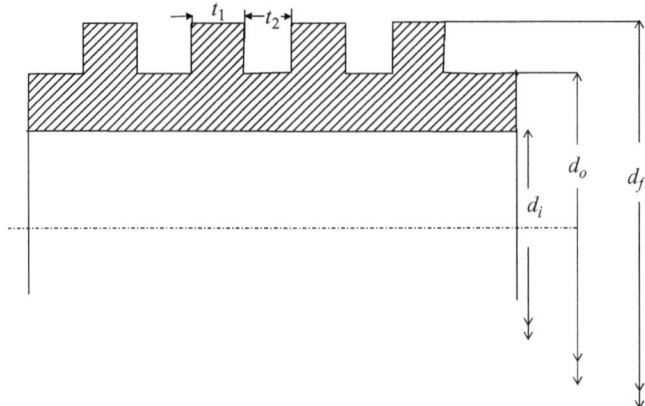

Fig. 9.17 Finned tubes dimensions.

Thus the new objective function is

$$F = \frac{T}{C_{tn}} = \left\{ \sum_{i=1}^{Np} \text{Tubes } (i) \right\} H + (8760 \ K) \cdot \theta_n \dot{m}_w \Delta P \tag{9.92}$$

However, the various optimum chillers corresponding to different fin designs cannot be compared on the basis of the above defined objective function in view of varying C_{tn}. This can be done by referring to the objective function defined with respect to a fixed value of C_t that is, cost of copper tubing per unit length. Using this fixed value as that corresponding to the existing design of the finned surface (C_{te}), we can do the comparison on the basis of function F' defined as:

$$F' = \frac{T}{C_{te}} = R * F \tag{9.93}$$

Dhar and Saraf [7] present the results of a detailed study of the influence of various dimensions of the finned surface considered one by one, as also by varying them in proportional manner. Some of these results are presented below.

Table 9.10 shows the results obtained for various values of tube inner diameter d_i with only d_o changed proportionally, that is, $\left[d_o = \left(\frac{d_o}{d_i} \right)_{exist} \cdot d_i \right]$ and d_f being kept the same as in the existing design. It is seen that there exists an optimum value of $d_i = 10.2$ mm for which the total cost is the minimum. This can be attributed to the variety of changes brought about in chiller design and total cost due to change in the tube inner diameter. Thus as d_i falls, so does the cost per unit length of tube (i.e., R falls); the coolant side pressure drop tends to increase rapidly and the optimization procedure tends to compensate for it by increasing the number of tubes and reducing their length; since d_f is

Table 9.10 Effect of tube inner diameter on optimal chiller design when only d_o is changed proportionally

| | | | | | No. of tubes in pass | | Length | Obj. function | | | Coolant |
S.no.	d_i (mm)	d_o (mm)	R (m y/kWh)	θ	1	2	(m)	First cost	Operating cost	Total cost	ΔP (kN/m²)
1	15.2	18.7	1.031	0.0296	154	168	4.54	1506.01	49.47	1555.48	7.35
2	13.8	17.0	1.0	0.0305	114	114	3.64	829.59	116.09	945.68	17.36
3	12.4	15.3	0.972	0.0314	130	133	2.75	702.1	114.9	817.0	17.26
4	10.2	12.5	0.932	0.0326	185	187	1.90	658.45	116.4	774.89	17.46
5	7.6	9.4	0.897	0.0338	334	337	1.22	734.56	106.18	840.74	15.89
6	5.1	6.3	0.872	0.0351	400	429	1.02	736.64	414.75	1149.39	61.98[a]

[a] Optimization stopped prematurely since the tube length reached the lower limit.

Table 9.11 Effect of tube inner diameter on optimal chiller design when both d_o and d_f are changed proportionally

| | | | | | | No. of tubes in pass | | Length | Obj. function | | | Coolant |
S.no.	d_i (mm)	d_o (mm)	d_f (mm)	R (m y/kWh)	θ	1	2	(m)	First cost	Operating cost	Total cost	ΔP (kN/m²)
1	15.2	18.7	21.0	1.209	0.0253	87	92	4.01	868.54	125.19	993.73	18.73
2	13.8	17.0	19.0	1.0	0.0305	114	114	3.64	829.59	116.09	945.68	17.36
3	12.4	15.3	17.0	0.811	0.0375	143	148	3.31	781.79	113.14	894.93	16.97
4	10.2	12.5	14.0	0.543	0.0561	237	244	2.72	708.99	99.66	808.05	14.91
5	7.6	9.4	10.5	0.305	0.100	463	476	2.12	608.11	94.13	702.24	14.12

kept constant, reduction in d_i causes appreciable increase in the finned surface area per unit length and there also is the impact of d_i on the heat transfer coefficients. The interplay of all these factors causes the total cost to fall as the d_i is decreased up to a limit beyond which the cost tends to increase.

The results change substantially if we do not keep d_f constant, but change it also in a manner to retain the overall geometrical proportions of the finned surface. Table 9.11 shows the results of this optimization exercise. It is now seen that the total cost decreases monotonously with d_i. The main reasons for this change from the earlier results is that when d_f also decreases proportionally (in contrast to the above case where it was held constant) the cost of copper tubing per unit length decreases very rapidly.

Thus from Tables 9.10 and 9.11 we can see that for $d_i = 7.6$ mm the R value drops from 0.897 in the former to only 0.305 in the latter case. Consequently, during optimization it is possible to obviate the effect of increasing running costs with decreasing d_i by selecting designs with much larger number of these cheaper tubes (939 tubes in Table 9.11 as against 671 in Table 9.10 for $d_i = 7.6$ mm).

Following a similar approach the influence of varying other parameters like the base value of relative cost factor θ, the operating parameters like the coolant inlet temperature, the coolant mass flow rate, etc., on the optimum chiller design have also been presented by Dhar and Saraf [7]; details can be seen in that monograph.

9.2.4 Refrigerant Condenser

Dhar and Saraf [7] also present the results of optimizing the design of a typical refrigerant condenser. The condenser taken up for detailed analysis, Fig. 6.5, is a component of an industrial 120 TR refrigeration plant.

Its simulation using the detailed process model developed was presented earlier in Chapter 6. To optimize its design without influencing the performance of other components, they follow a strategy similar to that adopted for chiller and described in Sections 9.2.2 and 9.2.3. Following similar steps, the condenser optimization problem can be stated as:

$$\text{Minimize } F = \frac{\text{Total cost}}{C_{ct}} = \sum_{i=1}^{Np} Tubes(i)H + (8760.K)\theta \dot{m}_w \Delta P \qquad (9.94)$$

Subject to constraints:

Condenser pressure, vapor entering temperature,
coolant mass flow rate and temperature at inlet, = Corresponding rated values
condensate flow velocity over subcooler

$$\text{Degree of subcooling} \geq \text{Rated value}$$

$$\text{Total heat transfer} \geq \text{Rated value}$$

$$\text{Velocity of coolant in tubes} \leq 3$$

$$L(i) \leq \text{Variables }(i) \leq H(i) \quad i = 1, 2, \ldots, n$$

The variables to be optimized are, as in the case of chillers, viz. number of tubes in various passes and in the subcooler, and the tube length (H). The other design parameters are considered as "logical" variables to be optimized by a case study approach.

As in the case of chillers, this problem is also a mixed–integer nonlinear programming problem and been solved by using modified Box's method. A few typical results illustrating the effect of energy costs are presented in Table 9.12.

Once again, it can be seen that the optimum designs are strongly dependent on the value of relative cost factor θ. Thus the optimum design for $\theta = 0.0305$ has 303 tubes of 2.01 m length, while that for $\theta = 0.061$ has 379 tubes of 1.73 m length. It is seen that as the relative cost factor θ rises (i.e., the cost of electric power relative to that of copper tubes rises), the optimal designs have:

(i) Lower coolant side pressure drop.
(ii) Increased number of tubes in each pass of reduced length [both of which enable (i) above].
(iii) Higher fixed and the pumping costs.

As expected, these results are qualitatively of the some kind as those obtained during flooded chiller optimization. The effect of other parameters like number of passes, finned tube dimensions, the coolant temperature and flow rate etc. have also been presented by Dhar and Saraf [7]. The results are generally of the same kind as obtained in chiller optimization, and the details can be seen in the monograph.

9.3 OPTIMUM DESIGN OF THERMAL SYSTEMS

In this section we shall present comprehensive studies on optimum design of typical thermal systems like a refrigeration system, a power plant and a desiccant cooling system.

9.3.1 Refrigeration System

In Chapter 6 we discussed in detail the simulation of a 120-TR (nominal) capacity liquid chilling plant (Fig. 6.1) manufactured by a prominent industry in India. Herein we shall present the methodology of optimizing the design of the whole plant, following the general approach outlined in Chapter 7.

The Objective Function
The total cost of the plant over its entire life span is taken as the objective function to be minimized. This involves both the initial capital cost of various components and the operating cost over the entire life span.

Table 9.12 Effect of energy cost on the optimum design of condenser

S.No.	θ (m y/kWh)	Op. cost: existing design	Optimum design variables								in sub cooler	Length (m)	Obj. function			Coolant ΔP (kN/m²)
			No. of tubes in pass										First cost	Op. cost	Total cost	
			A1	A2	A3	B1	B2	B3								
1	0.0305	174.63	49	48	48	59	38	42		19	2.01	511.24	92.3	663.54	16.9
2	0.0610	349.26	62	62	71	66	51	44		23	1.73	614.25	112.05	726.3	10.2
3	0.0914	523.89	85	81	82	67	59	48		27	1.55	653.09	116.84	769.94	7.1
4	0.1219	698.52	83	76	83	89	69	61		30	1.47	676.4	131.05	807.45	6.0
5	0.1524	873.15	86	85	92	100	82	70		33	1.37	706.55	131.7	838.2	4.8

Not much saving is expected in the initial cost of the compressor and the expansion device by optimizing their "thermal" design parameters. Further the operating cost of the compressor is influenced strongly by its operating parameters like suction and discharge pressures and temperatures, and the refrigeration capacity, but very weakly by its dimensions like bore/stroke/clearance, etc. So, the compressor design and the design of the expansion valve have been kept unaltered during the optimization of the liquid chilling plant.

The objective function thus includes the first and the operating costs of the chiller and the condenser, the fixed cost of the liquid vapor heat exchanger (LVHE) and the operating cost of the compressor. Further, the cost of the LVHE is extremely small in comparison to the costs of the chiller and the condenser. Moreover, since the use of LVHE in R-22 refrigeration systems results in a fall in the COP due to super heating of the vapors entering the compressor (see Table 9.1), it is obvious that to get the lowest compressor operating cost, we should actually do away with LVHE. However, LVHE is necessary for safe operation of the compressor during sudden drop in load to prevent refrigerant liquid droplets from entering the compressor. Therefore, we have allowed the existing LVHE to be used even in the optimized plant. The objective function has therefore been defined as (cf. Eqs. 9.84 and 9.94):

$$F = \frac{\text{Total cost}}{\text{Cost of heat exchanger tubing per unit length}} \quad (9.95)$$

$$= \frac{F_{evap} + F_{cond} + P_{evap} + P_{cond} + P_{comp}}{C_c}$$

$$= \left[\sum_{i=1}^{N_{pass}} \text{Tubes } (i), \ H \right]_{evap} + \left[\sum_{i=1}^{N_{pass}} \text{Tubes } (i) \ H \right]_{cond}$$

$$+ 8760 \cdot K\theta \left(\dot{m}_{evap} \cdot \Delta P_{evap} + \dot{m}_{cond} \cdot \Delta P_{cond} + P_{comp} \right) \quad (9.96)$$

where θ, the relative cost factor is defined by Eq. (9.85).

The Variables

As the designs of the compressor, expansion device, and the LVHE have not been altered, the design variables to be optimized are obviously those pertinent to the chiller and the condenser.

Following the same approach, as discussed above in Sections 9.2.3 and 9.2.4, 11 design variables $[x(i)]$ for the optimization of the complete system have been identified, viz. the chiller length (1), the number of tubes in each of its two passes (2–3), the condenser length (4), the number of tubes in its six passes (5–10), and the numbers of tubes in the subcooler (11). All other variables like the tube diameters, the number of passes, types of fins on the tubes etc. have been considered as "logical variables" (cf. Sections 9.2.3 and 9.2.4).

The Constraints

It is necessary that the optimized system should be capable of chilling the required flow rate of water over the specified range of temperatures. Further to decouple the refrigeration plant from the cooling tower, the temperature, and the flow rate of water cooling the condenser are constrained to be the same as in the original design.

As in the case of optimization of individual components the coolant flow velocities in both the chiller and the condenser are restricted to be less than 3 m/s, the recommended limit from erosion and corrosion considerations. For the sake of simplicity the condensate velocity over the subcooler tubes is also kept the same as in the original design.

Finally, we have the regional constraints to ensure that the values of various design variables are lying within the prescribed limits.

To sum up the constraints can be written in mathematical form as:

$$Q_{chiller}\{x(i), i = 1, 2, \ldots, 11\} \geq Q_{chiller, rated} \tag{9.97}$$

$$\dot{m}_{chiller} = \dot{m}_{chiller, rated} \tag{9.98}$$

$$\dot{m}_{cond} = \dot{m}_{cond, rated} \tag{9.99}$$

$$T_{i,cond} = T_{i,cond(rated)} \quad \{i: \text{water inlet}\} \tag{9.100}$$

$$Q_{cond}\{X(i), i = 1, 2, \ldots, 11\} \leq Q_{cond, rated} \tag{9.101}$$

$$V_{chiller, water} \leq 3 \text{ m/s} \tag{9.102}$$

$$V_{cond, water} \leq 3 \text{ m/s} \tag{9.103}$$

$$L_{(i)} \leq x(i) \leq H_{(i)} \quad i = 1, 2, \ldots, 11 \tag{9.104}$$

Optimization Results

As in optimization of the chiller and the condenser considered separately (Sections 9.2.3 and 9.2.4), this problem too falls in the category of mixed integer programming. Its solution has also been obtained by using the modified Box Complex method developed by Dhar [4]. The detailed results obtained for three values of θ are presented in Tables 9.13–9.15.

Once again we note the strong dependence of the optimum designs of the chiller and the condenser, as also of the operating temperatures, on relative cost factor θ. The difference in the total cost of the optimum and the original designs increases with increasing values of θ. Thus existing plant costs, respectively, 4% and 5.85% higher than the corresponding optimum designs for θ values of 0.0305 and 0.0915, respectively.

Table 9.13 The original and the optimum designs of the chilling plant for varying energy costs

		Chiller design variables			Condenser design variables								
		No. of tubes in pass		Length	No. of tubes in pass						in sub	Length	
S.no.	θ (m y/kWh)	1	2	(m)	1	2	3	4	5	6	cooler	(m)	
1	–	114	114	3.64	36	36	36	18	36	36	14	2.49[a]	
2	0.0305	95	92	3.31	33	34	33	47	54	36	14	4.31	
3	0.0610	108	83	3.2	62	114	84	27	60	46	24	3.60	
4	0.0914	109	82	3.14	69	115	80	34	64	53	31	3.79	

[a] The original design.

Table 9.14 The total and the constituent costs of the original and the optimum designs of the chilling plant

		Fixed, operat., and total cost (original design)						Fixed, operat., and total cost (optimal design)					
S.no.	θ	F_{ce}	F_{cc}	P_{ce}	P_{cc}	P_{comp}	T_{cs}	F_{ce}	F_{cc}	P_{ce}	P_{cc}	P_{comp}	T_{cs}
1	0.0305	829.59	527.71	117.04	154.16	23,753.74	25,382.2	619.22	1082.88	153.36	196.71	22,354.34	24,406.52
2	0.0610	829.59	527.71	234.10	308.34	47,507.48	49,407.2	611.08	1501.76	299.01	145.30	44,452.21	47,009.36
3	0.0914	829.59	527.71	351.11	462.47	71,261.22	73,432.1	599.73	1691.12	444.48	197.28	66,437.94	69,369.80

Table 9.15 Operating conditions in the original and the optimum designs

		Compressor	Coolant pressure drop[a]		Saturation temp.		Subcooling
S.no.	θ (m y/kWh)	BHP	Chiller	Condenser	Chiller	Condenser	in condenser
1	–[b]	119.25	0.177	0.298	3.51	42.83	7.23
2	0.0305	112.23	0.232	0.380	2.47	39.32	6.63
3	0.0610	111.58	0.226	0.141	2.34	38.99	6.14
4	0.0914	111.18	0.224	0.127	2.24	38.81	6.30

[a] In kgf/cm^2.
[b] Original design.

From these results it is seen that as θ increases:

1. The fixed cost of the condenser increases significantly while that of the chiller reduces slightly.
2. The operating costs of the chiller and the compressor increase while that of the condenser is minimum at $\theta = 0.0610$.
3. The saturation temperatures, both of the chiller and the condenser reduce.
4. The power consumption of the compressor and the coolant pressure drops, both in the chiller and the condenser reduce, the reduction being more significant in case of the condenser.
5. The number of tubes in the condenser increases significantly while there is very small change in the number of tubes in the chiller.

We can also get an insight into this optimization exercise on physical grounds. Thus as θ increases, the energy costs become more significant and the optimum design would be the one needing lesser power consumption even if it entails a modest increase in the fixed cost. Since the compressor power consumption accounts for more than 90% of the total power consumption, the focus is on reducing it. This is achieved by reducing significantly the condenser saturation temperature and reducing the chiller saturation temperature marginally. This demands a significant increase in the condenser surface area, and therefore in its cost, and a marginal reduction in the chiller surface area and the cost (see Table 9.13). Further it can be seen from this table that the increase in the surface area of the condenser is achieved by increasing the number of tubes, since this also reduces the coolant pressure drop, and hence the coolant pumping cost. On the other hand, the reduction in chiller surface area is achieved by reducing its length, rather than the number of tubes, since the latter strategy would have increased the coolant pressure drop and its pumping cost.

Finally let us see how the optimum designs of the chiller and the condenser differ from those obtained earlier while optimizing the components individually. Table 9.16 brings out this comparison for θ value of 0.0914.

As expected the two sets of optimum designs differ considerably. This is obviously due to the fact that during system optimization, since the largest contribution to the total cost (the objective function) arises from the compressor running cost, those designs (of both the chiller and the condenser) are optimum which reduce this cost; while during individual component optimization the compressor operating cost remains the same as in the original design, viz. $P_{com} = 71261.72$ (see Table 9.14), and the changes in the design are done only to minimize the total (fixed and running) cost of the components, which are much smaller than P_{comp}. While optimizing the chilling plant as a whole, the P_{comp} value decreases substantially to 66437.9 and the resulting increase in the costs of the condenser (see Table 9.14) is much smaller than the decrease in P_{comp}, resulting in a net reduction in the total "cost" of the complete chilling plant.

Table 9.16 Comparison of the system optimal design and individual component optimum designs

	Chiller design variables					Costs		
	Number of tubes in pass		Length	F_{ce}	P_{ce}	T_{ce}		
Case	1	2	(m)					
A[a]	109	82	3.14	599.7	444.5	1044.2		
B[b]	155	160	2.97	935.7	164.3	1100.0		

	Condenser design								Costs		
	Number of tubes in pass						Sub cooler	Length (m)	F_{cc}	P_{cc}	T_{cc}
Case	1	2	3	4	5	6					
A	69	115	80	34	64	53	31	3.79	1691.1	197.3	1888.4
B	85	81	82	67	59	48	27	1.55	653.1	116.8	769.9

[a] System optimization.
[b] Component optimization.

This clearly shows that much larger savings in the total cost of a liquid chilling plant are possible when it is a considered as a complete system, than the savings obtained by optimizing the chiller and the condenser individually.

9.3.2 Combined Cycle Power Plant (CCPP)

In Chapter 6, Section 6.2, we have already discussed the case study of simulation of the combined cycle power plant (Fig. 6.9), using the given design details of its various components. In this section we shall discuss the optimization of this plant.

In a CCPP the air compressor, the gas and the steam turbines, various pumps and valves are standard equipments which are manufactured in various standard sizes only. The optimization of their design is done by the manufacturers using detailed thermal-hydraulic modeling coupled with stress analysis and experimental studies. As far as the CCPP is concerned, the waste heat recovery boiler (WHRB) is the only key equipment which is specifically designed and built for every power plant. WHRB includes a number of heat exchangers and its cost is a significant fraction of the total cost of power plant. There is thus considerable scope for cost reduction of a CCPP by optimizing the design of its WHRB. We shall present the details of optimization exercise based on the paper by Seyedan et al. [8], and the Ph.D. thesis of Seyedan [9]. The details of this plant (Fig. 6.9), as also the method of simulating its performance, have already been presented in Section 6.2.

The Objective Function

Since we are optimizing a "component" of the whole CCPP through which fluids are flowing at rates governed externally, only the capital cost of the boiler has been taken as the objective function. This includes the costs of all the eight heat exchangers within the WHRB (Fig. 6.9). As in other heat exchangers (see Sections 9.2.2 and 9.2.3), here too, the cost of heat exchangers has been taken as a direct function of the total weight of the heat-transfer surfaces. Accordingly the objective function to be minimized is:

$$F = \sum_{i=1}^{N} [N_R N_T V \rho]_i \qquad (9.105)$$

where N_R is the total number of rows in exchanger i, N_T is the number of tubes in a row of exchanger i, V is the volume of tube material in each tube, ρ is the density of tube material (kg/m³), and N is the total number of heat exchangers in the WHRB.

The Variables

The WHRB has eight heat exchangers of different kinds—superheaters, economizers, evaporators, etc. Though theoretically we are free to choose tubes of different diameters, different fin types, and length for each, it would become very onerous to build and install such a boiler, and even more difficult to maintain it in view of wide variety of inventory requirements. Therefore, in conformity with the current engineering design practice, manufacturing considerations and the need for standardization, it has been presumed that the overall configuration of all the heat exchangers is similar. Accordingly, the variables to be optimized have been identified as:

- **(i)** The length of the tubes.
- **(ii)** The number of tubes in a row.
- **(iii)** The transversal pitch between the tubes.
- **(iv)** The longitudinal pitch between the tubes.
- **(v)** The number of fins per unit length.
- **(vi)** The height of fins.
- **(vii)** The thickness of fins.

The values of these design variables have been kept the same for all the heat exchangers, which differ from each other only in the number of tube rows. Again, for the sake of conformity with the existing configuration, the number of tube rows in various heat exchangers has not been altered. The other design variables like the tube diameter and thickness, the fin design, etc., can be considered as "logical" variables, whose "optimum" values are found by repeating the optimization exercise for a few "permissible values" of these variables and then choosing the best amongst the various optima thus obtained. The variable "number of tubes" can take only integer values while the remaining six variables can take any value greater than zero. Thus this optimization problem also is a mixed integer programming problem.

The Constraints

Since changes in the design of the heat exchangers in the WHRB could change the performance of the system, it is necessary to ensure that the net power output does not fall below the rated value, that is,

$$P_{cc} \geq P_{ccr} \tag{9.106}$$

Again, geometric compatibility of the tube bank configuration demands:

$$S_T \geq d_o + 2h_f \tag{9.107}$$

$$S_L \geq d_o + 2h_f \tag{9.108}$$

where d_o is the outer diameter of the heat exchanger tubes, h_f is the fin height, S_T and S_L are the transverse and longitudinal tube pitches.

There is also a need to restrict the velocity of flue gases (V_g) over the heat exchanger tubes to restrain vibrations, noise, and erosion:

$$V_{gmax} \leq (V_{gr} = 14\,\text{m/s}) \tag{9.109}$$

Finally to avoid the search of new designs in obviously infeasible variable-space, suitable upper and lower bounds have been prescribed for all the variables.

Results

Since this problem too has a variable which can have only integer values, it has been solved by using the modified Box's complex method developed by Dhar [4]. The existing design of the heat exchangers is given as the base design, and the algorithm gives "better" system designs which progressively reduce the overall cost of the WHRB.

Table 9.17 shows the distinctive design features of the final optimum design obtained through this search procedure, vis-a-vis the existing design.

It is seen that the optimum design uses more number of tubes of smaller length. The fin thickness is also reduced (as expected) to the limiting value of 1 mm; but the number of fins per meter length is increased. The optimum design results in a reduction of total weight of heat exchangers by about 25%. In absolute terms, for the complete plant of 800 MW capacity with a total of four WHRB's, this implies a reduction of total weight by 918 tons.

Table 9.17 Comparison of the optimum design of WHRB with the existing design

S.no.	Variables	design value	Optimal design value	optimum design with fixed fin thickness
1	Length of tube (m)	18.5	13.4	16.8
2	No. of tubes in a row	98	112	85
3	Transverse pitch (mm)	80	84.4	85
4	Longitudinal pitch (mm)	70	60.0	62.4
5	No. of fins/meter	150	168.2	170.0
6	Height of fins (mm)	15	14.91	15.3
7	Thickness of fin (mm)	1.3	1.0	1.3
	Total weight (tons)	922.3	692.8	801.3

In order to determine how much of this weight reduction was contributed by reduction in fin thickness (which may be more governed by other considerations like strength and erosion) the procedure was also used to optimize the plant while keeping the fin thickness constant at its existing value of 1.3 mm. The results are shown in the last column of Table 9.17. As expected, the magnitude of reduction in the weight of heat exchangers is lesser, but still at a significant value of $\sim 13\%$.

The effect of tube diameter—a "logical variable"—has been studied by repeating the optimization exercise with four different tubes of "standard dimensions" usually employed in boilers. Table 9.18 shows the results. It is seen that a further weight reduction of about 2% is possible by using smaller diameter ($OD = 25.5$ mm) tubes in the heat exchangers.

Table 9.18 Optimum design of WHRB with four possible tube diameters

S.no.	Variables	$OD = 31.8$ mm $TH = 2.6$ mm	25.5 2.6	38.1 3.406	51 3.403
1	Length of tube (m)	13.4	13.9	18.1	10.8
2	No. of tubes in a row	112	101	81	108
3	Transverse pitch (mm)	34.4	93.8	89.3	119.7
4	Longitudinal pitch (mm)	60.0	65.9	64.9	88.6
5	No. of fins/meter	168.2	169.6	169.9	152.2
6	Height of fins (mm)	14.9	18.5	14.3	17.9
7	Thickness of fin (mm)	1.0	1.0	1.0	1.0
	Total weight	692.8	677.8	876.7	977.3

TH, tube thickness.

9.3.3 Liquid Desiccant-Based Air Conditioning System

In Section 6.3 we had presented the strategy for simulation of a proposed novel air conditioning system based on liquid desiccants like LiBr, LiCl, Glycols, etc. In this section we shall discuss the salient features of optimizing its design as presented by Jain et al. [10]. The main purpose of this exercise was to assess whether this new technology would be commercially feasible and how it compared with conventional vapor compression refrigeration (VCR)-based air conditioning systems.

Referring to its schematic diagram, Fig. 6.19, we note that the proposed system basically consists of a large number of heat and mass exchangers and heat-exchangers through which fluids are being circulated. Jain et al. [10], have therefore taken the total cost (capital + running) of the system over its life span as the objective function to be minimized. The capital costs are calculated, as in the above two cases (Sections 9.3.1 and 9.3.2), on the basis of the cost of the material (basically heat-exchanger tubes/plates) used in the whole system. The operating costs consist of the cost of circulating air and the liquid desiccant through various components. Since the system needs considerable amount of thermal energy to regenerate the desiccant solution after it absorbs moisture in the "absorber" (Fig. 6.19) the cost of this energy should also be incorporated in the objective function unless the system uses waste heat available from the local sources. The cost of thermal energy (per kJ) has been taken as 1/3rd of the cost of electrical energy needed to operate the liquid pumps and the air blowers.

Since this was an exercise in finding the best possible design of a new system, not only the design variables of various equipment but also the operating variables like various flow rates have been included in the list of variables to be optimized. These are:

Absorber
Configuration
Falling film shell-and-tube design with air and solution in parallel flow and cooling water flowing in counter flow direction.

Variables
Number of tubes, length of tubes, mass flow rate of air, mass flow rate of liquid-desiccant solution. Tube diameter is fixed depending on the system capacity and not included in the set of variables to be optimized.

Regenerator
Falling film counter flow plate type configuration.

Variables
Number of plates, width of the plates, gap between the plates, length of the plates, Mass flow rate of air.

The mass flow rate of liquid solution in the regenerator cannot be taken as independent variable since this is the same as the liquid flow rate at the exit of the absorber.

Solution-Solution Heat Exchanger
Double pipe counter flow configuration.

Variable
Total length of the tubes.

Solution Heater
Double pipe counter flow heat exchanger.

Variable
Total length of the tubes.

Air-Air Heat Exchangers in the Absorber and Regenerator Circuits
Configuration
Continuous fin tube heat exchanger.

Variables (for Each)
Number of tubes in a row, tube length.

Tube diameter, number of rows, fin type are kept as "logical" variables whose values are fixed depending on the capacity.

We thus have 15 design variables out of which 4 can have only integer values.[3]

Finally we need to fix the extent of "cooling" that the new system should provide. Thus two constraints arise to ensure that the sensible and latent loads of the space are met. Besides this we specify upper bounds on the air velocities to prevent splash entrainment and flooding of the solution in the absorber and the regenerator.

The optimization problem has been solved by using the modified Box Complex method developed by Dhar [4]. The simulation procedure discussed in Section 6.3 is used to predict the performance of the various designs generated in the search procedure. The environmental condition has been taken as the 1% outside design condition for monsoon months of Delhi, viz. 38.6°C DBT/26.0°C WBT. Table 9.19 gives the details of the optimized design of a 20 *TR* unit.

The initial cost of the plant was estimated (in 1994) to be Rs. 3,49,000 and the life cycle cost as Rs. 6,51,000, which compares favorably with that of a 20 TR VCR system, even after incorporating the effect of improper wetting which would increase the cost by 30–40%. The operating cost (of electric power) however still remains quite small in comparison to that of a VCR plant which is expected to be around 20 kW. This situation would of course change drastically if the heating costs were also to be included in LDAC operating costs. Thus, this liquid desiccant-based system seems to be feasible only if ample amount of waste heat is available.

This theoretical exercise also reveals a potential operating problem. The concentration of the solution at the absorber outlet is 62.9%, which is quite high. Operating at such high concentration increases the risk of crystallization of the solution at low temperatures during the plant shut down. The solubility of lithium Bromide [the desiccant selected by Jain [11]] in water at 30°C is 62%, and therefore it is advisable to keep the solution concentration below this value. Thus, an additional constraint should be put in the optimization program and new results obtained.

---

[3] Namely, the number of tubes in the absorber and the two air-air heat exchangers, and the number of plates in the regenerator.

Table 9.19 Optimized design of a 20 TR LDAC system

Absorber	Regenerator
Tubes size (mm): 36.9 ID, 42.3 OD	Plate width: 0.95 m
No. of tubes: 300	Number of plates: 38
Length: 3.5 m	length: 2.28 m
Water flow rate: 15 kg/s	Gap between plates: 14 mm
Mass flow rate of solution: 1.78 kg/s	
Mass flow rate of air: 3.85 kg/s	Mass flow rate of air: 1.5 kg/s
Solution-solution HX	**Solution heater**
Inner tube: ID/OD (mm): 28.2/33.5	Inner tube: ID/OD: 28.2/33.5
Outer tube: ID/OD (mm): 42.3/48.1	Outer tube: ID/OD: 42.3/48.1
Total length: 80 m	Total length: 54 m
	Oil flow rate: 1.5 kg/s

Air-air heat exchanger
Tube ID/OD (mm): 19.94/22.22
No. of rows: 8
Fins: 0.33 mm thick; 315 fin/m
No of tubes per row: 72 in absorber HX
: 41 in regenerator HX
Length: 1.89 m in absorber HX
: 2.0 m in regenerator HX

Operating conditions
Solution conc. at absorber outlet: 62.9%
Supply air conditions: 15.3°C/0.0067 kg/kg da
Room conditions: 25°C/0.010 kg/kg da
Ventilation air: 10%
Electrical power consumption: 5.83 kW
Heating requirement: 215 kW

REFERENCES

[1] P.L. Dhar, C.P. Arora, On optimum suction state for vapour compression cycle, in: Proceedings of the National Symposium on Refrigeration and Airconditioning, Paper I-302, CMERI, Durgapur, 1972.

[2] P.L. Dhar, C.P. Arora, Optimum interstage temperature for cascade systems, in: Proceedings of the Second National Symposium on Refrigeration and Airconditioning, Roorkee, 1973, pp. 211–215.

[3] P.L. Dhar, C.P. Arora, Optimum design of finned surfaces, J. Franklin Inst. 301 (4) (1976) 379–392.

[4] P.L. Dhar, Optimization in refrigeration systems, PhD thesis, IIT, Delhi, 1974.

[5] P.L. Dhar, C.P. Arora, Design of direct expansion chillers, in: Proceedings of the Fourth National Symposium on Refrigeration and Airconditioning, IIT, Delhi, 1975, pp. 230–239.

[6] G.R. Saraf, P.L. Dhar, Optimization of flooded chiller design, Indian J. Technol. 18 (1980) 214–218.

[7] P.L. Dhar, G.R. Saraf, Computer Simulation and Design of Refrigeration Systems, Khanna Publishers, Delhi, 1987.

[8] B. Seyedan, P.L. Dhar, R.R. Gaur, G.S. Bindra, Optimization of waste heat recovery boiler of a combined cycle power plant, J. Eng. Gas Turbines Power 118 (1996) 561–564.

[9] B. Seyedan, Simulation and optimization studies of a combined cycle power plant, PhD thesis, Department of Mechanical Engineering, IIT, Delhi, 1995.

[10] J. Sanjeev, P.L. Dhar, S.C. Kaushik, Optimal design of liquid desiccant cooling systems, ASHRAE Trans. 106 (Pt 1) (2000) 79–87.

[11] S. Jain, Studies on Desiccant Augmented Evaporative Cooling Systems, PhD thesis, Department of Mechanical Engineering, IIT, Delhi, 1995.

CHAPTER 10

Dynamic Response of Thermal Systems

Hitherto, we have focused our attention on simulating the steady-state performance of thermal systems. It forms the basis of their steady-state behavior at part loads. Thus we can get an idea of the part load operating conditions and efficiency using the steady-state simulation procedures. However, to understand what happens during the transition from full load to part load operation we need dynamic simulation. Sometimes the operating conditions encountered during transient conditions may even cause damage to the system. For example, while pulling down the temperature of a cold storage, the initial temperature of the air surrounding the evaporator coil is typically around 30–35°C which is much higher than the steady-state design value (varying for 2°C to −30°C). This results in extremely high load on the system, and the compressor motor, and even the condenser, can get overloaded. Both these conditions can harm the system. The dynamic analysis helps us to identify such potentially harmful transient situations so that control systems can be developed to prevent their occurrence. Another important application of dynamic analysis is in investigating the stability of such control systems to ensure safe and efficient operation.

In this chapter we shall study the basic principles of dynamic analysis of simple thermal systems, and the solution of the resulting differential equations using Laplace transforms which have been explained at length in Section 2.5.[1] The utility of the concept of transfer function, explained in Section 2.5.1, in predicting the dynamic response to any type of stimulus, as also in developing block diagrams of complex systems, shall be illustrated by a few examples. This shall be followed by analysis of the performance and stability of typical control systems.

10.1 DYNAMICS OF THE FIRST-ORDER SYSTEMS

The systems for which the governing equation is a first-order differential equation are called first-order systems. The simplest example of such a thermal system is a typical temperature measuring device like a mercury/alcohol thermometer, a thermocouple

---

[1] The student is advised to brush up the concepts of Laplace transformation, as explained in Section 2.5, before reading this chapter.

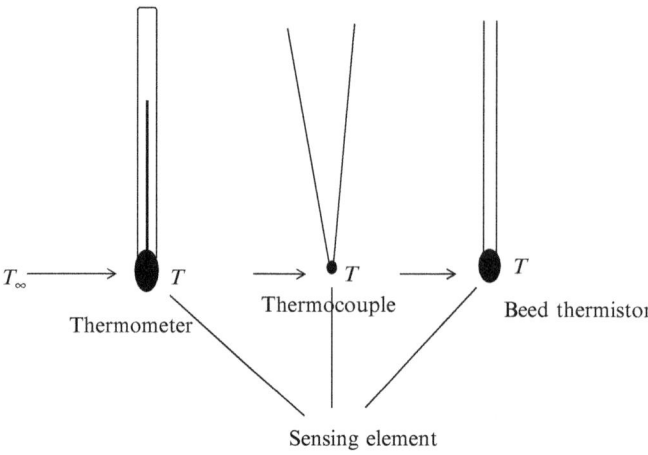

$T_\infty \longrightarrow$ T Thermometer $\longrightarrow$ $\bullet\, T$ Thermocouple $\longrightarrow$ $\bullet\, T$ Bead thermistor   Sensing element

Fig. 10.1 First-order systems.

or even a thermistor, Fig. 10.1. Considering the temperature sensing element to be at a uniform temperature T and, in the case of thermometer, neglecting the thermal capacity and thermal resistance of glass, we can write the energy equation as:

$$mc\frac{dT}{d\tau} = h \cdot A(T_\infty - T) \qquad (10.1)$$

where m is the mass of sensing element, c its specific heat, and h is the convective heat transfer coefficient between the sensing element and the surrounding fluid which is at temperature T_∞.

Eq. (10.1) is a first-order differential equation and so these devices are first-order systems. It is a simple differential equation whose solution is known to be:

$$T = T_\infty + Ce^{\frac{-hA}{mc}\tau} \qquad (10.2)$$

where C is a constant of integration whose value is found using the initial/boundary condition.

The quantity $\left(\frac{mc}{hA}\right)$ has the units of time and is called the time constant of the system. Putting $\frac{mc}{hA} = \tau_c$, the above equation becomes

$$T = T_\infty + Ce^{-\frac{\tau}{\tau_c}} \qquad (10.3)$$

We can also solve Eq. (10.1) using Laplace transforms (Section 2.5). Taking Laplace transform of Eq. (10.1) we get

$$\frac{mc}{hA} \cdot [s\text{Ł}(T) - T(0)] = \text{Ł}\{T_\infty\} - \text{Ł}\{T\} \qquad (10.4)$$

which can be rearranged putting $\frac{mc}{hA} = \tau_c$, to get

$$\frac{Ł(T)}{Ł(T_\infty)} = \frac{1 + \frac{T(0)\cdot\tau_c}{Ł\{T_\infty\}}}{1 + \tau_c s} \tag{10.5}$$

The left-hand side (LHS) of Eq. (10.5), being the ratio of Laplace transforms of the output and the input, it is clearly the transfer function for this system. As discussed in Section 2.5.1, this would become helpful in the dynamic analysis only if $T(0) = 0$. We therefore normalize the equation by defining the deviation variables as follows:

$$T' = T - \overline{T} \tag{10.6}$$

$$T'_\infty = T_\infty - \overline{T}_\infty \tag{10.7}$$

where the symbols with bar $(\overline{T}, \overline{T}_\infty)$ indicate values of these temperatures corresponding to steady state at $\tau = 0$. Now, from Eq. (10.1) it follows that at steady state, the LHS should be zero. Therefore

$$0 = hA(\overline{T}_\infty - \overline{T})$$

$$\text{or} \quad \overline{T} = \overline{T}_\infty \tag{10.8}$$

Using these deviation variables (which indicate the deviation in their values from the steady-state value) in Eq. (10.1) we get:

$$mc\frac{d}{d\tau}(T' + \overline{T}) = hA(\overline{T}_\infty + T'_\infty - \overline{T} - T')$$

Since $\frac{d\overline{T}}{d\tau} = 0$ and $\overline{T}_\infty = \overline{T}$, we can simplify above equation as:

$$\frac{dT'}{d\tau} = \frac{(T'_\infty - T')}{\tau_c} \tag{10.9}$$

Taking Laplace transform, and recalling that $T'(0)$ since at $\tau = 0$, $T = \overline{T}$, we get the transfer function as:

$$\frac{Ł(T')}{Ł(T'_\infty)} = \frac{T'(s)}{T'_\infty} = \frac{1}{1 + \tau_c s} \tag{10.10}$$

Using this transfer function we can get the response of the system [i.e., $T'(\tau)$] to various types of perturbations in the surrounding temperature, that is, $T'_\infty(\tau)$.

Thus considering first the case of a sudden change in the surrounding temperature, as indicated in Fig. 10.2, we put

$$T'_\infty = AU(\tau) \tag{10.11}$$

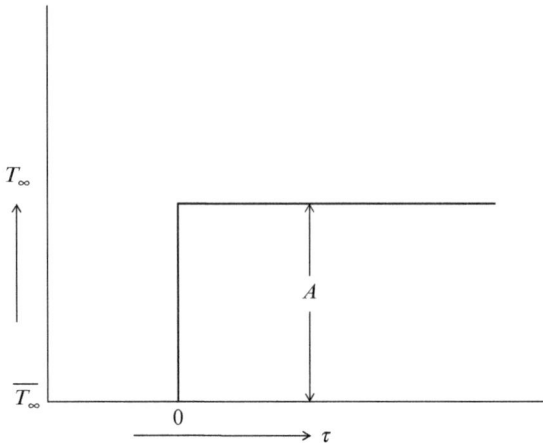

Fig. 10.2 Sudden changes in T_∞.

where $U(\tau)$ is a unit step function which has a value of zero for $\tau < 0$ and of 1 for $\tau > 1$ (Section 2.5). Its Laplace transform is:

$$\text{Ł}(T'_\infty) = A\text{Ł}(U(\tau))$$

$$= \frac{A}{s} \qquad \text{(Table 2.5)} \tag{10.12}$$

Putting this in Eq. (10.10) gives:

$$T'(s) = \frac{A}{s} \cdot \frac{1}{1 + \tau_c s} = \frac{A}{s} - \frac{A}{s + 1/\tau_c}$$

Taking inverse transform we get:

$$T'(\tau) = A(1 - e^{-\tau/\tau_c}) \qquad \tau \geq 0 \tag{10.13}$$

Eq. (10.13) shows the dependence of the response on the time constant τ_c. Thus:

$$\text{for} \quad \tau = \tau_c \quad T'(t) = 0.632A$$
$$\tau = 2\tau_c \quad T'(t) = 0.865A$$
$$\tau = 3\tau_c \quad T'(t) = 0.950A$$
$$\text{and for} \quad \tau \to \infty \quad T'(\tau) \to A$$

We note that the thermometer would take much greater time than τ_c to read the correct changed value of the surrounding temperature. Substituting other values of τ, we note that for $\tau = 5\tau_c$, $T'(\tau) = 0.9933A$; so a time $> 5\tau_c$ would be needed to measure the surrounding temperature with an error of $<1\%$.

Example 10.1

A thermocouple being used to measure the temperature inside a furnace has a time constant of 20 s. Determine its response to a linear rise in the temperature of the furnace at the rate of 5°C/min.

Solution

$$\text{Here} \quad T'_{\infty}(\tau) = 5\tau$$

where time τ is in minutes. This is also called a ramp function.

$$\text{Now from Table 2.5} \quad \mathcal{L}\{T'_{\infty}(\tau)\} = \frac{5}{s^2}$$

$$\text{Hence} \quad T'(s) = \frac{5}{s^2(1 + s/3)} \quad \text{Since } \tau_c = 20\,\text{s} = \frac{1}{3}\,\text{minutes}$$

Rearranging using partial fractions, we can write:

$$T'(s) = \frac{15}{s^2(s + 3)} = \frac{5}{s^2} - \frac{5}{3s} + \frac{5}{3(s + 3)}$$

Taking inverse transform we get the thermocouple temperature as:

$$T'(\tau) = 5\tau - \frac{5}{3}(1 - e^{-3\tau})$$

Fig. 10.3 shows the plot of the stimulus $T'_{\infty}(\tau)$ and the response $T'(\tau)$.

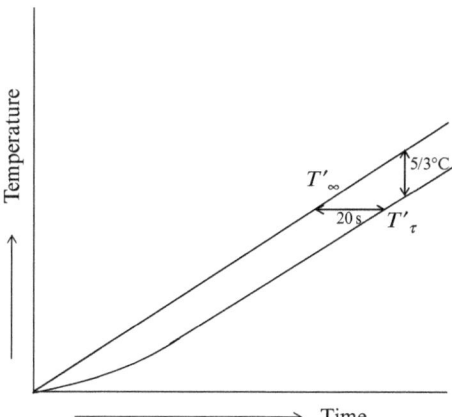

Fig. 10.3 Response to a ramp input.

We see that after an initial transient period, the response is almost parallel with the stimulus, but the thermocouple temperature never equals to the furnace temperature. It lags the furnace temperature by $\frac{1}{3}$ minutes, that is, its time constant. In steady state, after the initial transient, the thermocouple temperature is lower than the furnace temperature by $5/3°C$.

Example 10.2

Consider the pneumatic system shown in Fig. 10.4A. Gas flowing through a header has a tapping connecting it to a vessel of volume V through an orifice. Determine the transfer function for change in the vessel pressure due to small changes in the gas pressure. The orifice characteristic (pressure difference ΔP versus mass flow rate $\dot{m}$) is as shown in Fig. 10.4B. Assume the gas inside the vessel undergoes a polytropic process due to gas entering/leaving it.

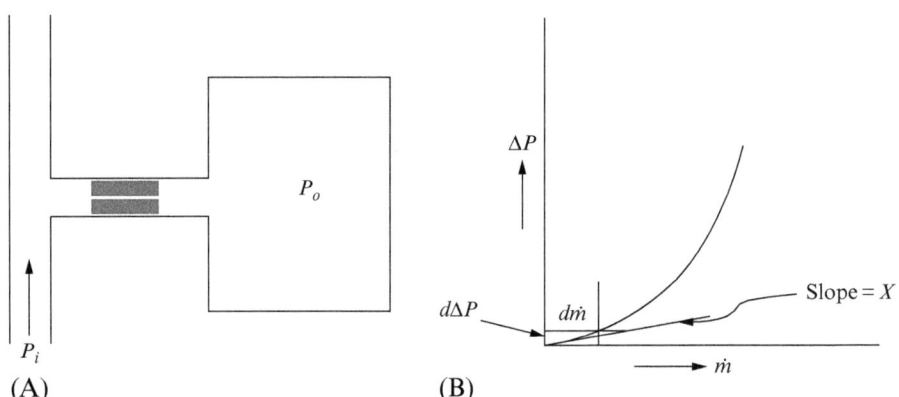

Fig. 10.4 Figure for Example 10.2.

Solution

It is clear that at steady state:

$$\overline{P}_i = \overline{P}_o$$

We have to determine how would the pressure in the vessel change due to any change in the gas header pressure, that is, relationship between P'_o and P'_i.

Now since we are confining our attention to small changes around the steady state, the relationship between ΔP and $\dot{m}$ can be written as

$$\Delta P = (P'_i - P'_o) = X \cdot \dot{m}$$

where X is the slope of the orifice characteristic curve, shown in Fig. 10.4B. Thus the mass of gas which will enter the vessel in time $d\tau$ would be:

$$dm = \dot{m} \cdot d\tau = \frac{(P_i' - P_o')}{X} \cdot d\tau$$

Now, this change in pressure P_o', being the consequence of entry of additional gas, can be related to the mass that has entered by the gas laws and the type of process occurring within the vessel. The gas law gives (assuming ideal gas):

$$m = \frac{P_o V}{R_g T}$$

$$\Rightarrow \quad dm = \frac{dP_o V}{R_g T} - \frac{P_o V}{R_g T^2} \cdot dT$$

where R_g is the gas constant and T is the gas temperature.

Now for polytropic process we know that pressure and temperature change are related as:

$$\left(\frac{P_o + dP_o}{P_o}\right) = \left(\frac{T \mid dT}{T}\right)^{\frac{n}{n-1}}$$

$$\Rightarrow \quad \frac{dP_o}{P_o} = \frac{n}{n-1} \cdot \frac{dT}{T}$$

Substituting it in the above expression for dm we get:

$$dm = \frac{1}{n} \cdot \frac{dP_o \cdot V}{R_g T} = \frac{1}{n} \cdot \frac{dP_o'}{R_g T} \cdot V \quad \{\text{since } P_o = \overline{P} + P' \quad \text{and} \quad \overline{P} \text{ is constant}\}$$

Combining it with the expression for dm obtained above from the orifice properties, we get:

$$\frac{P_i' - P_o'}{X} = \frac{V}{nR_g T} \cdot \frac{dP_o'}{d\tau}$$

$$\left(\frac{XV}{nR_g T}\right) \frac{dP_o'}{d\tau} + P_o' = P_i'$$

which is the governing equation for a first-order system.

Taking Laplace transform we get the transfer function:

$$\frac{P_o'(s)}{P_i'(s)} = \frac{1}{1 + s\tau'}$$

where $\tau' = \frac{X \cdot V}{nR_g T}$ is the time constant.

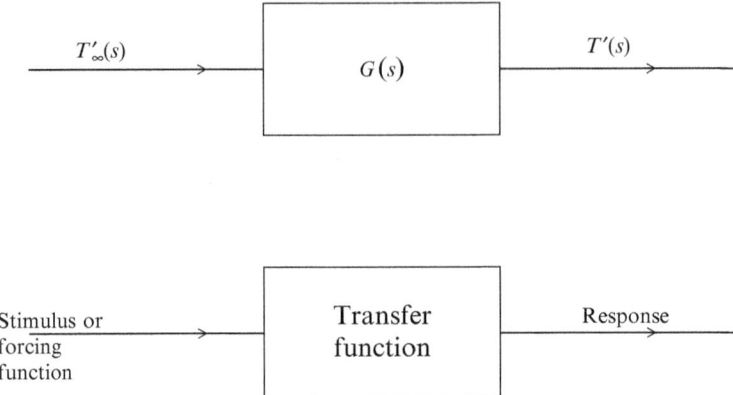

Fig. 10.5 Block diagram of a first-order system.

The relationship between the stimulus and the response is often indicated by a block diagram (Section 2.5.1) shown in Fig. 10.5, where the transfer function $G(s)$ is given by Eq. (10.10). The standard transfer function of a first-order system is:

$$G(s) = \frac{K}{1 + \tau_c s}$$

where K and τ_c are constants; τ_c being the time constant.

As another example of a first-order system let us consider a simple arrangement to heat water, shown in Fig. 10.6. Water entering at temperature T_i, is heated by an electric heater immersed in the water collected in a tank. There is a mixer in the tank to ensure uniformity of its temperature, T, at which water is withdrawn. Assuming the flow rate of the two streams to be equal $(= \dot{m})$, we can write the energy balance equation as:

$$\dot{m}cT_i + Q = \dot{m}cT + \rho Vc\frac{dT}{d\tau} \tag{10.14}$$

where c is the specific heat of the fluid (assumed independent of temperature), Q is the rate of heat input, V is the volume of liquid in the tank, and ρ its density.

At steady state, Eq. (10.14) gives:

$$\dot{m}c\overline{T}_i + \overline{Q} = \dot{m}c\overline{T} + \rho Vc\left(\frac{d\overline{T}}{d\tau}\right) \tag{10.15}$$

where the bar over a symbol indicates its steady-state value; the derivative value being clearly equal to zero. Subtracting Eq. (10.15) from Eq. (10.14) we get:

$$\dot{m}c(T_i - \overline{T}_i) + (Q - \overline{Q}) = \dot{m}c(T - \overline{T}) + \rho Vc\left(\frac{d(T - \overline{T})}{d\tau}\right) \tag{10.16}$$

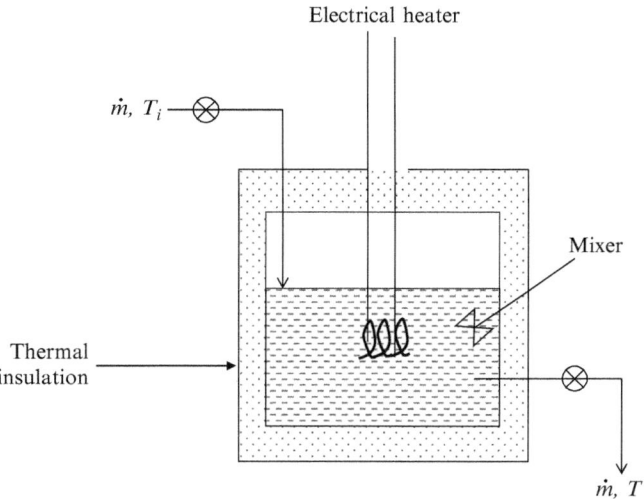

Fig. 10.6 A simple water heating system.

Thus we get the governing equation in terms of the deviation variables as:

$$\dot{m}cT_i' + Q' = \dot{m}cT' + \rho Vc\frac{dT'}{d\tau} \tag{10.17}$$

If we assume that the water inlet temperature is maintained constant, then we have one stimulus (Q') and one response (T'). The above equation simplifies to:

$$Q' = \dot{m}cT' + \rho Vc\frac{dT'}{d\tau} \tag{10.18}$$

Taking Laplace transform we get:

$$Q'(s) = \dot{m}cT'(s) + \rho Vc \cdot sT'(s) \tag{10.19}$$

Which on rearrangement gives the transfer function as:

$$\frac{T'(s)}{Q'(s)} = \frac{1/\dot{m}c}{1 + (\rho V/\dot{m})s} = \frac{K}{1 + s\tau_c} \tag{10.20}$$

where $\tau_c = \frac{\rho V}{\dot{m}}$ is the time constant, and $K = 1/\dot{m}c$ is another constant called the "gain." Thus this system also exhibits first-order dynamics.

We could also analyze the above system in absence of heating and study the response to change in the inlet temperature. Eq. (10.17) now becomes:

$$\dot{m}cT_i' = \dot{m}cT' + \rho Vc\frac{dT'}{d\tau} \tag{10.21}$$

which after taking Laplace transform, can be simplified to get the new transfer function as:

$$\frac{T'(s)}{T_i'(s)} = \frac{1}{(\rho V/\dot{m})s + 1} = \frac{1}{1 + \tau_c s} \tag{10.22}$$

which is again the transfer function of a first-order system.

Example 10.3

Water entering at 25°C is being heated in a tank with an electric heater (Fig. 10.6). The electrical heater is provided current in such a manner that its heat output (kW) is given by the equation

$$Q = 2.5 + 2.5\sin(\tau/180)$$

where time τ is in seconds. If the flow rate of both the entering and leaving water streams is 2 kg/min determine the temperature profile of the outgoing stream. Given the volume of water in the tank is 2 L.

Solution

The transfer function for this heating arrangement is given by Eq. (10.20) with

$$K = \frac{1}{\dot{m}c} = \frac{1}{(2/60)(4.18)} = 7.177 (\text{k/kW})$$

$$\tau_c = \frac{\rho V}{\dot{m}} = \frac{(1000)(2 \times 10^{-3})}{(2/60)} = 60\,\text{s}$$

Now the change in the rate of heat input Q' for $t > 0$ is given as:

$$Q'(\tau) = 2.5\sin(\tau/180)$$

Taking Laplace transform we get:

$$Q'(s) = \frac{2.5(1/180)}{s^2 + (1/180)^2}$$

Using this in Eq. (10.20) gives:

$$T'(s) = \left(\frac{7.177}{1 + 60s} \right) \left(\frac{2.5/180}{s^2 + (1/180)^2} \right)$$

$$= \left(\frac{(17.9425) \times 1/60}{s + 1/60} \right) \left(\frac{1/180}{s^2 + (1/180)^2} \right)$$

To take the inverse transform of $T'(s)$, we first expand the right-hand side (RHS) by partial fractions:

$$T'(s) = \frac{1}{s^2 + (1/180)^2} \cdot \frac{17.9425(1/180)}{1 + (1/180)^2(60)^2} + \frac{(17.9425)(1/180)(60)}{1 + (1/180)^2 \cdot (60)^2} \cdot \frac{1}{s + (1/60)}$$

$$- \frac{(17.9425)(1/180)(60)}{1 + (1/180)^2 \cdot (60)^2} \cdot \frac{s}{s^2 + (1/180)^2}$$

Now we can get the inverse transform using Table 2.5 as:

$$T'(\tau) = \frac{(17.9425) + (1/180)(60)}{1 + (1/180)^2(60)^2}e^{-\tau/60} - \frac{17.9425(1/180)(60)}{1 + (1/180)^2(60)^2} \cdot cos(\tau/180)$$
$$+ \frac{17.9425}{1 + (1/180)^2 \cdot (60)^2} \cdot sin(\tau/180)$$

The two trigonometric terms can be combined using the identity:

$$a\ cos\theta + b\ sin\theta = c\ sin(\theta + \phi)$$
$$\text{where } c = \sqrt{a^2 + b^2} \quad tan\phi = \frac{a}{b}$$

to get:

$$T'(\tau) = \frac{(17.9425)(1/180)60}{1 + (1/180)^2(60)^2}e^{-\tau/60} + \frac{17.9425}{\sqrt{1 + \left(\frac{1}{180}\right)^2 60^2}}sin(\tau/180 + \phi)$$

$$\text{where} \quad \phi = tan^{-1}\left(-\frac{1}{180} \cdot 60\right) = -18.445° \text{ or } -0.3218 \text{ radians}$$

Simplifying, this can be written as:

$$T'(\tau) = 5.38275e^{-\tau/60} + 17.0216\ sin(\tau/180 - 0.3218)$$

To find the actual water temperature, we note that at $\tau = 0$, at steady state (Eq. 10.15):

$$\left(\frac{2}{60}\right)(4.18)\overline{T}_i + 2.5 = \left(\frac{2}{60}\right)(4.18)\overline{T}$$
$$\text{Since } \overline{T}_i = 25°C, \text{ this gives } \overline{T} = 42.94°C$$

Thus the actual temperature T, of water leaving the tank is given by the equation:

$$T = T' + \overline{T} = 42.94 + 5.38275e^{-\tau/60} + 17.0216sin\left(\frac{\tau}{180} - 0.3218\right)$$

The contribution of the exponential term to the temperature would become very small within 5 min and thereafter the water temperature would also very sinusoidally, lagging the heat input by 18.4°.

10.1.1 Linearization

In all the examples considered hitherto, the equations modeling the process were linear. Actually, most physical systems are nonlinear and therefore not suitable for characterization through transfer functions. Since the use of transfer functions is extremely useful in dynamic analysis, nonlinear problems are usually "linearized" before Laplace transformation. This is done by approximating the nonlinear functions in proximity of the steady-state operating point by a Taylor series expansion and neglecting all terms after the first partial derivatives.

Thus, if $f(x_1, x_2)$ is a nonlinear function and $(\bar{x}_1, \bar{x}_2)$ are the steady-state values of the two variables, we can write:

$$f(x_1, x_2) = f(\bar{x}_1, \bar{x}_2) + (x_1 - \bar{x}_1) \left(\frac{\partial f}{\partial x_1} \right)_{\bar{x}_1, \bar{x}_2} + (x_2 - \bar{x}_2) \left(\frac{\partial f}{\partial x_2} \right)_{\bar{x}_1, \bar{x}_2} + \dots \text{ higher order terms}$$

(10.23)

Linearization implies:

$$f(x_1, x_2) \simeq f(\bar{x}_1, \bar{x}_2) + (x_1 - \bar{x}_1) \left(\frac{\partial f}{\partial x_1} \right)_{\bar{x}_1, \bar{x}_2} + (x_2 - \bar{x}_2) \left(\frac{\partial f}{\partial x_2} \right)_{\bar{x}_1, \bar{x}_2}$$

(10.24)

Thus any nonlinear function can be converted into a linear function.

We shall illustrate this with the help of an example.

Example 10.4

Consider a tank containing water (Fig. 10.7) with a water stream entering at the top and another leaving at the bottom. The flow rate of the outgoing stream, V_0, depends on the instantaneous height of water column in the tank:

$$V_o = kh^{\frac{1}{2}}$$

where k is a constant.

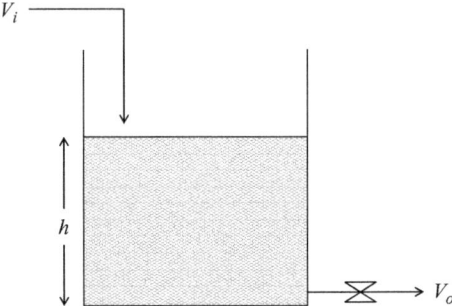

Fig. 10.7 Dynamics of a water tank.

How would the water level respond to a sudden change in the flow rate of the incoming steam, V_i?

Solution

Considering mass balance around the tank, we can write

$$V_i - V_o = A \frac{dh}{d\tau}$$

where A is the cross-sectional area of the tank.

Substituting the given relationship of V_o and h, the equation becomes:

$$V_i - kh^{1/2} = A\frac{dh}{d\tau}$$

Before proceeding further, we linearize the $h^{1/2}$ term using Eq. (10.24) to get:

$$V_o = kh^{1/2} = k\left[\bar{h}^{1/2} + (h - \bar{h})\left(\frac{1}{2}h^{-1/2}\right)_{\bar{h}}\right]$$

$$= k\left[\bar{h}^{1/2} + \frac{1}{2}\frac{h - \bar{h}}{\bar{h}^{1/2}}\right]$$

where $\bar{h}$ is the steady-state value of the water level in the tank.

At steady state, at $\tau = 0$

$$\frac{\bar{h}}{d\tau} = 0 \quad \bar{V}_i = \bar{V}_o = k\sqrt{\bar{h}}$$

Introducing derivation variables:

$$V_i' = V_i - \bar{V}_o \quad \text{and} \quad h' = h - \bar{h}$$

we get from the above equations:

$$V_i' - \frac{kh'}{2\sqrt{\bar{h}}} = A\frac{dh'}{d\tau}$$

or

$$A\frac{dh'}{d\tau} + \left(\frac{k}{2\sqrt{\bar{h}}}\right) \cdot h' = V_i'$$

which is a linear equation and thus easily amenable to Laplace transformation, which gives:

$$Ash'(s) + \left(\frac{k}{2\sqrt{\bar{h}}}\right)h'(s) = V_i'(s)$$

$$\Rightarrow \quad \frac{h'(s)}{V_i'(s)} = \frac{1}{\frac{k}{2\sqrt{\bar{h}}} + As} = \frac{2\sqrt{\bar{h}/k}}{1 + s\tau_c}$$

where $\tau_c = \dfrac{2A\sqrt{\bar{h}}}{k}$ is the time constant.

The transfer function is thus a typical TF of a first-order system and the response to a sudden change in $V_i'(s)$ by an amount ΔV can be found as before to be of the form of Eq. (10.13), that is,

$$h'(\tau) = \frac{2\Delta V\sqrt{\bar{h}}}{K}(1 - e^{-\tau/\tau_c})$$

It should be kept in mind that this linearization is valid only for small deviations from the steady state since only then can the terms involving higher order derivatives in the Taylors series expansion (Eq. 10.23) be neglected to get a linear representation of a nonlinear function.

10.1.2 First-Order System in Series

We often come across complex systems which can be visualized as a set of "first-order systems" in series. For example, considering again the simple mercury thermometer of Fig. 10.1, if the thermal resistance of the glass bulb is also taken into account, the representation of the system by Eq. (10.1) would not be valid. Now, the surroundings are interacting with the glass bulb and the bulb is, in turn, interacting with the mercury inside it (Fig. 10.8). We could consider both these interactions as simple first-order systems and represent their transfer functions as shown in Fig. 10.9.

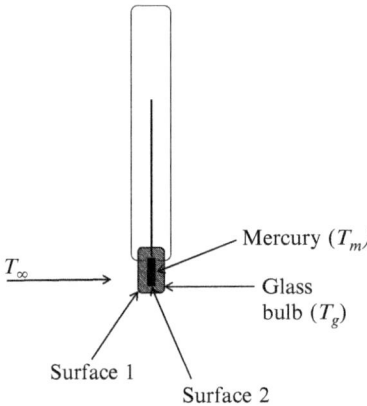

Fig. 10.8 Mercury thermometer with thick glass bulb.

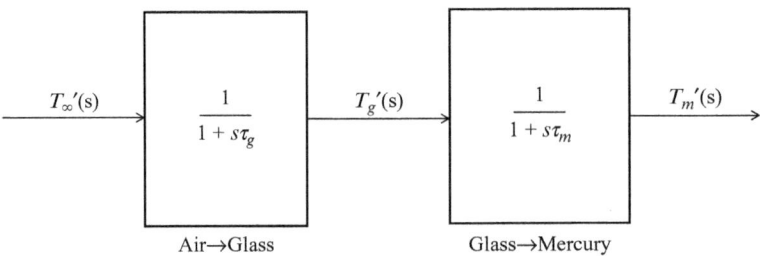

Fig. 10.9 Mercury thermometer as two first-order systems in series.

This however is not an accurate representation of this system. It has an implicit assumption that the glass bulb temperature T_g is influenced only by the temperature of the surrounding air T_∞ and the mercury temperature T_m only by the glass bulb temperature T_g. This is clearly not true since the glass bulb experiences heat transfer at its both surfaces—with air at the outer surface (1) and with mercury at the inner surfaces (2), as is evident from its complete heat balance equation:

$$h_1 A_1 (T_\infty - T_g) + h_2 A_2 (T_m - T_g) = \frac{dT_g}{d\tau} \cdot m_g c_g \tag{10.25}$$

Of course, the temperature of the mercury changes only because of heat transfer from the glass surface:

$$h_2 A_2 (T_g - T_m) = \frac{dT_m}{d\tau} \cdot m_m c_m \tag{10.26}$$

In Eqs. (10.25) and (10.26) h_1 and A_1 refer to the heat transfer coefficient and the surface area at the outer surface of the glass bulb, and h_2, A_2 refer to these quantities at its surface in contact with mercury; m_g, c_g are, respectively, the mass and the specific heat of the glass bulb, and m_m, c_m that of the mercury inside the bulb.

If we neglect the second term on the LHS of Eq. (10.25), then both the glass bulb and the mercury can be approximated as independent, noninteracting first-order systems and their block diagrams can be represented by Fig. 10.9, and the overall transfer function can be found out directly from the figure as:

$$\frac{T'_m(s)}{T'_\infty(s)} = \frac{1}{(1 + s\tau_m)(1 + s\tau_g)}$$

$$= \frac{1}{(1 + s(\tau_m + \tau_g) + s^2 \tau_m \tau_g)} \tag{10.27}$$

$$\text{where} \quad \tau_m = \frac{m_m c_m}{h_2 A_2}; \quad \tau_g = \frac{m_g c_g}{h_1 A_1} \tag{10.28}$$

However, if both the terms on the LHS of Eq. (10.25) are significant, the two subsystems (glass and mercury) interact with each other and the overall transfer function would have to be determined ab initio.

We define the deviation variables $T'_\infty(\tau)$, $T'_g(\tau)$, and $T'_m(\tau)$ as before (cf. Eqs. 10.6 and 10.7) as:

$$T'_\infty(\tau) = T_\infty(\tau) - \overline{T}_\infty$$
$$T'_g(\tau) = T_g(\tau) - \overline{T}_g$$
$$T'_m(\tau) = T_m(\tau) - \overline{T}_m \tag{10.29}$$

Eqs. (10.25) and (10.26) can now be converted in terms of these deviation variables, by following a similar procedure as above, to get:

$$h_1 A_1 (T'_\infty - T'_g) + h_2 A_2 (T'_m - T'_g) + h_1 A_1 (\overline{T}_\infty - \overline{T}_g) + h_2 A_2 (\overline{T}_m - \overline{T}_g)$$

$$= \frac{dT'_g}{d\tau} \cdot m_g c_g \tag{10.30}$$

$$h_2 A_2 (T'_g - T'_m) + h_2 A_2 (\overline{T}_g - \overline{T}_m) = \frac{dT'_m}{d\tau} \cdot m_m c_m \tag{10.31}$$

Further it follows for Eqs. (10.25) and (10.26) that, at steady state:

$$h_1 A_1 (\overline{T}_\infty - \overline{T}_g) + h_2 A_2 (\overline{T}_m - \overline{T}_g) = 0 \tag{10.32}$$

$$h_2 A_2 (\overline{T}_g - \overline{T}_m) = 0 \tag{10.33}$$

Using these results in Eqs. (10.30) and (10.31), we get:

$$h_1 A_1 (T'_\infty - T'_g) + h_2 A_2 (T'_m - T'_g) = m_g c_g \frac{dT'_g}{d\tau} \tag{10.34}$$

$$h_2 A_2 (T'_g - T'_m) = m_m c_m \frac{dT'_m}{d\tau} \tag{10.35}$$

Taking the Laplace transforms we get:

$$h_1 A_1 (T'_\infty(s) - T'_g(s)) + h_2 A_2 (T'_m(s) - T'_g(s)) = m_g c_g s T'_g(s) \tag{10.36}$$

$$h_2 A_2 (T'_g(s) - T'_m(s)) = m_m c_m s T'_m(s) \tag{10.37}$$

Eq. (10.37) gives us:

$$T'_g(s) = \left(\frac{m_m c_m}{h_2 A_2} \cdot s + 1 \right) T'_m(s)$$

$$= (\tau_m s + 1) T'_m(s) \tag{10.38}$$

where the time constant τ_m is given by Eq. (10.28). Substituting this result in Eq. (10.36), and simplifying we get:

$$\frac{T'_m(s)}{T'_\infty(s)} = \frac{1}{1 + s^2 \tau_m \tau_g + s(\tau_m + \tau_g + R\tau_m)} \tag{10.39}$$

$$\text{where} \quad R = \frac{h_2 A_2}{h_1 A_1}$$

Comparing Eq. (10.39) with Eq. (10.27) we note that the effect of "interaction" between the two sub-system is to introduce an additional term $R\tau_m s$ in the denominator of the transfer function. We shall study the effect of this on the dynamic response through an example.

Example 10.5

A mercury-in-glass thermometer is kept inside an air duct in a laboratory experiment. To start with the air and the thermometer are both at 30°C. Then suddenly the air temperature is raised to 50°C. Determine the response of the thermometer to this sudden change in air temperature using the following data.

Glass bulb dimensions: $OD/ID : 4.5/3.5$ mm;

Length : 15 mm

Specific heat of glass : 0.84 kJ/kg K

Its density : 2.4×10^3 kg/m^3

Specific heat of mercury : 0.14 kJ/kg K

Its density : 13.6×10^3 kg/m^3

Heat transfer coefficient at the outside surface of the glass bulb : 60 W/m^2 K

at its inner surface : 30 W/m^2 K

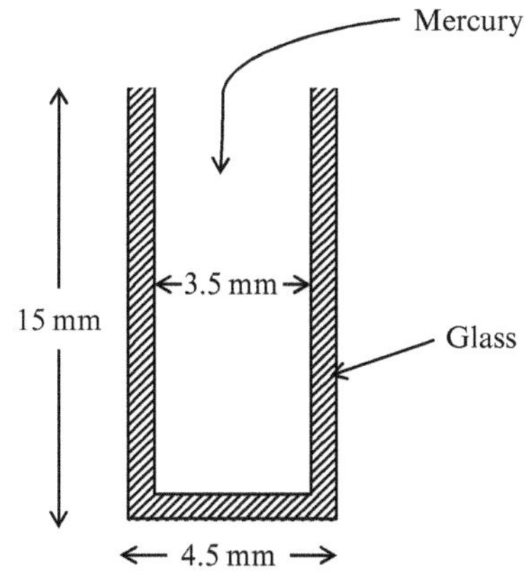

Fig. 10.10 The thermometer bulb of Example 10.5.

Solution

We first find the two time constants τ_m and τ_g (Eq. 10.28), and for this we need to find the values of h_1, A_1, m_g, c_g and h_2, A_2, m_m, c_m from the given data using the bulb geometry (Fig. 10.10).

$$A_1 = \pi \times 4.5 \times 15 + \frac{\pi \times 4.5^2}{4} \text{ mm}^2$$

$$= 227.96 \text{ mm}^2$$

$$A_2 = \pi \times 3.5 \times 14.5 + \frac{\pi \times 3.5^2}{4} \text{ mm}^2$$

$$= 169.06 \text{ mm}^2$$

$$m_g = \rho_g \times \text{(volume of glass bulb)}$$

$$= (2.4 \times 10^3)(10^{-9}) \left(\frac{\pi}{4} \cdot (4.5^2 - 3.5^2) \times 15 + \frac{\pi}{4} \cdot 3.5^2 \times 0.5 \right) \text{kg}$$

$$= 0.2377 \text{ g}$$

$$m_m = \rho_m \times \text{(volume of mercury)}$$

$$= (13.6 \times 10^3)(10^{-9}) \left(\frac{\pi}{4} \cdot 3.5^2 \times 14.5 \right) \text{kg}$$

$$= 1.897 \text{ g}$$

$$\therefore \tau_g = \frac{m_g c_g}{h_1 A_1} = \frac{(10^{-3} \times 0.2377) \times (0.84 \times 10^3)}{60 \times (227.96 \times 10^{-6})} = 14.6 \text{ s}$$

$$\tau_m = \frac{m_m c_m}{h_2 A_2} = \frac{(1.897 \times 10^{-3}) \times (0.14 \times 10^3)}{30 \times (169.06 \times 10^{-6})} = 52.4 \text{ s}$$

If we neglect the effect of interaction between the glass bulb and mercury on the energy balance of glass, we can treat Fig. 10.9 as a representation of the thermometer. This gives the overall transfer function as (Eq. 10.27):

$$\frac{T'_m(s)}{T'_\infty(s)} = \frac{1}{(1 + s \times 14.6)} \cdot \frac{1}{(1 + s \times 52.4)}$$

For a sudden change in T_∞ from 30 to 50°C:

$$T'_\infty(\tau) = 20 \ U(\tau) \Rightarrow T'_\infty(s) = \frac{20}{s}$$

Substituting it in the above equation gives:

$$T'_m(s) = \frac{20}{s} \cdot \frac{1}{1 + 14.6s} \cdot \frac{1}{1 + 52.4s}$$

$$= \frac{20}{s} - \frac{20(14.6)^2}{14.6 - 52.4} \cdot \frac{1}{1 + 14.6s}$$

$$- \frac{20(52.4)^2}{52.4 - 14.6} \cdot \frac{1}{1 + 52.4s}$$

$$= \frac{20}{s} + \frac{112.783}{1 + 14.6s} - \frac{1452.78}{1 + 52.4s}$$

Taking inverse transform we get the thermometer response:

$$T'_m(\tau) = 20 + \frac{112.783}{14.6} e^{-\tau/14.6} - \frac{1452.78}{52.4} \cdot e^{-\tau/52.4}$$

$$= 20 + 7.7249 e^{-\tau/14.6} - 27.725 e^{-\tau/52.4}$$

If we consider the interaction between the two subsystems, we have to use Eq. (10.39) to get the overall transfer function:

$$T'_m(s) = \frac{20}{s} \cdot \frac{1}{1 + s^2 \cdot 14.6 \times 52.4 + s(14.6 + 52.4 + 0.3708 \times 52.4)}$$

where $R = \dfrac{h_2 A_2}{h_1 A_1} = \dfrac{30 \times 169.06}{60 \times 227.96} = 0.3708$

We can simplify above equation as

$$T'_m(s) = \frac{20}{s} \cdot \frac{1}{1 + 765.04 s^2 + 86.43 s} = \frac{20}{s} \cdot \frac{1}{(1 + 76.4185 s)(1 + 10.0115 s)}$$

The RHS can be expressed in partial fractions as:

$$T'_m(s) = \frac{20}{s} - \frac{20 \times 76.4185^2}{76.4185 - 10.0115} \cdot \frac{1}{1 + 76.4185 s}$$
$$+ \frac{20 \times 10.0115^2}{76.4185 - 10.0115} \cdot \frac{1}{1 + 10.0115 s}$$
$$= \frac{20}{s} - \frac{1758.79}{1 + 76.4185 s} + \frac{30.19}{1 + 10.0115 s}$$

Taking the inverse transform, we get the temperature profile:

$$T'_m(\tau) = 20 - \frac{1758.79}{76.4185} e^{-\tau/76.4185} + \frac{30.19}{10.0115} \cdot e^{-\tau/10.0115}$$
$$= 20 - 23.015 e^{-\tau/76.4185} + 3.015 e^{-\tau/10.0115}$$

The temperature profiles as predicted by these two equations have been calculated and plotted using Excel software. The results are shown in Table 10.1 and Fig. 10.11. It is seen that the effect of interaction is to slow down the thermometer response. This is a general result applicable to all interacting systems. To appreciate the effect of excluding the thermal capacity of glass in determination of the dynamic response of the thermometer, we have also calculated the response predicted by Eq. (10.13), which gives for this case

Table 10.1 Results of Example 10.4

Time	No interact	Interact	Only mercury	Time	No interact	Interact	Only mercury
0.000	0.000	0.000	0.000	10.000	0.986	0.919	3.475
20.000	3.035	2.694	6.346	30.000	5.350	4.608	8.718
40.000	7.576	6.420	10.678	50.000	9.574	8.057	12.298
60.000	11.304	9.512	13.636	70.000	12.774	10.794	14.741
80.000	14.009	11.922	15.655	90.000	15.039	12.912	16.410
100.000	15.896	13.781	17.034	110.000	16.606	14.544	17.549
120.000	17.195	15.213	17.975	130.000	17.681	15.800	18.327
140.000	18.084	16.316	18.617	150.000	18.417	16.767	18.858
160.000	18.692	17.164	19.056	170.000	18.919	17.512	19.220
180.000	19.107	17.817	19.356	190.000	19.262	18.085	19.468

Continued

Table 10.1 Results of Example 10.4—cont'd

Time	No interact	Interact	Only mercury	Time	No interact	Interact	Only mercury
200.000	19.390	18.320	19.560	210.000	19.496	18.526	19.636
220.000	19.584	18.707	19.700	230.000	19.656	18.865	19.752
240.000	19.716	19.004	19.795	250.000	19.765	19.127	19.831
260.000	19.806	19.234	19.860	270.000	19.840	19.328	19.884
280.000	19.868	19.410	19.904	290.000	19.891	19.483	19.921
300.000	19.910	19.546	19.935	310.000	19.925	19.602	19.946
320.000	19.938	19.651	19.955	330.000	19.949	19.693	19.963
340.000	19.958	19.731	19.970	350.000	19.965	19.764	19.975
360.000	19.971	19.793	19.979	370.000	19.976	19.818	19.983
380.000	19.980	19.841	19.986	390.000	19.984	19.860	19.988
400.000	19.987	19.877	19.990	410.000	19.989	19.892	19.992
420.000	19.991	19.906	19.993	430.000	19.992	19.917	19.995
440.000	19.994	19.927	19.995	450.000	19.995	19.936	19.996
460.000	19.996	19.944	19.997	470.000	19.996	19.951	19.997
480.000	19.997	19.957	19.998	490.000	19.998	19.962	19.998
500.000	19.998	19.967	19.999	510.000	19.998	19.971	19.999

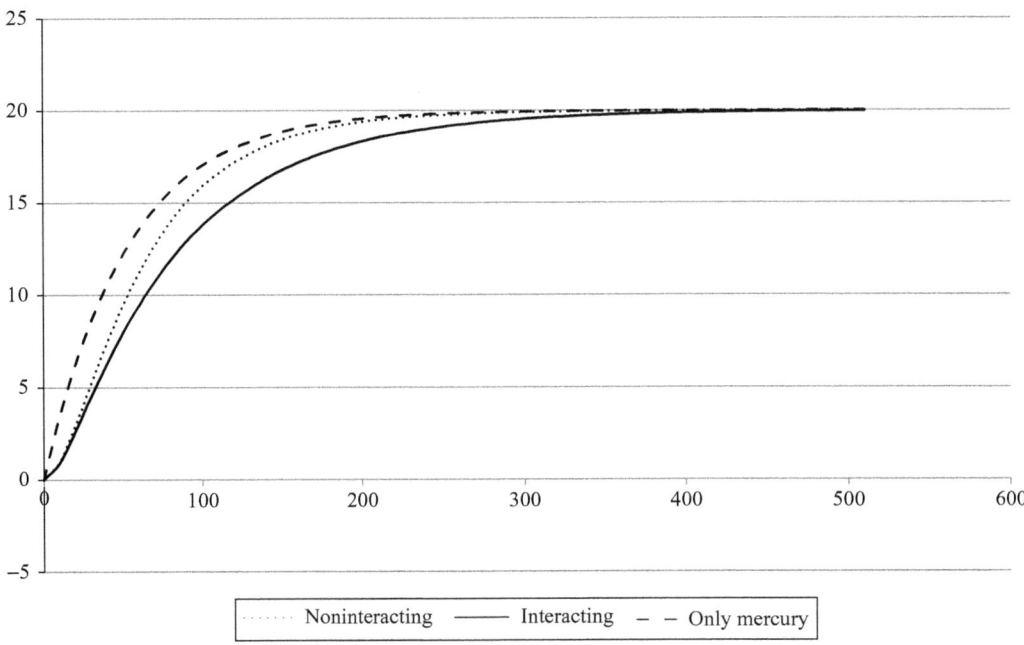

Fig. 10.11 Temperature profile of Example 10.4.

$$T(\tau) = 20[1 - e^{-\tau/52.4}]$$

These results have also been tabulated along with the above two in Table 10.1. The considerable deviation in response predicted by this method becomes obvious from Fig. 10.11, where all these results have been plotted. Thus the time required to reach $19.9°C$ (i.e., within $0.1°C$ of the actual value of $20°C$) is given by the three methods as:

1. Considering only mercury: 275 s
2. Considering both glass and mercury but no interaction: 295 s
3. Considering both glass and mercury with interaction: 415 s

10.2 HIGHER ORDER SYSTEMS

The systems in which the process model is a differential equation of order two or more are termed as higher order systems. Such systems can typically arise from two or more first-order systems in series. As we saw in Section 10.1.2, the equation relating the stimulus (T'_∞) to the final response (T'_m) is a second-order differential equation (indicated by s^2 term in the Laplace transform). The thermowell used to measure the temperature of a stream flowing through a pipe, Fig. 10.12, is a third-order system and can be visualized as three interacting systems in series involving interaction from (i) fluid to thermowell, (ii) thermowell to oil, and finally (iii) oil to the thermometer. If in the thermometer analysis we consider glass and mercury separately, the overall system will become a fourth-order system with the third interaction above replaced by two interactions, that is, from oil to glass and glass to mercury.

Many systems are intrinsically higher order systems and not just because of series connection of a few first-order systems. One such system which we have already discussed in some detail in Section 4.3 is the spring loaded ring-valve used in large refrigeration compressors (Fig. 4.36). The equation of motion for this ring-type plate value is (cf. Eq. 4.214):

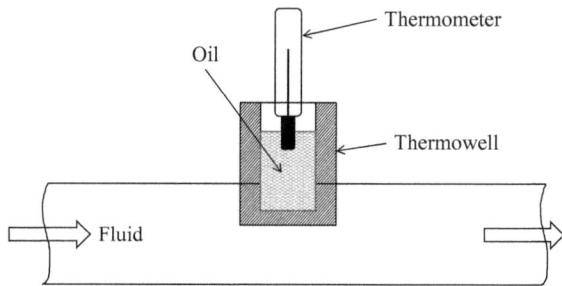

Fig. 10.12 Measuring temperature of a fluid flowing through a pipe.

$$M_r \frac{d^2 h}{dt^2} + \xi \frac{dh}{dt} + N_{sp} \cdot S_k h = P_i A_P + \frac{P_i + P}{2}(A_{vp} - Ap) - PA_{vp} - N_{sp}S_{pl} \quad (10.40)$$

This is the governing equation for a typical second-order system whose general form is often written as:

$$\tau_c^2 \frac{d^2 Y}{d\tau^2} + 2\xi \tau_c \frac{dY}{d\tau} + Y = X(\tau) \quad (10.41)$$

where τ_c is called the time constant, ξ the damping coefficient, $X(\tau)$ the forcing function (or the stimulus), and $Y(\tau)$ is the system response.

Taking the Laplace transform, starting from steady-state conditions, we get the transfer function in terms of the deviation variables as:

$$\frac{Y'(s)}{X'(s)} = \frac{1}{\tau_c^2 s^2 + 2\xi T_c s + 1} \quad (10.42)$$

Clearly two parameters τ_c and ξ characterize the dynamics of second-order systems in contrast to only one parameter for the first-order systems. The nature of response depends strongly on the relative values of τ_c and ξ. To illustrate this dependence let us consider the response to a unit step forcing function, for which

$$X'(s) = \frac{1}{s} \quad (10.43)$$

$$\Rightarrow \quad Y'(s) = \frac{1}{s(\tau_c^2 + 2\tau_c \xi s + 1)}$$

$$= \frac{1/\tau_c^2}{s(s - a)(s - b)} \quad (10.44)$$

where a, b, the roots of the quadratic equation $\tau_c^2 s^2 + 2\tau_c \xi s + 1 = 0$, are:

$$a, b = \frac{-\xi}{\tau_c} \pm \frac{\sqrt{\xi^2 - 1}}{\tau_c} \quad (10.45)$$

It is evident from Eq. (10.45) that if $\xi < 1$, the roots will be imaginary, while if $\xi \geq 1$, the roots will be real. Accordingly the inverse transforms of Eq. (10.44) will be quite different.

The solution for $\xi < 1$ can be obtained by breaking the expression for the transfer function, Eq. (10.44), into parts, as was done in earlier examples. The distinguishing feature here is that both a and b are complex numbers (Eq. 10.45), which we write for simplification in algebra as:

$$a = \alpha_1 + i\alpha_2; \quad b = \alpha_1 - i\alpha_2 \quad (10.46)$$

$$\text{where} \quad \alpha_1 = -\frac{\xi}{\tau_c} \quad \text{and} \quad \alpha_2 = \frac{\sqrt{1 - \xi^2}}{\tau_c} \quad (10.47)$$

The coefficients of the partial fraction terms would also have both real and imaginary parts, and so we write,

$$\frac{1/\tau^2}{s(s-a)(s-b)} = \frac{C_1}{s} + \frac{C_2 + iC_3}{s - \alpha_1 - i\alpha_2} + \frac{C_4 + iC_5}{s - \alpha_1 + i\alpha_2} \tag{10.48}$$

As before, we equate the coefficients of various terms, keeping the real and imaginary parts separate. This gives:

$$C_1 = 1; \quad C_2 = C_4 = -\frac{1}{2}; \quad C_3 = -C_5 = -\frac{1}{2}\frac{\alpha_1}{\alpha_2} \tag{10.49}$$

The inverse transform of Eq. (10.48), using the rules given in Section 3.5, is:

$$Y'(\tau) = 1 - \frac{1}{2}\left(1 + i\frac{\alpha_1}{\alpha_2}\right)e^{(\alpha_1 + i\alpha_2)\tau} - \frac{1}{2}\left(1 - \frac{\alpha_1}{\alpha_2}i\right)e^{(\alpha_1 - i\alpha_2)\tau} \tag{10.50}$$

Eq. (10.50) can be simplified using the trigonometric rule:

$$e^{i\theta} = cos\theta + i\,sin\theta; \quad e^{-i\theta} = cos\theta - i\,sin\theta$$

to get:

$$Y'(\tau) = 1 - e^{\alpha_1 \tau}\left[cos(\alpha_2 \tau) - \frac{\alpha_1}{\alpha_2}sin(\alpha_2 \tau)\right] \tag{10.51}$$

This can further be simplified using the trigonometric identity:

$$a\,cos\theta + b\,sin\theta = \sqrt{a^2 + b^2}\,sin(\theta + tan^{-1}a/b)$$

to get the solution for $\xi < 1$, the under-damped system as:

$$Y'(\tau) = 1 - \frac{1}{\sqrt{1 - \xi^2}}e^{-\xi\tau/\tau_c}\;sin\left(\sqrt{1 - \xi^2}\frac{\tau}{\tau_c} + \phi\right) \tag{10.52}$$

$$\text{where} \quad \phi = tan^{-1}\left(\frac{\sqrt{1 - \xi^2}}{\xi}\right) \tag{10.53}$$

Proceeding in a similar manner the solutions for $\xi > 1$ and $\xi = 1$ can be obtained as:

$$\underline{\xi > 1:} \tag{10.54}$$

$$Y'(\tau) = 1 - e^{-\xi\tau/\tau_c}[cosh\sqrt{\xi^2 - 1}\frac{\tau}{\tau_c}$$

$$+ \frac{\xi}{\sqrt{\xi^2 - 1}}sinh\sqrt{\xi^2 - 1}\frac{\tau}{\tau_c}] \tag{10.55}$$

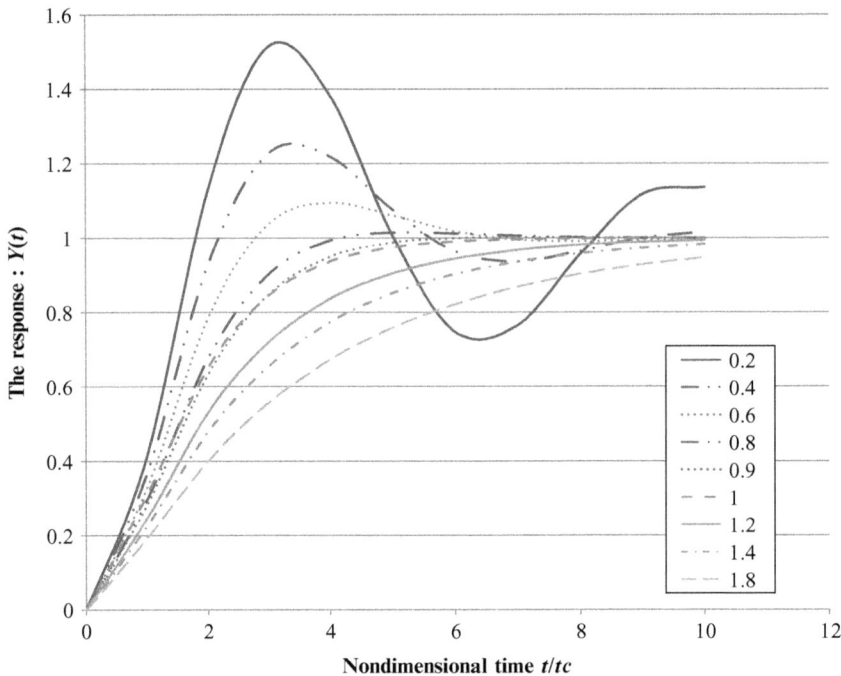

Fig. 10.13 Response of second-order systems to a unit step stimulus.

$$\xi = 1 :$$

(10.56)

$$Y(\tau) = 1 - \left(1 + \frac{\tau}{\tau_c}\right) e^{-\tau/\tau_c}$$

(10.57)

Fig. 10.13 shows a plot of these three equations (viz. 10.52, 10.55, 10.57) indicating the values of the response $Y'(\tau)$ as a function of time (τ/τ_i) and for various values of the damping factor ξ. It is seen that for under-damped systems, the response is oscillatory, smaller the value of damping coefficient larger the amplitude of oscillation about the steady-state solution. For damping coefficient values ≥ 1 there are no oscillations and the response approaches the steady-state solution exponentially. For values of $\xi \gg 1$, the overdamped systems, the approach to steady state becomes slower; therefore $\xi = 1$ is termed as critical damping coefficient.

10.3 TRANSPORTATION LAG

Most thermal systems have fluids transported through pipes. This introduces delay in transmission of stimulus given at one end to the equipment connected at the other end.

This is termed as the transportation lag or Dead time. Fig. 10.14 shows a typical situation where a fluid is flowing through an insulated pipe of length L, at a constant volumetric flow rate Q. Initially the system is in steady state so the temperature of the fluid entering the pipe equals that the exit, that is,

$$\bar{y} = \bar{x} \tag{10.58}$$

Now, if a change were made in the temperature $x(\tau)$ at $\tau = 0$, the change would not be felt at exit till the first packet of fluid with changed temperature reaches the exit. This will take time $\tau_d = LA/Q$ seconds as shown in Fig. 10.15. We can thus indicate the relationship between $x(\tau)$ and $y(\tau)$ as

$$y(\tau) = x(\tau - \tau_d) \tag{10.59}$$

subtracting Eq. (10.58) from Eq. (10.59) and introducing deviation variables, we get

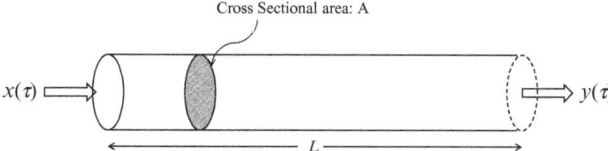

Fig. 10.14 System with transporation lag.

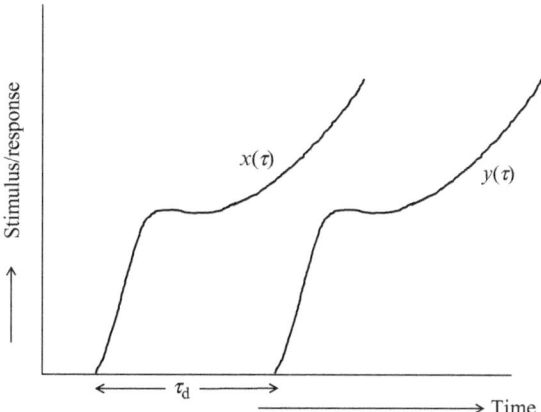

Fig. 10.15 Response to transporation lag.

$$y'(\tau) = x'(\tau - \tau_d) \tag{10.60}$$

We know from the rules for Laplace transforms that:

$$\text{if} \quad Ł(x(\tau)) = X(s)$$

$$Ł(x(\tau - \tau_d)) = e^{-s\tau_d} X(s) \tag{10.61}$$

Using this to transform Eq. (10.60), we get:

$$\frac{Y'(s)}{X'(s)} = e^{-s\tau_d} \tag{10.62}$$

Thus the transfer function of transportation lag is $e^{-s\tau_d}$. Since $e^{-s\tau_d}$ is not a rational function, it is often approximated (for stability analysis) as

$$e^{-s\tau_d} = \frac{1}{e^{s\tau_d}} = \frac{1}{1 + \tau_d s + \tau_d^2 s^2/2 + \dots\dots\dots} \tag{10.63}$$

Using only the first-order term, we can write:

$$e^{-s\tau_d} \cong \frac{1}{1 + \tau_d s} \tag{10.64}$$

A better approximation, often preferred is obtained by writing

$$e^{-\tau_d s} = \frac{e^{-s\tau_d/2}}{e^{s\tau_d/2}} \cong \frac{1 - s\tau_d/2}{1 + s\tau_d/2} \tag{10.65}$$

To find the inverse transform of $Y'(s)$ given by Eq. (10.62), the shifting property of Laplace transforms is helpful. This gives:

$$Ł^{-1}\{e^{-as} F(s)\} = u_a(\tau) f(\tau - a) \tag{10.66}$$

$$\text{where} \quad Ł\{f(\tau)\} = F(s) \tag{10.67}$$

$$\text{and} \quad u_a(\tau) = 0 \text{ for} \quad \tau < a$$

$$= 1 \text{ for} \quad \tau > a \tag{10.68}$$

Transport lag can play a significant role in design of controls for thermal systems, as illustrated by Example 10.6.

Example 10.6

Consider the simple water heating system of Example 10.3 (Fig. 10.6). If in the exit pipe of diameter 25.0 mm a thermocouple is being used to measure the temperature of the outgoing stream, what will be the difference in the temperature as measured by it and the tank temperature after a sudden rise in the incoming water temperature from 25°C to 45°C. Assume that the heater input remains constant at 2.5 kW. The thermocouple is placed 60 cm away from the tank exit.

Solution

The transfer function relating change in the inlet temperature to the change in the tank temperature is given by Eq. (10.22) as:

$$\frac{T'(s)}{T'_i(s)} = \frac{1}{1 + \tau_c s} = \frac{1}{1 + 60\,s}$$

The time constant for transportation lag between the tank and the thermocouple location is $[Q = 2\,L/min]$:

$$\tau_d = \frac{LA}{Q} = \frac{(60)(\frac{\Pi}{4} \times 2.5^2)}{(2000/60)} = 8.84 \text{ s}$$

Assuming the time constant of the thermocouple to be very small, we can say that its temperature T_t equals the temperature of the fluid in contact with it. It is related to the tank deviation temperature T' by Eq. (10.62), that is,

$$\frac{T'_t(s)}{T'(s)} = e^{-8.84\,s}$$

Combining the two transfer functions (Fig. 10.16) we get:

$$T'_t(s) = \frac{e^{-8.84\,s}}{1 + 60\,s} \cdot T'_i(s)$$

For a step change in the inlet temperature from 25°C to 45°C:

$$T'_i(s) = \frac{20}{s}$$

Substituting, we get

$$T'_t(s) = \frac{20}{s} \cdot \frac{e^{-8.84\,s}}{1 + 60\,s}$$

This can also be written as:

$$T'_t(s) = e^{-8.84\,s} \frac{20}{s(1 + 60)}$$

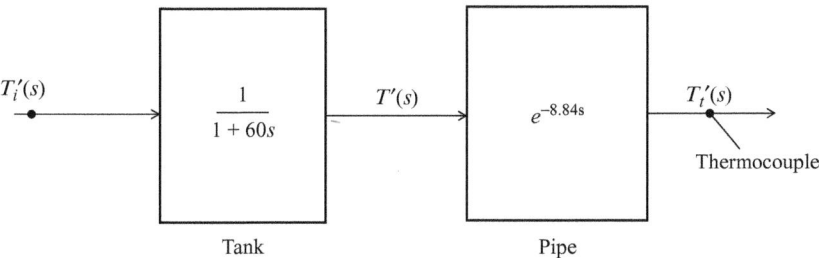

Fig. 10.16 Block diagram for Example 10.6.

Using the shifting property of Laplace transforms, the inverse transform can be obtained (Eqs. 10.66 and 10.67).

$$T'_t(\tau) = u(8.84)f(\tau - 8.84)$$

where

$$f(\tau) = \pounds^{-1}\left\{\frac{20}{s(1 + 60s)}\right\}$$

$$= \pounds^{-1}\left\{\frac{20}{s} - \frac{20}{(s + 1/60)}\right\}$$

$$= 20(1 - e^{-\tau/60})$$

This gives

$$T'_t(\tau) = 0 \text{ for } 0 < \tau < 8.84$$

$$= 20(1 - e^{-(\tau - 8.84)/60}) \text{ for } \tau > 8.84$$

To compare it with the tank temperature we find $T'(s)$ and then $T'(\tau)$.

$$T'(s) = \frac{20}{s} \cdot \frac{1}{1 + 60\ s} = \frac{20}{s} - \frac{20}{s + 1/60}$$

$$\Rightarrow \quad T'(\tau) = 20 - 20e^{-\tau/60}$$

Fig. 10.17 shows a plot of the two temperatures based on these equations. It can be seen that the thermocouple temperature lags behind the tank temperature by 8.84 s. This

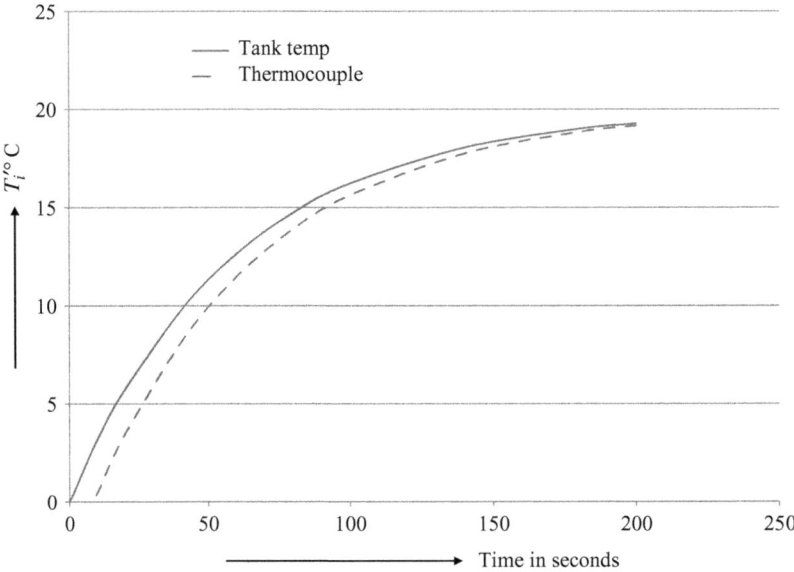

Fig. 10.17 Effect of transporation lag.

will clearly have a bearing on the design of any control system for the tank based on the thermocouple temperature.

10.4 PRINCIPLE OF SUPERPOSITION

An important advantage of the linearized analysis is that the effect of various stimuli on the system is additive. Thus if a system has two (or more) possible stimuli, x_1' and x_2', with G_1 and G_2 being the individual transfer functions for them when considered in isolation, the system response when both these stimuli are simultaneously present can be obtained as:

$$Y' = Y_1' + Y_2' = x_1' G_1 + x_2' G_2 \tag{10.69}$$

This is indicated in the block diagram shown in Fig. 10.18. The result is termed as the principle of superposition.

To appreciate this principle let us consider again the illustration of a simple water heating system (Fig. 10.6) analyzed earlier in Section 10.1. There are two possible stimuli in this system, viz. change in the (i) incoming water temperature T_i and (ii) electrical power input to the heater. In Section 10.1, we obtained the transfer function for both of these stimuli considered in isolation, viz. Eq. (10.20) for heater power and Eq. (10.22) for water inlet temperature. In case both these stimuli are simultaneously present, the governing equation is Eq. (10.17). Taking its Laplace transform we get:

$$\dot{m}c T_i'(s) + Q'(s) = \dot{m}c T'(s) + \rho V c_s T'(s)$$

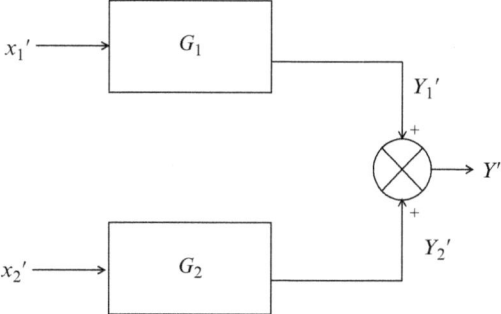

Fig. 10.18 Block diagram showing superposition.

which on rearrangement gives:

$$T'(s) = \frac{\dot{m}c T_i'(s)}{\dot{m}c + \rho V c s} + \frac{Q'(s)}{\dot{m}c + \rho V c s}$$

$$= \frac{T_i'(s)}{1 + \left(\frac{\rho V}{\dot{m}}\right) s} + \frac{Q'(s) \cdot 1/\dot{m}c}{1 + \left(\frac{\rho V}{\dot{m}}\right) s}$$

$$= \frac{T_i'(s)}{1 + \tau_c s} + \frac{K Q'(s)}{1 + \tau_c s} \tag{10.70}$$

This transfer function can be written in terms of the transfer function for each individual stimuli, Eqs. (10.20) and (10.22), as:

$$T'(s) = G_1 T_i'(s) + G_2 Q'(s)$$

$$\text{where} \quad G_1 = \frac{1}{1 + \tau_c s} \quad \text{and} \quad G_2 = \frac{K}{1 + \tau_c s} \tag{10.71}$$

This clearly is in accordance with the principle of superposition as expressed mathematically by Eq. (10.69).

An interesting application of the principle of superposition is in studying the dynamics of a boiler drum (Fig. 10.19). The boiler drum liquid level is influenced by two factors, viz. the admission of cold feed water and the ingress of heat from the burning fuel surrounding the drum. The cold water condenses the entrained vapor bubbles and thus tends to *decrease* the water level following first-order behavior, with transfer function $\frac{K_1}{1+s\tau_1}$. On the other hand the heat input causes steam production which tends to increase the water level in an integral form (since total steam generated is directly governed by *total* heat input), which leads to transfer function K_2/s. The cumulative effect of these two transfer functions is that the overall output transfer function is (see Fig. 10.20):

$$\frac{K_2}{s} - \frac{K_1}{1 + \tau_1 s} = \frac{(K_2 \tau_1 - K_1)s + K_2}{s(1 + \tau_1 s)} \tag{10.72}$$

It is clear from Eq. (10.72) that

$$\text{for} \quad K_2 \tau_1 < K_1 \tag{10.73}$$

the second term will dominate initially and we will have a negative response, that is, the liquid level will decrease for some time before it starts increasing. This can also be seen from Fig. 10.21 which shows graphically the response of boiler drum with time. The possibility of inverse response is evident. Clearly, if K_2 is sufficiently high, that is, the rate of heat input is high, then Eq. (10.73) will not be satisfied and we will not have inverse response.

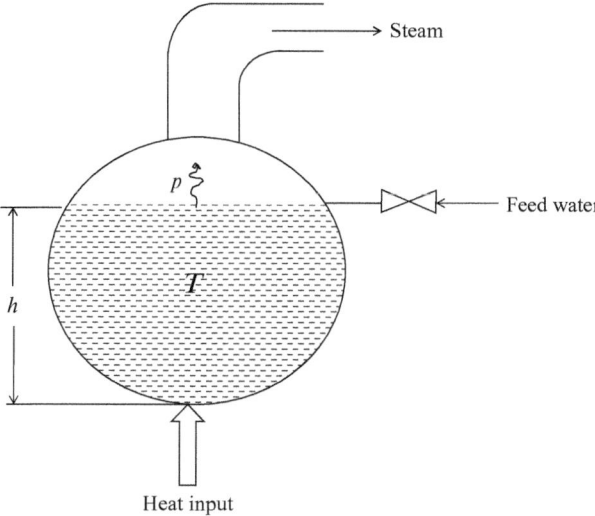

Fig. 10.19 Boiler drum dynamics.

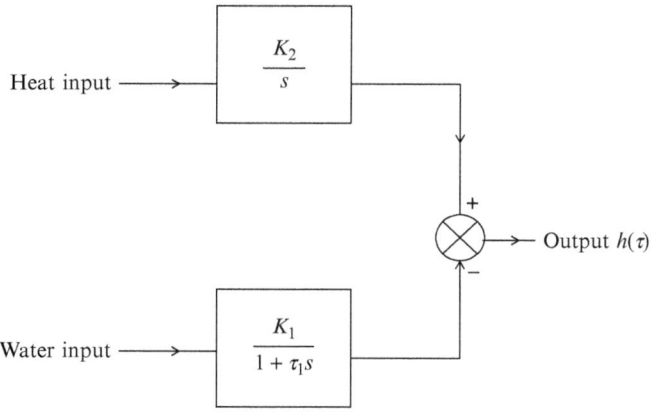

Fig. 10.20 Block diagram showing transfer functions for a boiler drum.

Such inverse response can occur in any system where there exist two (or more) opposing influences. Another commonly encountered example is that of mercury in glass thermometer. Both the glass bulb and mercury would expand on temperature rise and have opposing effect on the liquid position in the thermometer stem (i.e., the temperature as indicated by the thermometer). If the glass expansion predominates initially, the thermometer would show an inverse response. While designing the control systems for such systems, special attention has to be paid to this possibility.

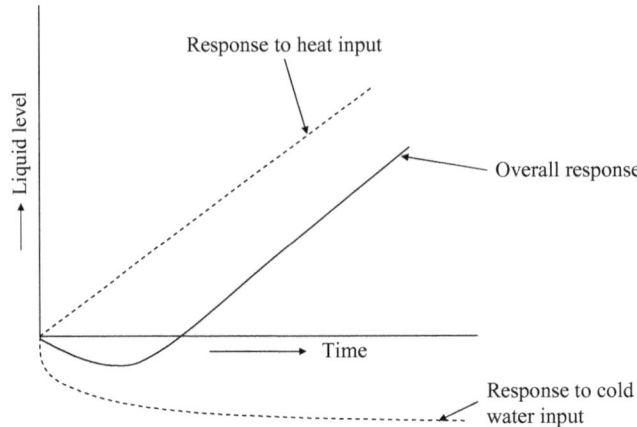

Fig. 10.21 Liquid level change showing inverse response.

10.5 CONTROL SYSTEM ANALYSIS

Thermal systems experience frequent load variations. To ensure that these systems continue to operate efficiently and safely during the transients, suitable control systems have to be designed and incorporated within them. In this section we shall discuss, in brief, the basic principles of analysis and design of such control systems.

We begin our discussions from the simplest possible control system (discussed in brief earlier in Section 2.5.1) to regulate the temperature of a liquid being heated (electrically or through steam) in a tank as shown in Fig. 10.22. The liquid enters the tank at temperature T_i and leaves at the temperature T. For simplicity, we take the mass flow rates of the two streams to be the same at $\dot{m}$. It is desired to regulate the tank temperature at value T_R by means of a suitable control system, whose basic elements are indicated in Fig. 10.22. The controlled variable (tank temperature) is measured by an appropriate device (e.g., here it could be a thermocouple or a thermistor). This signal is transmitted to the controller system consisting mainly of two components (Fig. 10.23), viz. a comparator which compares the measured value with the desired value (i.e., the set point) T_R and the controller which, based on their difference (or error ϵ) changes the heat input (Q) to the tank in such a way as to reduce the error ϵ. This it does through an appropriate equipment called the final control element (FCE) which, in this example, could be a variac controlling the power input to the electric heater or a control valve that adjusts the flow of steam.

The controller decides the relationship between ϵ and the magnitude of change in the heat input ΔQ. One of the simplest relationships is direct proportionality (i.e., $\Delta Q \alpha \epsilon$). This is termed as the proportional control. The overall control system thus has four main components, viz. the process, the measuring element, the controller system

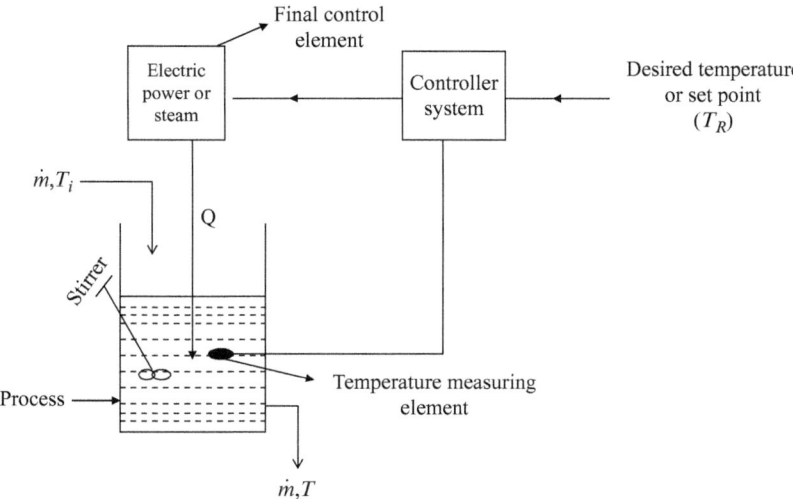

Fig. 10.22 A simple control system for a hot water tank.

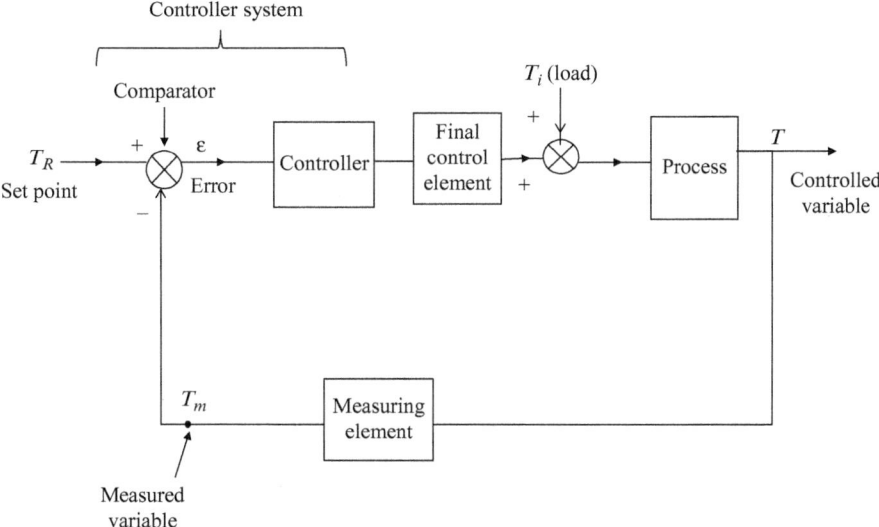

Fig. 10.23 Block diagram of simple control system.

and the FCE (here including the heating device). This is generally true of most control systems that we would be discussing.

For analytical purposes it is helpful to represent the interactions occurring in a control system through a block diagram Fig. 10.23. The term "load" in this diagram refers to that variable any change in which imposes "load" on the control systems by attempting to influence the value of the controlled variable. Thus in this example the temperature of the incoming liquid T_i is a load variable. Had we allowed change in its mass flow rate, that would have been another load variable.

In view of the nature of this block diagram, such a control system is also called a closed–loop system or a feedback system. More specifically, it is a negative feedback system since the value of measured temperature fed back to the comparator is subtracted from the set point value, that is,

$$\text{Negative feedback:} \quad \epsilon = T_R - T_M \tag{10.74}$$

It is easy to see intuitively that a negative feedback system is inherently stable. This is explained by the symbolic diagram of Fig. 10.24 which shows what happens whenever there is a disturbance from steady-state condition $(T = T_m = T_R)$, $\epsilon = 0$. We shall study a rigorous method of stability analysis in Section 10.5.7.

In contrast to negative feedback we can also have the positive feedback which is characterized by the equation:

$$\text{Positive feedback:} \quad \epsilon = T_R + T_m \tag{10.75}$$

It is easy to check that a positive feedback system would always be unstable, and is therefore not used. However, it may arise naturally in a complex system and then suitable safeguards are provided to avoid instability.

10.5.1 Two Kinds of Control Problems

In thermal engineering practice we come across two kinds of situations to be handled by the control system. The first, and the most commonly encountered situation, demands that the controlled variable (here the tank temperature) be maintained constant in spite of load changes. Such problems are called *regulator* problems. The other category, termed the *servo* problems, involve changing the controlled variable (i.e., the set point itself) according to some prescribed function of time. For example, there may be need to

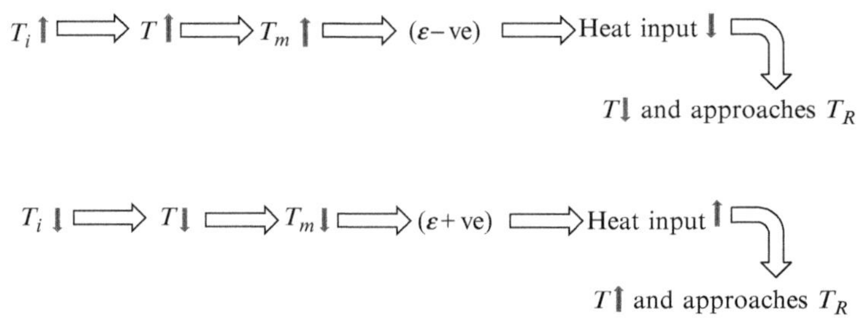

Fig. 10.24 Symbolic diagram explaining stability of negative feedback.

change the temperature of a furnace with time in a particular manner (e.g., during heat treatment). Usually, in servo problems we presume that there is no external load. Of course, in view of the principle of superposition, the response of a system to change in load can be superimposed on the response to set point change to obtain the response to any linear combination of set point and load changes.

10.5.2 Developing the Block Diagram

The simple block diagram of Fig. 10.23 showing only the manner of interaction between various components of the control system can be developed into a comprehensive "quantitative" block diagram by populating each of the boxes by the appropriate transfer function. We can do so directly by making use of the results obtained earlier in Sections 10.1 and 10.2 while analyzing the components individually.

Starting from the extreme left, we have the comparator whose governing equation for negative feedback is Eq. (10.74). By writing its steady-state and transient versions we can easily get the equation in terms of the deviation variables as:

$$c' = T'_R - T'_m \tag{10.76}$$

whose Laplace transformation gives:

$$\epsilon'(s) = T'_R(s) - T'_m(s) \tag{10.77}$$

This error signal actuates the controller which, in turn, adjusts the heat input to the tank through the FCE. These two equipment are usually combined during analysis since we are only interested in the change in heat input to the tank and not how it is achieved through the controller and the FCE. The relationship between the heat input Q and the error ϵ depends upon the nature of these controllers. Assuming proportional control we can write:

$$Q = K_c \epsilon + A \tag{10.78}$$

where K_c is the controller gain and A the heat input when $\epsilon = 0$ (also called bias value), that is, under steady state. This implies

$$A = \overline{Q} \tag{10.79}$$

Substituting this result in Eq. (10.78) we get:

$$Q - \overline{Q} = K_c \epsilon = K_c(\epsilon - \epsilon_s)$$

since $\epsilon_s = 0$ at steady state.

$$\text{or} \quad Q' = K_c \epsilon' \Rightarrow Q'(s) = K_c \epsilon'(s) \tag{10.80}$$

The next element in Fig. 10.23 is the "process," that is, the process happening in the tank as water enters and heat is input. We have already analyzed this in detail in Sections 10.1 and 10.4 to get the transfer function, Eq. (10.71), reproduced here for completeness:

$$T'(s) = \frac{T'_i(s)}{1 + \tau_c s} + \frac{KQ'(s)}{1 + \tau_c s} \tag{10.81}$$

$$\text{where} \quad \tau_c = \frac{\rho v}{\dot{m}} \tag{10.82}$$

$$K = \frac{1}{\dot{m}c} \tag{10.83}$$

The last component of the block diagram of Fig. 10.23 is the measuring element which is in contact with the liquid at temperature T and transmits the measured value T_m to the comparator. This was the very first device considered by us in Section 10.1 and its transfer function is given by Eq. (10.10). Indicating its time constant by τ_m, to distinguish it from the process time constant τ_c mentioned above, we can rewrite this equation as:

$$\frac{T'_m(s)}{T'(s)} = \frac{1}{1 + s\tau_m} \tag{10.84}$$

Combining all these individual component transfer functions we can get the block diagram for the whole control system shown in Fig. 10.25.

This diagram can be simplified by using the rules for combining transfer functions discussed in Section 2.5.1. We first break the TF for Q' into two parts viz. K and $\frac{1}{1+\tau_c s}$; and then combine the two blocks with the same transfer function of $\frac{1}{1+\tau_c s}$ by pushing it ahead of the "adder," to get the final block diagram shown in Fig. 10.26.

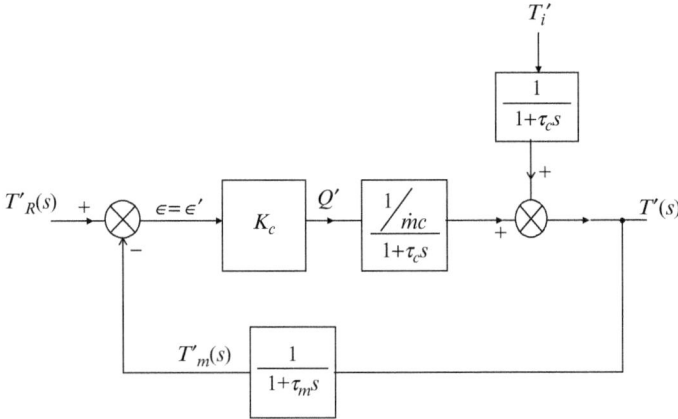

Fig. 10.25 Block diagram of the control system.

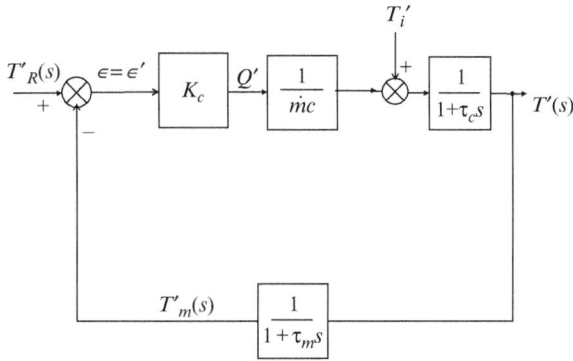

Fig. 10.26 Final block diagram of the control system.

10.5.3 Analyzing Servo Problems

The general block diagram of Fig. 10.26 can be used to analyze the impact of various kinds of stimuli on the performance of the control system. We consider first the servo problem for which $T_i'(s) = 0$. We start from the stimulus $T_R'(s)$ and note that in the forward direction there are three transfer functions which can be combined (using Rule (a), Fig. 2.9) to get the block diagram of Fig. 10.27. The feedback loop in Fig. 10.27 can now be eliminated by using Rule (c), Fig. 2.9, to get the overall transfer function for the servo problem as shown in Fig. 10.28. Here

$$G_s = \frac{K_c/\dot{m}c}{1 + \tau_c s} \quad \text{and} \quad H_s = \frac{1}{1 + \tau_m s}$$

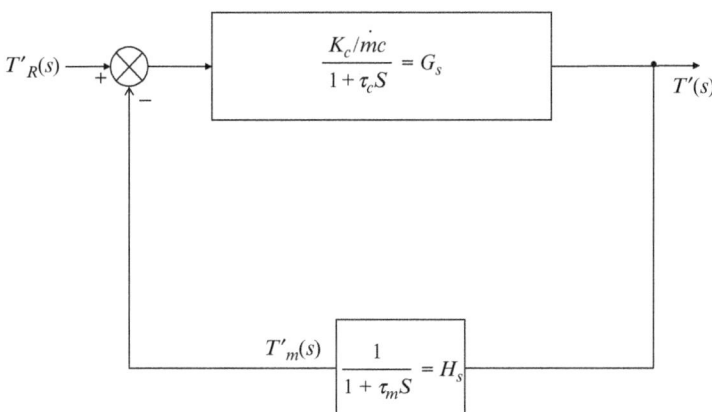

Fig. 10.27 Block diagram of servo problem.

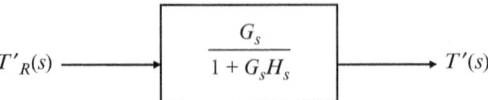

Fig. 10.28 Final block diagram of servo problem.

The subscript s in the above transfer functions indicates these are servo problem values. The overall transfer function thus becomes:

$$TF_s = \frac{T'(s)}{T_R'(s)} = \frac{\frac{K_c/\dot{m}c}{1+\tau_c s}}{1 + \frac{K_c/\dot{m}c}{1+\tau_c s} \cdot \frac{1}{1+\tau_m s}} \tag{10.85}$$

Using this overall transfer function we can predict the response of the control system to various types of changes in the set point. For the sake of simplicity, and to focus more on the influence of the feedback and the type of control on system performance, we carry out further analysis presuming that the time constant of the measuring element is very small. This is practically realizable by using a thermocouple with small bead for temperature measurement. Now, for $\tau_m \to 0$, we can simplify the overall TF of Eq. (10.85) to;

$$TF_s = \frac{G_s}{1 + G_s \cdot 1} = \frac{K_c/\dot{m}c}{1 + \tau_c s + K_c/\dot{m}c} = \frac{K_c/\dot{m}c}{\left(1 + \frac{K_c}{\dot{m}c} + \tau_c s\right)}$$

$$= \frac{\left(\frac{K_c/\dot{m}c}{1+K_c/\dot{m}c}\right)}{1 + \left(\frac{\tau_c}{1+K_c/\dot{m}c}\right) s} = \frac{K_c^s}{1 + \tau_c' s} \tag{10.86}$$

$$\text{where} \quad K_c^s = \frac{K_c/\dot{m}c}{1 + K_c/\dot{m}c} \quad \text{and} \quad \tau_c' = \frac{\tau_c}{1 + K_c/\dot{m}c} \tag{10.87}$$

We note that the TF of the control system is also a first-order transfer function but with a reduced time constant $\tau_c'(< \tau_c)$, This implies that the effect of feedback is to speed up the response.

To appreciate another effect of feedback let us find the response to a sudden change in the set point, say by one unit. We get:

$$T'(s) = \frac{1}{s} \cdot \frac{K_c^s}{1 + \tau_c' s} = \frac{K_c^s}{s} - \frac{K_c^s \tau_c'}{1 + \tau_c' s} \tag{10.88}$$

whose inverse transform is:

$$T'(\tau) = K_c^s(1 - e^{-\tau/\tau_c'}) \tag{10.89}$$

Clearly as $\tau \to \infty$, $T' \to K_c^s$, which is less than 1. This shows that the control system will not be able to ensure that the tank temperature is able to track the set point since the desired value of T' as $\tau \to \infty$ is 1, the change in the set point. This error is termed as the offset and is defined as

$$\text{Offset} = T_R'(\infty) - T'(\infty) \tag{10.90}$$

$$\text{which in this case} = 1 - K_c^s = \frac{1}{1 + K_c/\dot{m}c} \tag{10.91}$$

Clearly, as K_c increases, the offset decreases, but very high values of K_c may introduce instability, as we shall study later in Section 10.5.7.

Example 10.7

Consider the water heating system of Example 10.2, now fitted with a proportional control system to maintain the tank temperature at the desired set point. The steady-state conditions are:

$$T_i = 25°C \quad T_R = T = 45°C \quad \dot{m} = 2\,\text{kg/min} \quad \text{Volume of tank} = 2\,\text{L}$$

Determine the response of the control system to a sudden change of $5°C$ in the set point. Given the controller gain is $50\,\text{W/K}$.

Solution

We first find the values of K_c^s and τ_c^i using Eq. (10.87)

$$\tau_c = \frac{\rho V}{\dot{m}} = \frac{1000 \times (2 \times 10^{-3})}{(2/60)} = 60\,\text{s}$$

$$\frac{K_c}{\dot{m}c} = \frac{50}{(2/60)(4186)} = 0.35834$$

$$\therefore K_c^s = \frac{0.35834}{1 + 0.35834} = 0.2638$$

$$\tau_c' = \frac{60}{1 + 0.35834} = 44.17\,\text{s}$$

The response to a unit step change is given by Eq. (10.89); so the response to a change in T_R by $5°C$ would be

$$T'(\tau) = 5 \times 0.2638(1 - e^{-\tau/44.17})$$

$$\text{Clearly} \quad T'(\infty) = 5 \times 0.2638 = 1.319°C$$

which implies that the tank temperature would always be offset from the set point by $5 - 1.319 = 3.681°C$.

Comment This offset could be reduced by increasing the value of controller gain K_c. For example, if $K_c = 1500\,\text{W/K}$ the offset would be only $0.42°C$.

10.5.4 Analyzing Regulator Problem

Proceeding in a similar manner we can also study the response of the control system to load changes when the set point is kept fixed, that is, $T'_R = 0$. Fig. 10.29 shows how the overall transfer function for this regulator problem can be obtained by simplifying the general block diagram of Fig. 10.26 using the block-combining rules given in Fig. 2.9.

Since $T'_R = 0$, the comparator output is $-T'_m$, and thus the product $(T'_m)\left(\frac{K_c}{\dot{m}c}\right)$ has to be subtracted from the load T'_i. In view of these observations, the block diagram of Fig. 10.29A can be transformed into that in Fig. 10.29B. Its transformation to the final diagram of Fig. 10.29C follows directly from Rule (c), Fig. 2.9. The overall transfer function is thus:

$$TF_r = \frac{T'(s)}{T'_i(s)} = \frac{\frac{1}{1+\tau_c s}}{1 + \frac{1}{1+\tau_c s} \cdot \frac{K_c/\dot{m}c}{1+\tau_m s}} \tag{10.92}$$

Considering the case when $\tau_m \approx 0$, as assumed while analyzing the servo problem above, we can simplify it to:

$$\frac{T'(s)}{T'_i(s)} = \frac{1}{1 + \tau_c s + K_c/\dot{m}c} \tag{10.93}$$

$$= \frac{1/(1 + K_c/\dot{m}c)}{1 + \left(\frac{\tau_c}{1+K_c/\dot{m}c}\right)s} = \frac{K_c^r}{1 + \tau'_c s} \tag{10.94}$$

$$\text{where} \quad K_c^r = \frac{1}{1 + \frac{K_c}{\dot{m}c}} \tag{10.95}$$

and τ'_c is given by Eq. (10.87).

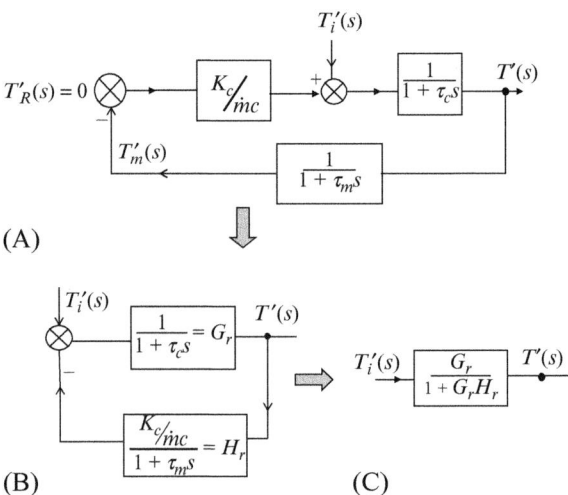

(A)

(B)   (C)

Fig. 10.29 Simplifying the block diagram for regulator problems.

As in the servo problem, here too, the control system behaves as a first-order system (under the assumed conditions) with a reduced time constant.

Let us also study its response to a step change in the inlet temperature, that is, $T'_i(s) = \frac{1}{s}$. It follows from Eq. (10.94) that:

$$T'(s) = \frac{1}{s} \cdot \frac{K^r_c}{1 + \tau'_c s} = K^r_c \left[\frac{1}{s} - \frac{\tau'_c}{1 + \tau'_c s} \right] \tag{10.96}$$

$$\Rightarrow T'(\tau) = K^r_c(1 - e^{-\tau/\tau'_c}) \quad \text{.... by taking inverse transform} \tag{10.97}$$

The steady-state value of T' is

$$T'(\infty) = K^r_c = \frac{1}{1 + K_c/\dot{m}c} \tag{10.98}$$

Since the set point is kept constant, an ideal situation would be that after any change in the load, the tank temperature should eventually return to the set point. This implies that ideally $T'(\infty) = T'_R(\infty) = 0$. Clearly this is not the case here and the offset is:

$$\text{Offset} = T'_R(\infty) - T'(\infty) = -\frac{1}{1 + K_c/\dot{m}c} \tag{10.99}$$

Had there been no feedback control, the offset would have been -1 (corresponding to $K_c = 0$). To reduce its magnitude, the controller gain K_c should be kept high.

Example 10.8

Determine the response of the water heating system of Example 10.6 to a sudden 5°C rise in the water inlet temperature, the set point remaining unaltered at 45°C.

Solution

As we saw above the new time constant τ'_c is the same as in servo problem, that is,

$$\tau'_c = 44.17\,\text{s}$$

$$\text{Further} \quad \frac{K_c}{\dot{m}c} = 0.35834 \quad \text{and} \quad \therefore K^r_c = \frac{1}{1 + 0.35834} = 0.7362$$

The transient response is given by Eq. (10.97) as:

$$T'(\tau) = 5 \times 0.7362(1 - e^{-\tau/44.17})$$

where we have multiplied the RHS by 5 since the change in T_i is 5 units.

Clearly as $\quad \tau \to \infty \quad T'(\tau) \to 3.6810 \quad$ or $\quad T(\tau) = \overline{T} + T'(\tau) = 48.681°C$

Thus the tank temperature is offset from the desired set point value by $T_R - T = 45 - 48.681 = -3.681°C$. As in the servo problem, here too, the offset could be reduced by increasing K_c. Thus for $K_c = 1500\,\text{W/K}$, the offset would be 0.42°C.

10.5.5 Proportional Integral (PI) and Proportional Integral Derivative (PID) Control

As we have seen above, both in the servo problem and in the regulator problem there remains a residual error (offset) with proportional control. To remove this error more complex modes of control have been devised. One such mode, which is quite popular in industry is the Proportional Integral or PI control. Here in addition to the linear term (Eq. 10.78) we have an additional integral in the controller response. Thus the PI control relationship for our system of Fig. 10.22 would be:

$$Q = K_c \epsilon + \frac{K_c}{\tau_I} \int_o^\tau \epsilon \, d\tau + A \qquad (10.100)$$

where the integration over time means that we sum up all the error terms up to the present time.

The term τ_I, called the integral time or the reset time, is an additional parameter (besides K_c) which can be adjusted to achieve the desired response.

To obtain its transfer function, we first convert the equation into deviation variables, as was done with Eq. (10.78). This gives

$$Q' = K_c \epsilon' + \frac{K_c}{\tau_I} \int_o^\tau \epsilon' \, d\tau \qquad (10.101)$$

Taking its transform and collecting terms we get the transfer function as:

$$\frac{Q'(s)}{\epsilon'(s)} = K_c \left(1 + \frac{1}{\tau_I s} \right) \qquad (10.102)$$

If this controller is used instead of a simple proportional controller in our water heating tank, the only change in the final block diagram of Fig. 10.26 will be that the controller gain K_c would be replaced by $K_c(1 + \frac{1}{\tau_I s})$. The transfer function for the set point change (with no measurement lag) would now be (cf. Eq. 10.86)

$$
\begin{aligned}
TF &= \frac{K_c/\dot{m}c \cdot \left(1 + \frac{1}{\tau_I s}\right)}{\left(1 + \frac{K_c}{\dot{m}c} \cdot \left(1 + \frac{1}{\tau_I s}\right) + \tau_c s\right)} \\
&= \frac{\frac{K_c}{\dot{m}c}(1 + \tau_I s)}{\tau_I s + \tau_c \tau_I s^2 + \frac{K_c}{\dot{m}c}(1 + \tau_I s)} \\
&= \frac{1 + \tau_I s}{\left(\frac{1 + \frac{K_c}{\dot{m}c}}{K_c/\dot{m}c}\right)\tau_I s + \frac{\tau_c \tau_I s^2}{K_c/\dot{m}c} + 1} \\
&= \frac{(1 + \tau_I s)}{\tau_1^2 s^2 + 2\xi \tau_1 s + 1} \qquad (10.103)
\end{aligned}
$$

where $\tau_1 = \sqrt{\dfrac{\tau_I \tau_c}{K_c/\dot{m}c}}$ and $\xi = \dfrac{1}{2}\left(1 + \dfrac{K_c}{\dot{m}c}\right)\sqrt{\dfrac{\tau_I}{\tau_c \cdot \left(\frac{K_c}{\dot{m}c}\right)}}$ (10.104)

Eq. (10.103) is a standard transfer function of quadratic form which we analyzed in detailed in Section 10.2. Depending upon the value of the damping coefficient the solution would be either oscillatory (Eq. 10.52) or moving exponentially towards the steady state (Eq. 10.55 or 10.57). In all cases however, the response to a unit step function is such that the solution approaches the stimulus in steady state. This can also be obtained directly from the final value theorem (Section 2.5) which states that for a finite valued function $f(\tau)$:

$$Lim_{\tau \to \infty}[f(\tau)] = Lim_{s \to \infty}[sf(s)] \tag{10.105}$$

Now when the stimulus is a unit step function, then for transfer function of Eq. (10.103) we get:

$$Lim_{\tau \to \infty}[T'(\tau)] = Lim_{s \to 0}\left[s \cdot \frac{1}{s} \cdot \frac{(1 + \tau_I s)}{\tau_1^2 s^2 + 2\xi \tau_1 s + 1} \right] \tag{10.106}$$

$$= 1$$

Thus there is no offset and the tank temperature approaches the new set point.

Similar analysis could also be carried out for load change. We again obtain a second-order transfer function, viz.

$$\frac{T'(s)}{T'_i(s)} = \frac{\tau_I/(\frac{K_c}{\dot{m}_c}) \cdot s}{\tau_1^2 s^2 + 2\xi \tau_1 s + 1} \tag{10.107}$$

where τ_1 and ξ are given by Eq. (10.104). For a unit step change in load, we get

$$T'(s) = \frac{1}{s} \cdot \frac{\tau_I/(\frac{K_c}{\dot{m}c})s}{\tau_1^2 s^2 + 2\xi \tau_1 s + 1} = \frac{\tau_I/(\frac{K_c}{\dot{m}c})}{\tau_1^2 s^2 + 2\xi \tau_1 s + 1} \tag{10.108}$$

which again would lead to oscillatory or exponentially varying solutions. The final limiting value would be, in accordance with Eq. (10.105) as

$$T'(\infty) = Lt_{s \to 0}\{s \cdot TF\} = Lim_{s \to 0}\left\{ \frac{s \cdot \tau_I/(\frac{K_c}{\dot{m}c})}{\tau_1^2 s^2 + 2\xi \tau_1 s + 1} \right\} = 0 \tag{10.109}$$

The "offset" is again zero since $T'_R(\infty) = 0$, there being no change in the set point during the change in load.

Elimination of the offset is the main distinguishing feature of PI control, however, it can introduce a long period of oscillatory response. To obviate this a derivative action is added to a PI controller to get what is called a PID controller. For PID controller relationship between the controller output and the error is (cf. Eq. 10.101)

$$\text{PID controller: } Q' = K_c \epsilon' + \frac{K_c}{\tau_I} \int_o^\tau \epsilon \, d\tau + K_c \tau_D \frac{d\epsilon}{d\tau} \qquad (10.110)$$

We now have an additional parameter τ_D to be adjusted, besides K_c and τ_I.

Its Laplace transform would be

$$\frac{Q'(s)}{\epsilon'(s)} = K_c \left(1 + \tau_D s + \frac{1}{\tau_I s} \right) \qquad (10.111)$$

Proceeding in a manner similar to PI controller analysis, the response to a PID controller can also be obtained. Qualitatively speaking the presence of derivative term in Eq. (10.110) implies that rapid changes in ϵ' would introduce large derivative responses which could contain the oscillatory response induced by the integral control. Thus, by proper choice of the parameters $K_c, \tau_D,$ and τ_I, a PID controller can quickly bring the system to steady state with minimum oscillations. This is also borne out by detailed analysis done through Laplace transforms. It is left as an exercise for the students to verify the above assertions.

10.5.6 Effect of Measurement Lag

In the above analysis of control systems for the sake of simplification, we neglected the measurement lag, that is, assumed $\tau_m \approx 0$. If we do not make this assumption, it is evident from Eqs. (10.85) and (10.92) that both the control situations—viz. servo problem and the regulator problem—lead to second-order transfer functions even for proportional control. Since we have already analyzed in detail the response of such systems, we will not discuss that again.

If we were to incorporate measurement lag in systems with PI control term K_c in the expressions for the two transfer functions (TF_s and TF_r), Eqs. (10.85) and (10.92), would have to be replaced by $K_c(1 + \frac{1}{\tau_I s})$ (see Eqs. 10.80 and 10.102). This would give

$$TF_s = \frac{\frac{K_c/\dot{m}_c}{1+\tau_c s} \left(\frac{1+\tau_I s}{\tau_I s} \right)}{1 + \frac{K_c/\dot{m}_c}{1+\tau_c s} \cdot \frac{1+\tau_I s}{\tau_I s} \cdot \frac{1}{1+\tau_m s}} \qquad (10.112)$$

$$TF_r = \frac{\frac{1}{1+\tau_s s}}{1 + \frac{1}{1+\tau_c s} \cdot \frac{K_c/\dot{m}_c}{1+\tau_m s} \left(\frac{1+\tau_I s}{\tau_I s} \right)} \qquad (10.113)$$

Thus both the transfer functions would have a third-order polynomial in the denominator and can be written in a general form as

$$TF_s = \frac{\alpha_s + \beta_s s}{1 + a_s s^3 + b_s s^2 + c_s s} \qquad (10.114)$$

$$TF_r = \frac{(\alpha_r + \beta_r s)s}{1 + a_r s^3 + b_r s^2 + c_r s} \qquad (10.115)$$

where the coefficient α, β, a, b, c for both these systems are functions of constants $\tau_I, \tau_c, \tau_m, K_c$, and $\dot{m}_c$,

To find the response of these systems to say a unit step change in the stimulus, we multiply these transfer functions by $\frac{1}{s}$.

$$\text{i.e.,} \quad T'(s) = \frac{1}{s} \cdot TF_s \tag{10.116}$$

$$T'(s) = \frac{1}{s} \cdot TF_r \tag{10.117}$$

To take the inverse transform, we resolve the RHS into components based on the roots of polynomials in the denominator of Eqs. (10.114) and (10.115). In general, the components would be written as

$$T'(s) = \frac{A}{s} + \frac{B}{s - b} + \frac{C}{s - c} + \frac{D}{s - d} \tag{10.118}$$

where b, c, and d are the roots of the cubic expression in the denominator.

Taking the inverse transform would give (refer Table 2.5)

$$T'(\tau) = A + Be^{b\tau} + Ce^{c\tau} + De^{d\tau} \tag{10.119}$$

Clearly, the nature of response $T'(\tau)$ would depend upon the nature of roots b, c, and d. If these are negative real numbers, or complex numbers with $-$ve real part, the exponential terms would decay with increasing time τ. However, if these roots are positive or complex numbers with positive real parts, the exponential terms would continuously increase with time and even a small stimulus would result in unbounded response. This is termed as instability and has to be avoided at all costs by adjusting various parameters. We shall study this in detail in the next section.

10.5.7 Stability Analysis

It follows from the above discussion that the analysis of stability of any control system boils down to finding the roots of the equation in the denominator of its transfer function. We would generalize this statement by noting that the denominator of all overall transfer functions can be written as $1 + G(s)$ where G is the product of individual transfer functions of all the blocks comprising the control system (see Figs. 10.26–10.29). Thus analysis of stability consists of determining the roots of the algebraic equation

$$1 + G(s) = 0 \tag{10.120}$$

This equation is called the characteristic equation.

$G(s)$ is also termed as the open loop transfer function since it is the overall transfer function when the feedback loop of Fig. 10.23 is disconnected after the measuring element. The characteristic equation is usually a polynomial, the main exception being

<u>Routh's Test for Stability</u>

1. Write the characteristic equation in the form

$$a_0 s^n + a_1 s^{n-1} + a_2 s^{n-2} + \cdots\cdots\cdots + a^n = 0 \quad a_0 > 0$$

2. If any of the coefficients $a_1, a_2, \ldots\ldots a_{n-1}, a_n$ is negative, the system is definitely unstable.

3. If all the coefficients are positive, the system may or may not be stable. To check apply following procedure:

4. Arrange these coefficients in the first two rows of the Routh array as indicated below for $n = 6$.

Row 1	a_0	a_2	a_4	a_6
Row 2	a_1	a_3	a_5	
Row 3	b_1	b_2	b_3	
Row 4	c_1	c_2	c_3	
Row 5	d_1	d_2		
Row 6	e_1	e_2		
Row 7	f_1			

5. The elements in row 3 and beyond are found using the following formulae:

$$b_1 = \frac{a_1 a_2 - a_0 a_3}{a_1} \quad b_2 = \frac{a_1 a_4 - a_0 a_5}{a_1} \ldots\ldots$$

$$c_1 = \frac{b_1 a_3 - a_1 b_2}{b_1} \quad c_2 = \frac{b_1 a_5 - a_1 b_3}{b_1} \ldots\ldots$$

6. The system is stable only if *none* of the elements in the first column is negative.

Fig. 10.30 Routh's test for stability.

systems with transportation lag which would need to be approximated as indicated in Section 10.3.

English mathematician Routh developed a simple algorithm to check whether the roots of a polynomial equation have positive real parts. The algorithm is summarized in Fig. 10.30. We shall illustrate the use of this method by an example.

Example 10.9

Consider the water heating system of Example 10.7, now fitted with a PI controller (instead of proportional controller) to maintain the tank temperature at the desired set point. The measuring system has a lag of 15 s and the reset time of PI controller is 10 s. Determine the maximum controller gain permissible to avoid instability.

Solution

Taking the data from solution of Example 10.7, we can summarize all the parameter values as

$$\tau_c = 60\,\text{s} \quad \tau_I = 10\,\text{s} \quad \tau_m = 15\,\text{s}$$

$$\frac{1}{\dot{m}c} = \frac{1}{\left(\frac{2}{60}\right)(4186)} = \frac{1}{139.53} \quad \text{K/W}$$

The characteristic equation is (representing controller gain by K_c)

$$1 + \left(\frac{K_c}{139.53}\right)\left(1 + \frac{1}{10s}\right)\left(\frac{1}{1 + 60s}\right)\left(\frac{1}{1 + 15s}\right) = 0$$

On simplification and collecting terms, this gives

$$900s^3 + 75s^2 + s\left(1 + \frac{K_c}{139.53}\right) + \frac{K_c}{1395.3} = 0$$

Since K_c is a positive number, all the coefficients of the above equation are positive. To establish the conditions for stability, we write down the Routh's array:

Row		
1	900	$\left(1 + \frac{K_c}{139.53}\right)$
2	75	$K_c/1395.3$
3	$\left(1 - \frac{K_c}{5 \times 139.53}\right)$	0
4	$\frac{K_c}{139.53}$	

Clearly, all the terms in the first column would be positive if

$$\text{if} \quad 1 - \frac{K_c}{5 \times 139.53} \geq 0$$

$$\text{or} \quad K_c \leq 697.65 \tag{10.121}$$

Thus to ensure stability the controller gain should be lesser than 697.65 W/K.

Example 10.10

Solve Example 10.9 if the controller is changed to PID controller with $\tau_D = 5\,\text{s}$ and $\tau_I = 10\,\text{s}$.

Solution

The transfer function of a PID controller is given by Eq. (10.111). Using it, instead of Eq. (10.102), the new characteristic equation can be written as:

$$1 + \left(\frac{K_c}{139.53}\right)\left(1 + 5s + \frac{1}{10s}\right)\left(\frac{1}{1 + 60s}\right)\left(\frac{1}{1 + 15s}\right) = 0$$

putting $\frac{K_c}{139.53} = K_c'$, the above equation can be simplified to

$$s(1 + 60s)(1 + 15s) + K_c'(s + 5s^2 + 0.1) = 0$$

Collecting terms it can be written in the standard polynomial form as

$$900s^3 + s^2(75 + 5K_c') + s(1 + K_c') + 0.1K_c' = 0$$

All the coefficients being positive (since K_c, $K_c' > 0$), we build the Routh's array to check for stability

Row		
1	900	$1 + K_c'$
2	$75 + 5K_c'$	$0.1K_c'$
3	$\dfrac{75 + 5K_c'^2 - 10K_c'}{75 + 5K_c'}$	0
4	$0.1K_c'$	

The system would be stable for K_c' values which give

$$75 + 5K_c'^2 - 10K_c' \geq 0$$

$$\text{Now} \quad 75 + 5K_c'^2 - 10K_c' = 5(K_c' - 1)^2 + 70$$

which is always positive, for all values of K_c'. Thus, there is no restriction on the value of K_c' from the perspective of stability.

10.6 DYNAMICS OF DISTRIBUTED SYSTEMS

In the analysis done hitherto we have focused on the temporal variation of various properties. Thus while analyzing the water heating system we assumed thorough mixing of water so that the spatial variation of temperature within the tank could be neglected. Such an analysis is termed as lumped system analysis.

There are numerous important thermal systems where spatial variation of temperature cannot be ignored, for example, in heat exchangers (Fig. 4.1) where transfer of heat occurs between two flowing fluids separated from each other by a conducting surface. Dynamic analysis of such distributed systems would have temperature (and other properties) as function of both distance and time, and would thus result in partial differential equations. As an illustration of the methodology of analyzing such distributed systems we shall first derive the equations governing the dynamic behavior of a concentric tube heat exchanger, Fig. 10.31.

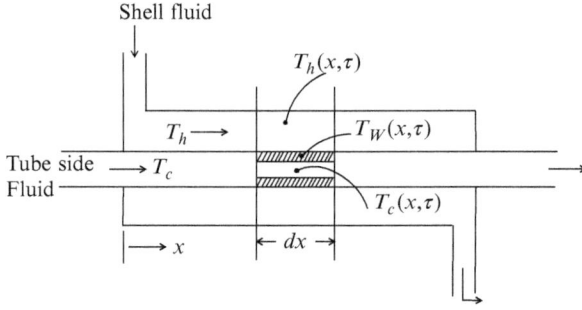

Fig. 10.31 Dynamic analysis of concentric tube heat exchangers.

In this heat exchanger both the hot fluid (properties indicated by subscript h) and the cold fluid (properties indicated by subscript c) flow parallel to each other in the same direction. Considering a small element of the heat exchanger of length dx situated at position x, we write down the energy balance equation for the two fluids (temperature T_c and T_h) and the tube walls (temperature T_w), using the general formulation of the first law of thermodynamics for open systems, Eq. (3.6) which, for the present case, becomes:

$$\frac{dE_{cr}}{d\tau} = \dot{Q}_{cv} + \dot{m}_i h_i - \dot{m}_e h_e \tag{10.122}$$

with $\dot{m}_i = \dot{m}_e$ for both the fluids.

Here we have neglected the effect of change in the elevation and the velocity of the fluids as they pass through the heat exchanger. We apply this equation first to the cold fluid flowing through the inner tube (Fig. 10.31). This gives

$$\frac{\partial}{\partial \tau}(dx \cdot A_c \cdot \rho_c \cdot c_{vc} \cdot T_c) = h_c(dxA_{sc})(T_w - T_c) + (A_c\rho_c \cdot V_c)c_{pc}(T_{ci} - T_{ce})$$

$$= \{h_c A_{sc}(T_w - T_c)\}dx + A_c\rho_c V_c c_{pc} \cdot \frac{\partial T_c}{\partial x}(-dx)$$

which can be simplified to

$$V_c\frac{\partial T_c}{\partial x} + \frac{c_{vc}}{c_{pc}} \cdot \frac{\partial T_c}{\partial \tau} = \frac{h_c A_{sc}}{A_c\rho_c C_{pc}}(T_w - T_c) \tag{10.123}$$

where A_{sc} is the tube inner surface area per unit length; V_c the velocity of the cold fluid, c_{vc}, c_{pc} being its specific heats at constant volume and constant pressure, respectively, and A_c is the cross-section area of its flow.

The ratio $\frac{c_{vc}}{c_{pc}}$ is significant only in gas flows; for liquids this value can be taken as 1.

Proceeding in a similar manner the energy balance equation for the hot fluid can be obtained as:

$$V_h\frac{\partial T_h}{\partial x} + \frac{c_{vh}}{c_{ph}} \cdot \frac{\partial T_h}{\partial \tau} = \frac{h_h A_{sh}}{A_h\rho_h c_{ph}}(T_w - T_h) \tag{10.124}$$

The energy equation for the metal walls can also be obtained from the general energy Eq. (10.122) by leaving the two mass flow rate terms on the RHS of the equation. This gives:

$$\frac{\partial}{\partial \tau}\{A_w \cdot \rho_w \cdot dx \cdot c_w T\} = -h_c dxA_{sc}(T_w - T_c) - h_h dxA_{sh}(T_w - T_h)$$

which can be simplified to

$$\frac{\partial T_w}{\partial \tau} = \frac{h_h A_{sh}}{A_w\rho_w c_w}(T_h - T_w) + \frac{h_c A_{sc}}{A_w\rho_w c_w}(T_c - T_w) \tag{10.125}$$

Defining the coefficients of these equations in terms of the time constants, we can write

$$\frac{A_c \rho_c c_{pc}}{h_c A_{sc}} = \tau_c; \quad \frac{A_h \rho_h c_{ph}}{h_h A_{sh}} = \tau_h; \quad \frac{A_w \rho_w c_w}{h_h A_{sh}} = \tau_{wh}; \quad \frac{A_w \rho_w c_w}{h_c A_{sc}} = \tau_{wc} \tag{10.126}$$

The governing equations for the dynamic analysis of this heat exchanger can now be compactly written as (Taking $c_v \approx c_p$)

$$V_c \frac{\partial T_c}{\partial x} + \frac{\partial T_c}{\partial \tau} = \frac{T_w - T_c}{\tau_c} \tag{10.127}$$

$$V_h \frac{\partial T_h}{\partial x} + \frac{\partial T_h}{\partial \tau} = \frac{T_w - T_h}{\tau_h} \tag{10.128}$$

$$\frac{\partial T_w}{\partial \tau} = \frac{T_h - T_w}{\tau_{wh}} + \frac{T_c - T_w}{\tau_{wc}} \tag{10.129}$$

Eqs. (10.127)–(10.129) are the three partial differential equations by solving which the dynamic analysis of a typical heat exchanger can be carried out.

We can use the above equations to express relationship between various variables in steady state. This gives, indicating the steady-state temperatures as $\overline{T}_c, \overline{T}_h, \overline{T}_w$,

$$V_c \frac{\partial \overline{T}_c}{\partial x} + \frac{\partial \overline{T}_c}{\partial \tau} = \frac{\overline{T}_w - \overline{T}_c}{\tau_c} \tag{10.130}$$

$$V_h \frac{\partial \overline{T}_h}{\partial x} + \frac{\partial \overline{T}_h}{\partial \tau} = \frac{\overline{T}_w - \overline{T}_h}{\tau_h} \tag{10.131}$$

$$\frac{\partial \overline{T}_w}{\partial \tau} = \frac{\overline{T}_h - \overline{T}_w}{\tau_{wh}} + \frac{\overline{T}_c - \overline{T}_w}{\tau_{wc}} \tag{10.132}$$

Subtracting Eqs. (10.130)–(10.132) from Eqs. (10.127)–(10.129) we get the governing equations in terms of the deviation variables (cf. Eq. 10.6)

$$V_c \frac{\partial T_c'}{\partial x} + \frac{\partial T_c'}{\partial \tau} = \frac{T_w' - T_c'}{\tau_c} \tag{10.133}$$

$$V_h \frac{\partial T_h'}{\partial x} + \frac{\partial T_h'}{\partial \tau} = \frac{T_w' - T_h'}{\tau_h} \tag{10.134}$$

$$\frac{\partial T_w'}{\partial \tau} = \frac{T_h' - T_w'}{\tau_{wh}} + \frac{T_c' - T_w'}{\tau_{wc}} \tag{10.135}$$

Assuming that at $\tau = 0$, the system is in steady state and the value of all deviation variables is zero, we can write the Laplace transform of the governing equations as

$$V_c \frac{dT_c'(s)}{dx} + sT_c'(s) = \frac{T_w'(s) - T_c'(s)}{\tau_c} \tag{10.136}$$

$$V_c \frac{dT_h'(s)}{\partial x} + sT_h'(s) = \frac{T_w'(s) - T_h'(s)}{\tau_h} \tag{10.137}$$

$$sT_w'(s) = \frac{T_h'(s) - T_w'(s)}{\tau_{wh}} + \frac{T_c'(s) - T_w'(s)}{\tau_{wc}} \tag{10.138}$$

We can use Eq. (10.138) to express $T_w'(s)$ in terms of $T_h'(s)$ and $T_c'(s)$, and then substitute this expression in Eqs. (10.136) and (10.137). This would give us two simultaneous ordinary differential equation in $T_c'(s)$ and $T_h'(s)$. An exact analytical solution of these equations cannot be obtained and so further analysis using Laplace transform is not possible without making some approximations.

However, if we consider simpler specific examples, complete transient response can be obtained using Laplace transforms. We shall illustrate this by considering the case where the hot fluid is saturated steam which is condensing over the cold tube. Thus the hot fluid temperature T_h is not a function of distance, though it may still vary with time. This results in considerable simplification since Eq. (10.137) becomes redundant. Thus, from Eq. (10.138) we can write

$$T_w'(s) = \frac{T_h'(s)\tau_{wc}}{(\tau_{wh}\tau_{wc}s + \tau_{wh} + \tau_{wc})} + \frac{T_c'(s)\tau_{wh}}{(\tau_{wh}\tau_{wc}s + \tau_{wh} + \tau_{wc})} \tag{10.139}$$

Substituting this in Eq. (10.136) gives

$$V_c\frac{dT_c'(s)}{dx} + sT_c'(s) + \frac{T_c'(s)}{\tau_c} = \frac{T_w'(s)}{\tau_c} = \frac{T_h'(s)\tau_{wc}}{(\tau_{wh}\tau_{wc}s + \tau_{wh} + \tau_{wc})\tau_c} + \frac{T_c'(s)\tau_{wh}}{(\tau_{wh}\tau_{wc}s + \tau_{wh} + \tau_{wc})\tau_c}$$

on rearranging we can write it as

$$V_c\frac{dT_c'(s)}{dx} + aT_c'(s) = bT_h'(s) \tag{10.140}$$

$$\text{where} \quad a = s + \frac{1}{\tau_c} - \frac{\tau_{wh}}{(\tau_{wh}\tau_{wc}s + \tau_{wh} + \tau_{wc})\tau_c} \tag{10.141}$$

$$b = \frac{\tau_{wc}}{(\tau_{wh}\tau_{wc}s + \tau_{wh} + \tau_{wc})\tau_c} \tag{10.142}$$

Eq. (10.140) is a first-order linear differential equation in $T_c'(s)$ whose solution for the boundary condition of $T_c'(x, s) = T_c'(0, s)$ at $x = 0$ is:

$$T_c'(x, s) = T_c'(0, s) + (1 - e^{-\frac{a}{v_c}\cdot x})\left(\frac{b}{a}T_h'(s) - T_c'(0, s)\right)$$

$$= T_c'(0, s)e^{-\frac{a}{v_c}x} + \frac{b}{a}(1 - e^{-\frac{a}{v_c}x})T_h'(s) \tag{10.143}$$

Eq. (10.143) gives the response of the cold fluid to any changes in the inlet water temperature and the steam temperature, in terms of the Laplace transforms of the temperatures. It can be converted into the two possible transfer functions by holding one constant.

Thus if the steam temperature is held constant, Eq. (10.143) reduces to

$$\frac{T_c'(x, s)}{T_c'(0, s)} = e^{-\frac{a}{v_c}x} \tag{10.144}$$

which is similar to the equation for transportation lag obtained earlier (Eq. 10.62), though with additional terms indicated in the expression for term "a" in Eq. (10.141).

On the other hand if the inlet water temperature is held constant, we get from Eq. (10.143):

$$\frac{T_c'(x, s)}{T_h'(0, s)} = \frac{b}{a}(1 - e^{-\frac{a}{v_c}x}) \tag{10.145}$$

substituting $x = L$, the length of the heat exchanger in Eqs. (10.144) and (10.145) we get the two transfer functions of this heat exchanger. The transient response to either of the two possible stimuli can be obtained by multiplying the Laplace transform of the stimulus with the transfer function and then taking the inverse transform.

Example 10.11

Consider the steam condenser shown in Fig. 10.32. Its design data are:
   Total surface area of contact between steam and water:

$$A_{sh} = 5\,\text{m}^2 \quad A_{sc} = 4.5\,\text{m}^2$$

Total cross-sectional area of tubes through which water is flowing : $50\,\text{cm}^2$

Total cross-sectional area of metal walls : $20\,\text{cm}^2$

Velocity of water through the tubes : $1\,\text{m/s}$

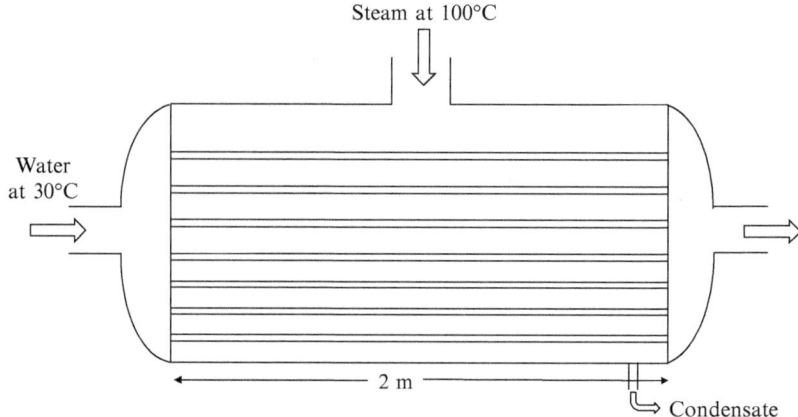

Fig. 10.32 Steam condenser of Example 10.10.

$$\text{Density of copper}: 8900\,\text{kg/m}^2; \quad \text{specific heat} = 385\,\text{J/kgK}$$

$$\text{Water side heat transfer coefficient}: 200\,\text{W/m}^2\text{K}$$

$$\text{Steam side heat transfer coefficient}: 1500\,\text{W/m}^2\text{K}$$

(a) Determine the transfer functions of this condenser for changes in the water temperature and the steam temperature.

Solution

We first determine various time constants using the given data.

$$\tau_c = \frac{A_c \rho_c c_{pc}}{h_c A_{sc}} = \frac{(50 \times 10^{-4})(1000)(4186)}{200 \times 4.5} = 23.26\,\text{s}$$

$$\tau_{wh} = \frac{A_w \rho_w c_w}{h_h A_{sh}} = \frac{(20 \times 10^{-4})(8900)(385)}{1500 \times 5} = 0.914\,\text{s}$$

$$\tau_{wc} = \frac{A_w \rho_w c_w}{h_c A_{sc}} = \frac{(20 \times 10^{-4})(8900)(385)}{200 \times 4.5} = 7.614\,\text{s}$$

$$a = s + \frac{1}{23.26} - \frac{0.914}{(s.0.914 \times 7.614 + 0.914 + 7.614)23.26}$$

$$= s + 0.04237 - \frac{0.005646}{(s + 1.2254)}$$

$$b = \frac{\tau_{wc}}{(T_{wh} T_{wc} s + \tau_{wh} + \tau_{wc})\tau_c} - \frac{7.618}{(6.9592 s + 8.528)} = \frac{1.09467}{s + 1.2254}$$

The two transfer functions are (Eqs. 10.140 and 10.141) ($V_c = 1 m/s$, $L = 2 m$)

$$\frac{T'_c(L, S)}{T_c(o, s)} = e^{-\left(s + 0.4237 - \frac{0.005646}{s + 1.2254}\right)L} = e^{-\left(2s + 0.08474 - \frac{0.011292}{s + 1.2254}\right)}$$

$$\frac{T'_c(L, S)}{T_h(s)} = \frac{1.09467}{(s + 0.04237)(s + 1.2254) - 0.005646}\left(1 - e^{-(2s + 0.08474 - \frac{0.011292}{s + 1.2254})}\right)$$

Considerable further simplification in the analysis is possible if we neglect the wall thermal capacity. The energy equation for the metal walls would now become

$$0 = h_h A_{sh}(T_h - T_w) + h_c A_{sc}(T_c - T_w) \qquad (10.146)$$

and in terms of the deviation variables, Eq. (10.138) would becomes

$$0 = \frac{T'_h(s) - T'_w(s)}{\tau_{wh}} + \frac{T'_c(s) - T'_w(s)}{\tau_{wc}}$$

$$\Rightarrow \quad T'_w(s) = \frac{T'_h(s) \cdot \tau_{wc}}{\tau_{wh} + \tau_{wc}} + \frac{T'_c(s) \cdot \tau_{wh}}{\tau_{wh} + \tau_{wc}}$$

$$= \frac{T'_h(s)\frac{1}{h_c A_{sc}}}{\frac{1}{h_c A_{sc}} + \frac{1}{h_h A_{sh}}} + \frac{T'_c(s) \cdot \frac{1}{h_h A_{sh}}}{\frac{1}{h_c A_{sc}} + \frac{1}{h_h A_{sh}}}$$

$$= \frac{T'_h(s)h_h A_{sh}}{h_c A_{sc} + h_h A_{sh}} + \frac{T'_c(s)h_c A_{sc}}{h_c A_{sc} + h_h A_{sh}} \tag{10.147}$$

Defining the overall heat transfer coefficient U based on the hot surface area as:

$$\frac{1}{UA_{sh}} = \frac{1}{h_h A_{sh}} + \frac{1}{h_c A_{sc}} \tag{10.148}$$

We can write the equation for wall temperature as

$$T'_w(s) = \frac{UA_{sh}}{h_c A_{sc}} \cdot T'_h(s) + \frac{U}{h_h} \cdot T'_c(s) \tag{10.149}$$

Substituting this in Eq. (10.136) we get

$$V_c \frac{dT'_c(s)}{dx} + sT'_c(s) = \frac{1}{\tau_c} \cdot \left[\frac{UA_{sh}}{h_c A_{sc}} \cdot T'_h(s) + \left(\frac{U}{h_h} - 1 \right) T'_c(s) \right]$$

which, on rearrangement gives

$$V_c \frac{dT'_c(s)}{dx} + \left(s + \frac{1}{\tau_c} - \frac{U}{\tau_c h_h} \right) T'_c(s) = \frac{1}{\tau_c} \cdot \frac{UA_{sh}}{h_c A_{sc}} \cdot T'_h(s)$$

This can also be put in the format of Eq. (10.140) with a and b now defined as

$$a = s + \frac{1}{\tau_c} \left(1 - \frac{U}{h_h} \right) \tag{10.150}$$

$$b = \frac{UA_{sh}}{\tau_c h_c A_{sc}} \tag{10.151}$$

Its solution would therefore be similar to Eq. (10.143) and the two transfer functions would be given by Eqs. (10.144) and (10.145) with $x = L$. This gives

$$\frac{T'_c(L, s)}{T'_c(0, s)} = e^{-\frac{a}{v_c} \cdot L} = e^{-[s + \frac{1}{\tau_c}(1 - \frac{U}{h_h})]\tau_d}$$

$$= e^{-s\tau_d} \cdot K_1 \tag{10.152}$$

where $\quad \tau_d = \dfrac{L}{V_c} \quad$ is the transportation lag $\tag{10.153}$

and $\quad K_1 = e^{-\frac{\tau_d}{\tau_c}(1 - \frac{U}{h_h})} \quad$ is a constant $\tag{10.154}$

The second transfer function would become

$$\frac{T'_c(L, s)}{T'_h(s)} = \frac{b}{a}(1 - K_1 e^{-s\tau_d}) \tag{10.155}$$

$$= \frac{b}{s + K_2}(1 - K_1 e^{-s\tau_d}) \tag{10.156}$$

$$\text{where} \quad K_2 = \frac{1}{\tau_c}(1 - \frac{U}{h_h}) = b \tag{10.157}$$

and "b" is given by Eq. (10.151).

These two transfer functions can be used to determine the response to changes in the cold water temperature and the steam temperature. Thus for step changes in these two stimuli by magnitudes ΔT_c and ΔT_h, respectively, we can write

$$T'_c(0, s) = \frac{\Delta T_c}{s}; \quad T'_h(s) = \frac{\Delta T_h}{s} \tag{10.158}$$

For the first case we get

$$T'_c(L, s) = \frac{\Delta T_c}{s} e^{-s\tau_d} \cdot K_1$$

$$= (K_1 \Delta T_c)\frac{1}{s}e^{-s\tau_d}$$

Its inverse transform is

$$T'_c(L, \tau) = K_1 \Delta T_c \cdot u_{\tau_d}(\tau) \tag{10.159}$$

which shows that the change in the outlet temperature of the cold fluid is delayed by τ_d seconds and the amplitude is K_1 times the change in the stimulus.

Similarly for the second case (change in the steam temperature by ΔT_h) we get

$$T'_c(L, s) = \frac{\Delta T_h}{s} \cdot \frac{b}{s + K_2} \cdot (1 - K_1 e^{-s\tau_d})$$

$$= \frac{\Delta T_h b}{K_2}\left[\frac{1}{s} - \frac{1}{s + K_2} - \frac{K_1}{s}e^{-s\tau_d} + \frac{K_1}{s + K_2}e^{-s\tau_d}\right] \tag{10.160}$$

Its inverse transform, using Eq. (10.66) is ($b = K_2$)

$$T'_c(L, \tau) = \Delta T_h\left[(1 - e^{-K_2\tau})u_o(\tau) - K_1(1 - e^{-K_2(\tau-\tau d)})u_{\tau d}(\tau)\right] \tag{10.161}$$

To get the actual values of the temperatures we need to add to the deviation value, the steady-state temperature at that location. For this we need the steady-state temperature distribution in the condenser. This can be found by solving Eq. (10.130), making use of the simple relation between $\overline{T}_w(x)$ and $\overline{T}_c(x)$ given by Eq. (10.146). This gives on simplification

$$\tau_c V_c\frac{d\overline{T}_c}{dx} + \left(1 - \frac{U}{h_h}\right)\overline{T}_c = \frac{\overline{T}_h UA_{sh}}{h_c A_{sc}} \tag{10.162}$$

Solution of Eq. (10.162) is (cf. Eq. 10.143)

$$\overline{T}_c(x) = \left(\overline{T}_c(0)\right) e^{-(1-\frac{U}{h_h})\frac{x}{V_c\tau_c}} + \frac{\overline{T}_h UA_{sh}}{h_c A_{sc}(1-\frac{U}{h_h})}\left(1 - e^{-(1-\frac{U}{h_h})\frac{x}{V_c\tau_c}}\right) \qquad (10.163)$$

This can be put in a compact form using the constants τ_d, and K_2 defined by Eq. (10.153), and (10.157) to get

$$\overline{T}_c(x) = \left(\overline{T}_c(0)\right) e^{-K_2\cdot\tau_d\cdot\frac{x}{L}} + \overline{T}_h(1 - e^{-K_2\tau_d\frac{x}{L}})$$
$$= \overline{T}_h + \left(\overline{T}_c(0) - \overline{T}_h\right) e^{-K_2\tau_d\frac{x}{L}} \qquad (10.164)$$

Combining Eqs. (10.164) and (10.161) the actual temperature profile during the transient operation of the heat exchanger can be found.

Example 10.12

The steam condenser of Example 10.11 is operating in steady state with water entering at 30°C and the steam temperature of 200°C.

(a) What is the water exit temperature?

(b) If the water inlet temperature is suddenly increased to 35°C, how would the water exit temperature change with time, steam temperature remaining constant.

(c) If the water temperature remains the same but the steam temperature increases by 10°C, how would the heat exchanger respond?

Solution

We first determine the U value for the heat exchanger using Eq. (10.148).

$$\frac{1}{U} = \frac{1}{h_h} + \frac{A_{sh}}{A_{sc}} \cdot \frac{1}{h_c} = \frac{1}{1500} + \frac{5}{4.5} \cdot \frac{1}{200} = 0.006222$$

$$\Rightarrow \quad U = 160.71 \text{ w/m}^2 \text{ K}$$

This gives $K_2 = \dfrac{1}{\tau_c}\left(1 - \dfrac{U}{h_h}\right) = \dfrac{1}{23.26}\left(1 - \dfrac{160.71}{1500}\right)$

$$= 0.03839$$

Also $\tau_d = \dfrac{L}{V_c} = \dfrac{2}{1} = 2 \text{ s}$

(a) Substituting these in the equation for cold fluid temperature profile, Eq. (10.164), gives

$$\overline{T}_c(x) = 200 + (30 - 200)e^{-(0.03839 \times 2 \times \frac{x}{L})}$$
$$= 200 - 170e^{-0.07678\frac{x}{L}}$$

The temperature of the cold water leaving the heat exchanger would therefore be

$$\overline{T}_c(L) = 200 - 170e^{-0.07678\frac{L}{L}} = 42.564°C$$

(b) If the water temperature is suddenly raised by 5°C, the response to this change is given by Eq. (10.159) where

$$\Delta T_c = 5°C \quad K_1 = e^{-\frac{\tau_d}{\tau_c}\left(1-\frac{U}{h_h}\right)}$$
$$= e^{-\frac{2}{23.26}\left(1-\frac{160.71}{1500}\right)}$$
$$= e^{-0.07677} = 0.9261$$

Therefore, the change in water outlet temperature is:

$$T'_c(L,\tau) = 5 \times 0.9261 \times u_2(\tau)$$
$$= 0 \quad \text{for} \quad \tau < 2$$
$$= 4.6305°C \quad \text{for} \quad \tau > 2$$

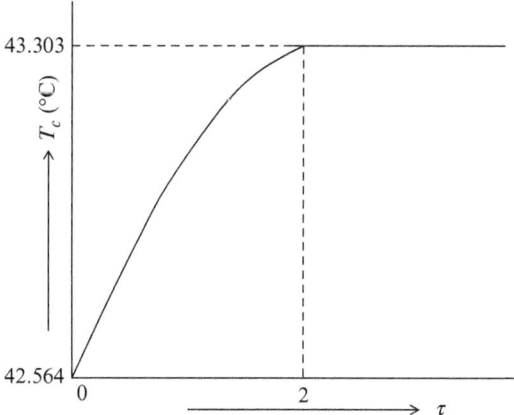

Fig. 10.33 Dynamic response of the steam condenser.

Adding to it the steady-state value obtained above, we can say that

$$T_c(L,\tau) = 42.564°C \quad \text{for} \quad \tau < 2$$
$$= 47.1945°C \quad \text{for} \quad \tau > 2$$

(c) The response to sudden change in the steam temperature is given by Eq. (10.161). To use it we first find the value of constant b which is given by Eq. (10.151).

$$b = \frac{UA_{sh}}{\tau_c h_c A_{sc}} = \frac{(160.71)5}{23.26 \times 200 \times 4.5} = 0.03839$$

Substituting values of the constants in Eq. (10.161) we get with $\Delta T_h = 10°C$

$$T_c'(L, \tau) = 10[(1 - e^{-0.03839\tau})u_o\tau - 0.9261(1 - e^{-0.03839(\tau-2)})u_2(\tau)]$$

$$= 0 \quad \tau < 0$$

$$= 10(1 - e^{-0.03839\tau}) \quad 0 \leq \tau < 2$$

$$= 10(1 - e^{-0.03839\tau}) - 9.261(1 - e^{-0.03839(\tau-2)}) \quad \text{for } \tau \geq 2$$

The actual temperature variation is thus

$$T_c(L, \tau) = 42.564 \quad \tau < 0$$

$$= 52.564 - 10(1 - e^{-0.03839\tau}) \quad 0 \leq \tau < 2$$

$$= 43.303 - 10e^{-0.03839\tau} + 9.261e^{-0.03839(\tau-2)} \quad \tau \geq 2$$

Fig. 10.33 shows the nature of this temperature profile. The outlet temperature becomes steady after 2 s.

PROBLEMS

10.1 A thermocouple installed inside a furnace has a time constant of 10 s. The furnace is initially at a steady temperature of 500°C. Determine the response of the thermocouple to change in the furnace temperature in accordance with Fig. 10.34.

10.2 Consider the flow of gas through an orifice of cross-sectional area A. The volumetric flow rate is given by the equation

$$Q = C_d \cdot A \cdot \sqrt{(P_1 - P_2)/\rho}$$

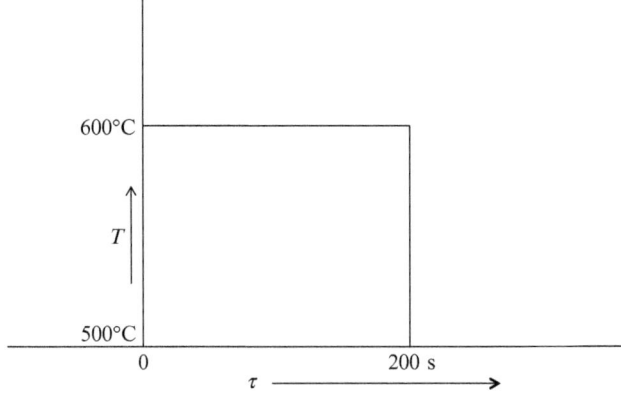

Fig. 10.34 Profile of temperature variation in Problem 10.1.

where P_1 and P_2 are the upstream and downstream pressures, ρ is the fluid density, and C_d is the coefficient of discharge of the orifice.

Determine the transfer functions of the orifice for change in the flow rate due to change in either of the two pressures.

10.3 A water tank, 0.5 m in diameter is in steady state at a water level of 1 m, with water flowing out from a 5 cm pipe at its bottom (Fig. 10.35).

(a) If the coefficient of discharge in 0.6, what is the steady-state water flow rate?

(b) Determine the transfer function for water level with respect to change in the incoming water flow rate.

(c) If the water flow rate is suddenly stopped, how would the level of water fall with passage of time?

10.4 A thermocouple is being employed to measure the temperature of a furnace maintained at a temperature which is high enough to cause significant heat transfer by radiation. Show that the dynamic response of the thermocouple to variation in furnace temperature T_∞ can be represented by the equation

$$\frac{dT}{d\tau} = C_1(T_\infty^4 - T^4) + C_2(T_\infty - T)$$

What are the values of the constants in terms of the parameters governing the transient response? Linearizing the above equation determine the transfer function of the thermocouple relating the change in its temperature to change in the furnace temperature.

10.5 The waste heat recovery boiler attached to a DG set has a large number of tubes carrying the flue gas immersed in a large pool of water. Fresh water enters the boiler at a flow rate $(\dot{m}_w)$ and the steam is also withdrawn at the same rate.

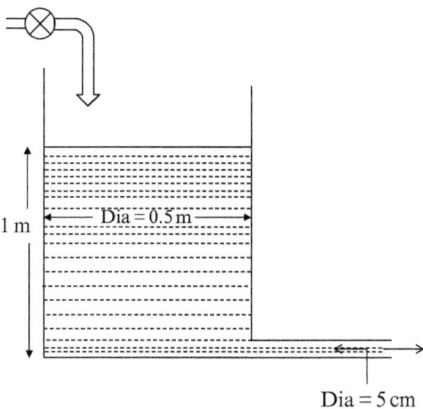

Fig. 10.35 Water tank of Problem 10.3.

The temperature of the feedwater can be assumed to be equal to the saturation temperature at the boiler pressure. The temperature of the flue gases entering the boiler changes with load on DG set. As a result the water evaporation rate changes, which alters the steam pressure. Determine the transfer function P'_s/T'_s using the following simplifying assumptions.

Mass of vapor in the boiler $= C_1 P$

Effectiveness of the boiler $= \epsilon$

Mass flow rate of the flue gas $= \dot{M}_{gas}$

Specific heat of flue gas $= C_{pgas}$

$T_{steam} \approx C_2 P$

C_1, C_2, ϵ: constants

10.6 Fig. 10.36 shows the arrangement for measuring the temperature of a fluid flowing through a pipe. The thermometer is dipped in oil kept in a thermowell which is welded to an opening in the pipe. Show that the dynamic behavior of the thermometer with respect to changes in the fluid temperature follows that of a third-order system. Determine the transfer function.

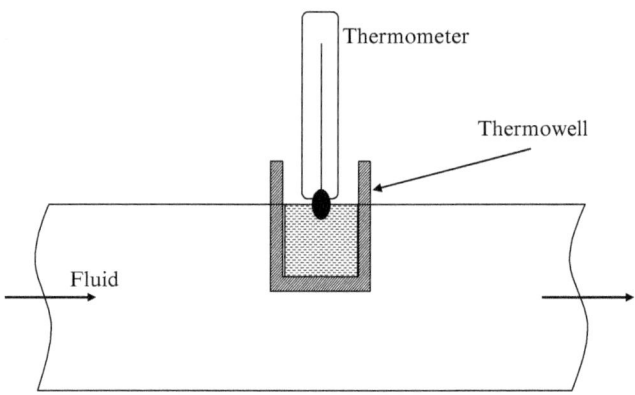

Fig. 10.36 Thermometer dipped in a well, Problem 10.6.

10.7 Fig. 10.37 shows a simple water manometer used to measure pressure difference. The total length of water column in both the tubes is L and the diameter of the tubes is D. If at time $t = 0$, a pressure difference occurs suddenly, determine the manner in which the water column will move, and the transfer function governing the response. Assume that the flow remains laminar, and the equations governing the flow in steady state are applicable during the transient operation also.

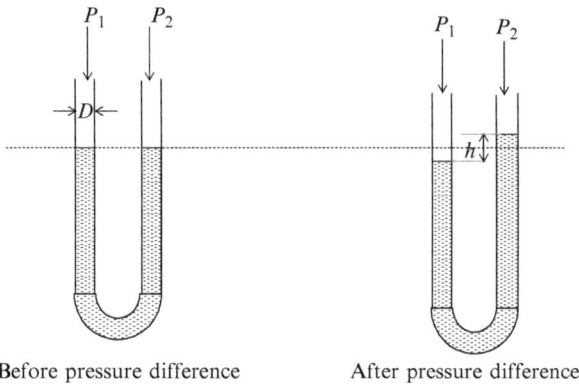

Before pressure difference   After pressure difference

Fig. 10.37 Manometer of Problem 10.7.

Solve the above problem for following data:

$$L = 1.50 \text{ m} \quad \text{Viscosity of water} = 0.8 \times 10^{-3} \text{ N s/m}^2$$
$$D = 0.05 \text{ m} \quad \text{Density of water} = 1000 \text{ kg/m}^3$$
$$P_1 = P_2 \text{ for } t < 0 \quad \text{and} \quad P_1 - P_2 = 100 \text{ kPa for } t \geq 0$$

10.8 Hot water is entering a tank at a temperature of 50°C with a flow rate of 1.5 kg/min, and flows out through a 5-m long pipe (Fig. 10.38). The tank and the pipe are well insulated and a mixer ensures uniformity of the tank temperature. Due to some problem in the water heating system, the incoming water temperature suddenly drops to the ambient temperature of 30°C.

(a) How would the temperature of the tank change with time?

(b) A thermocouple with a time constant of 5 s is placed at the exit of the pipe to measure the temperature of outgoing water. Determine its temperature response and compare it with the actual tank-water temperature.

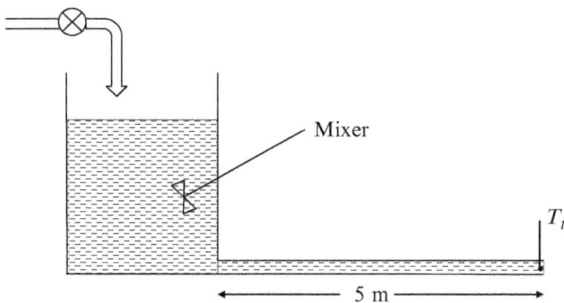

Mixer

T_t

5 m

Fig. 10.38 Hot water tank and pipe.

10.9 Solve Problem 10.8 for the case when the tank is not insulated and loses heat to the surroundings at 30°C. Take the overall heat-transfer coefficient between the tank water and the surroundings as 100 W/m²K. The tank surface area is 2 m².

10.10 Hot water from a boiler at 80°C is mixed with cold water at 30°C to get water at 60°C which is transported to a bathroom, 10 m away (Fig. 10.39). The pipe is uninsulated and the overall heat transfer coefficient between water and atmosphere is 50 W/m²K.

(a) Determine the temperature of water at the entry to the bathroom.

(b) Due to some problem in the boiler, the hot water supply suddenly ceases and cold water at 30°C now enters the pipe. Determine the rate at which the water temperature at pipe exit would change. Assume, mass flow rate of the water in the pipe remains constant at 5 kg/min, and the velocity of water in it as 1.2 m/s.

10.11 A mercury in glass thermometer is being used to measure temperature. The volume of the glass bulb exhibits first-order dynamics to changes in the surrounding temperature with a time constant τ_g and gain K_g. The volume occupied by the mercury can also be assumed to exhibit first-order dynamics to any change in the temperature of the surrounding glass with $\tau_m = 0.1$ s and $K_m = 1$. For what values of τ_g and K_g would the thermometer show inverse response?

10.12 A room is initially at the ambient temperature T_∞. At time $t = 0$, cool conditioned air at a constant temperature T_i starts entering the room at a steady mass flow rate $\dot{m}$. The air can be assumed to mix instantly with the room air, and leave the room at the same mass flow rate, and the room temperature $T(t)$ at any

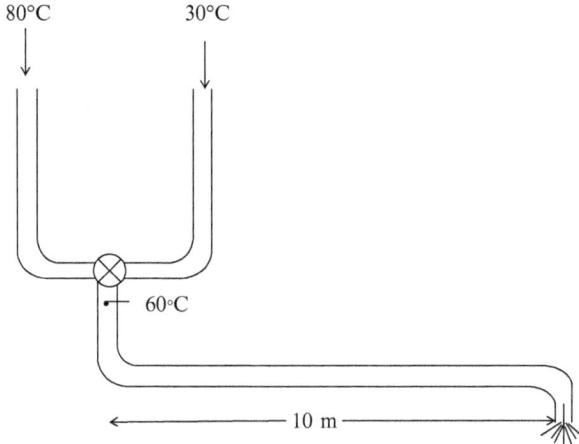

Fig. 10.39 Hot water supply to a bath room, Problem 10.10.

instant t. The overall heat transfer coefficient times the area (UA) of the room walls can be taken to be constant during the entire operation of temperature pull down, and the heat leaks into the room at a rate $UA(T_\infty - T(t))$ at any instant.

(a) Write the differential equation governing the energy balance of the room, expressing the room temperature T as a function of time t and other parameters listed above.

(b) Solve the above equation using Laplace transform to determine the variation in temperature T as a function of time.

(c) From this solution obtain the expression for the steady-state temperature of the room.

10.13 An electric heater is being used to heat ambient air at temperature T_i. Show that for small deviations in voltage from the rated value (200 V) the change in outlet temperature T_o is given by the transfer function

$$\frac{T'_o(s)}{V'(s)} = \frac{0.01 \; K/C_M}{1 + \left(\frac{C_M}{H}\right) s}$$

where

$$C_M = \text{mass flow rate of air} \times \text{specific heat of air}$$
$$C_H = \text{mass of heaters} \times \text{specific heat of heater material}$$
$$H = h \cdot A$$
$$h = \text{heat transfer coefficient between air and heater}$$
$$A = \text{surface area of the heater}$$
$$K = \text{kW rating of the heater}$$

Determine the transient response for a step change in voltage from 200 V to with 220 V with

$$H = 50 \, \text{W/K}; \quad C_H = 5 \, \text{kJ/K}; \quad C_M = 0.5 \, \text{kW/K}; \quad K = 10 \, \text{kW}; \quad T_i = 20°\text{C}$$

10.14 Fig. 10.40 shows the arrangement of safe heating of a liquid through a secondary fluid inside an insulated calorimeter. The electrical heater is immersed in the secondary fluid which evaporates and transfers heat to the coil carrying the primary fluid entering at temperature T_i. In turn, these vapors condense and fall down to the bottom of the container. The system is fitted with a control system which changes the electrical power supplied (P) to the heater in response to changes in the outlet temperature of the primary fluid (T), as measured by the thermocouple, T_m, (with a time constant τ_c) in accordance with the following relationship:

$$P = K(T_{set} - T_m)$$

where T_{set} is the set point temperature of the controller.

The rate of heat transfer (Q) in the coil of surface area A can be assumed to be governed by the relationship

$$Q = h_i A (T_i - T_{sat})$$

where the heat transfer coefficient h_i can be assumed to be constant.

T_{sat} the saturation temperature of the secondary fluid is a complex function of the power input. However, for the sake of simplification it can be assumed that the rate of change of deviation in T_{sat} is linearly related to the deviation in power P, from its steady-state value, that is,

$$\frac{dT'_{sat}}{dt} = C_4 P'$$

Draw the complete block diagram of the control loop in terms of the deviation variables with the primary fluid inlet temperature T_i as the load and its outlet temperature T as the controlled variable. Find out the transfer function for the system.

Using this transfer function find the response of the system to a sudden change in the fluid inlet temperature.

10.15 Consider a counterflow heat exchanger. Design a control system to maintain the cold-fluid exit temperature constant irrespective of the changes in the incoming stream temperature, without changing its flow rate. The piping arrangement permits by-passing a part of the hot fluid entering the heat exchanger. Draw

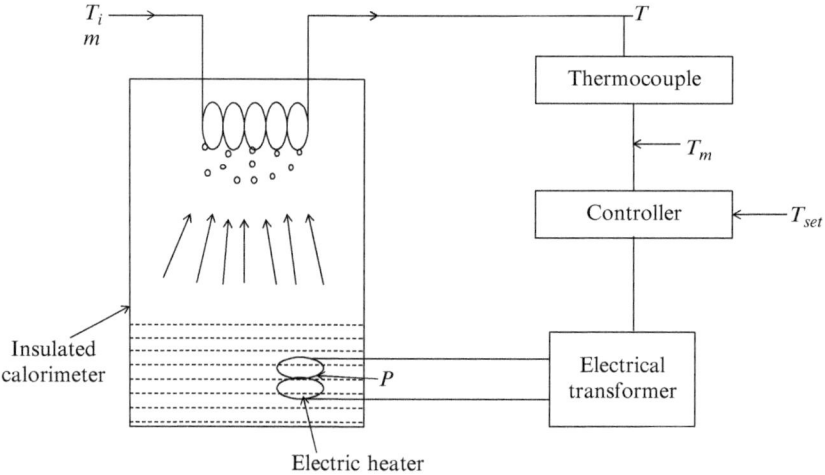

Fig. 10.40 Insulated calorimeter of Problem 10.14.

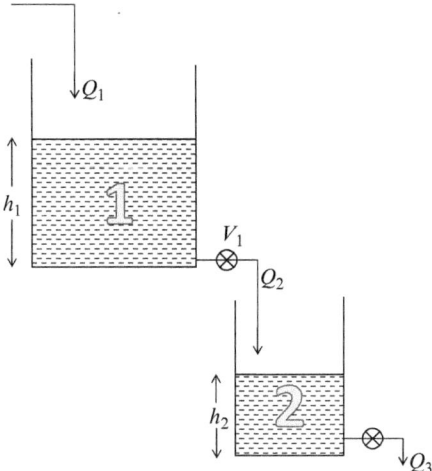

Fig. 10.41 Two-tank system of Problem 10.16.

its block diagram and determine the transfer function assuming proportional control.

10.16 Fig. 10.41 shows water flowing at rate $Q_1 m^3/3$ into tank 1 from outside. Depending upon the level of water (h_1) in this tank, and the adjustment of valve V_1, water leaves this tank with a flow rate Q_2 and flows into tank 2. It is desired to keep the level of water in tank 2 (h_2) at a specified value so that the rate of outflow from it remain constant. Design a control system for this two-tank system and determine its transfer function, assuming *PI* control.

10.17 Fig. 10.42 shows the block diagram of the control system of a thermostatic bath. Using the Routh's criterion determine the values of τ for which the output C is stable for all inputs R and U.

10.18 Consider the following transfer function of a system

$$TF = \frac{1}{(s+1)(s^2 - 3s + 2)}$$

Using Routh's criterion determine whether such a system would be stable or not. Verify the result by determining the dynamic response of the system to a step change in the input.

10.19 Solve Example 10.9 with a measurement lag of 5 s.

10.20 Solve Example 10.9 with the measuring device being a second-order system with transfer function $= 1/(1 + 5s + s^2)$.

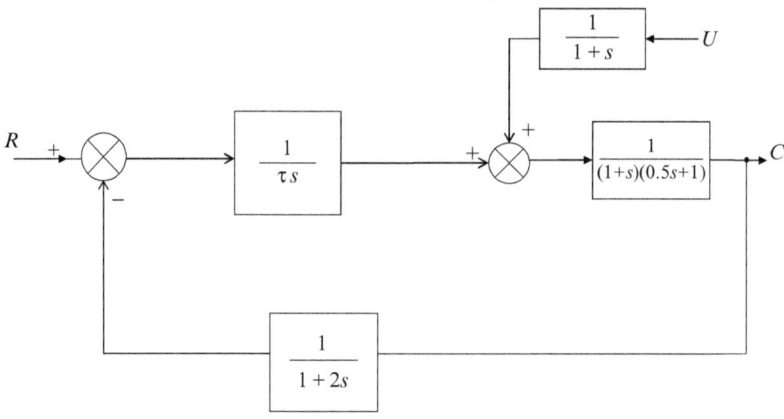

Fig. 10.42 Block diagram of the control system of Problem 10.17.

10.21 Consider a pin fin attached to a hot surface to increase its heat dissipation, which is taking place both by convection and radiation. Consider the energy balance of a small length of the fin and derive the equation expressing the fin temperature T as a function of time (τ) and distance along the fin as measured from its base attached to the hot surface. Assuming that at $\tau = 0$, the fin is already in steady state express the equation in terms of T', the deviation from the steady-state fin temperature $\overline{T}$ and T'_∞, the deviation from steady-state environment temperature $\overline{T}_\infty$.

10.22 Consider the problem of transient $1 - D$ conduction through the wall of a room. The temperature at the outer surface of the wall is taken as the stimulus and that at the inner surface as the response. Show that the transfer function is

$$\frac{T'(L, s)}{T'(0, s)} = e^{-\sqrt{s/\alpha}L}$$

where L is the wall thickness and α the thermal diffusivity, of the wall material.

Using this determine an expression for the change in the inner-wall temperature $T(L, \tau)$ due to a step change in the temperature at the outer wall.

10.23 Solve Example 10.12 for the case when both the water inlet temperature and the steam temperature rise by $10°C$. How would the heat exchanger respond?

10.24 Fig. 10.43 shows a power electronic unit approximated for thermal analysis as a box of dimensions $a \times b \times L$ losing heat to surroundings at a temperature T_∞.

Assuming both the base and the tip of the box to be at the surrounding temperature write down the energy balance at a typical location on the unit, assuming a uniform generation of heat at a volumetric rate Q W/m^3.

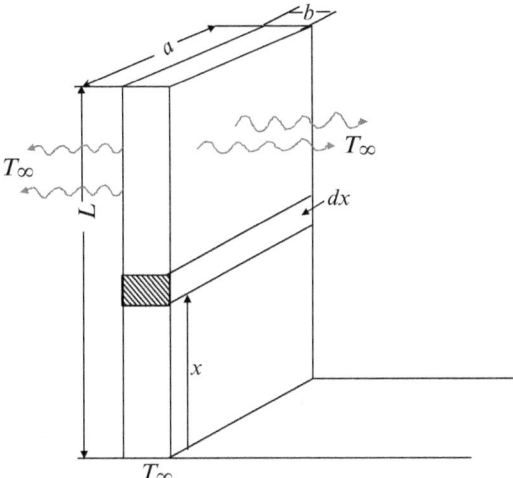

Fig. 10.43 The power-electronic unit of Problem 10.24.

(a) Determine the steady-state solution and the location where the temperature would be maximum.

(b) Assuming that at $\tau = 0$, the unit is in steady state, determine the Laplace transform for the change in the maximum temperature as a function of the change in the heat generation rate.

(c) Using this determine the response to a sudden change in the heat generation rate by an amount ΔQ.

CHAPTER 11

Additional Considerations in Thermal System Design

In our discussions on thermal system design in earlier chapters, we have focused primarily on thermofluid considerations, such as achieving the desired rates of heat transfer and/or pressure drop. In actual engineering practice many other considerations can become equally important. These include erosion and corrosion rates, fouling, the physical nature of the working fluid, vibrations, noise, safety, effect on environment, etc. In this chapter we shall study how these factors can impose additional constraints on the design of equipment.

Another important factor that can influence the optimal design is the uncertainties in the values of various parameters assumed to be constant in earlier chapters. The various types of uncertainties that are frequently encountered include those associated with various thermophysical properties and the empirical correlations used to estimate the heat transfer coefficient and pressure gradient, uncertainties in the environmental temperature and humidity, about the actual "thermal" load on the system and even about the cost data. A brief discussion on these stochastic considerations is also included in this chapter.

11.1 EROSION-CORROSION

Erosion refers to the wear of the metal surface due to high velocity of the fluid flowing over it. It typically occurs in soft metals such as copper and aluminum which are the preferred materials for heat exchanger tubes due to their high thermal conductivity. The most vulnerable positions are pipe bends (elbows), tube constrictions, and near the tube entrances. The rate of erosion depends on the fluid velocity. Since increasing the fluid velocity improves the rate of heat transfer, erosion considerations put a higher limit on the permission velocities to ensure sufficient equipment life. Typical recommended maximum velocities for specific materials are given in Table 11.1.

Fluid temperature also plays an important part in governing the erosion rate. Typically, the erosion rates increase with increase in temperature. Thus, while maximum recommended velocity in copper tubes is 2.4 m/s for cold water, if the average temperature of water is $\approx 60°C$, maximum recommended velocity is 1.5 m/s. If

Thermal System Design and Simulation
http://dx.doi.org/10.1016/B978-0-12-809449-5.00011-5

Table 11.1 Maximum water velocities for specific material

Low carbon steel	3 m/s
Stainless steel	4.5 m/s
Aluminum	1.8 m/s
Copper	2.4 m/s
90–10 Cupronickel	3.0 m/s
70–30 Cupronickel	4.5 m/s

the water temperature is usually higher than 60°C, the flow velocities should not exceed 1 m/s.

For fluids other than water, the maximum allowable fluid velocity (V_f) to ensure minimum erosion can be roughly estimated from the equation:

$$\frac{V_f}{V_w} = \left(\frac{\rho_w}{\rho_f}\right)^{1/2} \tag{11.1}$$

where ρ stands for density, V for velocity, subscript w for water, and subscript f for the fluid.

Erosion becomes a very serious issue when the fluids carry suspended particles (like ash in flue gases). These particles cause loss of material by abrasive wear and impact. This has resulted in development of special (nickel, chromium, and tungsten) alloys and coating for boiler tubes since failures entail huge repair and plant downtime costs. The material selection for boiler tubes is thus often governed by the need to have high-erosion resistance [1]. Still cost and erosion considerations impose an upper limit of 10–15 m/s on the velocity of the flue gas over the boiler tubes. This needs to be incorporated as a constraint in boiler design.

In some situations the rate of erosion is considerably enhanced due to the presence of water droplets in the fluid stream. Thus wet steam hitting diverter plates, and turbine blades can cause extensive pitting and erosion of the metallic surface due to the localized impact of high velocity water droplets. While use of special materials like alloys of steel containing nickel, chromium, and tungsten is helpful, the thermodynamic cycle is designed to ensure that the wetness fraction of steam does not exceed 10% at any position within the turbine. This becomes a crucial constraint in thermal power plant design. The erosion is very much enhanced in the presence of corrosion, that is, destruction of the surface due to chemical reaction which transforms it to another chemical compound. The most common cause of corrosion in the steel tubes of the condensers is the presence of chlorides in water especially when its pH is lower than 7. The presence of certain bacteria can also influence the corrosion rate. The control of corrosion therefore needs careful control of the quality of water circulating through the condenser tubes. However, it does not directly introduce any additional constraint in the design process.

The presence of any impurities, like salts in water, which may crystallize on the surface; or impurities, like sand or rust in water, which may deposit on the heat transfer surface, or biological organic growth material, has a strong effect on the extent of fouling. This may result in much greater reduction in the heat transfer rate than that accounted for by the fouling factor mentioned in Chapter 3. The design features which can minimize the effects of fouling include:

(a) maintaining high level of turbulence to prevent sedimentation (without increasing erosion);

(b) minimizing "dead spots" and velocity turns;

(c) ensuring easy accessibility to the heat transfer surfaces so that these can be easily cleaned mechanically.

11.2 VIBRATION AND NOISE

The thermal equipment can be classified into three categories from the perspective of vibration and noise: reciprocating devices such as the IC engines and compressors; rotating devices such as rotary compressors and fans; heat exchangers where noise and vibration are caused by the flow of fluid over tube bundles. The generation of vibration and noise in the IC engines and refrigeration compressors has been widely studied and appropriate measures devised to reduce it. Proper balancing of the inertial forces of the reciprocating and rotating masses is a prerequisite of all designs of reciprocating engines and compressors. In fact, various designs like cylinders arranged in V or W formations have been evolved primarily to reduce the vibration and noise. To reduce noise due to impulsive exhaust discharged from IC engines, mufflers are put in the exhaust pipe. These muffle the noise by making engine exhaust to flow through two or three large volume chambers in a circuitous manner. Again, to reduce transmission of engine vibrations (e.g., in an automobile), suitable rubber mounts are employed. The design of these mounts and mufflers is usually done by specialists in dynamics and noise control, and has no bearing on thermal system design. In contrast the noise control of rotating fans has a direct bearing on its design. Studies on noise in air cooled heat exchangers reveal that fan-generated noise is the major constituent of the total unit noise. This noise is strongly dependent on the fan tip speed; typically it is proportional to tip speed to a power of 5.6–5.8 [2]. Blade design is another crucial factor since this decides the flow rate obtained at a particular speed. The so-called low-noise fans have tapered twisted airfoil blades with wide chord which enables achieving same air flow rate at lower tip speeds. Thus the detailed aerodynamic design of fan blades is intimately related to the fan noise.

The importance of flow induced vibrations (FIVs) in heat exchangers (and the resulting noise) is now being increasingly appreciated. Though this noise is not so intense as the noise of reciprocating devices, incidents of boiler tube failure due to onset of heavy vibrations have drawn the attention of researchers and heat exchanger designers toward it. This problem has become more acute in modern compact heat exchangers, which are designed to operate at high fluid velocities. In the most popular design of the shell-and-tube heat exchangers, the trend has been toward fewer number of baffles to contain the shell side pressure drop at high fluid velocities. Both these factors, such as higher fluid velocities and fewer structural supports for the tube bank, exacerbate the FIV. Over last three decades the phenomenon of FIV has been studied in considerable detail with a view to evolve guidelines for heat exchanger design to minimize vibrations. A very brief review is presented here (for details see, e.g., Thulukkanam [2, Chapter 10]).

The vibrations can be induced in a heat exchanger due to four reasons, such as
1. vortex shedding;
2. turbulent buffeting;
3. acoustic resonance; and
4. fluid elastic instability.

The first three of these reasons are related to the resonance phenomena, which occurs when the frequency of excitations due to any of these reasons synchronizes with the natural frequency of vibration of the tubes. Fluid elastic instability occurs at a certain critical flow velocity when the amplitude of the tube vibration becomes large enough to cause collision of adjacent tubes, often resulting in tube failure.

11.2.1 Vortex Shedding

Vortex shedding refers to the formation of vortices due to vortex flow in the wake region behind the tube across which the flow is occurring (Fig. 11.1). This phenomenon is characterized by a nondimensional parameter known as the Strouhal number Su defined as

$$Su = \frac{f_v D}{U_\infty} \tag{11.2}$$

where D is the tube outer diameter, U_∞ the free stream velocity, and f_v the vortex shedding frequency. When f_v is very close to the natural frequency of vibration of the tube, resonance takes place resulting in large amplitude of vibration of the tube. The values of Su, as experimentally found for single tubes, are

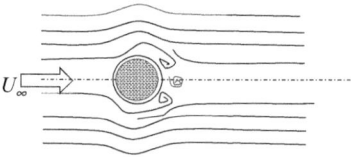

Fig. 11.1 Vortex shedding past a single tube.

$$300 < Re \le 2 \times 10^5 \qquad Su = 0.2$$
$$2 \times 10^5 < Re \le 3.5 \times 10^5 \qquad Su \text{ increases slightly}$$
$$3.5 \times 10^5 < Re \le 3.5 \times 10^6 \qquad \text{No vortex shedding}$$
$$Re > 3.5 \times 10^6 \qquad Su = 0.27 \qquad (11.3)$$

Heat exchangers have bundles of tubes arranged in a certain pattern and the value of Strouhal number depends on the two tube pitches and the tube layout. Many researchers have prepared maps showing plots of Su as a function of the pitch ratio (P/D) for different layouts; Thulukkanam [2] gives correlations for four tube layouts of Fig. 11.2, developed by Weaver.

$$Su = \frac{1}{1.73X_p} \quad \text{for 30 degree layout [Normal Triangle]}$$

$$= \frac{1}{1.16X_p} \quad \text{for 60 degree layout [Parallel Triangle]}$$

$$= \frac{1}{2X_p} \quad \text{for 90 degree and 45 degree layout [Normal \& Rotated square]} \quad (11.4)$$

$$\text{where} \quad X_p = P/D; \quad Su = \frac{f_v D}{U} = \frac{f_v D}{U_\infty [P/(P-D)]} \quad (11.5)$$

U being the velocity of flow in the inter-tubular gap. Thus, for any of the above configurations, using Eq. (11.4) the vortex shedding frequency can be estimated. The designer should then compare f_v with the natural frequency of the vibration of the tube, f_n. Ganapathy [3] suggests a simple equation to estimate f_n:

$$f_n = \frac{C}{2\pi} \sqrt{\left(\frac{E \cdot I \cdot}{M_e L^4} \right)} \quad (11.6)$$

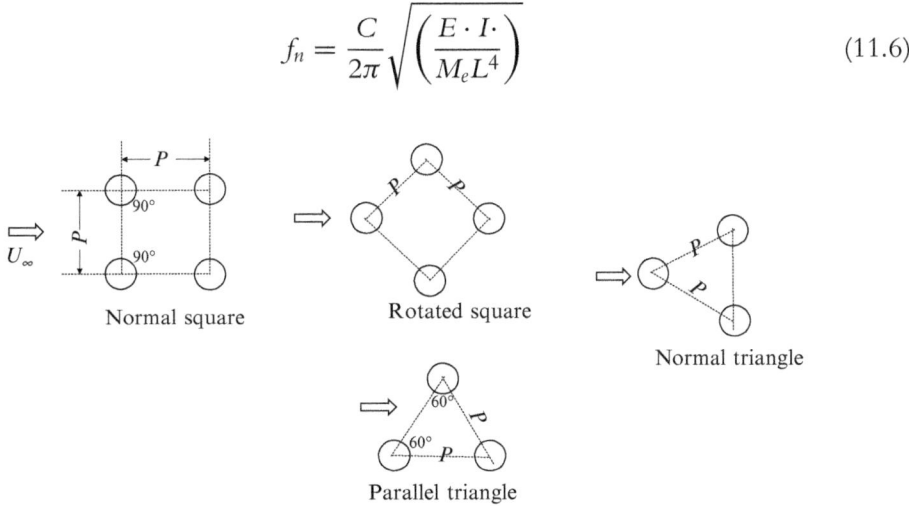

Fig. 11.2 Various tube layouts in Weaver's correlation.

where M_e is the total effective weight of the tube, which includes the contribution of the fluid weight inside and outside the tubes; I is the moment of inertia of the tubes; E the Young's modulus of elasticity for the material; L the tube length; and C a constant whose value varies over a wide range ($\sim$10–120) depending on the end support conditions. To avoid resonance f_v and f_n should not be within 20% of each other. Clearly by changing the tube pitch, the tube span or the end conditions, the heat exchanger can be designed to keep f_n and f_v far apart and avoid vibration problems.

For more accurate calculations the elaborate procedures suggested in the literature (e.g., Thulukkanam [2]) should be used. Since some of the factors influencing both f_n and f_v are difficult to quantify, for example, the effect of manufacturing tolerances and the end conditions, often experimental validation is also necessary before releasing the product for mass production.

11.2.2 Turbulence Included Vibrations

In modern compact heat exchangers the turbulence levels are generally very high. The fluctuating pressure fields present in such highly turbulent flows can also directly induce vibrations in the tube bundle, besides influencing it through vortex shedding. Since the perturbations in turbulent flow are associated with turbulent eddies of randomly varying frequencies, the turbulence induced vibrations need to be analyzed with probabilistic methods. The usual approach is to estimate the dominant control frequency in the flow field and the designer must ensure that it is not very close to the lower natural frequency of the tube. Thulukkanam recommends the following expression, based on study of gas flow normal to a tube bank (see Fig. 3.7), to find this frequency:

$$f_t = \frac{U}{DX_LX_T}\left[3.05\left(1 - \frac{1}{X_T}\right)^2 + 0.28\right]   \tag{11.7}$$

where

X_L = longitudinal pitch ratio = P_L/D (P_L is longitudinal pitch)
X_T = transverse pitch ratio = P_T/D (P_T is transverse pitch)
keeping f_t away from f_n (Eq. 11.6) should be objective of the designer.

11.2.3 Fluid Elastic Instability

Fluid elastic instability (FEI) is the most damaging situation resulting in uncontrolled increase in the amplitude of vibrating tubes, often leading to tube failure. This arises because of interplay between the excitation due to fluid dynamic forces exerted on the tube-bank by the fluid flowing across it, and the damping effect of the tube material and the fluid surrounding the tubes. As long as the damping effect predominates, the tube vibration is limited. However, as the fluid velocities increase, the energy imparted to the tube increases. Once this exceed the energy that can be dissipated by the damping, the instability sets in and the tube vibrations intensify to the point that adjacent tubes hit each other, resulting in tube failure. The fluid velocity at which this happens is termed the critical velocity U_{cr} and can be predicted by the Connors' equation:

$$\frac{U_{cr}}{f_n D} = K \left\{ \frac{m\delta}{\rho_s D^2} \right\}^{0.5} \tag{11.8}$$

where

   m = effective tube mass per unit length;
   $\delta = 2\pi \xi_n; \xi_n$ = the critical damping ratio;
   ρ_s = density of the shell side fluid;
   K is an empirical constant (called instability parameter) whose value depends on tube bundle configuration; typical value for arrays being $K = 4$.
Other correlations for U_{cr} can be seen in Thulukkanam [2].

   To avoid FEI, clearly the maximum fluid velocity within the tube bundle, U, should be less than U_{cr}. A conservative approach is to design the heat exchanger so that $U < 0.5 U_{cr}$.

11.2.4 Acoustic Resonance

Acoustic resonance can occur in heat exchangers where a gas (or steam) is flowing across a tube bundle. It is due to standing waves in the gas created within the shell (or any other walls confining the flow) by the fluctuating velocity component transverse to the mean flow direction, present in all turbulent flows. These standing waves can resonate with the vortex shredding or the turbulent eddies and create intense, low-frequency noise. If the standing wave frequency coincides with the natural frequencies of the structural components, such as the casing, the tubes, or even fin-plates, it can result in structural damage.

   As shown in Fig. 11.3, standing waves develop when the distance between the bounding walls, L_w, is $\frac{\lambda}{2}$ or a multiple of $\frac{\lambda}{2}$, that is $\frac{n\lambda}{2}$, where $n = 1, 2, 3, \ldots$; λ being the wavelength of the standing wave. Therefore, possible standing wave frequencies are given by the equation

$$f_a = \frac{nc}{2L_w} \quad \text{for } n = 1, 2, 3, \ldots \tag{11.9}$$

where c is the velocity of the sound in the fluid outside the tubes, and n is the mode number. The velocity of sound can be found using the standard equation

$$c = \sqrt{\frac{Z\gamma RT}{M_g}} \tag{11.10}$$

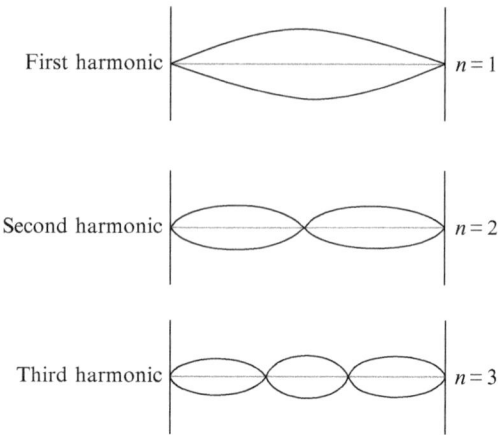

Fig. 11.3 Modes of acoustic standing waves in heat exchangers.

where Z is the gas compressibility factor, R the universal gas constant, T the absolute temperature of the gas, M_g its molecular mass, and γ the ratio of its two specific heats (cf. Eq. 3.146 for ideal gas).

Acoustic resonance would occur if $f_a = f_v$ or $f_a = f_t$.

11.2.5 Design for Minimizing Vibrations

It should be clear from the above brief discussion that after thermal design of a heat exchanger has been completed, a check for FIVs should be carried out, for all the four possible excitations. If the heat exchanger design is found to be susceptible to vibrations and noise, suitable changes in the design need to be done. Some of these changes are outlined below.

(i) *Increasing f_n the natural frequency vibration of the tubes.* It should be evident from Eq. (11.6) that f_n can be increased by reducing the tube span L (other variables being limited by standard tube dimensions and the material). Of course, reducing L will also increase the cross-flow velocity, which should not exceed the critical velocity U_{cr} (Eq. 11.8) to prevent fluid elastic instability.

(ii) *Decreasing cross flow velocity.* In case the design is found to be susceptible to fluid-elastic instability, reduction in cross flow velocity may be necessary. This could be achieved by reducing the flow rate, increasing the shell diameter, increasing the tube pitch, or even increasing the baffle spacing. These changes will also cause reduction in shell side heat transfer coefficient and an optimum design needs to be found.

(iii) In case the design is found to be susceptible to acoustic resonance, methods need to be devised to suppress the standing waves. This could be most conveniently done by placing an additional baffle inside the heat exchanger in a direction parallel to the flow and perpendicular to the direction of standing waves. Other methods can be seen in Thulukkanam [2, Section 10.7.3].

11.3 STOCHASTIC CONSIDERATIONS

During the design of thermal systems discussed hitherto, it has been implicitly assumed that the values of various parameters employed are fixed. However, we know that this is not strictly true. The environmental temperature and the humidity change continuously, but we design the systems for certain conditions. Thus the energy costs of operating an air-conditioning plant, or the power output of a thermal power plant, would be changing continuously throughout the year due to change in the load and the environmental conditions. There is typically ±15% uncertainty in the values of the heat transfer coefficients and the pressure gradients predicted by various empirical correlations. There is also a large uncertainty about the energy costs, inflation, interest rates, and so on. Numerous studies have demonstrated that the thermal system designed for certain fixed operating conditions may not achieve the desired performance at off-design conditions. Cho [4] has shown that if a heat exchanger is designed using mean values of design parameters, the probability of its meeting the requirements during actual operation is 50%. It is also evident that the systems designed to be optimal for certain operating conditions would not be optimal when these operate over widely varying conditions during their life time. Further, hitherto, we have focused on optimizing the design for a single objective such as minimal life-cycle cost, minimal energy costs, maximum heat recovery, etc. In many practical situation we encounter more than one objective, both of which are of equal importance (e.g., maximum heat transfer, minimum environmental pollution).

In this section we shall briefly discuss various approaches developed over last few decades to account for various kinds of uncertainties, part load operation, and multiple objectives while designing thermal systems.

11.3.1 System Design Under Uncertainty

The uncertainties that can influence the design of thermal systems can be classified into three categories.

Uncertainties Associated With Modeling

These include uncertainties in thermophysical and thermodynamic properties, in heat transfer coefficients, and in the ability of the model to exactly replicate the physical behavior of any equipment.

Uncertainties in the Values of the Input Variables

These arise from the random nature of certain process inputs such as the heating, cooling, or electrical loads; environmental conditions; prices; interest rates; and inflation.

Uncertainties Arising From Wear and Tear of Equipment

During the design process we often assume constant efficiencies of various components. Typical examples are mechanical and volumetric efficiencies of reciprocating devices, efficiencies of motors and batteries, isentropic efficiencies of turbines, and pumps and compressors. With passage of time the values of these efficiencies are bound to change randomly due to wear and tear. Traditional approaches to account for these uncertainties have been to design the equipment for certain nominal values of input variables and then incorporate "safety factors" to "enhance" the capacity of the equipment. In air-conditioning equipment a more "scientific" approach has become popular, that is, to design the system for most severe environmental conditions that have a certain probability of occurrence; for example, typically a system could be designed for 5% conditions which are obtained by statistical analysis of the hourly weather data of last 10 years in a manner to ensure that the probability of these being "exceeded" is only 5%, or in other words the probability of system satisfying the load requirements is 95%.

The modern approaches which have been evolved over last three decades can be divided into three categories, namely scenario approach, probabilistic approach, and chance constrained optimization. To appreciate the difference between these formulations, let us first write down the formal mathematical statement of system design under uncertainty.

$$\text{Optimize} \quad Z(x, d, u) \tag{11.11}$$
$$\text{subject to} \quad h(x, d, u) = 0$$
$$g(x, d, u) \leq 0$$
$$x^{\mathrm{L}} \leq x \leq x^{\mathrm{H}}; \quad d^{\mathrm{L}} \leq d \leq d^{\mathrm{H}}; \quad u^{\mathrm{L}} \leq u \leq u^{\mathrm{H}} \tag{11.12}$$

where $x, d,$ and u are the vectors of variables representing the state variables, design variables, and the uncertain variables, respectively. The uncertain variables are usually specified in terms of probabilistic distributions the most common being the normal Gaussian distribution. $Z(x, d, u)$ is the objective function, $h(x, d, u)$ and $g(x, d, u)$ are the vectors representing the equality and the inequality constraints, respectively. The superscript L and H represents the lower and higher bounds for the variables.

In the *scenario approach*, the stochastic optimization problem is converted into a set of deterministic problems. In each of these deterministic problems the uncertain variables are represented by a set of suitably sampled values of these variables, and the optimum obtained. Each of these sets of values of the uncertain variables constitutes a "scenario." The mathematical formulation of this method is given below

$$\text{Optimize} \quad Z(x, d, u^*) \tag{11.13}$$
$$\text{subject to} \quad h(x, d, u^*) = 0$$
$$g(x, d, u^*) \leq 0$$
$$x^L \leq x \leq x^H; \quad d^L \leq d \leq d^H; \quad u^L \leq u^* \leq u^H \tag{11.14}$$

Here u^* is the vector of values of the uncertain variables corresponding to each scenario. This optimization procedure is repeated for a large number of scenarios u^* and a probabilistic representation of the optimum Z is thus obtained.

To build scenarios of uncertain variables, a number of techniques have been developed. Monte Carlo technique is most commonly used to generate random values of variables u^* from specified probability distribution function $f(u)$ and the cumulative distribution function $F(u)$. The steps involved are

(i) generate a pseudo–random number r_j, uniformly distributed between numbers 0 and 1. Numerous computer programs are available for this.

(ii) determine a sample value u_j^* using

$$u_j^* = F^{-1}(r_j) \tag{11.15}$$

where F is the cumulative distribution function for the random variable u_j. The pertinent simplified equations for various common types of distributions can be seen in the paper by Badar et al. [5]. The complete strategy can be represented by the schematic diagram of Fig. 11.4 (following Badar et al. [5]).

In the *probabilistic approach*, the objective function itself is a probabilistic functional. The mathematical formulation can be written as:

$$\text{Optimize} \quad P1(z) = P1\{z(x, d, u)\} \tag{11.16}$$

subject to constraints

$$P2\{h(x, d, u)\} = 0$$
$$P3\{g(x, d, u)\} \leq 0 \tag{11.17}$$

where P's represent the probabilistic functionals. For example, if the goal is to optimize an expected value, we get

$$P \equiv E(F(u)) = \int_0^1 f(u) \, dp(u) \tag{11.18}$$

This integral can be calculated by sampling the function, as

$$E\{F(u)\} = \frac{\sum_{i=1}^{N_{samp}} F(u)}{N_{samp}} \tag{11.19}$$

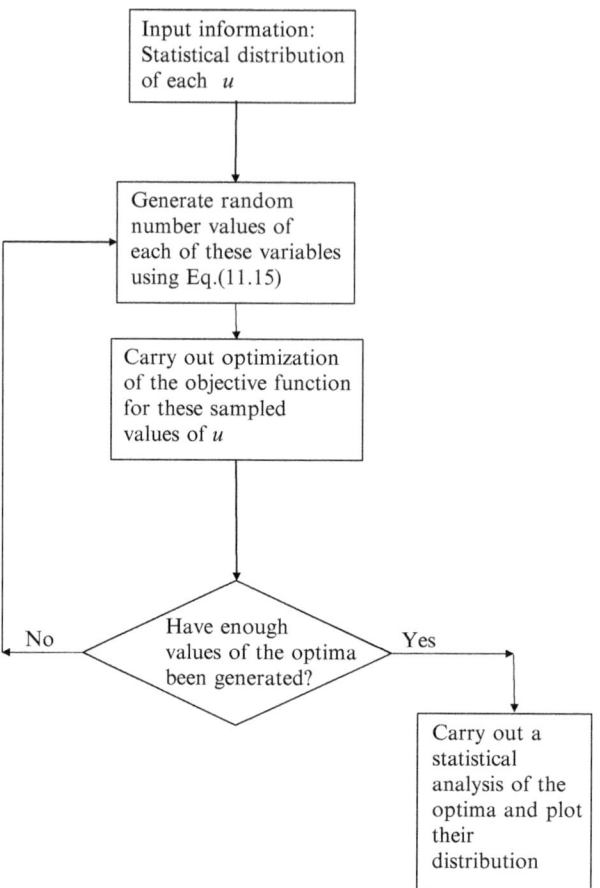

Fig. 11.4 Flow diagram of scenario approach for stochastic optimization.

The *chance constrained optimization* differs from this approach in that the constraints (Eq. 11.17) are replaced by

$$P\{h(x, d, u) > \beta\} \leq P_c \qquad (11.20)$$

Thus we are relaxing the constraints so that the these need not hold strictly, but the probability of exceeding a certain value is restricted to be less than an acceptable value. Diwekar et al. [6] illustrate the application of various stochastic optimization strategies to advanced coal–based thermal power systems for electric power generation. Che et al. [7] present a simple case study of applying Monte Carlo base approach for optimal design of a pressure vessel.

11.4 SYSTEM DESIGN CONSIDERING PART-LOAD OPERATION

The thermal systems—both power plants and refrigeration/air-conditioning systems—operate for most of their life or part loads. In the usual design process, these systems are, however, designed for certain nominal operating conditions. We thus often end up with costly over-designed systems since the actual operating costs are much lower than those considered during the design process. Clearly a rigorous approach would be to consider the actual loads that a system would experience throughout its operational life. Since the typical life of a thermal systems could be 15–30 years, and the loads and the environment conditions are continuously changing, such a rigorous analysis would need enormous computational effort. Simplified approaches have therefore been devised. Judes et al. [8] subdivided on the basis of actual load profile the whole yearly operational time into four broad subdivisions of full-load and three levels of part-load operations of the power plant. To evaluate every design generated during the optimization process, its performance at these four load values was assessed and incorporated in the calculation of the objective function.

In renewable energy systems such as solar and wind energy systems, the environmental conditions (e.g., solar irradiation and the wind speed) change very rapidly and a detailed analysis of the time-series data is necessary to identify the optimum design.

11.5 ENVIRONMENTAL CONSIDERATIONS

With growing concerns of climate change, minimization of the environmental impact of the system could also be the objective of the optimal design. ISO 14040–14043 gives details of the life cycle assessment (LCA) method to determine the total environmental impact of each component of the system from its "cradle" (i.e., where the raw materials needed to manufacture it are extracted from earth), to its production and use, to finally its "grave" (i.e., disposal after completion of its life). Several methods to quantify their impact have been suggested.

Life cycle climate performance assesses the cumulative global warming emission from "cradle" to "grave." This is usually quantified in terms of tons of carbon dioxide equivalent. A comprehensive LCA includes, besides the global warming potential (GWP), the eutrophication potential (reduction in oxygen in water bodies), acidification potential (calculated in terms of kilogram of SO_2 equivalent), and the ozone depleting potential (ODP). To incorporate all these factors into the optimal design we need to assign weighting factors to each. Often, in view of the huge impact of global warming on climate, and amelioration in ozone depletion due to various policies implemented over last two decades, these days only GWP is considered as a measure of environmental impact. Its minimization is incorporated as an additional objective, besides the usual economic objectives, in a multiobjective optimization problem. An alternative approach suggests putting suitable costs on these environmental factors and combining these with other costs such as initial capital cost, operational costs, cost of recycling, disposal, etc. The costs can be assigned on the basis of costs of policy interventions to mitigate their GWP.

11.6 SYSTEM DESIGN FOR MULTIPLE OBJECTIVES

As indicated in Section 11.5, we do encounter situations in system design where more than one objective is important. Often besides the economic objectives, there may be other objectives such as minimization of the use of fossil fuels, carbon dioxide emission, floor space, etc. In mathematical terms it implies that for an optimization problem with k objectives, the "general formulation" of Eqs. (8.5)–(8.6) changes to:

$$\text{Minimize} \quad f_1(\vec{X}), f_2(\vec{X}), \ldots, f_k(\vec{X})$$
$$\text{where} \quad \vec{X} = [x_1, x_2, \ldots, x_n]^T$$

subject to constraints

$$g_j(\vec{X}) \leq 0 \quad j = 1, 2, \ldots, m$$

It is highly unlikely that a solution vector would exist which would simultaneously minimize all the objectives. The best possible solution would be the one which minimizes most of the objectives while for others, it gives a sufficiently low value and there exists no other design which would give a smaller value of these objectives without increasing the value of at least one of the other objectives. Such a solution is termed as the *Pareto optimal* design. Most multiobjective optimization problems give a set of such Pareto optimal solutions, and then a choice among these is made by some other additional criteria.

Probably the most commonly employed approach for such problems is the *weighting function method*. Here suitable weights (w_i) are assigned to various objectives in accordance with the importance attached to them and a single-objective function F is defined as the weighted sum of all the objective functions, that is

$$F = \sum_{i=1}^{k} w_i f_i(X)$$

and the design vector $\vec{X}$ is found which minimizes F.

Another approach is that of first finding the optimum design vector for each of the objectives considered singly, and then finding the global optimal so as to minimize the total deviation of this global optimum from each individual optimum. Thus if $f_i^*(\vec{X}^*)$ is the value of the ith objective function when optimized individually subject to the same constraints, the global objective function $F(\vec{X})$ could typically be

$$F(\vec{X}) = \sum_{i=1}^{k} \left| \frac{f_i^*(\vec{X}^*) - f_i(\vec{X})}{f_i^*(\vec{X}^*)} \right| \tag{11.21}$$

or

$$F(\vec{X}) = \sum_{i=1}^{k} \left[\frac{f_i^*(\vec{X}^*) - f_i(\vec{X})}{f_i^*(\vec{X}^*)} \right]^2 \tag{11.22}$$

The design vector which minimizes $F(\vec{X})$ would be the optimal design vector of the multiobjective optimization problem.

Yet another approach is that of choosing the most important among the objectives as the (single)-objective function and converting all the rest into constraints by specifying the upper and the lower bounds of their acceptable values. *Goal programming* is a variant of this approach. For each objective certain goals are set and the optimum design is chosen which minimizes the deviations from these set goals.

11.7 COMMERCIAL SOFTWARES

Over the years numerous commercial software packages have been developed for simulation and design of equipment used in thermal systems. Most of these proprietary softwares enable selection of suitable equipment (heat exchanger, condenser, boiler, evaporator, and compressor) to meet the expected duty from among the equipment manufactured by the company. Major HVAC companies such as Carrier, Trane, York, etc. and even small companies making only a few components (say air-cooled heat exchangers) have developed such softwares. Carrier corporation has a complete suite of programs for meeting all the requirements of design of HVAC system for a modern building like hourly load calculation, optimizing the design of piping and life cycle economic analysis of these systems. For design of power plants too a number of softwares have been developed. Most prominent among them are the set of softwares being marketed by FIRECAD company. These include softwares for the design of fire-tube and water-tube boilers, economizers, superheaters, etc.

Another set of widely used softwares for heat exchangers design is marketed by HTRI, their latest set of programs being Xchanger suite 7. It includes modules for analysis and design of various types of heat exchangers and fired heaters. The calculations are done using local values of the heat transfer coefficient and the pressure gradients. It has the flexibility of using input files from other commercial softwares such as UniSim, HTFS, and Aspen EDR. Besides the thermal design, it also has modules for mechanical design and vibration analysis. A German company, TLK-Thermo GmbH offers TIL, suite for both steady-state and transient simulation of various kinds of thermal systems such as refrigeration systems, power plants, adsorption systems, fuel cells, etc. It also offers other suites of programs for facilitating connection of various simulation programs.

As thermal engineering equipment like heat exchangers are integral parts of most chemical process industries, commercial softwares for chemical process simulation and optimization invariably include modules for heat exchanger design. Some of the widely used such softwares are Aspen Plus, UniSim, Simsci, etc. Aspen EDR is the module of Aspen Plus for the design and rating of heat exchangers. It permits use of simplified procedures to build preliminary designs which could be compared against the design objectives, and the chosen design could then be the starting point for a more rigorous design exercise. UniSim and Simsci are two other competitor softwares with similar capabilities. TRNSYS is another software especially suitable for simulating the transient behavior.

These softwares differ from each other primarily in the type of graphic user interface provided, the models and equations used for estimation of heat transfer coefficients, pressure gradients, and various thermodynamic and thermophysical properties of the working fluids. Most softwares periodically provide update to remove the shortcomings pointed out by the users. Typical shortcomings could be difficulty of convergence of programs under certain working conditions, inability to decipher physically impossible operating conditions (e.g., a fluid flowing below its freezing point), etc.

Some free softwares for thermodynamic analysis and preliminary thermal design analysis are also available. Prominent among them are CATT, for evaluating the thermodynamic properties of a number of common fluids, and COOLPACK, for thermodynamic analysis of various types of refrigeration cycles, including multistage systems, and determination of UA values of the heat exchangers used in these cycles. A number of free softwares (such as SOLO, T★SOL, PV★SOL) with limited capabilities are also available for solar thermal and photovoltaic system design; their professional versions enable comprehensive simulation and design. The German company TLK-Thermo GmbH also offers a freely downloadable software ClaRa for simulation of the thermal power plant cycles.

Such softwares greatly facilitate the study of the influence of various parameters on the performance of the equipment or the system as a whole. The designer can easily change the working fluids, the operating conditions, designs of only some of the components while retaining existing design of others, etc. However, since most softwares do not provide the source code, the designer is also constrained in some crucial aspects. Thus, with advancement of knowledge regarding heat transfer and fluid flow in various equipments as new correlations are developed to predict the heat transfer coefficient and pressure gradients under various conditions, the designer cannot incorporate these in the software. Such constraints are not there if the designer builds his own customized software based on the basic principles enunciated in the book.

REFERENCES

[1] ASME Standards Technology LLC, Design Guidelines for Corrosion, Erosion and Steam Oxidation of Boiler Tubes in Pulverized Coal-Fired Boilers, ASME Standards Technology LLC, NY. USA, 2014.

[2] K. Thulukkanam, Heat Exchanger Design Handbook, second ed., CRC Press, Boca Raton, FL, USA, 2013.

[3] V. Ganapathy, Avoid heat transfer equipment vibration, Hydrocarbon Process. 66 (1987) 62.

[4] S.M. Cho, Uncertainty analysis of heat exchanger thermal-hydraulic designs, Heat Transfer Eng. 8 (1987) 63.

[5] M.A. Badar, S.M. Zubair, A.K. Sheikh, Uncertainty analysis of heat exchanger thermal designs using the MonteCarlo simulation technique, Energy 18 (8) (1993) 859–866.

[6] U.M. Diwekar, E.S. Rubin, H.C. Frey, Optimal design of advanced power systems under uncertainty, Energy Convers. Manage. 38 (1997) 1725.

[7] J. Che, J. Wang, K. Li, A Monte Carlo based robustness optimization method in new product design process: a case study, Am. J. Ind. Bus. Manage. 4 (2014) 360–369.

[8] M. Judes, S. Vigerske, G. Tsatsaronis, Optimization of the design and partial load operation of power plants using MINLP, in: Optimization in Energy Industry, Springer, Berlin, Heidelberg, 2009, http://www.math.hu-berlin.de/~stefan/JuTsVi08.pdf (accessed October 28, 2014).

INDEX

Note: Page numbers followed by *b* indicate boxes, *f* indicate figures and *t* indicate tables.